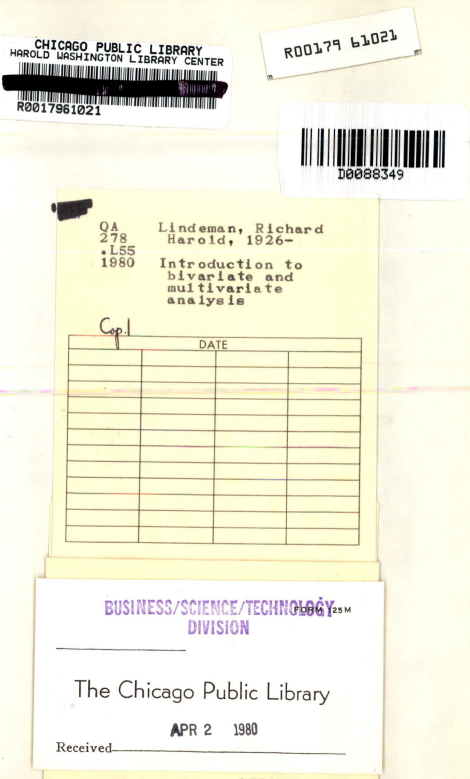

Introduction to Bivariate and Multivariate Analysis

Introduction to Bivariate and Multivariate Analysis

Richard H. Lindeman
Teachers College, Columbia University

Peter F. Merenda
University of Rhode Island

Ruth Z. Gold
Teachers College, Columbia University

Scott, Foresman and Company Glenview, Illinois
Dallas, Texas Oakland, New Jersey Palo Alto, California
Tucker, Georgia London, England

Acknowledgments

Chapter 2. [Figures 2.4.1, 2.4.3] Figures 4.1, 4.2 by Maurice M. Tatsuoka from *Multivariate Analysis: Techniques for Educational and Psychological Research*. Copyright © 1971, by John Wiley & Sons, Inc. Reprinted by permission of John Wiley & Sons, Inc./ [Figure 2.4.2] By permission. Copyright by Helen M. Walker and Joseph Lev, *Statistical Inference*, Figure 10.1, p. 247, Henry Holt and Company, 1953. [Figure 2.10.1] Helen M. Walker and Joseph Lev, *Elementary Statistical Methods*, Figure 14.2, p. 246. Copyright © 1969 by Holt, Rinehart & Winston, Inc.

Chapter 8. [Figure 8.2.1, Tables 8.2.3, 8.2.4, 8.2.5, 8.3.1, 8.4.2, 8.4.3, 8.4.4] Tables 6.1, 6.2, 7.3, 7.4, 7.6, 8.13, 13.7, 14.5, 14.10 by Harry H. Harman, from *Modern Factor Analysis*, 3rd Edition, revised, © 1960, 1967, 1976 by the University of Chicago. Published by the University of Chicago Press. Reprinted by permission. Quotations from *Modern Factor Analysis*, third edition revised, by Harry H. Harman. Copyright © 1960, 1967, 1976 by the University of Chicago Press. Reprinted by permission.

Chapter 9. [Table 9.6.8] Table 3, by Goodman in *Journal of the American Statistical Association*. Copyright 1970 by the American Statistical Association. Published quarterly by the American Statistical Association.

Appendix B. [Figure B.1] Figure 15, by Ross R. Middlemiss, from *Differential and Integral Calculus*. Copyright 1940 by the McGraw-Hill Book Company, Inc. Published by McGraw-Hill Company, Inc.

Appendix C. [Table C.1] Table B, Gene V. Glass, Julien C. Stanley, *Statistical Methods in Education and Psychology*, 1970, p. 513. Reprinted by permission of Prentice-Hall, Inc., Englewood Cliffs, New Jersey./[Table C.4] "Percentage Points of F Distribution" from *Biometrika Tables for Statisticians*, Vol. II, edited by E. S. Pearson and H. O. Hartley. Copyright © Biometrika Trustees, 1972, by permission of the publisher, Cambridge University Press./[Table C.5] "Critical Values of the Spearman Rank Correlation Coefficient, r_s" by Jerrold H. Zar, from "Significance Testing of the Spearman Rank Correlation Coefficient," in the *Journal of the American Statistical Association*. Published quarterly by the American Statistical Association./[Table C.6] Reproduced by permission of the publishers, Charles Griffen & Company Ltd., of London and High Wycombe, from Kendall, *Rank Correlation Methods*, 4th ed., 1970./[Table C.7] From Table III in "A Comparison of Alternative Tests of Significance for the Problem of *m* Rankings," by Milton Friedman, from the *Annals of Mathematical Statistics*, vol. XI, p. 91, 1940./[Tables C.8a, C.8b, C.8c] Tables by W. L. Jenkins, *Psychometrika*, vol. 20, no. 3, Sept. 1955. Published by the Psychometric Society./[Tables C.9a, C.9b] Reprinted with permission from *Handbook of Tables for Probability and Statistics*, second edition, William H. Beyer (editor). Copyright The Chemical Rubber Company Company, CRC Press, Inc.

Library of Congress Cataloging in Publication Data

Lindeman, Richard Harold, 1926–
Introduction to bivariate and multivariate analysis.

Bibliography: p. 432.
Includes index.
1. Multivariate analysis. I. Merenda, Peter Francis, 1922– joint author.
II. Gold, Ruth Z., 1921– joint author.
III. Title.
QA278.L55 1980 519.5'3 79-19133
ISBN 0-673-15099-2

Printed in the United States of America.

12345678910-RRC-85848382818079

BST cop. 1

To the Instructor

This book is designed for students in the social sciences who need to have a working knowledge of applied bivariate and multivariate statistics in order to understand and evaluate research reports in the literature of their fields and to select and use appropriate statistical methods in their own research. It has been written as a text for intermediate statistics courses that teach multivariate analysis. We have assumed that the reader has had some basic preparation in statistics, including descriptive methods, the fundamentals of probability theory, an introduction to random variables and their properties (such as expected value, variance, and covariance), and elementary inferential concepts and methods (such as t, χ^2, and F in estimation and in testing statistical hypotheses).

A working knowledge of high-school algebra, including an introduction to matrix operations, is sufficient mathematical background. In a few places we have used elementary differential calculus, principally to find a maximum or minimum. Readers who have a knowledge of calculus will find this useful and will be able to follow the presentation of derivations of necessary formulas. Those without a knowledge of calculus may have to take these results on "faith," but can skip the sections using calculus without loss of continuity.

16.95

Throughout the book we have relied quite heavily on the use of numerical examples to show the steps in the computations of various statistics and to illustrate the applications of the methods in research situations. In many cases we have used simple contrived data, especially to illustrate computational procedures, so that all of the steps involved could be shown and easily followed by the reader. A large number of the examples, as well as many of the end-of-chapter exercises, however, are based on real data, in some cases adapted to illustrate particular procedures or applications. In general, we have tried to provide not only a clear description of the methods treated, but also realistic examples of their application in practical research settings.

The first two chapters dealing with bivariate regression and correlation, as well as portions of Chapter 3 on special bivariate methods, cover material which many students will have encountered in an introductory course. We have included these chapters for three principal reasons. First, they give a more thorough coverage than introductory texts and can be used not only for review, but also to augment previous knowledge. Second, they provide a simpler context than later chapters in which to introduce both notation and important basic concepts, so that later chapters will be easier to read and comprehend. Third, the method discussed in later chapters can, in many cases, be regarded as extensions of the bivariate methods treated in

Chapters 1 through 3. Thus, the inclusion of the first three chapters permits a more integrated picture of correlational methods in general.

In Chapters 4 through 8, we treat the topics usually associated with multivariate analysis texts, namely, multiple regression and correlation, canonical correlation, discriminant analysis, multivariate analysis of variance, and factor analysis. Chapter 9 deals with the multivariate analysis of categorical data, a topic that is not typically found in multivariate analysis textbooks but is now quite important in social science research.

We are indebted to many students who have offered valuable suggestions for improvement in the presentation of this text. We are also indebted to several former teachers of ours, all of whom had important roles in stimulating our interest in multivariate analysis and who contributed to our understanding of statistical methods in general.

Our thanks go to Owen Whitby for his careful and thorough review of the manuscript and for his excellent suggestions for improvements of a technical nature. We also thank Joe Sonnefeld for his many valuable editorial comments and suggestions and Laurell Johnson for her careful checking of the arithmetical computations in the examples.

To the Student

This book has been designed for students in the social sciences who need to have a working knowledge of applied bivariate and multivariate statistics in order to understand and evaluate research reports in the literature of their fields and to select and use appropriate statistical methods in their own research. We have assumed that the reader has had some basic preparation in statistics, including descriptive methods, the fundamentals of probability theory, an introduction to random variables and their properties (such as expected value, variance, and covariance), and elementary inferential concepts and methods (such as t, χ^2, and F in estimation and in testing statistical hypotheses). Such a background can usually be obtained in a good one-semester (or two-quarter) course at either the undergraduate or graduate level.

A working knowledge of high-school algebra, including an introduction to matrix operations, is sufficient mathematical background. We mention the need for matrix algebra here in order to stress its importance in dealing with multivariate methods. In fact, about three-fourths of the methods treated in the book cannot be fully comprehended without an understanding of matrix operations. Many, perhaps most, high school mathematics curricula now include some attention to matrix algebra. Nearly all introductory college mathematics courses include it. However, for those readers who lack such background, or who need to review it, we have included an appendix that treats all of the matrix procedures required for the book itself. We urge students who may need such study or review to make use of Appendix A prior to reading the last section of Chapter 1.

In a few places, and as early as Chapter 1, we have used elementary differential calculus, principally to find a maximum or minimum. Readers who have a knowledge of calculus will be able to follow the presentation and see exactly how the formulas for obtaining the desired results were derived. Those without a knowledge of calculus may have to take these results on "faith," but can skip the sections using calculus without loss of continuity. For readers who may find it helpful, a brief review of the elements of differential calculus is provided in Appendix B.

We have endeavored to provide proofs wherever the methods of proof are consistent with the mathematical and statistical background assumed of the reader. The theory which underlies most of the methods discussed in this book, however, cannot be treated without extensive preparation in mathematical statistics. Many results throughout the book are therefore stated without proof, although in such cases we do give references to proof sources. Some readers will wish to read these, while others will not. A vertical rule in the margin next to these sections, as well as the sections using calculus, easily identifies them; at the same time, they can be skipped without loss of continuity. We urge those who have the necessary background to read these sections because we believe they help to provide a depth of understanding of important concepts that is ultimately desirable.

Throughout the book we have relied quite heavily on the use of numerical examples to show the steps in the computations of various statistics and to illustrate the applications of the methods in research situations. In many cases we have used simple contrived data, especially to illustrate computational procedures, so that all of the steps involved could be shown and easily followed by the reader. A large number of the examples, as well as many of the end-of-chapter exercises, however, are based on real data, in some cases adapted to illustrate particular procedures or applications. In general, we have tried to provide not only a clear description of the methods treated, but also realistic examples of their application in practical research settings.

The first two chapters dealing with bivariate regression and correlation, as well as portions of Chapter 3 on special bivariate methods, cover material that many students will have encountered in an introductory course. We have included these chapters for three principal reasons. First, they give a more thorough coverage than introductory texts and can be used not only for review, but also to augment previous knowledge. Second, they provide a simpler context than later chapters in which to introduce both notation and important basic concepts, so that later chapters will be easier to read and comprehend. Third, the methods discussed in later chapters can, in many cases, be regarded as extensions of the bivariate methods treated in Chapters 1 through 3. Alternatively, of course, the bivariate methods can be viewed as special cases of the more general multivariate methods of later chapters. Thus, the inclusion of the first three chapters permits a more integrated picture of correlational methods in general.

In Chapters 4, 5, 6, and 7, we treat topics usually associated with multivariate analysis texts, namely, multiple regression and correlation, canonical correlation, discriminant analysis, and multivariate analysis of variance. In these chapters, as well as in the first three chapters, we have tried to maintain a consistent notation, with which the reader should try to become familiar as early as possible. In Chapter 8, on factor analysis, we have found it best to depart to some extent from this notation in order to be more consistent with the somewhat different notational requirements of the topic and with the more or less standard notation that appears in most factor analysis literature. Similarly, in Chapter 9, which deals with multivariate analysis of categorical data, the notation is more consistent with that of the important journal literature on this topic than with that of the earlier chapters.

Finally, we should comment on the presentation of numerical values obtained by working out the examples. In carrying out computations, particularly those involving matrix operations, it is best to retain as large a number of digits as is permitted by the computer or calculator being used. In printing results for many of the examples, we have rounded off intermediate results in the interest of saving space. However, in most cases, the *un*rounded intermediate results were used in further computations to produce final results. Thus, any discrepancy that might occur from using rounded intermediate results does not necessarily indicate an error. In any case, the discrepancies should usually be small, occurring in the fourth or fifth digit.

Contents

1/Bivariate Regression Analysis

We assume that the reader of this book is already familiar with certain elementary concepts of statistical estimation and hypothesis testing in the univariate case. We might, for example, be interested in estimating the mean, μ_Y, of a population of N values of a variable Y. Unless the entire population of Y values were available to us, we could not compute the value of μ_Y directly, but would have to estimate it from a sample consisting of only a portion, n $(n<N)$, of the values in the population. A satisfactory estimator of μ_Y would then be the sample mean, $\bar{Y}=\Sigma_{i=1}^{n} Y_i/n$. Furthermore, the sample mean, $\bar{Y}$, would also be a satisfactory predictor of any unobserved *individual* Y value in the population, because there would be no way of knowing whether that individual value was larger or smaller than the population mean.

As an illustration, suppose that Y is the grade point average (GPA) of freshman students ($N=1{,}000$) at a given college, and that the mean of a sample of $n=50$ values of Y from the entire class is $\bar{Y}=2.07$. Not only is $\bar{Y}=2.07$ the best available estimate of μ_Y, but it is also the best prediction of Jane Adams' Y value, assuming that her actual GPA is unknown to us. Furthermore, if we assume no important change in grading policies from year to year, our best prediction of the GPA of John Smith, who will be a member of next year's freshman class, would also be $\bar{Y}=2.07$. The point is that in the absence of additional information about other related characteristics of individuals in the population, our best estimate of average performance is the mean of a sample from the same or a similar population of Y values.

In this chapter we consider the case in which values of an additional variable, X, are available to increase the precision of estimates of μ_Y and of predicted individual values of Y. Such additional information is frequently available. For example, we might expect to predict Jane Adams' GPA more precisely if we knew her score on a measure of academic aptitude, such as the verbal subtest of the Scholastic Aptitude Test (SAT-V) of the College Entrance Examination Board. If her SAT-V score were $X=700$, considerably above the mean score of 500 for freshmen at her college, we could expect her GPA to be well above 2.07, because experience has shown that there is a substantial positive relationship between SAT scores and college freshman GPA. On the other hand, if her SAT-V score were $X=500$, we would expect her GPA to be closer to the observed sample mean of 2.07. It would not necessarily equal that value, of course, since SAT and GPA are not perfectly related. Our aim would be to use data on a related variable, such as SAT-V, in a manner that yields predictions of GPA having the greatest possible precision. In this chapter we develop and illustrate procedures for accomplishing that aim.

1.1 LINEAR EQUATIONS

Some kinds of relationships encountered in everyday life can be expressed conveniently and accurately in terms of linear equations. The statements that "3 feet equal 36 inches" and that "4 feet equal 48 inches" are simply verbal translations of the equation $Y=12X$, in which Y is the number of inches and X is the number of feet. Similarly, the relationship between time and distance travelled at a constant rate of speed, say 30 miles per hour, is expressed by the equation $d=30t$, where d is the distance in miles and t is the time in hours. These linear equations represent *exact* relationships, that is, the exact value of either variable can be determined when the value of the other is known.

Suppose a company pays each employee X dollars per month and gives a year-end bonus of \$500. The annual income, Y, of any employee can then be written in the form of an equation, $Y=500+12X$. This expression can be generalized as $Y=a+bX$, in which Y is the total annual income, a is the bonus, b is the number of months employed, and X is the monthly salary. The variable Y to the left of the equal sign is regarded as the *dependent* variable because its value depends on that of X. The variable X is considered the *independent* variable.

Substitution of a set of numerical values for X in the equation $Y=a+bX$ produces a set of corresponding values for Y. For example, consider the equation $Y=3+2X$. If $X=0$, the equation becomes $Y=3$. The pair of values $X=0$ and $Y=3$ is said to "satisfy" the equation because it maintains the equality. Further substitution for X produces the pairs of values for X and Y given in Table 1.1.1. Each of these pairs of values can be represented by a point on a graph. If the pairs of values in Table 1.1.1 are plotted, as in Figure 1.1.1, it can be seen that all of the points lie precisely on a straight line. For this reason, the equation $Y=3+2X$ or, more generally, $Y=a+bX$ is called a *linear* equation.

The constants a and b in the equation $Y=a+bX$ determine the position of the line on the graph. Note in Table 1.1.1 that as X increases one

TABLE 1.1.1 VALUES OF *X* AND *Y* THAT SATISFY THE LINEAR EQUATION $Y=3+2X$

	Pair								
Variable	1	2	3	4	5	6	7	8	9
X	0	1	2	3	4	5	6	7	8
Y	3	5	7	9	11	13	15	17	19

unit, Y consistently increases two units. As X increases two units, Y increases 4 units. In general, as X increases k units, Y increases $2k$ units. The constant b in the general equation is called the *slope* of the line, and is equal to the increase in Y divided by the corresponding increase in X. For the line in Figure 1.1.1, b is equal to $2k/k$ or simply 2. The value of Y when $X=0$ is called the *Y intercept* and is equal to a in the general equation.

Thus, one can determine by an inspection of its equation that the straight line $Y=3+2X$ crosses the Y axis at $Y=+3$ and has a slope of $+2$.

FIGURE 1.1.1 Graph of the Linear Equation $Y = 3 + 2X$

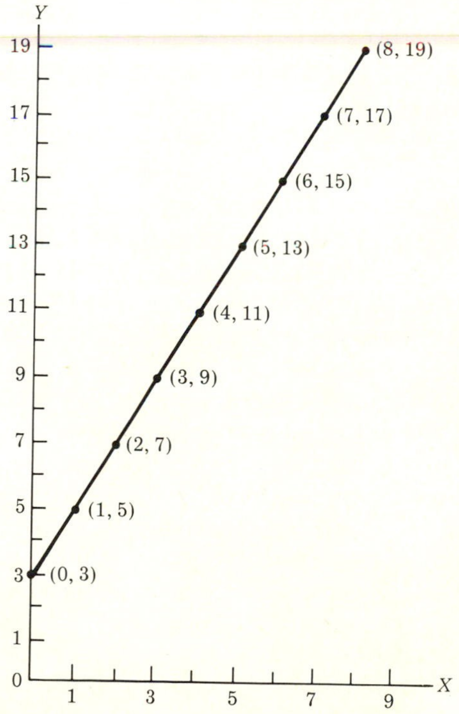

TABLE 1.1.2 VALUES OF X AND Y THAT SATISFY THE LINEAR EQUATION $Y=3-2X$

	Pair								
Variable	1	2	3	4	5	6	7	8	9
X	0	1	2	3	4	5	6	7	8
Y	3	1	−1	−3	−5	−7	−9	−11	−13

The slope of a line representing a linear equation may be negative or zero as well as positive. Consider the equation $Y=3-2X$, for example. Substituting various values for X in this equation yields the pairs of values for X and Y given in Table 1.1.2. Note here that as the value of X increases, Y decreases, that is, the change in Y is negative. Therefore, the slope is negative, as indicated by the sign of the term involving X in the equation. Consider also the equation $Y=3$. Here the slope is zero, and the line is parallel to the X axis at $Y=3$. It is important to note that the position of a line on a graph is completely determined by the values of the constants a and b. If either of these values is altered, the position of the line will be changed.

1.2 USING LINEAR EQUATIONS IN PREDICTION

In statistical work the relationships between variables are rarely exact. In dealing with psychological variables, inexactness arises from a number of sources, among which are measurement error and the influence of various moderator variables. Consider, for example, the midterm (X) and final examination (Y) scores of 43 students in an introductory statistics course (Table 1.2.1). There is clearly a relationship between the two variables because the largest X scores tend to be associated with the largest Y scores. Students who performed most poorly on the midterm (X) also tended to get the lower scores on the final (Y). However, the relationship is not perfect. Among the seven students who had a score of 35 on the midterm examination, scores on the final examination ranged from 28 to 38. Of those scoring 34 on the midterm, two had relatively low scores on the final (21 and 28) while one had a much higher score, 38. Thus, if we were to use X to predict scores on Y, there would be errors present in the predictions. However, because there is a relationship between X and Y, even though it is an imperfect one, we can obtain more accurate predictions using X than we could using no independent variable at all.

Even when measurement error is minimized, as in the realm of physical measurements, relationships may still be inexact. For example, height and weight are not perfectly related, even though each can be measured accurately. Not everyone of a given height has exactly the same weight,

TABLE 1.2.1 MIDTERM AND FINAL EXAMINATION SCORES OF 43 STUDENTS IN AN INTRODUCTORY STATISTICS COURSE

Student number	*Midterm score (X)*	*Final score (Y)*	*Student number*	*Midterm score (X)*	*Final score (Y)*
1	29	32	23	34	28
2	32	24	24	40	38
3	34	34	25	33	23
4	40	40	26	30	30
5	35	37	27	18	23
6	33	24	28	27	30
7	30	12	29	24	19
8	36	31	30	30	33
9	35	28	31	31	24
10	37	38	32	28	34
11	26	27	33	33	27
12	35	33	34	35	32
13	30	33	35	37	21
14	35	36	36	35	38
15	17	13	37	36	36
16	36	39	38	25	30
17	20	20	39	26	29
18	34	38	40	23	24
19	35	32	41	34	33
20	33	24	42	19	21
21	23	28	43	27	29
22	34	21			

because there are differences in body structure arising from both hereditary and environmental sources. Thus, we cannot exactly determine a person's weight from a knowledge of her height, although knowing her height increases the accuracy with which we can predict her weight.

1.2.1 Formulating the Linear Regression Equation

We shall make further use of the examination scores in Table 1.2.1 later, but at this point we shall use simpler data to develop and illustrate the application of the linear equation to the problem of predicting Y from X.

Suppose that two ten-item tests, X and Y are given to seven students with results as shown in Table 1.2.2. Each pair of values may be represented by a

TABLE 1.2.2 SCORES OF SEVEN STUDENTS ON EACH OF TWO TEN-ITEM TESTS, *X* AND *Y*

	Student no.						
Pair	1	2	3	4	5	6	7
X	7	6	5	4	3	2	1
Y	6	7	4	5	1	3	2

FIGURE 1.2.1 Scatter Diagram of Scores of Seven Students on Two Tests, X and Y

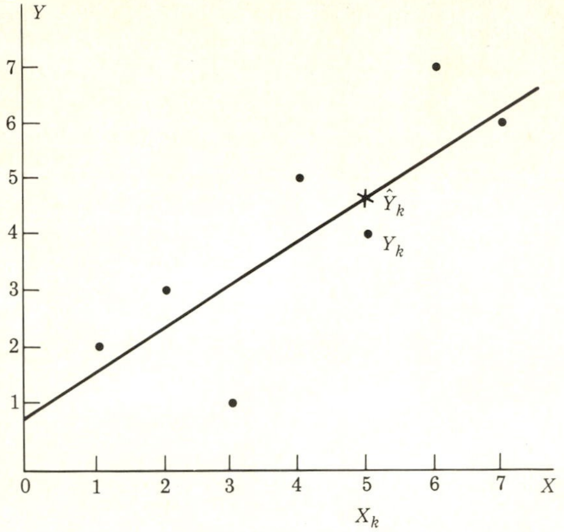

point on a graph and plotted in a *scatter diagram*, as in Figure 1.2.1. The small circles represent the seven points on the diagram. It is obvious that no straight line can be drawn through all of the points.

It is clear that any straight-line representation of the relationship between X and Y will involve some degree of error. The problem in developing the prediction equation is to locate the line so that this error is minimized.

The linear equation representing the relationship between X and Y will be of the general form $Y = a + bX$, where, as we indicated above, a is the Y intercept and b is the slope of the line. Substitution in this equation of a specific value for X, say X_k, will produce a corresponding value for Y, say $\hat{Y}_k$. The symbol $\hat{Y}$ (read "Y-hat") is used to denote a prediction of Y, instead of the actual value, because as we have pointed out above, some error will usually be made in predicting Y. The pair of values X_k and $\hat{Y}_k$ may be represented by a point that lies exactly on the straight line approximating the linear relationship between X and Y. Note in Figure 1.2.1 that the predicted value, $\hat{Y}_k$, and the actual value, Y_k, do not coincide. Such discrepancies between predicted and actual values of Y would also occur for other values of X, although not necessarily for all of them. The difference $Y_k - \hat{Y}_k$ thus represents the error we make in using the straight-line equation to predict Y from X. In recognition of the fact that the straight-line prediction equation does not provide perfect predictions, it is usually written $\hat{Y} = a + bX$, rather than $Y = a + bX$, to indicate that the prediction, $\hat{Y}$, would be the value of the variable Y if the linear relationship were perfect.

When formulating a prediction equation, the objective is to fit the line to the observed data in such a way as to minimize discrepancies between the actual and the predicted values. One might intuitively adopt

$$\sum_{i=1}^{n} (Y_i - \hat{Y}_i) = 0$$

as the criterion for a line of good fit, giving an average discrepancy of zero; but there are infinitely many lines that have this property. An alternative would be to minimize the discrepancies regardless of their direction, that is, to seek a line that minimizes

$$\sum_{i=1}^{n} |Y_i - \hat{Y}_i|.$$

However, this criterion involves the use of algebraically bothersome absolute values.

The criterion for "best" fit that has been found most useful is the "least-squares" criterion. In fitting a straight line to a set of data by this criterion, the values of a and b in the equation are chosen so as to minimize the sum of the squared discrepancies

$$\sum_{i=1}^{n} (Y_i - \hat{Y}_i)^2,$$

in which the summation is taken over all n pairs of observations to which the line is to be fitted. In addition to avoiding the use of absolute values, this criterion has other advantages which we shall see later.

Substitution for $\hat{Y}_i$ from the equation, $\hat{Y}_i = a + bX_i$, yields

$$G = \sum_{i=1}^{n} [Y_i - (a + bX_i)]^2 = \sum_{i=1}^{n} (Y_i - a - bX_i)^2. \tag{1.2.1}$$

Squaring the expression in parentheses on the right produces

$$G = \sum (Y^2 + a^2 + b^2X^2 - 2aY - 2bXY + 2abX)$$

$$= \sum Y^2 + na^2 + b^2 \sum X^2 - 2a \sum Y - 2b \sum XY + 2ab \sum X,$$

in which the subscripts and summation limits have been dropped to simplify notation. The symbol G to the left of the equality sign has no special meaning and is used only as a name for the function on the right side of the equation.

To choose values of a and b that minimize the function G, we shall employ the methods of differential calculus.

For the reader not familiar with this method, Appendix B contains a brief explanation of the basic concepts involved. Differentiating with respect to a and b, we have

$$\frac{\partial G}{\partial a} = 2na - 2\sum Y + 2b \sum X,$$

and

$$\frac{\partial G}{\partial b} = 2b\sum X^2 - 2\sum XY + 2a\sum X.$$

Setting each of these expressions equal to zero, dividing by 2, and rearranging, we have

$$na + \left(\sum X\right)b = \sum Y, \tag{1.2.2}$$

and

$$\left(\sum X\right)a + \left(\sum X^2\right)b = \sum XY. \tag{1.2.3}$$

These are the so-called "normal" equations, which must be solved for a and b. Some important consequences of these equations will be discussed later in this section.

Two equations in two unknowns can be solved in several ways. Here we use a method that involves subtracting one equation from the other. Multiplying Equation 1.2.2 by $\sum X$ and Equation 1.2.3 by n yields

$$\left(n\sum X\right)a + \left(\sum X\right)^2 b = \sum X \sum Y,$$

and

$$\left(n\sum X\right)a + \left(n\sum X^2\right)b = n\sum XY.$$

Now, subtracting the first of these equations from the second, we have

$$\left(n\sum X^2\right)b - \left(\sum X\right)^2 b = n\sum XY - \sum X \sum Y,$$

and solving for b gives

$$b = \frac{n\sum XY - \sum X\sum Y}{n\sum X^2 - \left(\sum X\right)^2} = \frac{\sum XY - \dfrac{\sum X\sum Y}{n}}{\sum X^2 - \dfrac{\left(\sum X\right)^2}{n}}. \tag{1.2.4}$$

An expression for a may be obtained easily by solving Equation 1.2.2 to produce

$$a = \bar{Y} - b\bar{X}. \tag{1.2.5}$$

The numerical value of a may then be obtained by substituting in Equation 1.2.5 the numerical value of b obtained from Equation 1.2.4.

To simplify notation in this and later sections, we define the following:

$$S_{xx}=\sum(X-\bar{X})^2=\frac{n\sum X^2-(\sum X)^2}{n}=\sum X^2-\frac{(\sum X)^2}{n} \qquad (1.2.6)$$

$$S_{yy}=\sum(Y-\bar{Y})^2=\frac{n\sum Y^2-(\sum Y)^2}{n}=\sum Y^2-\frac{(\sum Y)^2}{n} \qquad (1.2.7)$$

$$S_{xy}=\sum(X-\bar{X})(Y-\bar{Y})=\frac{n\sum XY-\sum X\sum Y}{n}=\sum XY-\frac{\sum X\sum Y}{n}. \qquad (1.2.8)$$

Using this notation, Equation 1.2.4 can be rewritten as

$$b=\frac{S_{xy}}{S_{xx}}. \qquad (1.2.9)$$

Consider again the distribution of X and Y scores of seven students given in Table 1.2.2. The steps necessary to formulate the regression equation based on those data are:

1. Calculate the values necessary for determination of b, namely S_{xy} and S_{xx}. For the data in Table 1.2.2 these are:

$$S_{xy}=135-\frac{(28)^2}{7}=23,$$

and

$$S_{xx}=140-\frac{(28)^2}{7}=28.$$

2. Substitute these values in Equation 1.2.9 to obtain

$$b=\frac{23}{28}=.82143.$$

3. Substitute the value obtained for b in Equation 1.2.5 to obtain

$$a=4-(.82143)(4)=4-3.28=.71429.$$

4. Substitute the numerical values for a and b in the linear prediction equation, $\hat{Y}=a+bX$, to obtain the equation

$$\hat{Y}=.71429+.82143X.$$

This equation can then be used to obtain predictions of the Y scores for members of the population for whom scores on only X are available. For example, the predicted Y score for an individual whose X score is 3 would be

$$\hat{Y}=.71429+(.82143)(3)=3.18.$$

The equation developed to predict Y from X is called a *regression equation*. The line that represents the regression equation in graphical form is called a *regression line* or *least-squares line*, and the constants a and b in the regression equation are called *regression coefficients*. The linear regression equation is only one of several types of such estimation or prediction equations commonly used in psychological and educational research. We shall consider other possible forms of regression equations later in this chapter.

1.2.2 Origin of the Term Regression

The concept of regression has some historical significance, which we should take note of here. Sir Francis Galton was one of the first persons to work with statistics dealing with relationships. At the turn of the century, Galton was conducting many investigations concerning the influence of heredity on human attributes, mental as well as physical, and several of his studies involved father-son relationships. In particular, Galton (1889) reported findings about the relationships between heights of fathers and sons. He observed that tall fathers tended to have tall sons and that short fathers tended to have short sons. However, he also observed what he called a *regression effect* in this relationship. He noticed, for instance, that the heights of sons tended to be "regressed" toward the mean of their group. Very tall fathers tended to produce tall sons. However, they were not as tall, on the average, as their fathers. Very short fathers tended to have short sons, who were, nevertheless, not as short, on the average, as their fathers. For those fathers in the medium range, the averages of the sons' heights corresponded more closely to those of the fathers. Hence, knowing the height of the father, one could predict reasonably well the height of his son and vice versa. Galton referred to the phenomenon of regression as "filial regression." He denoted the relationship between father's and son's heights by the symbol r (for regression), a symbol commonly used today to denote the coefficient of linear correlation (see Chapter 2). Some insight into the phenomenon of regression toward the mean will be gained later in this chapter when we have examined the mathematical model for linear regression.

Although the terms *regression line* and *regression equation*, as used in this and succeeding chapters, stem from Galton's interest and work, these terms apply in modern usage to a function that is employed in statistical *prediction*. Thus, the equation may also be properly referred to as the *prediction equation*.

1.2.3 Consequences of the Normal Equations

Aside from yielding solutions for a and b that can be used to compute the regression equation, $\hat{Y} = a + bX$, the normal equations provide several interesting and useful facts about the variables in regression analysis and the relationships between them. To state these facts it will be convenient to express Equations 1.2.2 and 1.2.3 in slightly different but algebraically equivalent forms. Considering first Equation 1.2.2, we note that it can be expressed as

$$\Sigma Y_i = na + b\Sigma X_i = \Sigma(a + bX_i) = \Sigma \hat{Y}_i.$$

Then it is clear that the sum of discrepancies between observed and estimated values of Y equals zero, that is,

$$\Sigma(Y_i - \hat{Y}_i) = 0. \tag{1.2.10}$$

Equation 1.2.3 can be expressed as

$$\Sigma X_i Y_i - a\Sigma X_i - b\Sigma X_i^2 = 0,$$

which is equivalent to

$$\Sigma X_i(Y_i - a - bX_i) = 0,$$

or

$$\Sigma X_i(Y_i - \hat{Y}_i) = 0. \tag{1.2.11}$$

These equations lead to the following results:

1. $\overline{Y} = \overline{\hat{Y}}$, that is, the means of the observed and predicted values of Y are equal. The proof of this statement is left to the reader (see Exercise 1.7).

2. The point $(\overline{X}, \overline{Y})$ is on the regression line, $\hat{Y} = a + bX$. The proof of this statement is also left to the reader (see Exercise 1.6).

3. $\Sigma(Y_i - \hat{Y}_i)^2 = \Sigma(Y_i - a - bX_i)^2$ is a minimum if $a = \overline{Y} - b\overline{X}$ and $b = S_{xy}/S_{xx}$. Of course, we used the method of differential calculus to accomplish exactly this. However, the statement can be verified using the normal equations. From Equation 1.2.10, we see that if c is any number, $\Sigma c(Y_i - \hat{Y}_i) = c\Sigma(Y_i - \hat{Y}_i) = 0$. Likewise, if d is any number, we see from Equation 1.2.11 that $\Sigma dX_i(Y_i - \hat{Y}_i) = d\Sigma X_i(Y_i - \hat{Y}_i) = 0$. Now, if $L = c + dX$ is the equation of *any* line, and if we let $L_i = c + dX_i$ be a prediction of Y_i for the ith individual, then,

$$\Sigma L_i(Y_i - \hat{Y}_i) = \Sigma(c + dX_i)(Y_i - \hat{Y}_i)$$
$$= c\Sigma(Y_i - \hat{Y}_i) + d\Sigma X_i(Y_i - \hat{Y}_i) = 0. \tag{1.2.12}$$

Furthermore, we may write

$$\Sigma(Y_i - L_i)^2 = \Sigma(Y_i - \hat{Y}_i + \hat{Y}_i - L_i)^2$$
$$= \Sigma(Y_i - \hat{Y}_i)^2 + 2\Sigma(Y_i - \hat{Y}_i)(\hat{Y}_i - L_i) + \Sigma(\hat{Y}_i - L_i)^2.$$

Using Equation 1.2.12, we can express the sum in the middle term on the right as

$$\Sigma(Y_i - \hat{Y}_i)(\hat{Y}_i - L_i) = \Sigma \hat{Y}_i(Y_i - \hat{Y}_i) - \Sigma L_i(Y_i - \hat{Y}_i) = 0.$$

Then $\Sigma(Y_i - L_i)^2$ becomes

$$\Sigma(Y_i - L_i)^2 = \Sigma(Y_i - \hat{Y}_i)^2 + \Sigma(\hat{Y}_i - L_i)^2,$$

which shows that $\Sigma(Y_i - \hat{Y}_i)^2 \leqq \Sigma(Y_i - L_i)^2$, and that these expressions are equal only if $L_i = \hat{Y}_i = a + bX_i$.

1.3 THE STANDARD ERROR OF ESTIMATE

As we indicated earlier, when a sample of paired observations on psychological or educational variables is sufficiently large, there are usually several different Y values associated with a particular X value. The distribution of Y values for a given value of X, say X_k, is called a *conditional distribution*, because it is the distribution of Y values on the condition that X has the value X_k. Associated with such a conditional distribution there is a conditional mean, which we denote by

$$\bar{Y}_k = \frac{\sum_{i=1}^{n_k} Y_{ik}}{n_k},$$

a conditional variance, which we denote by

$$s_{y \cdot k}^2 = \frac{\sum_{i=1}^{n_k} (Y_{ik} - \bar{Y}_k)^2}{n_k - 1},$$

and a regression-line value, $\hat{Y}_k = a + bX_k$. Because we find it useful to think of a separate regression-line value, $\hat{Y}_i$, for each individual in the sample, the regression equation is usually written $\hat{Y}_i = a + bX_i$, using the subscript i rather than k. Keep in mind, however, that this equation yields the same regression estimate for all individuals having $X_i = X_k$. In other words, there is only one regression-line value associated with each conditional distribution.

If there are several observations of Y for a given X and $s_{y \cdot k}^2 > 0$, then there must be discrepancies between the regression-line value and at least some of the observed values of Y for a given X. When there is only one observation of Y for a given X, so that $s_{y \cdot k}^2 = 0$, the observed Y and predicted Y may also be discrepant, because the relationship between X and Y in the sample is imperfect. A convenient measure of the size of these discrepancies, taken over all conditional distributions, is the *standard error of estimate*,

$$s_{y \cdot x} = \sqrt{\frac{\sum_{i=1}^{n} (Y_i - \hat{Y}_i)^2}{n - 2}}. \tag{1.3.1}$$

Note that the numerator of this expression is obtained by summing the squares of the discrepancies between the observed and predicted values of Y, for all n individuals in the sample. If $s_{y \cdot x}$ is large, then the error made in predicting Y from X is large. If $s_{y \cdot x}$ is small, then the error of prediction is small. If $s_{y \cdot x}$ is zero, each predicted value, $\hat{Y}_i$, is equal to the observed value, Y_i. In the latter case each of the conditional variances is equal to zero because all of the observed values of Y for any given X coincide with the regression-line value of Y.

The standard error of estimate may be interpreted in much the same way as a standard deviation. Both are measures of variation around a point of reference, the mean in the case of the standard deviation, and the regression-line values, $\hat{Y}_i$, in the case of the standard error of estimate. As can be seen from Figure 1.3.1(a), when the straight regression line represents well the kind of relationship between X and Y, the standard error of estimate provides a reasonable measure of the error of estimation for all values of X. However, if the straight line does not adequately represent the relationship, as in Figure 1.3.1(b), the standard error of estimate overesti-

FIGURE 1.3.1 Scatter Diagrams of Two X, Y Distributions Differing in Form of Relationship.

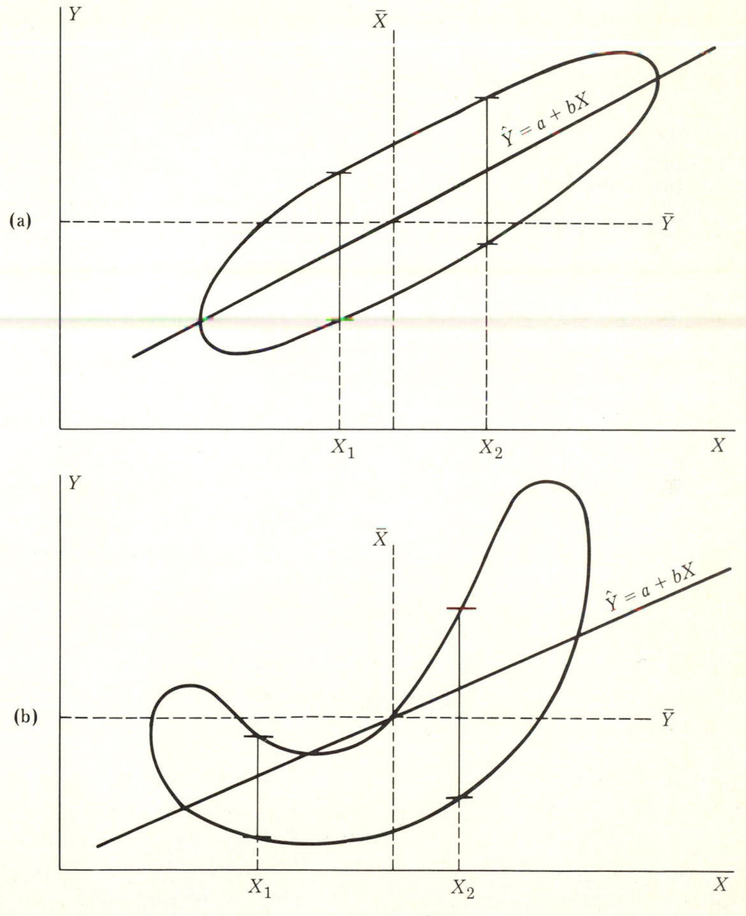

In (a), the variance of values around Y is equal at X_1 and X_2. In (b), the variance is larger at X_2 than at X_1.

mates the error for some values of X and underestimates the error for others. Thus, the proper interpretation of a computed standard error of estimate, as well as of other statistics in regression and correlation analysis, depends on how well the straight line represents the relationship between X and Y. We shall discuss this point further in a later section of this chapter.

A suitable computational formula for the standard error of estimate may be obtained algebraically by appropriate substitution for $\hat{Y}_i$ in Equation 1.3.1. Thus, by using the linear regression equation $\hat{Y}_i = a + bX_i$ and the fact that $a = \bar{Y} - b\bar{X}$, we can write

$$\begin{aligned}
\Sigma(Y_i - \hat{Y}_i)^2 &= \Sigma(Y_i - a - bX_i)^2 \\
&= \Sigma\left[Y_i - \bar{Y} - b(X_i - \bar{X})\right]^2 \\
&= \Sigma\left[(Y_i - \bar{Y})^2 + b^2(X_i - \bar{X})^2 - 2b(Y_i - \bar{Y})(X_i - \bar{X})\right] \\
&= S_{yy} + \frac{S_{xy}^2}{S_{xx}^2} S_{xx} - 2\frac{S_{xy}}{S_{xx}} S_{xy} = S_{yy} - \frac{S_{xy}^2}{S_{xx}}.
\end{aligned}$$

Then, substituting for $\Sigma(Y_i - \hat{Y}_i)^2$ in Equation 1.3.1,

$$s_{y \cdot x} = \sqrt{\frac{S_{yy} - \dfrac{S_{xy}^2}{S_{xx}}}{n-2}}. \tag{1.3.2}$$

To illustrate the application of Equation 1.3.2, we again use the data of Table 1.2.2. We know from previous calculations that $S_{xy} = 23$ and $S_{xx} = 28$. Using Equation 1.2.7, we obtain

$$S_{yy} = 140 - \frac{(28)^2}{7} = 140 - 112 = 28.$$

Then, substituting in Equation 1.3.2, we find that

$$s_{y \cdot x} = \sqrt{\frac{28 - \dfrac{(23)^2}{28}}{5}} = \sqrt{1.82143} = 1.35.$$

1.4 THE SECOND REGRESSION LINE

Just as Y may be predicted from a knowledge of X, the reverse is also possible. That is, X may be considered dependent and Y independent. The mathematical development of the regression equation is analogous to the case when Y is to be estimated. However, the regression coefficients, a and b, will *not*, in general, be equal to those in the former case, and some notational distinction must be made between them. When both regression lines are considered in a problem, they are usually written

$$\hat{Y} = a_{YX} + b_{YX}X, \tag{1.4.1}$$

and

$$\hat{X} = a_{XY} + b_{XY} Y. \tag{1.4.2}$$

Note that the second equation is obtained by interchanging X and Y in the first. The regression coefficients may be computed as follows:

$$b_{YX} = \frac{S_{xy}}{S_{xx}}; \tag{1.4.3}$$

$$a_{YX} = \bar{Y} - b_{YX}\bar{X}; \tag{1.4.4}$$

$$b_{XY} = \frac{S_{xy}}{S_{yy}}; \tag{1.4.5}$$

and

$$a_{XY} = \bar{X} - b_{XY}\bar{Y}. \tag{1.4.6}$$

Although it is useful to understand the distinction between the two regression lines and to be aware that either or both may be obtained in a given problem, one would rarely find it necessary to make use of both in an applied regression problem. Therefore, we shall omit the subscripts on a and b, unless it is necessary to make a distinction between the two lines. When subscripts are omitted, the reader should understand that Y is the dependent variable and X the independent variable.

1.5 THE LINEAR REGRESSION MODEL

The two variables involved in the linear regression model are defined quite differently. The predictor variable X is an *independent* variable, on which data are presumed to be available. No assumptions concerning the distribution of X are necessary in the model. In fact, statistically, X and functions of it are regarded as constants. Examples of independent variables are: (a) time or some function of it in a learning experiment, (b) pretest scores obtained prior to an experiment, and (c) entrance examination scores used to predict academic success.

The Y variable in the model is a *random* variable, the valves of which are assumed to *depend* on the values of X included in the sample. Examples of such a variable are: (a) scores on a postexperimental test of achievement, and (b) grade point averages earned by college freshmen. It is conceivable that a particular variable may be regarded as concomitant (independent) in one situation and as random (dependent) in another.

The linear regression model may be stated

$$Y_i = \alpha + \beta X_i + e_i. \tag{1.5.1}$$

This equation says that the ith observation on variable Y is a function of: (a) the two constants α and β, which are parameters of the model, (b) the value of the *i*th observation on X, and (c) some error, e_i. Assumptions about e_i for each conditional distribution are:

(1) e_i is a random variable with mean zero and variance $\sigma^2_{y \cdot x}$. That is, $E(e_i)=0$ and $\text{Var}(e_i)=\sigma^2_{y \cdot x}$. Note that $\sigma^2_{y \cdot x}$ is assumed to be equal in all conditional distributions. This property, equality of variances, is called *homoscedasticity*.

(2) e_i and e_j $(i \neq j)$ are not related. That is, knowing the error associated with one observation would tell us nothing about error in another. This implies that Y_i and Y_j $(i \neq j)$ are also unrelated.

(3) In applying estimation or hypothesis testing procedures, it is also necessary to assume that e_i is normally distributed. Under this assumption, together with assumption (2), e_i and e_j, as well as Y_i and Y_j $(i \neq j)$, are not only uncorrelated, but independent.

In this and in subsequent sections, it is important to distinguish between an entire population of Y values and two kinds of subpopulations of Y values that are determined by the X values included in a regression study. The distinction to be made will be clearer if we consider an example. Suppose we let Y be the grade point average (GPA), at the end of their first year of college, of those students from a particular school district who graduate from high school in a given year. Let X be the score (either Verbal or Quantitative) on the Scholastic Aptitude Test (SAT) taken in the senior year of high school by this population of students. The entire population of Y values would then be the GPAs of *all* graduating seniors in the district. The mean of this population is what we have denoted by μ_Y.

Of course, not all members of the population of graduating seniors go to college. In fact, many are not admitted because their SAT scores are not high enough to meet admission requirements. Therefore, when we consider the population of Y (GPA) values about which inferences can be drawn, we are really dealing with a subpopulation that is "defined" on the basis of the set of X (SAT) values that are high enough to satisfy college admission requirements. We denote the mean of *this* population of Y values by μ_y, to distinguish it from the mean μ_Y. Although our notation does not indicate it, μ_y is really a conditional mean, because it is the mean of a population of Y values that results from selecting a particular subset of X values. This distinction between μ_Y and μ_y is important, because in most regression studies it is μ_y, rather than μ_Y, about which we are able to draw inferences.

A third kind of Y mean is the mean of all Y values in the population *for a given X value*, denoted by $\mu_{y \cdot x}$. This would be the mean GPA, for example, of all graduating seniors from the district mentioned above who went to college *and* who had a particular SAT score, say 550. The relationship between μ_y and $\mu_{y \cdot x}$ is expressed by

$$\mu_{y \cdot x} = \mu_y + \beta(X_i - \overline{X}) = \alpha + \beta X, \tag{1.5.2}$$

where $\alpha = \mu_y - \beta\overline{X}$. Using this equation and the assumptions listed earlier, the linear regression model may also be stated as

$$E(Y_i)=\alpha+\beta X_i=\mu_{y\cdot x}=\mu_y+\beta(X_i-\bar{X}). \qquad (1.5.3)$$

Thus, Equations 1.5.2 and 1.5.3 may both be regarded as statements of the linear regression model. We should emphasize that the expectation operators used here and in subsequent sections indicate *conditional* rather than unconditional expectations. That is, they denote estimates of parameters of that subpopulation of Y values that is determined by the set of X values included in the regression problem under investigation. Thus, for example, the expected value of the sample mean, $\bar{Y}$, is $E(\bar{Y})=\mu_y$, and not μ_Y.

1.5.1 A Comment about Regression Toward the Mean

Having examined the linear regression model, we can now offer an explanation of the phenomenon known as regression toward the mean. Let us rewrite Equation 1.5.2 as

$$\mu_{y\cdot x}-\mu_y=\beta(X_k-\bar{X}),$$

where X_k is a particular value of X. The difference on the left is between the mean Y for a particular value of X and the mean Y for *all* values of X, while that on the right is between a particular X and the mean of all values of X. With reference to Galton's work, we let Y represent son's height and X father's height. Then $\mu_{y\cdot x}-\mu_y$ is the difference between the mean height of sons whose fathers have height X_k and the mean height of all sons, and $X_k-\bar{X}$ is the difference between father's height, X_k, and the mean height of all fathers.

Now it is easy to see that if β is between 0 and 1, that is, $0<\beta<1$, then $|\mu_{y\cdot x}-\mu_y|<|X_k-\bar{X}|$. This is in fact the regression of sons' heights toward their mean that Galton observed. In other words, if $X_k-\bar{X}$ is a relatively large positive value (very tall fathers), $\mu_{y\cdot x}-\mu_y$ is smaller than $X_k-\bar{X}$, but still positive (tall sons, but not as tall, on the average, as their fathers). If $X_k-\bar{X}$ is a relatively large negative value (very short fathers), $|\mu_{y\cdot x}-\mu_y|$ is less than $|X_k-\bar{X}|$ (short sons, but not as short, on the average, as their fathers). But how likely is it that $0<\beta<1$? As we shall see in Chapter 2, if: (1) the relationship between fathers' and sons' heights is not perfect (which it is not); (2) tall fathers tend to have tall sons and short fathers tend to have short sons (which Galton observed to be true); and (3) the variance of sons' heights is approximately equal to the variance of fathers' heights (which it is), then β is almost sure to be between 0 and 1.

It is interesting to note that if we merely interchange X and Y so that Y is father's height and X is son's height, the same argument shows that father's height also regresses toward the mean. Thus, father's height and son's height *appear* to be regressing toward the mean at the same time.

1.6 INFERENCES ABOUT PARAMETERS IN THE LINEAR REGRESSION MODEL

There are a number of parameters associated with the linear regression model, for example, μ_y, α, β, $\mu_{y \cdot x}$, and $\sigma^2_{y \cdot x}$. The Xs in this model are regarded as constants, so-called *known parameters*. They are not estimated, but are actually determined from the sample. The parameters α and β, on the other hand, are estimated from the sample by the regression coefficients, a and b. The conditional mean, $\mu_{y \cdot x}$, may also be estimated from the sample data. Interest in estimation and hypothesis testing is usually centered on μ_y, β, and $\mu_{y \cdot x}$.

In order to form confidence intervals or to test hypotheses regarding μ_y or β, it is necessary to identify suitable estimators of them, and to determine certain parameters of the sampling distributions of these estimates. Consider first the parameter μ_y. When data on a concomitant variable are available, $\bar{Y}$ is an unbiased estimator of μ_y, with variance $\sigma^2_{y \cdot x}/n$. The proofs follow.

(1) $E(\bar{Y}) = \mu_y$.

Since

$$Y_i = \alpha + \beta X_i + e_i = \mu_y + \beta(X_i - \bar{X}) + e_i,$$

$$\Sigma Y_i = n\mu_y + \beta\Sigma(X_i - \bar{X}) + \Sigma e_i,$$

and

$$\bar{Y} = \mu_y + \frac{\Sigma e_i}{n},$$

because

$$\Sigma(X_i - \bar{X}) = 0.$$

Therefore,

$$E(\bar{Y}) = \mu_y + \frac{E(\Sigma e_i)}{n} = \mu_y,$$

because

$$E(e_i) = 0.$$

(2) $\mathrm{Var}(\bar{Y}) = \dfrac{\sigma^2_{y \cdot x}}{n}$.

$$\mathrm{Var}(\bar{Y}) = \mathrm{Var}\left(\mu_y + \frac{\Sigma e_i}{n}\right) = \mathrm{Var}\left(\frac{\Sigma e_i}{n}\right)$$

$$= \frac{1}{n^2}[\mathrm{Var}(e_1) + \ldots + \mathrm{Var}(e_n)]$$

$$= \frac{n\sigma^2_{y \cdot x}}{n^2} = \frac{\sigma^2_{y \cdot x}}{n}. \tag{1.6.1}$$

Note that $\sigma^2_{y \cdot x}$ in this context is *not* the variance of Y values taken about μ_y, but instead is the variance around $\mu_{y \cdot x}$, that is, around the regression line. If there is some degree of linear relationship between X and Y, $\sigma^2_{y \cdot x}$ can be expected to be smaller than the variance of Y values taken from μ_y. The square of the standard error of estimate, $s^2_{y \cdot x}$, is an appropriate sample estimate of $\sigma^2_{y \cdot x}$.

If we estimate β by b, computed from sample values by means of Equation 1.2.9, then the expected value of b is β and its variance is $\sigma^2_{y \cdot x}/S_{xx}$.

The necessary proofs are:

(1) $E(b) = \beta$.

$$E(b) = E\left(\frac{S_{xy}}{S_{xx}}\right).$$

Because the values of the concomitant variable are regarded as constant in the model, S_{xx} is a constant in the above expression. Therefore,

$$E(b) = \frac{1}{S_{xx}}\left[E(S_{xy})\right].$$

$$\begin{aligned} E(S_{xy}) &= E\left[\Sigma(Y_i - \bar{Y})(X_i - \bar{X})\right] \\ &= E\left[\Sigma(X_i - \bar{X})Y_i - \Sigma(X_i - \bar{X})\bar{Y}\right] \\ &= E\left[\Sigma(X_i - \bar{X})Y_i\right], \end{aligned}$$

because $\bar{Y}\Sigma(X_i - \bar{X}) = 0$.

$$\begin{aligned} E\left[\Sigma(X_i - \bar{X})Y_i\right] &= E\left[(X_1 - \bar{X})Y_1\right] + \ldots + E\left[(X_n - \bar{X})Y_n\right] \\ &= E\left[(X_1 - \bar{X})(\mu_y + \beta(X_1 - \bar{X}) + e_1)\right] + \ldots \\ &\quad + E\left[(X_n - \bar{X})(\mu_y + \beta(X_n - \bar{X}) + e_n)\right] \\ &= E\left[(X_1 - \bar{X})\mu_y + \beta(X_1 - \bar{X})^2 + (X_1 - \bar{X})e_1\right] + \ldots \\ &\quad + E\left[(X_n - \bar{X})\mu_y + \beta(X_n - \bar{X})^2 + (X_n - \bar{X})e_n\right] \\ &= E\left[\mu_y\Sigma(X_i - \bar{X}) + \beta\Sigma(X_i - \bar{X})^2 + \Sigma(X_i - \bar{X})e_i\right] \\ &= E\left[\beta\Sigma(X_i - \bar{X})^2\right], \end{aligned}$$

because $\Sigma(X_i - \bar{X}) = 0$ and $E(e_i) = 0$. Therefore, since the Xs and functions of them are constants in the model,

$$E\left[\Sigma(X_i - \bar{X})Y_i\right] = E\left[\beta\Sigma(X_i - \bar{X})^2\right] = \beta S_{xx},$$

and

$$E(b) = \frac{\beta S_{xx}}{S_{xx}} = \beta. \qquad (1.6.2)$$

(2) $$\mathrm{Var}(b) = \frac{\sigma^2_{y \cdot x}}{S_{xx}}.$$

$$\mathrm{Var}(b) = \mathrm{Var}\left(\frac{S_{xy}}{S_{xx}}\right) = \frac{1}{S^2_{xx}}\mathrm{Var}(S_{xy}).$$

$$\begin{aligned}\mathrm{Var}(S_{xy}) &= \mathrm{Var}\left[\Sigma(X_i - \bar{X})(Y_i - \bar{Y})\right] \\ &= \mathrm{Var}\left[\Sigma(X_i - \bar{X})Y_i\right] \\ &= \mathrm{Var}\left[(X_1 - \bar{X})Y_1 + \cdots + (X_n - \bar{X})Y_n\right] \\ &= \mathrm{Var}\left[(X_1 - \bar{X})Y_1\right] + \cdots + \mathrm{Var}\left[(X_n - \bar{X})Y_n\right],\end{aligned}$$

because the $(X_i - \bar{X})$ are regarded as constants and $\mathrm{Cov}(Y_i, Y_j) = 0$ $(i \neq j)$. Therefore,

$$\begin{aligned}\mathrm{Var}(S_{xy}) &= (X_1 - \bar{X})^2 \mathrm{Var}(Y_1) + \cdots + (X_n - \bar{X})^2 \mathrm{Var}(Y_n) \\ &= \sigma^2_{y \cdot x} S_{xx},\end{aligned} \qquad (1.6.3)$$

and

$$\mathrm{Var}(b) = \frac{\sigma^2_{y \cdot x} S_{xx}}{S^2_{xx}} = \frac{\sigma^2_{y \cdot x}}{S_{xx}}. \qquad (1.6.4)$$

1.6.1 Interval Estimation in Linear Regression

Interval estimates in this chapter may be obtained from probability statements whose general form is

$$P\left[\hat{\theta} + t_{\frac{\alpha}{2}; n_e} s(\hat{\theta}) < \theta < \hat{\theta} + t_{1-\frac{\alpha}{2}; n_e} s(\hat{\theta})\right] = 1 - \alpha, \qquad (1.6.5)$$

where θ is the parameter to be estimated, $\hat{\theta}$ is the estimator of θ based on a sample of size n, $t_{\frac{\alpha}{2}; n_e}$ and $t_{1-\frac{\alpha}{2}; n_e}$ are percentile points in the t distribution with n_e degrees of freedom*, $s(\hat{\theta})$ is a sample estimator of the standard error of θ, and $1 - \alpha$ is the level of confidence. To obtain interval estimates of μ_y and β one simply has to obtain the necessary numerical values and substitute them in Equation 1.6.5.

The standard errors of the estimators of μ_y and β involve $\sigma^2_{y \cdot x}$, which is unknown and must be estimated from sample data. As we stated earlier, a reasonable estimator of $\sigma^2_{y \cdot x}$ is the square of the standard error of estimate,

*For example, if $\alpha = .05$ and $n_e = 15$, then $t_{\frac{\alpha}{2}; n_e} = t_{.025; 15} = -2.13$, and $t_{1-\frac{\alpha}{2}; n_e} = t_{.975; 15} = 2.13$.

defined by Equation 1.3.2. We now show that this estimator is unbiased, that is, that $E(s^2_{y \cdot x}) = \sigma^2_{y \cdot x}$.

$$E\left(s^2_{y \cdot x}\right) = \frac{1}{n-2} E\left(S_{yy} - \frac{S^2_{xy}}{S_{xx}}\right)$$
$$= \frac{1}{n-2}\left[E(S_{yy}) - \frac{1}{S_{xx}} E\left(S^2_{xy}\right)\right].$$
$$E\left(S^2_{xy}\right) = \text{Var}(S_{xy}) + \left[E(S_{xy})\right]^2,$$

which becomes, by Equations 1.6.2 and 1.6.3,

$$E\left(S^2_{xy}\right) = \sigma^2_{y \cdot x} S_{xx} + \beta^2 S^2_{xx}.$$

To find $E(S_{yy})$, we first write Equation 1.2.7 as

$$S_{yy} = \Sigma Y_i^2 - n\bar{Y}^2.$$

Then

$$E(S_{yy}) = E(\Sigma Y_i^2) - nE(\bar{Y}^2).$$

Since $\text{Var}(Y_i) = E(Y_i^2) - [E(Y_i)]^2$,

$$E(Y_i^2) = \text{Var}(Y_i) + [E(Y_i)]^2 = \sigma^2_{y \cdot x} + \left[\mu_y + \beta(X_i - \bar{X})\right]^2.$$

Summing over the n sample observations,

$$E(\Sigma Y_i^2) = n\sigma^2_{y \cdot x} + \Sigma\left[\mu_y + \beta(X_i - \bar{X})\right]^2$$
$$= n\sigma^2_{y \cdot x} + \Sigma\left[\mu_y^2 + \beta^2(X_i - \bar{X})^2 + 2\mu_y\beta(X_i - \bar{X})\right]$$
$$= n\sigma^2_{y \cdot x} + n\mu_y^2 + \beta^2 S_{xx}.$$

Now, since

$$E(\bar{Y}^2) = \text{Var}(\bar{Y}) + \left[E(\bar{Y})\right]^2 = \frac{\sigma^2_{y \cdot x}}{n} + \mu_y^2,$$

we can write

$$E(S_{yy}) = n\sigma^2_{y \cdot x} + n\mu_y^2 + \beta^2 S_{xx} - n\left(\frac{\sigma^2_{y \cdot x}}{n} + \mu_y^2\right)$$
$$= n\sigma^2_{y \cdot x} - \sigma^2_{y \cdot x} + \beta^2 S_{xx}$$
$$= \sigma^2_{y \cdot x}(n-1) + \beta^2 S_{xx}.$$

Now, substituting for $E(S_{yy})$ and $E(S^2_{xy})$ yields

$$E\left(s^2_{y \cdot x}\right) = \frac{1}{n-2}\left[\sigma^2_{y \cdot x}(n-1) + \beta^2 S_{xx} - \sigma^2_{y \cdot x} - \beta^2 S_{xx}\right]$$
$$= \sigma^2_{y \cdot x}. \tag{1.6.6}$$

We are now prepared to show probability statements that may be used to obtain interval estimates of μ_y and β. We make use of the general form of probability statement given in Equation 1.6.5.

Interval Estimation of μ_y. In this case, $\theta = \mu_y$, $\hat{\theta} = \bar{Y}$, $n_e = n-2$, and $s(\hat{\theta}) = s_{y \cdot x}/\sqrt{n}$. The probability statement is

$$P\left[\bar{Y} + t_{\frac{\alpha}{2};n-2}\frac{s_{y \cdot x}}{\sqrt{n}} < \mu_y < \bar{Y} + t_{1-\frac{\alpha}{2};n-2}\frac{s_{y \cdot x}}{\sqrt{n}}\right] = 1 - \alpha. \qquad (1.6.7)$$

To illustrate the application of Equation 1.6.7, we shall use the data of Table 1.2.1. For those data we have $\bar{Y} = 28.9767$, $n = 43$, and $s_{y \cdot x} = 5.5813$. If we wish to construct a 95% two-sided confidence interval, $t_{.975;41} = 2.02$, and the lower and upper limits are

$$\text{Lower: } 28.9767 - 2.02\left(\frac{5.5813}{\sqrt{43}}\right) = 27.2574$$

$$\text{Upper: } 28.9767 + 2.02\left(\frac{5.5813}{\sqrt{43}}\right) = 30.6960.$$

We conclude with 95% confidence that these limits include the value of μ_y.

Note that the width of this interval, $30.6960 - 27.2574 = 3.4386$, is smaller than the one that would have been obtained without using the information on X. In that case, the interval would be obtained by substituting in

$$P\left[\bar{Y} + t_{\frac{\alpha}{2};n-1}\frac{s_y}{\sqrt{n}} < \mu_y < \bar{Y} + t_{1-\frac{\alpha}{2};n-1}\frac{s_y}{\sqrt{n}}\right] = 1 - \alpha.$$

The necessary values are $\bar{Y} = 28.9767$, $n = 43$, and $s_y = 6.8678$. Limits for the 95% interval are

$$\text{Lower: } 28.9767 - 2.02\left(\frac{6.8678}{\sqrt{43}}\right) = 26.8611$$

$$\text{Upper: } 28.9767 + 2.02\left(\frac{6.8678}{\sqrt{43}}\right) = 31.0923.$$

Here the width of the interval is $31.0923 - 26.8611 = 4.2312$, which is noticeably larger than that obtained using Equation 1.6.7. Thus, using information on X produces a more precise estimate of μ_y than could otherwise be obtained.

Recall that we distinguished in Section 1.5 between μ_Y and μ_y. It is important to note that the interval obtained using Equation 1.6.7 estimates μ_y and not μ_Y. One should take this distinction into account when designing regression studies, particularly if one of the objectives is to estimate the mean of a population of Y values. The definition of the population having a mean μ_y depends on the design of the study. The selection of different sets of Xs would lead to different values of μ_y and, hence, to different estimates.

Interval Estimation of β. In this case, $\theta = \beta$, $\hat{\theta} = b$, $n_e = n-2$, and $s(\hat{\theta}) = \sqrt{\hat{\text{V}}\text{ar}(b)}$, in which $\text{Var}(b)$ is equal to $s_{y \cdot x}^2/S_{xx}$. The probability statement is

$$P\left[b+t_{\frac{\alpha}{2};n-2}\frac{s_{y\cdot x}}{\sqrt{S_{xx}}}<\beta<b+t_{1-\frac{\alpha}{2};n-2}\frac{s_{y\cdot x}}{\sqrt{S_{xx}}}\right]=1-\alpha. \quad (1.6.8)$$

Again using the data of Table 1.2.1 for illustration, we have $b=.7032$, $s_{y\cdot x}=5.5813$, and $S_{xx}=1423.1163$. For a 95% confidence interval, $t_{.975;41}=2.02$. The limits are

$$\text{Lower: } .7032-2.02\left(\frac{5.5813}{\sqrt{1423.1163}}\right)=.4043$$

$$\text{Upper: } .7032+2.02\left(\frac{5.5813}{\sqrt{1423.1163}}\right)=1.0021.$$

We conclude with 95% confidence that these limits include β.

Interval Estimation of the Mean Value of Y for a Given X. Earlier in this chapter we formulated the sample regression equation for obtaining a point estimate, $\hat{Y}_i$, of $\mu_{y\cdot x}$, the mean of Y for a given X. In some cases it is more useful to have an interval estimate than a point estimate. The form of the estimator is the same as that shown by Equation 1.6.5. However, we must determine the variance of the estimator $\hat{Y}_i$, and find a suitable estimator of the variance, in order to formulate the desired confidence interval.

Recall that the linear regression model may be stated $Y_i=\mu_y+\beta(X_i-\overline{X})+e_i$, where β and μ_y are constant regardless of the value of X_i. The variation in $\hat{Y}_i$ is due to

1. the sampling variation in $\overline{Y}$: $\text{Var}(\overline{Y})=\dfrac{\sigma^2_{y\cdot x}}{n}$, and
2. the sampling variation in b: $\text{Var}(b)=\dfrac{\sigma^2_{y\cdot x}}{S_{xx}}$.

To obtain an expression for the variance of $\hat{Y}_i$, we must first show that $\text{Cov}(\overline{Y},b)=0$.

Let $c=c_1Y_1+\cdots+c_nY_n=\Sigma c_iY_i$ and $d=d_1Y_1+\cdots+d_nY_n=\Sigma d_iY_i$, in which c_i and d_i $(i=1,\ldots,n)$ are constants. The covariance of c and d is

$$\text{Cov}(c,d)=E[(c-E(c))(d-E(d))]$$

in which

$$\begin{aligned}E(c)&=E(c_1Y_1+\cdots+c_nY_n)\\&=E(c_1Y_1)+\cdots+E(c_nY_n)\\&=c_1E(Y_1)+\cdots+c_nE(Y_n)\\&=\mu_y(c_1+\cdots+c_n).\end{aligned}$$

Similarly, $E(d)=\mu_y(d_1+\cdots+d_n)$. Then

$$\begin{aligned}\mathrm{Cov}(c,d) &= E\big[(c_1Y_1+\cdots+c_nY_n-\mu_y(c_1+\cdots+c_n))\\ &\qquad\times(d_1Y_1+\cdots+d_nY_n-\mu_y(d_1+\cdots+d_n))\big]\\ &= E\big[(c_1(Y_1-\mu_y)+\cdots+c_n(Y_n-\mu_y))\\ &\qquad\times(d_1(Y_1-\mu_y)+\cdots+d_n(Y_n-\mu_y))\big].\end{aligned}$$

Because Y_i and Y_j $(i\neq j)$ are assumed to be uncorrelated, all terms of the form $E[c_id_j(Y_i-\mu_y)(Y_j-\mu_y)]=c_id_jE[(Y_i-\mu_y)(Y_j-\mu_y)]$ $(i\neq j)$ are equal to zero. Hence,

$$\begin{aligned}\mathrm{Cov}(c,d) &= E\big[c_1d_1(Y_1-\mu_y)^2+\cdots+c_nd_n(Y_n-\mu_y)^2\big]\\ &= c_1d_1E(Y_1-\mu_y)^2+\cdots+c_nd_nE(Y_n-\mu_y)^2\\ &= \sigma_{y\cdot x}^2(c_1d_1+\cdots+c_nd_n)\\ &= \sigma_{y\cdot x}^2\Sigma c_id_i.\end{aligned}$$

Now, if we set $c=\bar{Y}=\frac{1}{n}Y_1+\cdots+\frac{1}{n}Y_n=\Sigma c_iY_i$, it is clear that $c_i=\frac{1}{n}$. Furthermore, if we set $d=b$, where $b=S_{xy}/S_{xx}$, then

$$\begin{aligned}d &= \frac{1}{S_{xx}}\Sigma(X_i-\bar{X})(Y_i-\bar{Y})\\ &= \frac{1}{S_{xx}}\big[\Sigma Y_i(X_i-\bar{X})-\bar{Y}\Sigma(X_i-\bar{X})\big]\\ &= \frac{X_1-\bar{X}}{S_{xx}}Y_1+\cdots+\frac{X_n-\bar{X}}{S_{xx}}Y_n\\ &= \Sigma\frac{X_i-\bar{X}}{S_{xx}}Y_i=\Sigma d_iY_i,\end{aligned}$$

because $\bar{Y}\Sigma(X_i-\bar{X})=0$. Thus, it is clear that $d_i=\frac{X_i-\bar{X}}{S_{xx}}$. Substituting in the above equation for $\mathrm{Cov}(c,d)$, we find that

$$\mathrm{Cov}(\bar{Y},b)=\sigma_{y\cdot x}^2\Sigma\left(\frac{1}{n}\right)\left(\frac{X_i-\bar{X}}{S_{xx}}\right)=\frac{\sigma_{y\cdot x}^2}{nS_{xx}}\Sigma(X_i-\bar{X})=0.$$

Then the variance of $\hat{Y}_i=\bar{Y}+b(X_i-\bar{X})$ is

$$\begin{aligned}\mathrm{Var}(\hat{Y}_i) &= \mathrm{Var}(\bar{Y})+\mathrm{Var}\big[b(X_i-\bar{X})\big]\\ &= \frac{\sigma_{y\cdot x}^2}{n}+\frac{(X_i-\bar{X})^2\sigma_{y\cdot x}^2}{S_{xx}}=\sigma_{y\cdot x}^2\left[\frac{1}{n}+\frac{(X_i-\bar{X})^2}{S_{xx}}\right].\end{aligned}$$

We are now prepared to show the probability statement for use in estimating $\mu_{y \cdot x}$. In this case $\theta = \mu_{y \cdot x}$, $\hat{\theta} = \hat{Y}_i = a + bX_i$, $n_e = n - 2$, and

$$s(\hat{\theta}) = s_{y \cdot x}\sqrt{\frac{1}{n} + \frac{(X_i - \bar{X})^2}{S_{xx}}} = s(\hat{Y}_i).$$

The probability statement is

$$P\left[\hat{Y}_i + t_{\frac{\alpha}{2};n-2}s(\hat{Y}_i) < \mu_{y \cdot x} < \hat{Y}_i + t_{1-\frac{\alpha}{2};n-2}s(\hat{Y}_i)\right] = 1 - \alpha. \quad (1.6.9)$$

Note that the width of this confidence interval depends, in part, upon the magnitude of the deviation $X_i - \bar{X}$. This implies that the accuracy of estimates of $\mu_{y \cdot x}$ is greater for values of X near $\bar{X}$ and decreases as values of X deviate more from $\bar{X}$.

Prediction Interval of Y for an Individual Having a Given X. In this case $\theta = Y_{i \cdot x}$, $\hat{\theta} = \hat{Y}_i = a + bX_i$, $n_e = n - 2$, and

$$s(\hat{\theta}) = s_{y \cdot x}\sqrt{1 + \frac{1}{n} + \frac{(X_i - \bar{X})^2}{S_{xx}}} = s_i(\hat{Y}_i),$$

where $s_i(\hat{Y}_i)$ is used to distinguish this estimator from $s(\hat{Y}_i)$, used when the mean $\mu_{y \cdot x}$ is being estimated. The probability statement is

$$P\left[\hat{Y}_i + t_{\frac{\alpha}{2};n-2}s_i(\hat{Y}_i) < Y_{i \cdot x} < \hat{Y}_i + t_{1-\frac{\alpha}{2};n-2}s_i(\hat{Y}_i)\right] = 1 - \alpha. \quad (1.6.10)$$

Note here that the estimator, $\hat{Y}_i$, is the same as that used in estimating $\mu_{y \cdot x}$, but the standard error of the estimator is different. The result is that the prediction interval of an individual value will always be wider than an interval estimate of the mean value, $\mu_{y \cdot x}$, for a given X. The standard error of the individual observation takes into account the variation of Y values around the estimated regression line.

To illustrate the use of Equations 1.6.9 and 1.6.10 with data from Table 1.2.1, we shall obtain 95% confidence intervals for $\mu_{y \cdot 35}$ and for the Y value of a person having $X = 35$. To estimate $\mu_{y \cdot 35}$, the mean Y value of all persons having $X = 35$, we need to compute $\hat{Y} = 7.3235 + .7032(35) = 31.9355$,

$$s(\hat{Y}_i) = 5.5813\sqrt{\frac{1}{43} + \frac{(35 - 30.7907)^2}{1423.1163}} = 1.0546.$$

The necessary t value is again $t_{.975;41} = 2.02$. The limits for $\mu_{y \cdot 35}$ are

Lower: $31.9355 - 2.02(1.0546) = 29.8052$

Upper: $31.9355 + 2.02(1.0546) = 34.0658$.

To obtain the 95% prediction interval of the Y value of an individual having $X = 35$, we substitute in Equation 1.6.10 with

$$s_i(\hat{Y}_i) = 5.5813\sqrt{1 + \frac{1}{43} + \frac{(35 - 30.7907)^2}{1423.1163}} = 5.6801.$$

The limits are

Lower: $31.9355-2.02(5.6801)=20.4617$

Upper: $31.9355+2.02(5.6801)=43.4093.$

Note that the interval for the individual score is much wider than that for the mean, $\mu_{y\cdot 35}$.

1.6.2 Tests of Hypotheses in Linear Regression

We may outline a general procedure for testing hypotheses about a parameter, θ, as follows:

(1) State the null hypothesis, H_0: $\theta=\theta^*$, where θ^* denotes a specific value of θ, and one of the alternatives, H_1: $\theta\neq\theta^*$, H_1: $\theta<\theta^*$, or H_1: $\theta>\theta^*$.

(2) Specify the test statistic.

(3) State the distribution of the test statistic under H_0.

(4) Specify α, the level of significance of the test.

(5) Set up the critical region corresponding to the alternative of interest.

(6) From the sample data, compute the value of the test statistic.

(7) Compare the value of the test statistic with the critical region. If the computed value of the test statistic falls within the critical region, reject H_0. Otherwise accept H_0.

This general procedure can be applied in carrying out a large number of different tests. Most of these differ from one another only in terms of the test statistic to be used. In the tests described below, only the test statistic and its distribution will be given.

Test of the Hypothesis H_0: $\beta=\beta^*$. This is a test of the hypothesis that the population regression coefficient, β, has some specific value, β^*. The test statistic is

$$t=\frac{(b-\beta^*)}{\sqrt{\dfrac{s_{y\cdot x}^2}{S_{xx}}}}=\frac{(b-\beta^*)\sqrt{S_{xx}}}{\sqrt{s_{y\cdot x}^2}}. \tag{1.6.11}$$

Note that the factor $\sqrt{s_{y\cdot x}^2/S_{xx}}$ is the standard deviation of the sampling distribution of b (see Equation 1.6.4), with $\sigma_{y\cdot x}^2$ estimated by $s_{y\cdot x}^2$. When H_0 is true, the statistic has the t distribution with $n-2$ degrees of freedom.

To illustrate this test we consider again the data of Table 1.2.1, where $b=.7032$, $s_{y\cdot x}=5.5813$, and $S_{xx}=1423.1163$. Suppose we wish to test the null hypothesis H_0: $\beta=.5$, against the alternative H_1: $\beta>.5$, with $\alpha=.05$. Such a test may have been motivated by some past experience that led us to believe that β might be very close to .5. Substituting in Equation 1.6.11, we have

$$t=\frac{(.7032-.5)\sqrt{1423.1163}}{5.5813}=1.3734.$$

Because $n=43$, and H_1 suggests a one-tailed test, we compare $t=1.3734$ with the critical region, $t>1.68$ ($t_{.95;41}=1.68$). Because the observed t is not in the critical region, we conclude that there is insufficient evidence to reject the null hypothesis. Thus, although the test does not prove that $\beta=.5$, it does not provide convincing evidence that β is *greater* than .5.

Test of the Hypothesis H_0: $\beta=0$. Although this test may be carried out by specifying $\beta^*=0$ in Equation 1.6.11, an alternative procedure is both useful and instructive. It involves a breakdown of S_{yy} into regression and error components. Consider the identity $Y_i-\bar{Y}=Y_i-\hat{Y}_i+\hat{Y}_i-\bar{Y}$. Squaring both sides yields

$$(Y_i-\bar{Y})^2=(Y_i-\hat{Y}_i)^2+(\hat{Y}_i-\bar{Y})^2+2(Y_i-\hat{Y}_i)(\hat{Y}_i-\bar{Y}).$$

If the last term on the right side is summed over the n sample observations, it becomes

$$\Sigma(Y_i-\hat{Y}_i)(\hat{Y}_i-\bar{Y})=\Sigma(Y_i-\hat{Y}_i)b(X_i-\bar{X}),$$

because $\hat{Y}_i=\bar{Y}+b(X_i-\bar{X})$, from Equations 1.4.1 and 1.4.4. Then, further substitution for $\hat{Y}_i$ yields

$$\begin{aligned}\Sigma(Y_i-\hat{Y}_i)(\hat{Y}_i-\bar{Y})&=b\Sigma\left[(Y_i-\bar{Y})-b(X_i-\bar{X})\right](X_i-\bar{X})\\&=b\Sigma\left[(X_i-\bar{X})(Y_i-\bar{Y})-b(X_i-\bar{X})^2\right]\\&=bS_{xy}-b^2S_{xx}\\&=S_{xy}^2/S_{xx}-S_{xy}^2/S_{xx}=0.\end{aligned}$$

Therefore,

$$\Sigma(Y_i-\bar{Y})^2=\Sigma(Y_i-\hat{Y}_i)^2+\Sigma(\hat{Y}_i-\bar{Y})^2. \qquad (1.6.12)$$

An examination of Equation 1.3.2 shows that we may write this equation as

$$S_{yy}=s_{y\cdot x}^2(n-2)+\frac{S_{xy}^2}{S_{xx}}.$$

This breakdown of S_{yy} is summarized in Table 1.6.1.

TABLE 1.6.1 SUMMARY OF ANALYSIS OF VARIANCE TEST OF H_0: $\beta=0$

Source of variation	*Sum of squares (SS)*	*Degrees of freedom*	*Mean square (MS)*	*F*
Regression	$SS_r=\frac{S_{xy}^2}{S_{xx}}$	1	$MS_r=SS_r$	$\frac{MS_r}{MS_e}$
Error	$SS_e=S_{yy}-\frac{S_{xy}^2}{S_{xx}}$	$n-2$	$MS_e=\frac{SS_e}{n-2}$	
Total	$SS_t=S_{yy}$	$n-1$		

It can be shown that $n-2$ degrees of freedom are associated with $\Sigma(Y_i - \hat{Y}_i)^2$, and that $n-1$ degrees of freedom are associated with S_{yy}. By subtraction, one degree of freedom is associated with S_{xy}^2/S_{xx}. It can also be shown that when H_0 is true, the test statistic,

$$F = \frac{(S_{xy}^2/S_{xx})(n-2)}{S_{yy} - S_{xy}^2/S_{xx}}, \tag{1.6.13}$$

has the F distribution with 1 and $n-2$ degrees of freedom.

We shall again use the data of Table 1.2.1 to illustrate the test. Table 1.6.2 is the summary table based on those data.

Since $F_{.95;\,1,41} = 4.07$, the critical region is $F > 4.07$. The observed result, $F = 22.5932$, falls within the critical region. Therefore, we reject H_0, and conclude that the value of β is greater than zero in the population of statistics students from which our sample was drawn.

Test for Lack of Fit of the Linear Model. Thusfar, when we have obtained interval estimates or tested hypotheses we have assumed that the straight-line or linear model is correct. Under that assumption, $s_{y\cdot x}^2$ is an unbiased estimate of $\sigma_{y\cdot x}^2$, as shown earlier.

In the sample, consider the conditional distribution of Y for a particular value of X, say X_k. Associated with this distribution we have two, possibly different, expressions for the variation of Y values around a measure of central position. One of these is

$$\sum_{i=1}^{n_k} (Y_{ik} - \hat{Y}_k)^2,$$

which is based on deviations from the linear regression estimate. The other is

$$\sum_{i=1}^{n_k} (Y_{ik} - \bar{Y}_k)^2,$$

which is based on deviations from the sample mean of the conditional distribution. It can be shown algebraically that the sum of squared deviations around a constant, that is, $\Sigma(Y_i - c)^2$, is a minimum if $c = \bar{Y}$. Therefore,

$$\sum_{i=1}^{n_k} (Y_{ik} - \hat{Y}_k)^2 \geqq \sum_{i=1}^{n_k} (Y_{ik} - \bar{Y}_k)^2.$$

TABLE 1.6.2 SUMMARY OF ANALYSIS OF VARIANCE BASED ON THE DATA OF TABLE 1.2.1

Source of variation	*Sum of squares*	*Degrees of freedom*	*Mean square*	*F*
Regression	703.7949	1	703.7949	22.5932
Error	1277.1818	41	31.1508	
Total	1980.9767	42		

TABLE 1.6.3 SUMMARY OF LACK-OF-FIT TEST OF THE LINEAR REGRESSION MODEL

Source of variation	*Sum of squares*	*Degrees of freedom*	*Mean square*	*F*
Lack of fit	SS_L	$K-2$	$MS_L = \dfrac{SS_L}{K-2}$	$\dfrac{MS_L}{MS_p}$
Pure error	SS_p	$n-K$	$MS_p = \dfrac{SS_p}{n-K}$	
Error	SS_e	$n-2$		

Summing the left side of the above inequality over all K conditional distributions (one for each value of X), we have

$$\sum_{k=1}^{K} \sum_{i=1}^{n_k} (Y_{ik} - \hat{Y}_k)^2.$$

Similarly, summing the right side of the above inequality, we have

$$\sum_{k=1}^{K} \sum_{i=1}^{n_k} (Y_{ik} - \bar{Y}_k)^2,$$

which we call the sum of squares for "pure" error (SS_p). The difference between these two sums of squares is

$$SS_L = \sum_{k=1}^{K} \sum_{i=1}^{n_k} (Y_{ik} - \hat{Y}_k)^2 - \sum_{k=1}^{K} \sum_{i=1}^{n_k} (Y_{ik} - \bar{Y}_k)^2,$$

in which SS_L is the sum of squares for lack of fit. When SS_L is divided by the appropriate number of degrees of freedom, it can be used in a test for lack of fit of the linear model. The test is summarized in Table 1.6.3.

Note that this test cannot be carried out unless there are at least two Y values for at least one value of X in the sample. A significant F value means that the linear model is not entirely satisfactory, even though a significant linear effect may be operating. A nonsignificant F means that there is no reason to doubt the adequacy of the linear model. However, a nonsignificant F does not prove that the linear model is the correct one.

To illustrate this test, and to clarify the distinction between cases where the linear model is appropriate and those where it is not, we consider the following simple data sets:

(a)	X:	1	1	2	2	3	3	4	4	5	5	6	6	7	7
	Y:	2	6	3	5	4	8	5	8	7	8	7	11	10	12
(b)	X:	1	1	2	2	3	3	4	4	5	5	6	6	7	7
	Y:	1	2	4	8	7	11	10	12	9	10	4	6	1	4
(c)	X:	1	1	2	2	3	3	4	4	5	5	6	6	7	7
	Y:	5	7	3	5	1	2	3	7	6	9	7	10	10	12

FIGURE 1.6.1 Scatter Diagrams of Data Sets Having Different Degrees of Curvilinearity

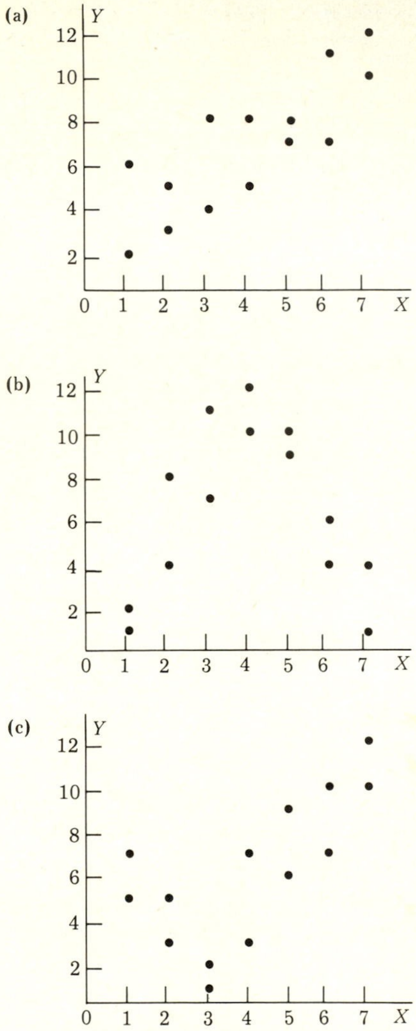

TABLE 1.6.4 SUMMARY OF LACK-OF-FIT TEST FOR DATA SET (b)

Source of variation	*Sum of squares*	*Degrees of freedom*	*Mean square*	*F*
Lack of fit	157.5536	5	31.5107	8.65
Pure error	25.5000	7	3.6429	
Error	183.0536	12		

Scatter diagrams of these data sets are shown in Figure 1.6.1. Note that Figure 1.6.1(a) shows a nearly linear relationship, for which the model $Y_i = \alpha + \beta X_i + e_i$ would be quite satisfactory. Figures 1.6.1(b) and 1.6.1(c) show different types of *curvilinearity*, for which the linear model may not be appropriate.

We shall illustrate the test for linearity described above by applying it to data set (b). The steps are as follows:

(1) Compute for each value of X the sum of squared deviations of Y values about their conditional mean, $\bar{Y}_k$. For example, if $X=1$,

$$\Sigma(Y_{ik} - \bar{Y}_k)^2 = (1-1.5)^2 + (2-1.5)^2 = .5.$$

(2) Sum the values obtained in (1) over all values of X to get the sum of squares for pure error,

$$SS_p = \sum_{k=1}^{K} \sum_{i=1}^{n_k} (Y_{ik} - \bar{Y}_k)^2 = 25.5.$$

(3) Compute the error sum of squares,

$$SS_e = \sum_{k=1}^{K} \sum_{i=1}^{n_k} (Y_{ik} - \hat{Y}_k)^2 = S_{yy} - \frac{S_{xy}^2}{S_{xx}}$$

$$= 183.2143 - \frac{(3)^2}{56} = 183.0536.$$

(4) Compute the sum of squares for lack of fit,

$$SS_L = SS_e - SS_p = 157.5536.$$

Substitution of the above values in Table 1.6.3 yields Table 1.6.4. For these data, there is a lack of fit significant beyond the .01 level. This finding, as well as the appearance of the scatter diagram, suggests that the linear model is not correct, and that one should look for an alternative. A model that is quadratic in X, such as $Y_i = \alpha + \beta X_i + \gamma X_i^2 + e_i$, might be considered.

1.7 LINEAR REGRESSION IN MATRIX TERMS

The linear regression model was stated in Equation 1.5.1 as $Y_i = \alpha + \beta X_i + e_i$. This statement implies a set of n linear equations, as follows:

$$Y_1 = \alpha + \beta X_1 + e_1$$

$$Y_2 = \alpha + \beta X_2 + e_2$$

$$\vdots$$

$$Y_n = \alpha + \beta X_n + e_n,$$

where each equation represents the Y value of a given individual in terms of the parameters of the model, the individual's X value, and an error component. (Note: before reading further in this section, the reader may wish to refer to Appendix A on matrix algebra.)

The above set of linear equations can be represented in matrix form by defining

$$\underset{(n\times 1)}{\mathbf{Y}=\begin{bmatrix} Y_1 \\ Y_2 \\ \vdots \\ Y_n \end{bmatrix}}, \quad \underset{(n\times 2)}{\mathbf{X}=\begin{bmatrix} 1 & X_1 \\ 1 & X_2 \\ \vdots & \vdots \\ 1 & X_n \end{bmatrix}}, \quad \underset{(2\times 1)}{\boldsymbol{\beta}=\begin{bmatrix} \alpha \\ \beta \end{bmatrix}}, \text{ and } \underset{(n\times 1)}{\mathbf{e}=\begin{bmatrix} e_1 \\ e_2 \\ \vdots \\ e_n \end{bmatrix}}.$$

Then, in matrix form, the linear regression model is

$$\mathbf{Y}=\mathbf{X}\boldsymbol{\beta}+\mathbf{e}. \tag{1.7.1}$$

With $\mathbf{X}$ and $\mathbf{Y}$ defined as above, it can easily be shown that

$$\mathbf{X'X}=\begin{bmatrix} n & \Sigma X \\ \Sigma X & \Sigma X^2 \end{bmatrix} \quad \text{and} \quad \mathbf{X'Y}=\begin{bmatrix} \Sigma Y \\ \Sigma XY \end{bmatrix}.$$

Now, if the sample estimate, $\mathbf{b}=\begin{bmatrix} a \\ b \end{bmatrix}$, is substituted for $\boldsymbol{\beta}$, the normal equations (see Equations 1.2.2 and 1.2.3) can be written in matrix form as

$$\mathbf{X'Xb}=\mathbf{X'Y}. \tag{1.7.2}$$

This equation can be solved for $\mathbf{b}$ by premultiplying both sides by the inverse of $\mathbf{X'X}$:

$$(\mathbf{X'X})^{-1}(\mathbf{X'X})\mathbf{b}=(\mathbf{X'X})^{-1}(\mathbf{X'Y})$$

$$\mathbf{Ib}=(\mathbf{X'X})^{-1}(\mathbf{X'Y})$$

$$\mathbf{b}=(\mathbf{X'X})^{-1}(\mathbf{X'Y}). \tag{1.7.3}$$

The solution of this matrix equation provides the estimates a and b, defined by Equations 1.2.4 and 1.2.5. These estimates may be substituted in the regression equation, $\hat{Y}_i=a+bX_i$, which itself may be expressed in matrix terms as $\hat{\mathbf{Y}}=\mathbf{Xb}$, where

$$\hat{\mathbf{Y}}=\begin{bmatrix} \hat{Y}_1 \\ \hat{Y}_2 \\ \vdots \\ \hat{Y}_n \end{bmatrix}.$$

Equations 1.7.1, 1.7.2, and 1.7.3 are important to remember because they are general in form and apply to many different types of regression problems. As we shall see later, their application depends on the appropriate definition of $\mathbf{X}$ and $\boldsymbol{\beta}$.

1.8 EXERCISES

1.1. Given the equation $Y=1+2X$.

(a) Complete the following table by determining the value of Y for each given value of X.

X:	0	1	2	3	4	5
Y:						

(b) Plot each pair of points (X, Y) on graph paper and draw the straight line representing the equation in (a).

1.2. Determine the slope and the Y intercept for each of the following equations:

(a) $Y=-2+X$ (c) $Y=-\dfrac{X}{2}$

(b) $3Y=12+1.5X$ (d) $-Y=X-2$

1.3. Use the results of Exercise 1.2 to graph the equations in 1.2 (a) through (d).

1.4. On the graphs drawn for Exercise 1.3, show by plotting new lines what change in location occurs if the equations in Exercise 1.2 (a) through (d) are changed as follows:

(a) $Y=-2-X$ instead of $Y=-2+X$

(b) $3Y=12+3X$ instead of $3Y=12+1.5X$

(c) $Y=-\dfrac{X}{2}+4$ instead of $Y=-\dfrac{X}{2}$

(d) $-Y=X+4$ instead of $-Y=X-2$

1.5. The following table contains the scores on a variable Y for seven individuals, as well as several possible alternative sets (a, b, c, and d) of X scores for the same seven individuals. For each combination of Y with X:

(a) Compute the regression coefficients, a_{YX} and b_{YX}, and formulate the regression equation, $\hat{Y}_i=a_{YX}+b_{YX}X_i$.

(b) Compute the regression coefficients, a_{XY} and b_{XY}, and form the other regression equation, $\hat{X}_i=a_{XY}+b_{XY}Y_i$.

(c) Sketch the two regression lines on graph paper, along with lines representing the X and Y means.

		X Scores			
Individual	*Y Score*	(a)	(b)	(c)	(d)
1	7	7	6	1	7
2	6	6	7	2	4
3	5	5	4	3	2
4	4	4	5	4	1
5	3	3	2	5	3
6	2	2	1	6	5
7	1	1	3	7	6

In each case, note the relationship betwen the positions of the two regression lines on the graph.

1.6. Show that the regression line, $Y_i = a_{YX} + b_{YX}X_i$, always passes through the point $(\bar{X}, \bar{Y})$.

1.7. Prove that $\bar{Y} = \bar{\hat{Y}}$.

1.8. Using the data on midterm and final examination scores in Table 1.2.1:

(a) Compute the regression coefficients, a_{YX} and b_{YX}, and formulate the regression equation for estimating Y from X.

(b) Draw a scatter diagram by plotting the 43 points representing the students in the class and draw the regression line obtained in (a) on the graph.

(c) Use the regression equation obtained in (a) to predict the final examination scores, $\hat{Y}_i$, for individuals having midterm scores of 20, 25, 30, and 35. Compare the estimates with the actual Y scores in each case.

(d) Compute the standard error of estimate, $s_{y \cdot x}$, for these data using Equation 1.3.2.

(e) Suppose that it is known that two students differ by 10 points on X. What estimate could be made of their difference on Y?

1.9. Prove that $\Sigma(Y_i - c)^2$ is a minimum if $c = \bar{Y}$.

1.10. Varying doses of poison were given to ten groups of mice each consisting of 20 animals. The following results were observed:

	Group									
Variable	1	2	3	4	5	6	7	8	9	10
Dose in mg (X)	6	8	10	12	14	16	18	20	22	24
Number of deaths (Y)	1	3	5	6	8	9	14	11	12	16

(a) Obtain the equation for the least-squares line which would enable one to predict the number of deaths per 20 mice (Y) for a given dose of poison (X).

(b) Use the equation obtained in (a) to predict how many of a group of 20 mice would be killed by a dose of 19 milligrams of poison.

(c) Obtain a 95% confidence interval for μ_y, the mean number of deaths per 20 mice.

(d) Obtain an interval estimate for β, based on the above data. Use $\alpha = .05$.

(e) Obtain a 95% confidence interval for the mean number of deaths per 20 mice resulting from a dose of poison equal to 14 milligrams.

(f) Obtain a 95% prediction interval for the number of deaths in a particular group of 20 mice administered 18 milligrams of poison.

1.11. The following table contains scores earned by 20 students on two tests, X and Y:

Student	X	Y	*Student*	X	Y
1	52	62	11	56	55
2	57	50	12	52	56
3	56	57	13	57	57
4	36	38	14	57	48
5	47	55	15	36	36
6	49	43	16	42	50
7	60	59	17	60	63
8	56	57	18	49	47
9	52	55	19	42	44
10	42	47	20	47	49

Given these data,

(a) Test the hypothesis H_0: $\beta = 0$, using the test statistic of Equation 1.6.11. Set $\alpha = .05$ and use H_1: $\beta \neq 0$ as the alternative hypothesis.

(b) Test the hypothesis H_0: $\beta = 0$ again, but use the F statistic of Equation 1.6.13 (or of Table 1.6.1). Set $\alpha = .05$ and use H_1: $\beta \neq 0$ as the alternative hypothesis. What do you observe on comparing this F with the t statistic in (a)?

(c) Carry out a test of the hypothesis that the linear model, $Y_i = \alpha + \beta X_i + e_i$, is correct for this problem. Use $\alpha = .05$.

2/ Bivariate Linear Correlation

In Chapter 1 we considered in some detail the problem of estimating the value of a dependent variable, Y, when the value of an independent (concomitant) variable, X, was known. When the primary purpose of research is to obtain such estimates, then the procedures described in Chapter 1 are appropriate. However, the research objective may be merely to determine whether there are linear *relationships* between variables, and the question of prediction or estimation may be of no more than secondary importance. In such cases, the primary concern is with the *correlation* between variables.

In this chapter we shall consider first how the concept of correlation is related to the concept of linear regression, and how the *correlation coefficient* may be defined. We then discuss estimation and tests of hypotheses in correlation analysis; applications of correlational methods in research; the interpretation of results of correlational analysis; and factors that must be considered in the design of correlational studies. Again the discussion is restricted to the bivariate case, as in Chapter 1.

2.1 THE CORRELATION COEFFICIENT

As we pointed out in Chapter 1, the standard error of estimate, $s_{y \cdot x}$, is a measure of the error made in estimating Y from X. If its value is relatively small, then estimates are quite accurate. We would expect estimates to be accurate if there were a close linear relationship between X and Y. Thus, a small value of $s_{y \cdot x}$ would imply a relatively strong relationship, while a relatively large value of $s_{y \cdot x}$ would imply a somewhat weaker relationship.

Since the magnitude of $s_{y \cdot x}$ depends, at least in part, on the strength of relationship between X and Y, one might be tempted to use it as a measure of relationship. However, the size of the standard error of estimate depends not only on the strength of relationship, but also on the variances of X and Y. We can see this dependence in the computational formula for $s_{y \cdot x}$, namely,

$$s_{y\cdot x}=\sqrt{\frac{S_{yy}-\dfrac{S_{xy}^2}{S_{xx}}}{n-2}}.$$

Note that the numerator of the fraction under the radical sign depends in part on the values of S_{xx} and S_{yy}. If S_{yy} and S_{xx} are both large, then $s_{y\cdot x}$ may be large even when the relationship is relatively strong. Similarly, if S_{yy} and S_{xx} are both small, then $s_{y\cdot x}$ may be small even when the relationship is relatively weak. Therefore, $s_{y\cdot x}$ is not entirely satisfactory as a measure of relationship. A better measure would be one which does not depend on the variances of X and Y.

Recall from Chapter 1 that the sum of squared deviations of Y values from $\bar{Y}$ can be partitioned as follows (see Equation 1.6.12):

$$\Sigma(Y_i-\bar{Y})^2=\Sigma(Y_i-\hat{Y}_i)^2+\Sigma(\hat{Y}_i-\bar{Y})^2. \tag{2.1.1}$$

Each of the terms in this equation must be greater than or equal to zero, because each is formed by summing squared quantities. Therefore, it is obvious that

$$\Sigma(Y_i-\bar{Y})^2\geqslant\Sigma(Y_i-\hat{Y}_i)^2,$$

and that

$$\Sigma(Y_i-\bar{Y})^2\geqslant\Sigma(\hat{Y}_i-\bar{Y})^2.$$

In other words, $\Sigma(Y_i-\bar{Y})^2$ is the maximum possible value of each of the other two terms in Equation 2.1.1. Noting that $\Sigma(Y_i-\hat{Y}_i)^2$ is the numerator of $s_{y\cdot x}^2$, the square of the standard error of estimate, and that $\Sigma(Y_i-\bar{Y})^2$ is the numerator of s_y^2, the variance of Y, we may now obtain a measure of relationship that does not depend on the measurement scales of X and Y. First we divide both sides of Equation 2.1.1 by $\Sigma(Y_i-\bar{Y})^2$ to obtain

$$1=\frac{\Sigma(Y_i-\hat{Y}_i)^2}{\Sigma(Y_i-\bar{Y})^2}+\frac{\Sigma(\hat{Y}_i-\bar{Y})^2}{\Sigma(Y_i-\bar{Y})^2}. \tag{2.1.2}$$

Subtracting the first term on the right from both sides of the equation yields

$$1-\frac{\Sigma(Y_i-\hat{Y}_i)^2}{\Sigma(Y_i-\bar{Y})^2}=\frac{\Sigma(\hat{Y}_i-\bar{Y})^2}{\Sigma(Y_i-\bar{Y})^2}. \tag{2.1.3}$$

Consider first the expression on the left of the equality sign. If there were *no* error of estimation, then $\Sigma(Y_i-\hat{Y}_i)^2=0$, and the left side of Equation 2.1.3 would equal exactly 1.00. In such a case there would be a perfect linear relationship between X and Y. On the other hand, if there were *no* linear relationship between X and Y, then the best available linear estimate of Y would be $\bar{Y}$, as we noted in Chapter 1. In such a case $\hat{Y}_i = \bar{Y}$, and the expression to the left of the equality sign would equal zero.

Now consider the term on the right. If there were no linear relationship between X and Y, and each estimate of Y were equal to $\bar{Y}$, then this term would equal zero. If each estimate were exactly equal to the actual value of Y, that is, equal to Y_i, then the numerator of this term would be $\Sigma(Y_i - \bar{Y})^2$ and the expression on the right would equal exactly 1.00, indicating a perfect linear relationship between X and Y.

Each side of the expression given in Equation 2.1.3 defines the square of the *Pearson product-moment correlation coefficient, r*. Thus,

$$r^2 = 1 - \frac{\Sigma(Y_i - \hat{Y}_i)^2}{\Sigma(Y_i - \bar{Y})^2} = \frac{\Sigma(\hat{Y}_i - \bar{Y})^2}{\Sigma(Y_i - \bar{Y})^2}, \tag{2.1.4}$$

and

$$r = \pm\sqrt{1 - \frac{\Sigma(Y_i - \hat{Y}_i)^2}{\Sigma(Y_i - \bar{Y})^2}} = \pm\sqrt{\frac{\Sigma(\hat{Y}_i - \bar{Y})^2}{\Sigma(Y_i - \bar{Y})^2}}. \tag{2.1.5}$$

Note that Equations 2.1.4 and 2.1.5 show that the correlation coefficient, r, has the following properties:

1. The value of r is zero when there is no linear relationship between X and Y. In this case $\Sigma(Y_i - \hat{Y}_i)^2$ has its maximum value, $\Sigma(Y_i - \bar{Y})^2$, and $\Sigma(\hat{Y}_i - \bar{Y})^2$ has its minimum value, zero.

2. The possible values of r range from -1.00 to $+1.00$. The values $r = -1.00$ and $r = +1.00$ occur when there is a perfect linear relationship (negative or positive) between X and Y. When there is a perfect linear relationship, $\Sigma(Y_i - \hat{Y}_i)^2 = 0$ and $\Sigma(\hat{Y}_i - \bar{Y})^2 = \Sigma(Y_i - \bar{Y})^2$. Then $r^2 = 1.00$, and r equals either $+1.00$ or -1.00.

3. Because r^2 is a ratio of a sum of squares to its maximum value, $\Sigma(Y_i - \bar{Y})^2$, it may be interpreted, according to Equation 2.1.4, as the proportion of the total Y variance accounted for, or "explained," by the linear relationship between X and Y. For example, if r were computed and found to be .7, r^2 would be .49, and we could conclude that 49%, or about half, of the Y variance could be "explained" on the basis of the linear relationship with the variable X. Of course, $1 - r^2$ is the proportion of Y (or X) variance *un*explained on the basis of X (or Y). Note also that r^2 represents the proportionate *reduction* in the variance of Y when it is measured from the regression line, $\hat{Y}_i = a + bX_i$, rather than from $\bar{Y}$.

2.2 ALTERNATIVE FORMULAS FOR THE CORRELATION COEFFICIENT

The formula for r given by Equation 2.1.5 is useful for showing how r is related to $\Sigma(Y_i - \hat{Y}_i)^2$, but is inconvenient for computational purposes, because it would require the computation of the regression estimate, $\hat{Y}_i$, for each of n individuals in the sample. Because $\hat{Y}_i = a + bX_i$, we may write

$$\Sigma(Y_i - \hat{Y}_i)^2 = \Sigma(Y_i - a - bX_i)^2.$$

Substitution for a from Equation 1.2.5 yields

$$\Sigma(Y_i - \hat{Y}_i)^2 = \Sigma(Y_i - \bar{Y} + b\bar{X} - bX_i)^2$$
$$= \Sigma\left[(Y_i - \bar{Y}) - b(X_i - \bar{X})\right]^2.$$

Squaring the expression in brackets on the right gives

$$\Sigma(Y_i - \hat{Y}_i)^2 = \Sigma\left[(Y_i - \bar{Y})^2 + b^2(X_i - \bar{X})^2 - 2b(X_i - \bar{X})(Y_i - \bar{Y})\right]$$
$$= \Sigma(Y_i - \bar{Y})^2 + b^2\Sigma(X_i - \bar{X})^2 - 2b\Sigma(X_i - \bar{X})(Y_i - \bar{Y}).$$

Using the notation of Chapter 1, where $S_{xx} = \Sigma(X_i - \bar{X})^2$, $S_{yy} = \Sigma(Y_i - \bar{Y})^2$, and $S_{xy} = \Sigma(X_i - \bar{X})(Y_i - \bar{Y})$, we have

$$\Sigma(Y_i - \hat{Y}_i)^2 = S_{yy} + b^2 S_{xx} - 2bS_{xy}.$$

Because $b = S_{xy}/S_{xx}$ (see Equation 1.2.9), we can write

$$\Sigma(Y_i - \hat{Y}_i)^2 = S_{yy} + \frac{S_{xy}^2 S_{xx}}{S_{xx}^2} - 2\frac{S_{xy}^2}{S_{xx}}$$
$$= S_{yy} - S_{xy}^2/S_{xx}. \tag{2.2.1}$$

Now, using Equation 2.2.1 and the fact that $\Sigma(Y - \bar{Y})^2 = S_{yy}$, we may substitute for $\Sigma(Y_i - \hat{Y}_i)^2$ and $\Sigma(Y_i - \bar{Y})^2$ in Equation 2.1.5, to obtain

$$r = \pm\sqrt{1 - \frac{S_{yy} - S_{xy}^2/S_{xx}}{S_{yy}}} = \pm\sqrt{\frac{S_{xy}^2}{S_{xx}S_{yy}}}$$
$$= \frac{S_{xy}}{\sqrt{S_{xx}S_{yy}}}. \tag{2.2.2}$$

Equation 2.2.2 provides a computing formula for the correlation coefficient that requires only the three values S_{xy}, S_{xx}, and S_{yy}. As indicated in Chapter 1, these are readily available from the sample data. Note that whether the sign of r is + or − depends entirely on the sign of S_{xy}, which can be either + or −. A positive sign indicates that large values of X tend to be associated with large values of Y, and that small values tend to be associated with small values. A negative sign indicates that small values of either X or Y tend to be associated with large values of the other variable.

Other formulas for r, equivalent algebraically to Equations 2.1.5 and 2.2.2, show its relationship to the regression coefficients, b_{XY} and b_{YX}. Recall that if one wishes to estimate Y from X, the value of the regression coefficient b_{YX} will usually be different from b_{XY}, used in estimating X from Y. The formulas for these two coefficients are $b_{YX} = S_{xy}/S_{xx}$ and $b_{XY} = S_{xy}/S_{yy}$. Comparing these expressions with Equation 2.2.2 for r, we see that the signs of r, b_{YX}, and b_{XY} must all be identical, because all have the same numerator and all have positive denominators. Furthermore, if we obtain the product of the two regression coefficients, we see that r equals its square root. Thus,

$$r = \pm\sqrt{b_{YX}b_{XY}} = \pm\sqrt{\frac{S_{xy}^2}{S_{xx}S_{yy}}} = \frac{S_{xy}}{\sqrt{S_{xx}S_{yy}}}. \qquad (2.2.3)$$

The correlation coefficient may also be expressed as a function of either b_{YX} or b_{XY} alone. If we multiply both numerator and denominator of Equation 2.2.2 by $\sqrt{S_{xx}}$, we have

$$r = \frac{S_{xy}\sqrt{S_{xx}}}{S_{xx}\sqrt{S_{yy}}} = b_{YX}\sqrt{\frac{S_{xx}}{S_{yy}}} = b_{YX}\frac{s_x}{s_y}, \qquad (2.2.4)$$

where s_x and s_y are the sample standard deviations of X and Y, respectively. This shows that r and b_{YX} differ only by a factor equal to the ratio of the standard deviations of X and Y. If these standard deviations are equal, then r is equal to b_{YX}. Of course, r may also be expressed in terms of b_{XY} as

$$r = b_{XY}\frac{s_y}{s_x}. \qquad (2.2.5)$$

These alternative formulas, relating r and the regression coefficients, are presented primarily as an aid to understanding how various concepts in regression and correlation analysis are related. In some cases, however, they may serve as useful computational devices as well.

2.3 A NUMERICAL EXAMPLE

To illustrate the computation of the correlation coefficient, we shall use the data given in Table 1.2.1, on midterm (X) and final examination (Y) scores in a course in introductory statistics. One would expect to find a positive relationship between X and Y, because both general aptitude and quantitative ability probably account for substantial portions of the variance in both X and Y. It is unlikely, though possible, that a student who had received one of the highest scores on X would receive one of the lowest on Y. We would expect, therefore, to find some degree of consistency in performance and thus would expect r to be between zero and one.

Inspecting the scatter diagram in Figure 2.3.1 reveals that there is a positive relationship between X and Y. The dotted lines showing the

FIGURE 2.3.1. SCATTER DIAGRAM SHOWING DISTRIBUTION OF MIDTERM AND FINAL EXAMINATION SCORES IN TABLE 1.2.1

$X_{.50} = 32.83$

$Y_{.50} = 29.50$

Final exam score (Y) \ Midterm score (X)	17–18	19–20	21–22	23–24	25–26	27–28	29–30	31–32	33–34	35–36	37–38	39–40
39–40										1		1
37–38									1	2	1	1
35–36										2		
33–34						1	2		2	1		
31–32							1			3		
29–30					2	2	1					
27–28				1	1				2	1		
25–26												
23–24	1			1				2	3			
21–22		1							1		1	
19–20		1		1								
17–18												
15–16												
13–14	1											
11–12							1					

location of the median values of X and Y divide the diagram into four quadrants. Relatively few pairs of observations fall in the lower right and upper left quadrants; most fall in the upper right and the lower left. Such a configuration suggests a positive relationship. That is, students tended to score at approximately the same position relative to other members of the group on both tests. There were exceptions, however.

For example, one student had 37 on the midterm, but dropped to 21 on the final. Another had 30 on the midterm, just below the median, and 12 on the final. Thus, although there *is* a positive relationship between X and Y, it is obviously not perfect. The correlation coefficient indicates quantitatively how strong the relationship actually is.

Computations based on the ungrouped data of Table 1.2.1 yielded the following values:

$$S_{xx} = 1423.12 \qquad S_{xy} = 1000.79$$

$$S_{yy} = 1980.98 \qquad n = 43.$$

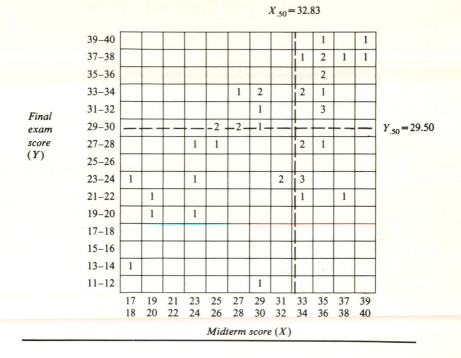

From these sums of squares and the cross-product sum we find that

$$r = \frac{S_{xy}}{\sqrt{S_{xx}S_{yy}}} = \frac{1000.79}{\sqrt{(1423.11)(1980.97)}} = .596,$$

$$b_{YX} = \frac{S_{xy}}{S_{xx}} = \frac{1000.79}{1423.11} = .703,$$

$$b_{XY} = \frac{S_{xy}}{S_{yy}} = \frac{1000.79}{1980.97} = .505,$$

$$s_x^2 = \frac{S_{xx}}{n-1} = \frac{1423.11}{42} = 33.88, s_x = \sqrt{33.88} = 5.82,$$

$$s_y^2 = \frac{S_{yy}}{n-1} = \frac{1980.97}{42} = 47.17, \text{ and } s_y = \sqrt{47.17} = 6.87.$$

Equation 2.2.2 was used here to obtain r. Note that b_{YX} and b_{XY} have the same sign as r, but are different in magnitude. However, the square root of the product of b_{YX} and b_{XY}, that is,

$$\sqrt{b_{YX}b_{XY}} = \sqrt{(.703)(.505)} = \sqrt{.355} = .596,$$

is equal to the value of r obtained from Equation 2.2.2. The same value of r also results from

$$r = b_{YX}\frac{s_x}{s_y} = .703\frac{5.82}{6.87} = .596,$$

and from

$$r = b_{XY}\frac{s_y}{s_x} = .505\frac{6.87}{5.82} = .596.$$

These equations may also be used to obtain b_{YX} and b_{XY} if r, s_y, and s_x are known. Thus,

$$b_{YX} = r\frac{s_y}{s_x} = .596\frac{6.87}{5.82} = .703,$$

and

$$b_{XY} = r\frac{s_x}{s_y} = .596\frac{5.82}{6.87} = .505.$$

Note that these are the same values obtained for b_{YX} and b_{XY} by using Equations 1.4.3 and 1.4.5.

We shall consider the question of the proper interpretation of r in a later section. However, we may note at this point that $r^2 = .355$, which means that we may consider 35.5% of the variance of the final scores, Y, to be "explained" or "accounted for" by the midterm, X. Another interpretation of the fact that $r^2 = .355$ is that the variation of Y values around the regression line, $Y_i = a + bX_i$, is 35.5% less than the variation of Y values around $\overline{Y}$. Note that these same statements could be made with X and Y

interchanged. The correlation coefficient merely expresses a relationship between two variables. It does not imply a cause-and-effect relationship between them.

2.4 A MATHEMATICAL MODEL FOR BIVARIATE LINEAR CORRELATION

In problems concerning the relationship between two variables, X and Y, the mathematical model used for estimation and tests of significance requires that observations be regarded as pairs, where each pair consists of one observation on X and one on Y. The two variables in this model are both regarded as random variables. It is assumed that the sample of pairs of observations is chosen by a random procedure, and that the values of each vary randomly from sample to sample. Thus, this model is different from the regression model, in which the values of X were regarded as fixed and only Y was considered a random variable. A population consisting of paired values on two random variables is called a bivariate population.

Estimation and tests of significance in correlation analysis are based on the assumption of a *normal bivariate* population. Such a population is completely determined by the values of its five parameters, μ_x, μ_y, σ_x^2, σ_y^2, and ρ, the population correlation coefficient. The bivariate normal density function is

$$f(X,Y) = \frac{1}{2\pi\sigma_x\sigma_y\sqrt{1-\rho^2}} \exp\left\{ -\frac{1}{2(1-\rho^2)} \left[\frac{(X-\mu_x)^2}{\sigma_x^2} + \frac{(Y-\mu_y)^2}{\sigma_y^2} - 2\rho \frac{(X-\mu_x)(Y-\mu_y)}{\sigma_x\sigma_y} \right] \right\}. \tag{2.4.1}$$

Graphing this function produces a three-dimensional figure such as that shown in Figure 2.4.1. The precise shape of the figure depends on the values of σ_x, σ_y, and ρ, and its location with respect to the axes depends on the values of μ_x and μ_y. The point (μ_x, μ_y) is called the *centroid* of the distribution. Note that it is no longer necessary to distinguish between μ_Y and μ_y, as in Chapter 1, because in this model, X and Y are both random variables.

The value of the above function at the point (X_i, Y_i) is the height of the surface above the point (X_i, Y_i) on the plane formed by the X and Y axes. Probabilities are represented by volumes instead of by areas as in the univariate normal distribution (see Figure 2.4.2).

A number of facts about the bivariate normal distribution are important in correlation analysis.

1. If the three-dimensional solid in Figure 2.4.1 is cut by any plane perpendicular to the plane formed by the X and Y axes and parallel to the

FIGURE 2.4.1 A Bivariate Normal Density Surface

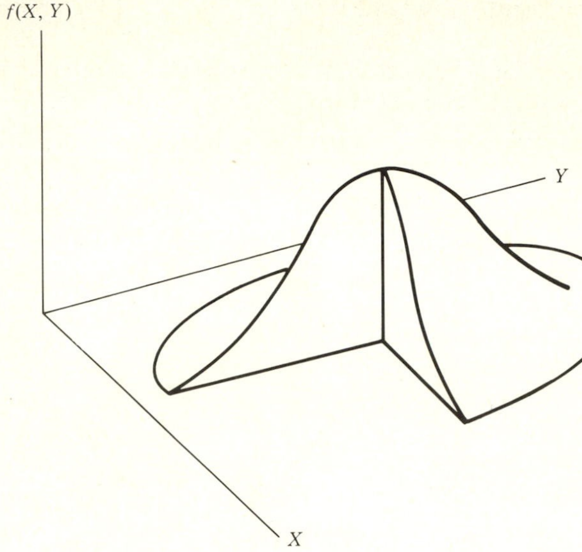

A portion of the three-dimensional solid has been removed to indicate the form of the marginal distributions of X and Y.

Y axis, the curve formed by the intersection of the plane and the surface of the solid will be normal with mean $\mu_{y \cdot x}$ and variance $\sigma^2_{y \cdot x}$. Each of these distributions is a conditional (univariate) distribution of Y for a given X. They all have the same variance, $\sigma^2_{y \cdot x}$, and their means fall on the same straight line, namely,

$$\mu_{y \cdot x} = \mu_y + \beta_{YX}(X - \mu_x), \tag{2.4.2}$$

FIGURE 2.4.2 Portion of Volume of a Bivariate Normal Distribution, Showing the Joint Probability that X Is Between a and b, and that Y Is Between c and d

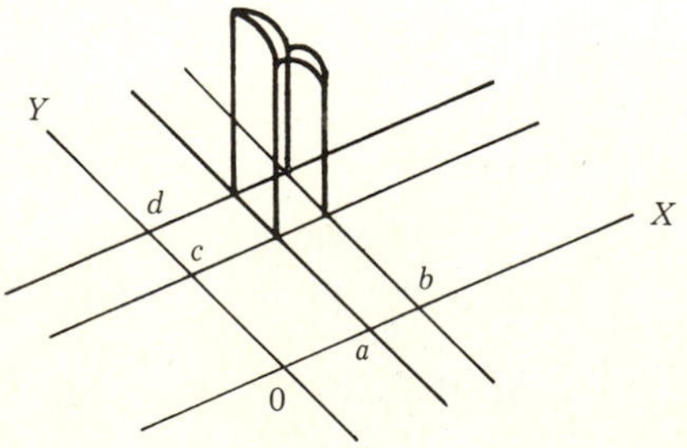

in which $\beta_{YX} = \rho(\sigma_y/\sigma_x)$. It is this property that gives precise meaning to the statement that X and Y have a linear relationship. If X and Y have a bivariate normal distribution, then the relationship, if any, between them must be linear.

2. An analogous set of statements can be made about the conditional X distributions, when the plane cutting the solid is parallel to the X axis. Then each such distribution is normal with variance $\sigma_{x \cdot y}^2$ and with means that are on the line

$$\mu_{x \cdot y} = \mu_x + \beta_{XY}(Y - \mu_y), \tag{2.4.3}$$

in which $\beta_{XY} = \rho(\sigma_x/\sigma_y)$.

3. The conditional variances, $\sigma_{y \cdot x}^2$ and $\sigma_{x \cdot y}^2$, are related to the unconditional variances, σ_y^2 and σ_x^2, and to ρ by the expressions

$$\sigma_{y \cdot x}^2 = \sigma_y^2(1 - \rho^2) \tag{2.4.4}$$

and

$$\sigma_{x \cdot y}^2 = \sigma_x^2(1 - \rho^2). \tag{2.4.5}$$

Solving each of these for ρ^2, we see that

$$\rho^2 = 1 - \frac{\sigma_{y \cdot x}^2}{\sigma_y^2} = 1 - \frac{\sigma_{x \cdot y}^2}{\sigma_x^2}. \tag{2.4.6}$$

This expression provides the same interpretation of ρ^2 as Equation 2.1.4 did for r^2. From Equation 2.4.6 we see that ρ^2 is a measure of the extent of reduction in variance of either X or Y due to the relationship between X and Y. If $\sigma_{y \cdot x}^2 = \sigma_y^2$ (or $\sigma_{x \cdot y}^2 = \sigma_x^2$), then $\rho^2 = 0$. If $\sigma_{y \cdot x}^2 = \sigma_{x \cdot y}^2 = 0$, then $\rho^2 = 1$, indicating a perfect linear relationship between X and Y.

4. If the three-dimensional solid of Figure 2.4.1 is cut by any plane parallel to the plane formed by the X and Y axes, the curve formed at the intersection is an ellipse with center at the point (μ_x, μ_y). This fact has led to the practice of representing the bivariate normal distribution by means of a graph such as that shown in Figure 2.4.3. This figure shows a family of ellipses that results from cutting, at several points, a bivariate normal distribution with $\mu_x = 15$, $\mu_y = 20$, $\sigma_x = \sigma_y = 5$, and $\rho = .6$.

5. If X and Y have a bivariate normal distribution, then the marginal distributions of both X and Y are normal. However, if both X and Y have univariate normal *marginal* distributions, it is *not* necessarily true that X and Y have a *bivariate* normal distribution.

6. If X and Y have a bivariate normal distribution, then X and Y are independent *if and only if* the correlation ρ_{XY} equals zero. While it is true for *any* joint distribution of X and Y that if X and Y are independent, $\rho_{XY} = 0$, the converse is also true for the bivariate normal distribution. In other words, there are bivariate distributions in which X and Y are *not* independent, but $\rho_{XY} = 0$; but for the bivariate normal distribution, $\rho_{XY} = 0$ implies that X and Y are independent, and vice versa. If the assumption of a bivariate normal joint distribution of X and Y can be supported, then ρ_{XY} is

FIGURE 2.4.3 Several Members of the Family of Ellipses Representing a Bivariate Normal Distribution with $\mu_x = 15$, $\mu_y = 20$, $\sigma_x = \sigma_y = 5$, and $\rho = .6$

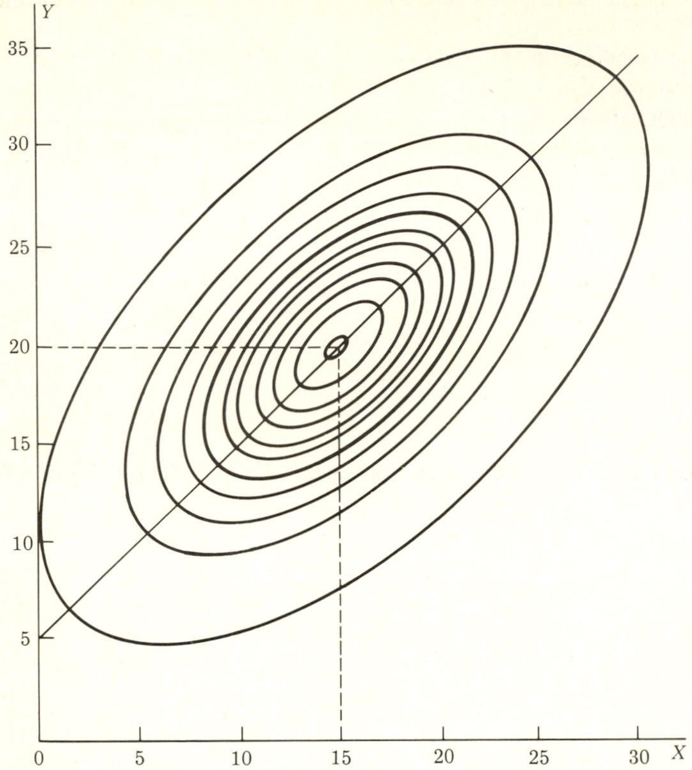

a direct measure of the dependence of X and Y. If such an assumption *cannot* be supported, then ρ_{XY} indicates only the degree of *linear* relationship between X and Y, and a nonlinear relationship may also be present.

2.5 THE SAMPLING DISTRIBUTION OF *r*

Since r ranges only from -1.00 to $+1.00$, it cannot have a truly normal distribution for any value of ρ. However, when $\rho=0$ and the sample is relatively large, the departure from normality does not seriously interfere with tests of significance based on the assumption of normality.

When $\rho \neq 0$, the sampling distribution of r is skewed and the test used when $\rho=0$ is not satisfactory. The logarithmic transformation of r (due to Fisher),

$$Z = \frac{1}{2}[\ln(1+r) - \ln(1-r)] = \tanh^{-1} r, \qquad (2.5.1)$$

which ranges from $-\infty$ to $+\infty$, has three important advantages over r for purposes of estimation and tests of significance:

1. Whereas the standard error of r, $\sigma_r \cong \frac{1-\rho^2}{\sqrt{n-1}}$, depends on the value of ρ, the standard error of Z, $\sigma_Z \cong \frac{1}{\sqrt{n-3}}$, does not depend on ρ, but only on the sample size.

2. The distribution of r is not even approximately normal for small samples, nor even for large samples if ρ departs substantially from zero. The distribution of Z is nearly normal even for a sample size as small as $n=8$.

3. The distribution of r changes its form as ρ changes, while that of Z is practically constant in form whatever the size of ρ.

For these reasons the logarithmic transformation, Z, is used in most of the statistical tests and interval estimates involving the correlation coefficient. Values of Z for various values of r are tabled in Appendix C, Table C.10.

2.6 INTERVAL ESTIMATION IN CORRELATION ANALYSIS

An interval estimate for ρ is obtained by first determining the interval estimate of Z_ρ, the population value of Z, and then reading from a table the corresponding values for ρ. The estimate may be obtained from the following probability statement:

$$P\left[Z+\frac{z_{\frac{\alpha}{2}}}{\sqrt{n-3}}<Z_\rho<Z+\frac{z_{1-\frac{\alpha}{2}}}{\sqrt{n-3}}\right]=1-\alpha, \qquad (2.6.1)$$

where Z is the value of Fisher's logarithmic transformation corresponding to the sample value, r, and $z_{\frac{\alpha}{2}}$ and $z_{1-\frac{\alpha}{2}}$ are percentiles of the standard normal distribution.

Our example is again based on the data of Table 1.2.1. Suppose we wish to obtain a 95% confidence interval for ρ, the population value of the correlation between the midterm (X) and the final examination (Y) scores of students in an introductory statistics course. Such an estimate might be of interest to a statistics professor who wants to assess the consistency of performance of students on the types of tests he usually administers at the introductory level.

The data required for obtaining the 95% confidence interval ($\alpha=.05$) are that $r=.596$ and $n=43$. Using Table C.10 we find that the Fisher Z transformation of $r=.596$ is $Z=.6869$. From the table of the standard normal distribution in Appendix C we find that $z_{1-\frac{\alpha}{2}}=z_{.975}=1.96$ and $z_{\frac{\alpha}{2}}=z_{.025}=-1.96$. Then the limits for Z_ρ are

$$Z+\frac{z_{\frac{\alpha}{2}}}{\sqrt{n-3}}=.6869-\frac{1.96}{\sqrt{40}}=.3770,$$

and

$$Z + \frac{z_{1-\frac{\alpha}{2}}}{\sqrt{n-3}} = .6869 + \frac{1.96}{\sqrt{40}} = .9968.$$

Transforming these limits*, again using Table C.10 to limits for ρ, we find that the interval for ρ is .360 to .760. The risk that such an interval does not include ρ is 5 percent.

An alternative procedure for determining a 95% confidence interval for ρ is based on Charts C9a and C9b in Appendix C. To use these charts, we first locate the value of $r=.596$ on the horizontal scale of r. The closest value is $r=.6$. Then we read straight up on the vertical line marked $r=.6$ to its intersections with the curved lines labeled as close as possible to $n-2=41$ degrees of freedom. In this case, those lines are labeled 50 degrees of freedom. The intersection with the lower line occurs at $\rho=.4$, read on the vertical scale at the right or left side of the chart. The intersection with the upper line is at $\rho=.75$. These values agree quite closely with the limits .36 and .76, obtained by the procedure using Fisher's Z. If one wished to be somewhat conservative, it would be best to select the lines labeled 25 instead of 50. The interval in that case would be longer.

Note that if α is fixed, the only way to reduce the length of the confidence interval is to increase the sample size. We may easily obtain a formula for determining the sample size required to construct a confidence interval for Z_ρ of any desired length. Let L be the length of the interval desired. Then

$$L = 2\frac{z_{1-\frac{\alpha}{2}}}{\sqrt{n-3}}. \tag{2.6.2}$$

Solving for n, we have

$$n = \frac{4z^2_{1-\frac{\alpha}{2}}}{L^2} + 3. \tag{2.6.3}$$

To illustrate the use of this formula, suppose we wish to obtain a 95% confidence interval for Z_ρ of length .20. Substitution in equation 2.6.3 yields

$$n = \frac{4(1.96)^2}{(.20)^2} + 3 = 387.16,$$

or, rounding upwards, $n=388$. Therefore, a sample size of 388 would give us a 95% confidence interval for Z_ρ of length .20. Because the range of values of Z_ρ is greater than for ρ, the corresponding interval for ρ would have a length somewhat less than .20. Thus, if $r=.596$ and $n=388$, the

*Transforming Z to r may be done by use of the equation $r = \tanh Z = 1 - \frac{2}{e^{2Z}+1}$. Some hand calculators have trigonometric function keys, such as tanh and $\tanh^{-1}$ (see Equation 2.5.1), which give the same figures as those in Table C.10.

limits for Z_ρ are

$$Z+\frac{z_{\frac{\alpha}{2}}}{\sqrt{n-3}}=.6869-\frac{1.96}{19.6}=.5869,$$

and

$$Z+\frac{z_{1-\frac{\alpha}{2}}}{\sqrt{n-3}}=.6869+\frac{1.96}{19.6}=.7869,$$

which gives $L=.2$, as desired. The corresponding limits for ρ, obtained by using Table C.10, are .528 and .657, giving a length for the interval of $.657-.528=.129$. Thus, the interval for ρ is considerably shorter than for Z_ρ. Note that the length of the interval for ρ depends not only on the sample size, but also on the value of ρ.

2.7 TESTS OF HYPOTHESES IN CORRELATION ANALYSIS

The preceding sections have dealt with ways of estimating the population correlation coefficient, either by a single-valued or "point" estimate, r, or by interval estimation. In addition to estimating the population value of ρ, one may sometimes be interested in testing the hypothesis that ρ has some specific value, ρ_0. The null hypothesis that ρ equals zero is most often tested. The usual purpose of such a test is to determine whether a positive or negative linear relationship exists between two variables, X and Y. The research hypothesis is one of the possible alternatives to H_0: $\rho_0=0$, namely, $\rho<0$, $\rho>0$, or $\rho\neq 0$. Of course, the null hypothesis may also be that ρ equals some value other than zero. Then the possible alternatives are that $\rho<\rho_0$, $\rho>\rho_0$, or $\rho\neq\rho_0$.

The question often arises, When should one estimate ρ, and when should one test some hypothesis about ρ? There is no general answer which is correct in all circumstances. However, it is safe to say that estimation of ρ is especially appropriate in the exploratory stages of research, whereas hypothesis testing is most often carried out during the more refined stages in conjunction with testing for comfirmation of educational and psychological theories. In other words, estimation answers the question, What is the degree of linear relationship, *if any*, between X and Y?, whereas hypothesis testing answers the question, Am I correct in hypothesizing that ρ equals some specified value, say ρ_0? The hypothesized value, ρ_0, is usually based on previous empirical results obtained by other researchers or on established theory which is being tested under new conditions or with new populations. Thus, one may deduce from theory that X and Y should have a positive relationship and test the null hypothesis, H_0: $\rho=0$, against the alternative, H_1: $\rho>0$, in an attempt to test the theory. Alternatively, one may have found from previous research that in population A, $\rho = .5$, and may wish to test the hypothesis that in population B, ρ also equals .5.

The tests described below follow the seven-step procedure for testing hypotheses that was used in Chapter 1. In each case the test statistic and its distribution are given first, and these are followed by a numerical example.

2.7.1 Test of the Hypothesis H_0: $\rho=0$

To test this hypothesis we may use either

$$F = \frac{r^2(n-2)}{1-r^2}, \tag{2.7.1}$$

which has the F distribution with 1 and $n-2$ degrees of freedom, when H_0 is true, or

$$t = \frac{r\sqrt{n-2}}{\sqrt{1-r^2}}, \tag{2.7.2}$$

which, under H_0, has the t distribution with $n-2$ degrees of freedom. The t statistic should be used if the alternative hypothesis is either H_1: $\rho > 0$ or H_1: $\rho < 0$. Both of these alternatives imply a one-tailed test.

Again referring to the data of Table 1.2.1, suppose we wish to test the hypothesis H_0: $\rho = 0$. We found previously that the correlation coefficient *in the sample* was $r = .596$. While that value is our point estimate of ρ, it may be considerably in error because of sampling variability. The purpose of the test is to determine whether we may conclude that the *population* correlation coefficient is greater than zero, as we might expect theoretically and as the sample evidence suggests. Therefore, we shall use as alternative H_1: $\rho > 0$, and we shall arbitrarily choose $\alpha = .05$.

The steps in the test are as follows:

1. H_0: $\rho = 0$.
   H_1: $\rho > 0$.
2. The test statistic is $t = \dfrac{r\sqrt{n-2}}{\sqrt{1-r^2}}$.
3. The test statistic has the t distribution, under H_0, with $n-2=41$ degrees of freedom.
4. $\alpha = .05$.
5. The critical region is $t > 1.68$.
6. $t = \dfrac{.596\sqrt{41}}{\sqrt{1-(.596)^2}} = 4.75$.
7. Because t is in the critical region, we reject H_0 and conclude that ρ is greater than zero.

2.7.2 Test of the Hypothesis H_0: $\rho = \rho_0$

In this test, ρ_0 is some specific value of ρ. The test statistic is

$$z = (Z - Z_0)\sqrt{n-3}\,, \tag{2.7.3}$$

in which Z is Fisher's logarithmic transformation of r and Z_0 is Fisher's logarithmic transformation of ρ_0. The test statistic, z, has approximately the standard normal distribution. Although this statistic is more often used when ρ_0 is some value other than zero, it may be used when the null hypothesis is H_0: $\rho = 0$. However, it is the only appropriate statistic when ρ_0 is some value other than zero.

As a numerical example of the application of this test, we choose, quite arbitrarily, to test the null hypothesis H_0: $\rho = .40$, again using the data of Table 1.2.1. If α is chosen as .05, and if the alternative is H_1: $\rho > .40$, the steps in the test are:

1. H_0: $\rho = .40$.
   H_1: $\rho > .40$.
2. The test statistic is $z = (Z - Z_0)\sqrt{n-3}$.
3. The test statistic has approximately the standard normal distribution.
4. $\alpha = .05$.
5. The critical region is $z > 1.645$.
6. $z = (.6869 - .4236)\sqrt{40} = 1.67$.
7. Because z is in the critical region, we reject H_0 and conclude that ρ is greater than .40.

2.7.3 Test of the Hypothesis H_0: $\rho_1 = \rho_2$

In this test one wishes to determine whether the correlation coefficients, ρ_1 and ρ_2, between X and Y in two different populations are equal. One might hypothesize, for example, that the correlation between the statistics midterm (X) and final (Y) would differ in different sections of the course. The test statistic is

$$z = \frac{Z_1 - Z_2}{\sqrt{\dfrac{1}{n_1 - 3} + \dfrac{1}{n_2 - 3}}}, \tag{2.7.4}$$

where Z_1 and Z_2 are Fisher's logarithmic transformations of r_1 and r_2, respectively. The statistic, z, has approximately the standard normal distribution.

To illustrate this test, we suppose that the statistics midterm (X) and the final examination (Y) in Table 1.2.1 had been administered to two sections taught by professors whose teaching styles differed. We might wish to test the hypothesis H_0: $\rho_1 = \rho_2$, where ρ_1 and ρ_2 are the correlations in the populations of students taught by professors 1 and 2, respectively. In this example, we shall use as alternative H_1: $\rho_1 \neq \rho_2$ and set α at .05. If the data are $r_1 = .596$, $n_1 = 43$, $r_2 = .352$, and $n_2 = 52$, then the steps in the test are:

1. H_0: $\rho_1 = \rho_2$.
   H_1: $\rho_1 \neq \rho_2$.
2. The test statistic is $z = \dfrac{Z_1 - Z_2}{\sqrt{\dfrac{1}{n_1 - 3} + \dfrac{1}{n_2 - 3}}}$.
3. The test statistic has approximately the standard normal distribution.
4. $\alpha = .05$.
5. The critical region is $z > 1.96$ and $z < -1.96$.
6. $z = \dfrac{.6869 - .3677}{\sqrt{\dfrac{1}{40} + \dfrac{1}{49}}} = 1.50.$
7. Because z does not fall in the critical region, we accept H_0 and conclude that we do not have sufficient evidence that ρ_1 and ρ_2 are different.

2.7.4 Test of the Hypothesis H_0: $\rho_{XZ} = \rho_{YZ}$

It is sometimes of interest to determine whether, for a particular population, one variable, X, has a larger correlation with some criterion, Z, than does another variable, Y. For example, we may wish to determine whether achievement in a statistics course, Z, is predicted better by a test of quantitative aptitude, X, or by the score on the Miller Analogies Test, Y. A test statistic given by Hotelling (1940) may be used:

$$t = (r_{XZ} - r_{YZ})\sqrt{\frac{(n-3)(1 + r_{XY})}{2(1 - r_{XY}^2 - r_{XZ}^2 - r_{YZ}^2 + 2r_{XY}r_{XZ}r_{YZ})}} \quad (2.7.5)$$

This statistic has the t distribution with $n - 3$ degrees of freedom, under the assumption that the conditional distributions of Z are normal, with the same variance, for all values of X and Y.

To illustrate this test, let us suppose that for a sample of size $n = 33$, we have found $r_{XY} = .55$, $r_{XZ} = .65$, and $r_{YZ} = .49$. Then the steps in the test of H_0: $\rho_{XZ} = \rho_{YZ}$ against H_1: $\rho_{XZ} \neq \rho_{YZ}$ with $\alpha = .05$ are:

1. H_0: $\rho_{XZ} = \rho_{YZ}$.
   H_1: $\rho_{XZ} \neq \rho_{YZ}$.
2. The test statistic is t, defined by Equation 2.7.5.
3. The test statistic has the t distribution, under H_0, with $n - 3$ degrees of freedom.
4. $\alpha = .05$.
5. The critical region is $t < -2.04$ and $t > 2.04$.

6. $t = (.65 - .49)\sqrt{\dfrac{30(1.55)}{2(.385)}} = 1.24.$

7. Because $t = 1.24$ is not in the critical region, we would conclude that neither X nor Y is superior to the other in predicting Z.

2.7.5 Testing Hypotheses about an Average Correlation

Sometimes one has available several estimates of the correlation between X and Y, each based on a small sample, and desires to combine them and test some hypothesis about the combined estimate. Such a procedure is appropriate if it is known that the individual estimates are based on samples drawn from the same population or if it can be assumed that the samples have been drawn from populations having equal correlations. The latter assumption might be made if a test based on Equation 2.7.4 yielded a nonsignificant result.

The average correlation, r_{av}, may be obtained by first transforming each estimate to Z using Fisher's transformation (see Table C.10), obtaining a weighted average, Z_{av}, and then transforming this to r_{av}. For a total of K different estimates, the weighted average, Z_{av}, is obtained by

$$Z_{av} = \frac{(n_1 - 3)Z_1 + \cdots + (n_K - 3)Z_K}{n_1 + \cdots + n_K - 3K} \qquad (2.7.6)$$

To test the hypothesis H_0: $\rho_{av} = \rho_{0(av)}$, where $\rho_{0(av)}$ is some hypothesized value of ρ_{av}, the statistic is

$$z = (Z_{av} - Z_{0(av)})\sqrt{n_1 + \cdots + n_K - 3K}. \qquad (2.7.7)$$

This statistic is analogous to Equation 2.7.3, that is, Z_{av} and $Z_{0(av)}$ are Fisher's transformations of r_{av} and $\rho_{0(av)}$, respectively.

We illustrate the above procedure using an example of Fisher (1954, page 204). In two independent samples of sizes $n_1 = 20$ and $n_2 = 25$, drawn from populations in which the correlations between X and Y were assumed equal, the values of r were $r_1 = .6$ and $r_2 = .8$. The corresponding values of Z were $Z_1 = .6931$ and $Z_2 = 1.0986$. Substitution in Equation 2.7.6 with $K = 2$, yields

$$Z_{av} = \frac{17(.6931) + 22(1.0986)}{20 + 25 - 6} = .9218.$$

Transformation to r gives $r_{av} = .7267$. Note that this is not necessarily the same value which would have been obtained by calculating the corresponding weighted average of the r's themselves.

As an illustration of the use of Equation 2.7.7, we shall test H_0: $\rho_{av} = .65$ against H_1: $\rho_{av} \neq .65$ with $\alpha = .05$. The steps are:

1. H_0: $\rho_{av} = .65$.
   H_1: $\rho_{av} \neq .65$.
2. The test statistic is z defined by Equation 2.7.7.

3. The test statistic has approximately the standard normal distribution.
4. $\alpha = .05$.
5. The critical region is $z < -1.96$ and $z > 1.96$.
6. $z = (.9218 - .7753)\sqrt{39} = .915$.
7. Because z is not in the critical region, we accept H_0.

2.8 THE DESIGN OF REGRESSION AND CORRELATIONAL STUDIES

This section is not intended to be a comprehensive treatise on the design of research involving regression and correlation methods. There are many principles of good research design that apply to nearly all types of research in psychology and education. Entire books have been devoted to the problems and principles of research design (for example, Kerlinger, 1973). Our purpose is only to point out certain design considerations that apply particularly to regression and correlational studies.

2.8.1 Selection of the Sample

Because of differences between the mathematical models employed in regression and correlation analysis, the sampling methods differ. In the regression model, the concomitant variable, X, is regarded as a constant, and only Y is a random variable. One is therefore free to choose the values of X in any way desired; it is not necessary to select them randomly. How, then, should the values of X be selected? A part of the answer comes from recalling that the standard error of the regression coefficient, $\sigma_b = \sigma_{y \cdot x} / \sqrt{S_{xx}}$, depends on the variance of X. Thus, we might decide to choose the X values to maximize S_{xx}, thereby minimizing σ_b and attaining a more precise extimate of β. To maximize S_{xx} we should select all of the X values from the extremes of the possible range of values of X, excluding values of X in the middle of that range. But in order to achieve good estimates in the entire range of X, as well as to obtain a sample which permits determination of the nature of the relationship between X and Y, we should obviously include values in the middle of the range of X as well as the extremes. Therefore, a reasonable compromise is to choose X values at various equidistant points in the entire range of X and *randomly* select an equal number of Y values at each selected value of X. This procedure permits us to examine and test for linearity of regression and provides a relatively accurate estimate of β.

In the correlation model, *both* X and Y are random variables and values of both variables must be selected randomly. Actually, the sampling unit in correlational studies is a pair of values, one of X and one of Y, so that sampling is carried out on a population of such pairs. The distribution in the sample obtained thus depends upon the bivariate distribution of X and Y in the population.

2.8.2 Size of the Sample

In both correlation and regression studies, the samples typically used by researchers in psychology and education are often too small. As we showed earlier in this chapter a sample size of nearly 400 was required to obtain a 95% confidence interval for ρ of length about .13. A sample of 100, if $r = .596$ as in Table 1.2.1, would yield a 95% confidence interval of length .26. Although this interval length may be tolerable, it leaves a substantial degree of doubt concerning the true value of ρ. Although larger samples are desirable, we shall recommend $n = 100$ as a minimally acceptable sample size for regression and correlation studies.

2.8.3 Choice of Model

Most regression and correlation methods are based on the assumption of a linear model. Therefore, it is the responsibility of the researcher to justify the assumption that the relationship between X and Y in the population is indeed linear. Scatter diagrams should be used to show the bivariate distribution in the sample. If visual inspection of such figures suggests that a nonlinear trend may be present, then an appropriate test of nonlinearity, as described in Chapter 1, should be carried out. Too frequently, researchers compute linear regression and correlation coefficients without performing such a check. In cases in which the linear model is not appropriate, the linear statistics computed may lead to erroneous conclusions. As an example, consider the following illustrative pairs of observations:

X:	8	10	14	6	2	12	4
Y:	4	3	1	3	1	2	2

The scatter diagram representing these data is shown in Figure 2.8.1. An inspection of this figure shows clearly that there is a relationship between X and Y. However, the product-moment correlation coefficient computed for these data is exactly zero. Of course, for such simple data, one is quite likely to detect the nonlinear characteristic of the relationship merely by inspecting the list of X and Y values. But when the sample size is larger than, say 100, it is usually necessary to construct a scatter diagram in order to detect nonlinear trends. The failure to detect such trends when they exist can lead to serious misinterpretation of the data.

When tests of significance are to be carried out in regression and correlation studies, it is important to check on assumptions other than linearity. In the linear regression model, for example, it is assumed that the dependent variable, Y, is normally distributed for each value of X included

FIGURE 2.8.1 Scatter Diagram of Seven Pairs of Observations on X and Y

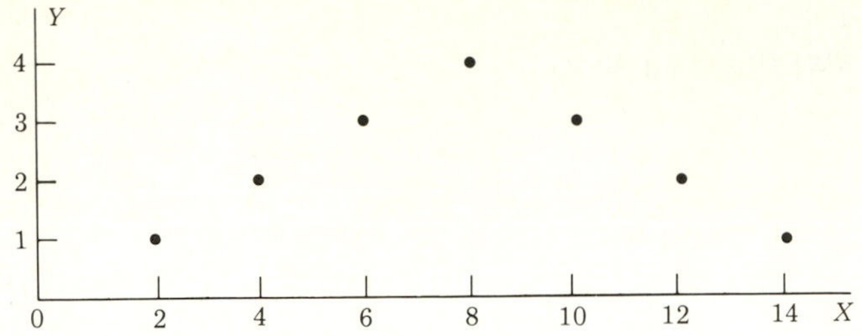

in the sample. In the model we assume that the variance of Y is the same for all conditional distributions. In the linear correlation model, a normal bivariate distribution is assumed, as we noted in Section 2.4.

In general, regression and correlation studies require stricter adherence to assumptions than do statistical methods involving comparisons among means, such as t tests and analyses of variance. However, for unimodal symmetric distributions, non-normality has relatively little effect on the distribution of the correlation coefficient. Nevertheless, researchers should give justifications for making the assumptions of the model.

2.8.4 Control of Extraneous Variables

It is important in *all* research to control in some way the influence of variables that may affect the results of one's research, but that are not of primary importance in the immediate research problem. For example, one may wish to study the relative effectiveness of two different instructional methods, recognizing that the age of the subjects may also have an important effect on the outcome. If one wishes to *detect* such an effect, one should apply the instructional methods to subjects of different ages, and determine whether the relative effectiveness differs from one age to another. On the other hand, if one wished to *eliminate* the effects of age differences in the experiment, one might select subjects who are all approximately the same age. Such devices for control of extraneous variables are well known in experimental design, where one is testing hypotheses about means. Control of such variables in regression and correlation studies is just as important from the point of view of research design, but, unfortunately, is generally done with less care, if at all.

Failure to control for extraneous variables can lead to serious misinterpretation of results. For example, suppose that one obtained data on size of shoe (X) and size of vocabulary (Y) for a large sample of children between the ages of 3 and 8 years. The product-moment correlation between X and Y in such a sample would be highly positive. Does a large positive correlation suggest that a meaningful relationship exists between shoe size and vocabulary in such children? Of course, the answer is no. Both X and Y are developmental variables, and are, therefore, highly related to age. If we restricted the sample to a single age, the correlation would probably be close to zero. The point is that if one were seriously interested in determining whether shoe size and vocabulary were related, the study should be designed so that the effects of age were eliminated. The effects of extraneous variables are not always as obvious as in this example. However, their control is an extremely important design consideration in regression and correlation studies, as well as in other research in the behavioral sciences.

2.8.5 Use of Extreme Groups

We have commented above on the freedom which the researcher has in selecting values of X in regression studies. We emphasize here again that one does *not* have such freedom in correlational studies. In particular, we caution the reader against using extreme groups (for example, the upper and lower quarters of a distribution) when designing correlational studies. The usual effect of such a practice is to inflate the value of the correlation coefficient artificially, so that it looks as if the variables involved are related more strongly than they actually are. There are some kinds of *experimental* designs in which the use of extreme groups can be justified, but we advise strongly against the practice in *correlational* studies.

2.9 INTERPRETATION OF RESULTS OF CORRELATIONAL STUDIES

After computing a correlation coefficient, one must decide what it really means in terms of the strength of relationship between X and Y. One often tends to think of r as indicating the "proportion of perfection" in the relationship. For example, a correlation of $r = .5$ might be thought to indicate a "fifty percent perfect" relationship. Such an interpretation is *not* correct. As pointed out earlier in this chapter, the *square* of the correlation coefficient may properly be interpreted as a proportion, namely, the proportion of the total variance of Y accounted for, or explained, by its linear relationship to X. Therefore, r should be interpreted in terms of its square. A correlation of .5, for example, would mean that one fourth ($r^2 = .25$) of the variability among the Y values could be explained on the basis of the linear relationship between X and Y.

There is also a tendency among researchers to talk about some values of r as "large" and others as "small." Too often it happens that the basis for such a judgment is nothing more than a test of significance of r, that is, a test of the null hypothesis H_0: $\rho = 0$. Remember that rejection of this hypothesis simply means that the population correlation coefficient is not equal to zero. Given a large enough sample size, a sample correlation of

$r = .10$, or even $r = .05$, would be statistically significant, but would indicate a very weak linear relationship between the variables involved.

Unfortunately, there is little agreement on the meaning of the terms "large" and "small." A value of r that seems large in one context may seem small in another. In making judgments about the size of an observed correlation one might consider the following questions:

1. Is the correlation larger than correlations *usually* obtained between the variables involved? For example, if one expected, on the basis of past research, to find a correlation of .6 between X and Y, than .4 would be considered small. If, on the other hand, one *expected* to find a correlation of .3, than .4 might be considered large.

2. Is it large enough so that regression-line predictions of Y from X, or vice versa, have satisfactory standard errors, that is, so that $s_{y \cdot x}$ does not exceed an upper limit based on the requirements of the research problem? In other words, could X be used to predict Y with sufficient accuracy so that it could be considered useful in the practical sense?

3. Is the observed correlation consistent in sign and magnitude with that expected on theoretical grounds? For example, if theory suggested a moderate negative correlation between X and Y, the observed value should be evaluated accordingly.

It seems clear that any observed correlation should be interpreted within the context in which it was obtained. Therefore, no general statement can be made concerning what values are "large" and which are "small."

2.10 FACTORS THAT AFFECT THE SIZES OF CORRELATION COEFFICIENTS

Suppose one computes a correlation coefficient and finds that its value is either smaller or larger than the one expected. Of course, the reason might be (and often is) that one's expectations were in error. Before drawing that conclusion, however, the researcher should consider whether any of several other factors might have been responsible. Some of these are:

1. *Unreliability of measures*. If either X or Y or both have unsatisfactory reliability, the result is a reduction in the size of r compared with what one might expect. Having reliable measures is just as important in regression and correlation studies as in other types of research.

2. *Sample size*. An observed correlation may be smaller or larger than expected because the sample size was too small. In small samples, the correlation coefficient is unstable and may be much larger or smaller than the population value it estimates. Increasing the sample size neither necessarily increases nor decreases the correlation. However, the value of r resulting from a larger sample is a more accurate estimate of the population value.

3. *Nonlinear relationships*. As we indicated previously, sometimes the linear model does not accurately represent the relationship between X and

Y. A correlation then may be smaller than expected because the true relationship is really nonlinear. A scatter diagram helps to detect such a trend.

4. *Transformations of data.* Adding a constant to either or both X and Y or multiplying either or both X and Y by a constant has no effect on the correlation coefficient, unless the constant multiplier changes all the signs of one variable, but not the other's. In the latter case, only the sign of r is changed. These transformations may, however, change the values of the regression coefficients. Other kinds of transformations, such as finding logarithms of all scores, *may* change the values of r, *and* the values of the regression coefficients.

5. *Grouping data.* Converting ungrouped to grouped data will usually change the values of r and the regression coefficients, to an extent that depends on how the grouping is done.

6. *Combining groups.* If two or more groups are combined, and r is computed for the combined group, it may differ greatly from the correlation in either group alone. If the groups have approximately equal means on X and Y, the effect is likely to be small. However, if the means differ substantially on either X or Y or both, there can be a substantial change in the correlation coefficient when groups are combined. See Figure 2.10.1 for some illustrations.

7. *Restriction in Range.* When the variance of either X or Y or both is small, so that the sample is fairly homogeneous, the correlation between X and Y is likely to be small. Such a restriction of range in the sample can occur because the population itself is restricted. For example, college students should have a smaller variance of SAT-V scores than high school students. Restriction in range can also occur as a result of faulty sampling procedures; as a result of measurement error arising from poor test construction or selection; and from mistakes in test administration or scoring.

The reason for the reduction of the size of the correlation coefficient is apparent when we consider the extreme case in which the variability on either X or Y is zero. Recall from Equation 2.2.2 that the numerator of r_{XY} is $S_{xy} = \Sigma(X_i - \bar{X})(Y_i - \bar{Y})$. Now suppose that all sample values of X were equal. Then all n values of $X_i - \bar{X}$ would be zero, so that S_{xy} would also be zero. (Technically, r in this case is really indeterminate, because S_{xx} in the denominator would also be zero. However, the effect on the covariance, $S_{xy}/(n-1)$, is clear.)

When it is known that restriction in the range of one of the variables, say X, has occurred, it is possible to estimate the correlation that would have been obtained had there been no restriction. If we denote this estimate by $\hat{\rho}_{XY}$ and let

s_X = standard deviation of the restricted distribution,

σ_X = standard deviation of the unrestricted distribution, and

r_{XY} = observed correlation of Y with restricted X,

FIGURE 2.10.1 Illustrations of the Effect on the Correlation Coefficient of Combining Two or More Groups with Unequal Means

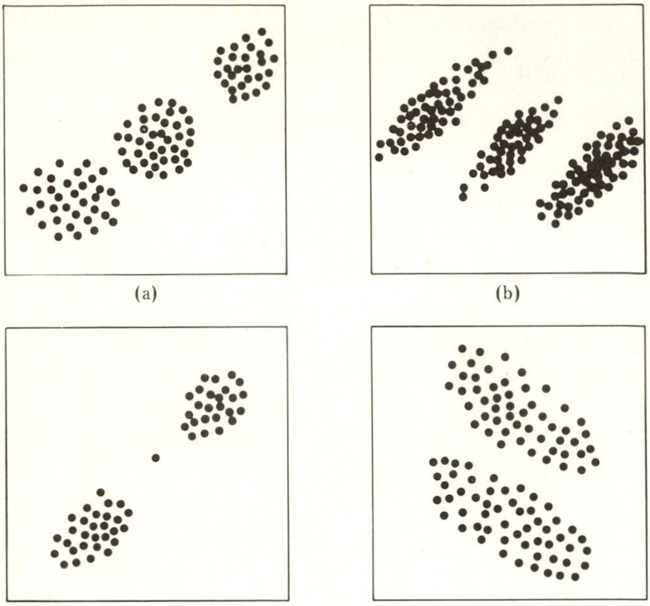

(a) In each group, r is near zero. In the composite group, r is large and positive.
(b) In each group, r is positive. In the composite group, r is negative.
(c) In each group, r is positive but small. In the composite group, r is positive and much larger.
(d) In each group, r is negative. In the composite group, r is near zero.

then

$$\hat{\rho}_{XY} = \frac{r_{XY}(\sigma_X / s_X)}{\sqrt{1 - r_{XY}^2 + r_{XY}^2\left(\frac{\sigma_X}{s_X}\right)^2}}, \qquad (2.10.1)$$

under the assumptions of linearity and homoscedasticity in the unrestricted bivariate population of X and Y values. Of course, the use of Equation 2.10.1 requires knowledge, specifically of σ_X and of the validity of the above assumptions, that may not always be available. However, the same kind of information that would lead one to suspect a restriction in range might also provide a basis for the use of this equation. In any case, the discussion here should serve to emphasize the importance of reporting not only r_{XY}, but also s_X and s_Y (as well as n, of course) when giving the results of a correlational study.

8. *Consistency among Coefficients*. Suppose we have data available for each of n individuals on each of three variables, say X_1, X_2, and X_3. We have computed the correlations between X_1 and X_2 and between X_1 and X_3 and found them to be $r_{12}=.5$ and $r_{13}=.3$. Given these two values, it can be

shown that the limits on the possible values of r_{23}, the correlation between X_2 and X_3, are no longer -1.00 and $+1.00$. Instead, the upper limit on r_{23} is

$$r_{12}r_{13}+\sqrt{1-r_{12}^2-r_{13}^2+r_{12}^2r_{13}^2}=.976,$$

and the lower limit is

$$r_{12}r_{13}-\sqrt{1-r_{12}^2-r_{13}^2+r_{12}^2r_{13}^2}=-.676.$$

Thus, for the values $r_{12}=.5$ and $r_{13}=.3$, we find that $-.676\leqq r_{23}\leqq .976$. A value of r_{23} within these limits would be considered *consistent* with r_{12} and r_{13}. A value outside would be *inconsistent*.

The point here is that one of the factors that influences the size of the correlation coefficient, r_{XY}, is the correlations that X and Y have with *other* variables. When such correlations are unknown, we must consider the possible range of r_{XY} to be $-1.00\leqq r_{XY}\leqq +1.00$. However, when additional information is available about relationships between X and Y and other variables, our interpretation of r_{XY} should take account of such information.

2.11 EXERCISES

2.1. An investigator determined that in a certain high school (School 1) there was a relationship between high school grade point average (GPA) at graduation of 188 students and the score on an aptitude test (APT) administered to them at the beginning of the 10th grade, as expressed by a Pearson product-moment correlation of $r_1=.62$. In a different high school (School 2) the correlation between final GPA and APT, also administered at the beginning of the 10th grade, was $r_2=.52$, based on 228 students.

(a) For each school, test the significance of the correlation between APT and GPA. Use $\alpha=.01$.

(b) Test the hypothesis that the correlation between APT and GPA was larger in School 1 than in School 2. Again use $\alpha=.01$.

(c) Obtain an estimate of the correlation between APT and GPA by combining the estimates from the two schools.

2.2. Given the following scatter diagram based on 900 independent observations on two variables, X and Y:

Y	X						
	40–44	45–49	50–54	55–59	60–64	65–69	f_Y
35–39					1	2	3
30–34				1	21	7	29
25–29			8	164	15		187
20–24		21	368	202			591
15–19	5	61	24				90
f_X	5	82	400	367	37	9	900

(a) Calculate the Pearson product-moment correlation coefficient based on these data.

(b) Calculate the two regression equations, one for estimating Y from X and one for estimating X from Y.
(c) Calculate the 95% confidence limits for $\mu_{y \cdot x}$ if $X=62$ and for $\mu_{x \cdot y}$ if $Y=33$.

2.3. Below are the heights, in inches, of 12 pairs of fathers and their adult sons:

Pair	*Father's height* (X)	*Son's height* (Y)
1	65	68
2	63	66
3	67	68
4	64	65
5	68	69
6	62	66
7	70	68
8	66	55
9	68	71
10	67	67
11	69	68
12	71	70

(a) Calculate the Pearson product-moment correlation for these data.
(b) Test the significance of the result obtained in (a) using $\alpha=.05$.
(c) Is there evidence from the data of the kind of regression effect discussed by Galton? Explain.

2.4. Use the data given in Exercise 2.4, and assume that the fathers' heights are restricted in range. If σ_X were actually 4.0, what would be the estimate of the corrected correlation between X and Y?

3/Further Methods of Bivariate Correlation

The product-moment correlation coefficient, r, discussed in Chapter 2, is applicable only when both X and Y can be regarded as continuous variables. Its utility as a descriptive statistic depends heavily on the assumption of linearity of regression of Y on X and of X on Y. Inferences based on r require the assumption of a bivariate normal distribution. When one or more of these assumptions is not met, it is necessary to use a different technique for determining the degree of relationship between two variables. In this chapter we shall discuss several methods of correlation that differ from the Pearson product-moment r, particularly in terms of assumptions about the joint distribution of X and Y. Some of these methods are variations of the Pearson r, whereas others use a quite different basis for expressing relationship.

3.1 RANK CORRELATION METHODS

Sometimes the basic data obtained in a correlational study are ranks rather than continuous variables. In other cases the original measures are continuous, but are transformed to ranks prior to analysis. In either case, if the objective is to express the degree of relationship between sets of ranks, the procedures discussed in Chapter 2 are not applicable. In this section we discuss several methods that are appropriate in such cases.

3.1.1 Spearman's Rank-Order Correlation

One of the earliest and most widely used methods of correlation when data are in the form of ranks is due to Charles Spearman, the British psychologist and statistician. Spearman's (1904a) rank-order correlation coefficient, which we shall denote by r', is based entirely on ranks and does not require assumptions regarding population parameters. Hence, it is an example of a nonparametric statistic.

The use of the Spearman rank-order correlation method should ordinarily be considered only when the observed data are in the form of ranks. However, there are occasions when data obtained on continuous variables do not meet the assumptions for proper interpretation of r and they are transformed to ranks for analysis; but changing interval data to ordinal form merely for the sake of using the computationally simpler rank correlation technique results in a considerable loss of information, and should be avoided.

The rank-order correlation, r', is simply the product-moment correlation between X and Y, where X and Y are expressed as ranks and both have distributions consisting of the consecutive integers, $1,2,\ldots,n$.

To derive a computational formula for r', we recall that the Pearson product-moment correlation coefficient was defined by Equation 2.2.2 as

$$r=\frac{S_{xy}}{\sqrt{S_{xx}S_{yy}}}$$

When X and Y consist of the same set of consecutive integers, S_{xx} and S_{yy} in this expression must be equal. Therefore, if we can derive an expression for S_{xx}, it will give us the value of S_{yy} as well.

We know from elementary algebra that the sum of the set of consecutive integers from 1 to n, and hence the value of ΣX, is

$$\Sigma X=1+2+\cdots+n=\frac{n(n+1)}{2}.$$

We know also that the sum of the squares of these same integers, and hence ΣX^2, is

$$\Sigma X^2=1^2+2^2+\cdots n^2=\frac{n(n+1)(2n+1)}{6}.$$

Because, by definition, $S_{xx}=\Sigma X^2-\dfrac{(\Sigma X)^2}{n}$, substitution yields

$$S_{xx}=\frac{n(n+1)(2n+1)}{6}-\frac{\left[\dfrac{n(n+1)}{2}\right]^2}{n}$$

$$=\frac{n(n+1)(2n+1)}{6}-\frac{n(n+1)^2}{4}=\frac{n^3-n}{12}.$$

Now, let us consider the numerator, S_{xy}, of Equation 2.2.2. Because $\bar{X}=\bar{Y}$ when both X and Y consist of the same set of ranks, we can write the identity $X-Y=(X-\bar{X})-(Y-\bar{Y})$. If we let $d=X-Y$, then

$$\Sigma d^2=\Sigma\left[(X-\bar{X})-(Y-\bar{Y})\right]^2$$

$$=\Sigma(X-\bar{X})^2+\Sigma(Y-\bar{Y})^2-2\Sigma(X-\bar{X})(Y-\bar{Y})$$

$$=S_{xx}+S_{yy}-2S_{xy}.$$

Thus,

$$S_{xy} = \frac{S_{xx} + S_{yy} - \Sigma d^2}{2} = \frac{n^3 - n}{12} - \frac{\Sigma d^2}{2}.$$

Substituting for S_{xx}, S_{yy}, and S_{xy} in Equation 2.2.2, we have

$$r' = \frac{\frac{n^3 - n}{12} - \frac{\Sigma d^2}{2}}{\frac{n^3 - n}{12}} = 1 - \frac{6\Sigma d^2}{n(n^2 - 1)}. \tag{3.1.1}$$

Thus, we see that the product-moment correlation based on ranks can be computed using the differences between paired ranks. Of course, r' could also be computed from the paired ranks by use of Equation 2.2.2, but Equation 3.1.1 facilitates the computation when it is done without the aid of a computer.

To illustrate the use of Spearman's rank correlation method, we shall consider the following example. Suppose a department head has asked two professors to rank eight textbooks in order to determine the best one for use in a general psychology course. The department head is interested in how well the two professors agree in their rankings, as well as in their evaluations of the texts. The best textbook is to receive a rank of one, the second best, two, etc. The data are given in Table 3.1.1.

Although the only sum needed for computation of r' is $\Sigma d^2 = 26$, the others are useful to obtain as checks. For example, not only should $\Sigma X = \Sigma Y$, but each should be equal to $n(n+1)/2$. Furthermore, Σd should equal zero, since $\Sigma d = \Sigma(X - Y) = \Sigma X - \Sigma Y = 0$.

The value of the rank-order correlation coefficient based on these data is

$$r' = 1 - \frac{6\Sigma d^2}{n(n^2 - 1)} = 1 - \frac{6(26)}{8(64 - 1)} = .69.$$

Hence, the degree of agreement, as expressed by the correlation between the ranks assigned to the texts by the two professors, appears to be moderately high. It would be reasonable to conclude, from an inspection of the ranks assigned, that Textbook H is the preferred book.

TABLE 3.1.1. RANKINGS OF EIGHT TEXTBOOKS BY TWO PROFESSORS

	Professor			
Textbook	*X*	*Y*	*d*	d^2
A	3	5	−2	4
B	7	8	−1	1
C	6	6	0	0
D	4	3	1	1
E	1	4	−3	9
F	8	7	1	1
G	5	2	3	9
H	2	1	1	1
Sum	36	36	0	26

When there are ties in rankings of persons or objects, then Equation 3.1.1 cannot be used for computing r'. In this case the mean rank of the tied individuals or objects is assigned to each of them, and the value of r' is computed using Equation 2.2.2. For example, if textbooks D and G had been considered equally good by Professor X, then each would have received a rank of 4.5 in Table 3.1.1, and the computed value of the rank order coefficient would have been $r' = .71$.

To test the statistical significance of r', the following statistic may be used when $n \geqslant 10$:

$$t = r'\sqrt{\frac{n-2}{1-(r')^2}} \tag{3.1.2}$$

This statistic has approximately a t distribution with $n-2$ degrees of freedom under the null hypothesis that ρ', the population parameter corresponding to r', equals zero. An exact test, for $n \leqslant 100$, may be carried out using a table prepared by Zar (1972), which is reproduced as Table C.5 in Appendix C.

3.1.2 Kendall's Tau

Another widely used measure of the relationship between pairs of variables in the form of ranks is the Tau coefficient proposed by Kendall (1948, 1955). Tau (τ) is based on the extent of agreement between judges in their relative orderings of all possible pairs of individuals or objects ranked. An agreement occurs when both judges give the same ordering to the objects in a pair. For example, in Table 3.1.1, both professors regarded textbook A superior to B; thus, they agreed in their ordering of the objects in that pair. A disagreement occurs when the ordering of objects in a pair is not the same for the two judges. Again referring to Table 3.1.1, we see that there is a disagreement on the pair (E, G), because Professor X regarded textbook E superior to G, while Professor Y thought that G was better than E.

The Tau coefficient may be computed by counting the number of agreements (n_{agree}) and the number of disagreements (n_{disagree}) among all possible pairs, finding the difference, and dividing by the number of pairs. Thus,

$$\tau = \frac{n_{\text{agree}} - n_{\text{disagree}}}{\frac{n(n-1)}{2}}, \tag{3.1.3}$$

in which n is the number of objects ranked. Because $n_{\text{disagree}} = \frac{n(n-1)}{2} - n_{\text{agree}}$, an alternative formula for Tau may be obtained by substituting for n_{disagree}. Thus,

$$\tau = \frac{n_{\text{agree}} - \left[\dfrac{n(n-1)}{2} - n_{\text{agree}}\right]}{\dfrac{n(n-1)}{2}} = \frac{2n_{\text{agree}}}{\dfrac{n(n-1)}{2}} - 1$$

$$= \frac{4n_{\text{agree}}}{n(n-1)} - 1. \qquad (3.1.4)$$

It can be seen from either of these equations that τ will have the value $+1.00$ if there is perfect agreement in rankings and the value -1.00 when there is complete disagreement, that is, when one set of rankings is the exact inverse of the other.

To simplify counting the number of agreements, it is helpful to arrange the individuals or objects in order according to their ranks as assigned by each judge. Such an arrangement for the data of Table 3.1.1 is shown in Table 3.1.2.

The number of agreements for each object is then determined by counting the objects that *both* judges placed below it in their rankings. For example, for textbook E, both judges ranked textbooks A, C, F, and B lower than E. Therefore, the number of agreements among pairs involving E is 4, as indicated at the top of the right-hand column in Table 3.1.2. For textbook H, both professors ranked textbooks A, D, G, C, B, and F lower than H, so the number of agreements is 6. The total number of agreements for these data is 21, and substitution in Equation 3.1.4 yields

$$\tau = \frac{4(21)}{56} - 1 = 1.5 - 1 = .5.$$

Tau may also be expressed in terms of the number of disagreements as

$$\tau = 1 - \frac{4n_{\text{disagree}}}{n(n-1)}. \qquad (3.1.5)$$

The number of disagreements may easily be determined by again ordering

TABLE 3.1.2 AGREEMENTS IN RANKINGS OF EIGHT TEXTBOOKS BY TWO PROFESSORS

Rank	*Ranking by Professor X*	*Ranking by Professor Y*	*Number of agreements*
1	E	H	4
2	H	G	6
3	A	D	3
4	D	E	3
5	G	A	3
6	C	C	2
7	B	F	0
8	F	B	0
			$n_{\text{agree}} = 21$

the individuals or objects according to their ranks given by each judge and connecting the same objects in the two rankings by straight lines. The number of disagreements is then the number of intersections of the lines. The number of disagreements for the data of Table 3.1.1 is shown in Figure 3.1.1. The numbers show that there are seven intersections, so $n_{\text{disagree}}=7$. Then by Equation 3.1.5, we have

$$\tau = 1 - \frac{4(7)}{56} = 1 - .5 = .5,$$

the same result obtained from Equation 3.1.4.

Equations 3.1.4 and 3.1.5 are useful only when there are no tied ranks. When ties exist, the data should be rearranged in a two-way frequency table with columns representing ranks assigned by one judge and rows representing those assigned by the other. To illustrate, suppose that the data of Table 3.1.1 had been as shown in Table 3.1.3.

On rearrangement, these data would be presented as shown in Table 3.1.4. To compute Tau, we carry out the following steps:

1. For each object (textbook, in this case) count the entries in all cells below and to the right of the cell containing that object. For example, below and to the right of textbook H, we have six entries, namely, G, D, A, C, F, and B. (Note that when there are *no* tied ranks, this number is simply the number of agreements, as defined above, associated with that object.) Obtain the sum of these, n_R. Here, $n_R = 6+3+4+3+3+2+0+0=21$.

2. For each object, count the entries in all cells below and to the *left* of the cell containing the object. For example, below and to the left of object H, we have one entry, namely, E. (Note that when there are *no* tied ranks,

FIGURE 3.1.1 Disagreements in Rankings of Eight Textbooks by Two Professors

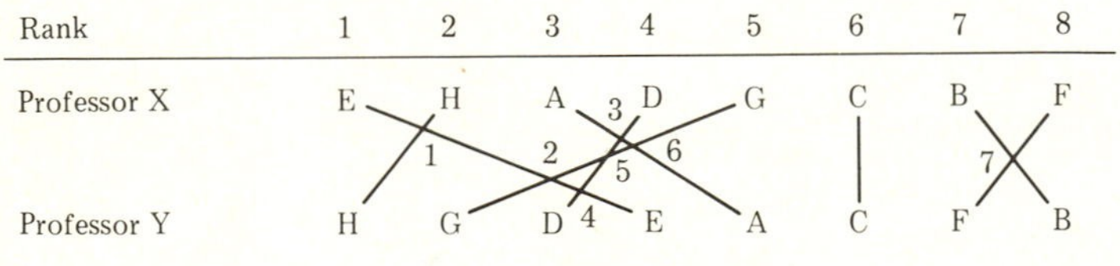

TABLE 3.1.3 TIES IN RANKINGS OF EIGHT TEXTBOOKS BY TWO PROFESSORS

	Ranking	
Textbook	*Professor X*	*Professor Y*
A	3	5
B	7	8
C	6	6
D	4.5	3.5
E	1	3.5
F	8	7
G	4.5	2
H	2	1

TABLE 3.1.4 REARRANGEMENT OF THE DATA IN TABLE 3.1.3 AS A TWO-WAY FREQUENCY TABLE

	X rank							
Y rank	1	2	3	4.5	6	7	8	*Total* (t_k)
1		H						1
2				G				1
3.5	E			D				2
5			A					1
6					C			1
7							F	1
8						B		1
Total (t_j)	1	1	1	2	1	1	1	8

this is the number of disagreements associated with the object.) Obtain the sum of these, n_L. Here, $n_L = 1+2+0+1+0+0+1+0=5$.

3. Compute

$$\tau = \frac{n_R - n_L}{\sqrt{\left(\frac{n(n-1)}{2} - T_X\right)\left(\frac{n(n-1)}{2} - T_Y\right)}}, \tag{3.1.6}$$

where

$$T_X = \frac{\sum_j t_j(t_j - 1)}{2},$$

$$t_j = \text{total for column } j,$$

$$T_Y = \frac{\sum_k t_k(t_k - 1)}{2}, \text{ and}$$

$$t_k = \text{total for row } k.$$

For our example,

$$\tau = \frac{21-5}{\sqrt{\left(\frac{8(7)}{2} - 1\right)\left(\frac{8(7)}{2} - 1\right)}} = .59,$$

which differs, of course, from the value obtained when there were no tied ranks.

When there are no ties, the statistical significance of Tau may be tested by computing the statistic

$$z = \frac{\tau}{\sigma_\tau} = \frac{\tau}{\sqrt{\frac{2(2n+5)}{9n(n-1)}}}, \tag{3.1.7}$$

which has approximately a standard normal distribution under the null hypothesis that Tau equals zero in the population. For $n > 10$, the standard normal distribution yields a satisfactory approximation. When ties are present, the denominator of Equation 3.1.7 should be replaced by the square root of

$$\sigma_\tau^2 = \frac{\left[n(n-1)(2n+5) - \sum_j t_j(t_j-1)(2t_j+5) - \sum_k t_k(t_k-1)(2t_k+5)\right]}{18}$$

$$+ \frac{\left[\sum_j t_j(t_j-1)(t_j-2)\right]\left[\sum_k t_k(t_k-1)(t_k-2)\right]}{9n(n-1)(n-2)}$$

$$+ \frac{\left[\sum_j t_j(t_j-1)\right]\left[\sum_k t_k(t_k-1)\right]}{2n(n-1)},$$

in which t_j and t_k are defined as in Equation 3.1.6. When no ties are present, this expression reduces to the square of the denominator of Equation 3.1.7.

When $n \leqslant 10$ and there are no ties, a test of significance of τ may be carried out using Table C.6 in Appendix C. This table gives critical values of the numerator of Equation 3.1.3, rather than of τ itself. Therefore, to use the table it is necessary to compute

$$\begin{aligned} S &= n_{\text{agree}} - n_{\text{disagree}} \\ &= 2n_{\text{agree}} - \frac{n(n-1)}{2} = \frac{n(n-1)}{2} - 2n_{\text{disagree}}. \end{aligned} \tag{3.1.8}$$

When there are ties present, Table C.6 may be used if S is defined as

$$S = n_R - n_L, \tag{3.1.9}$$

in which n_R and n_L are defined as in Equation 3.1.6.

To illustrate the use of Table C.6, we consider the example of the professors' ratings of eight textbooks. Based on the data of Table 3.1.1, we found that $n_{\text{agree}} = 21$. Therefore, $S = 2(21) - 8(7)/2 = 14$. Referring to the column for $n = 8$ in Table C.6, we find that a value of at least 16 would be required for significance at the .05 level. Therefore, we cannot consider the value of $\tau = .5$ to be significantly different from zero.

For the example in which ties were present, we find that $S = n_R - n_L = 16$. From Table C.6 with $n = 8$ we find that this value is equal to the critical value of S for a one-tailed test at the .05 level of significance. Therefore, we conclude that our observed value of $\tau = .59$ is significant at the .05 level.

Note that τ and r' express agreement between rankings in quite different ways. In computing r' the ranks are treated in the same way that pairs of test scores would be treated if we wished to express the correlation between them; hence, the concept of agreement expressed by r' is essentially the same as the concept of correlation expressed by the product-moment coefficient, r. The concept of agreement expressed by Tau can be seen by considering Equation 3.1.3. If we rewrite this equation as

$$\tau = \frac{n_{\text{agree}}}{\frac{n(n-1)}{2}} - \frac{n_{\text{disagree}}}{\frac{n(n-1)}{2}},$$

we see that it is merely the difference between two proportions whose sum is one. The first, $p_{\text{agree}} = n_{\text{agree}}/n(n-1)/2$, is the proportion of the total number of pairs of objects, $n(n-1)/2$, in which the relative ordering of the ranks is the same in the two sets of rankings. The second, $p_{\text{disagree}} = n_{\text{disagree}}/n(n-1)/2$, is the proportion of the total number of pairs in which the relative ordering is not the same. Thus, the value $\tau = .5$, based on Table 3.1.1, means that for the two professors involved, the difference, $p_{\text{agree}} - p_{\text{disagree}}$, was .5 and was in the direction of agreement, as indicated by the positive sign. If we think of p_{agree} and p_{disagree} as probabilities, then the value $\tau = .5$ means that, in selecting a pair of objects at random, the chance of agreement between these two professors in their ordering of the objects in the pair is considerably better (.5 more) than the chance of disagreement.

It is interesting to note that although r' and τ differ conceptually and in method of computation, they are highly correlated. As Kendall states, "It may be shown that for the population in which all rankings occur equally frequently... the product moment correlation between $[r']$ and τ is $2(n+1)/\sqrt{2n(2n+5)} \cong 1 - 1/4n$" (Kendall, 1952, p. 407). It follows that, when n is large, the correlation between these two statistics will be close to one. Furthermore, values of Tau will usually be closer to zero than those of r'; as n increases, the ratio of τ to r' approaches approximately 2/3.

We shall make no recommendation here concerning which of these two coefficients, r' or τ, it is better to use. As Kendall (1952, p. 393) states, "...one appears to be as good as the other so far as providing a measure of ranking concordance is concerned. [Spearman's r'] is, however, easier to calculate and is probably the most convenient to use. Against this must be set certain difficulties in its sampling distribution...and the fact that τ can be generalized to the case of partial rank correlation." The latter comment, concerning the properties of the sampling distribution of Spearman's r', suggests the only possible advantage of τ over r'. The sampling distribution of τ tends to normality quite rapidly as n increases; in fact, as Kendall (1952, p. 404) points out, "...the tendency is so rapid that for values of n greater than 10 the normal distribution provides an adequate approximation." Although this comment is worth noting, the advent of high-speed computers has made available more accurate tests of significance for Spearman's r'. Thus, the decision as to which statistic to use depends principally on one's preference for one or the other of the two definitions of agreement upon which the two statistics are based.

3.1.3 Kendall's Coefficient of Concordance

Before leaving the subject of correlation between rankings, we should indicate that a procedure analogous to those discussed above is available for expressing the degree of agreement among rankings given by *several* judges. Suppose, for example, that four professors, rather than just two, had been asked to rank the eight psychology texts of Table 3.1.1. The data might then be as shown in Table 3.1.5.

A statistic appropriate for expressing the degree of similarity among rankings given by m judges is Kendall's *coefficient of concordance*, W. This coefficient is simply the ratio of the variance of the rank sum, S_j, to its maximum possible variance. If all judges agreed exactly in their rankings, then the rank sum, S_j, would have its maximum possible variance and W would equal one. If all values of S_j were identical, then the variance of S_j would be zero, as would the value of W.

The value of W can be computed from

$$W = \frac{s}{\frac{1}{12} m^2 (n^3 - n)}, \tag{3.1.10}$$

in which m is the number of judges, n is the number of individuals or objects ranked, and

$$s = \frac{n \sum_{j=1}^{n} S_j^2 - \left(\sum_{j=1}^{n} S_j \right)^2}{n} \tag{3.1.11}$$

is the variance of the rank sums, that is, the S_js. For the data of Table 3.1.5, $m=4$, $n=8$, and $s=460$, so that

$$W = \frac{460}{\frac{1}{12}(4)^2(8^3 - 8)} = .68.$$

This result suggests a moderately high degree of similarity in the professors' rankings of the eight textbooks.

In addition to computing W in these problems, it is often useful to compute Spearman's rank-order correlation coefficient, r', for each pair of

TABLE 3.1.5 RANKINGS OF EIGHT TEXTBOOKS BY FOUR PROFESSORS

	Rankings				*Sum of*
Textbook	*Prof. W*	*Prof. X*	*Prof. Y*	*Prof. Z*	*Ranks,* S_j
A	4	3	5	4	16
B	6	7	8	7	28
C	5	6	6	8	25
D	2	4	3	3	12
E	1	1	4	5	11
F	7	8	7	6	28
G	8	5	2	2	17
H	3	2	1	1	7

judges. The mean of r' for all possible pairs, $\bar{r}'$, is related to W by

$$\bar{r}' = \frac{mW-1}{m-1}. \tag{3.1.12}$$

We tend to prefer W over $\bar{r}'$ for expressing overall similarity, because $\bar{r}'$ can have negative values that may be hard to interpret $(-1/m-1 \leqslant \bar{r}' \leqslant +1$, while $0 \leqslant W \leqslant +1)$. However, the values of r' for each pair can aid in identifying a judge who consistently disagrees with all other judges. Such a judge might be eliminated from the group or be required to take further training in the interest of increasing the reliability of the judgments obtained.

An approximate test of the significance of W can be carried out using the statistic

$$\chi^2 = m(n-1)W, \tag{3.1.13}$$

which, under the null hypothesis, approaches a chi-square distribution with $n-1$ degrees of freedom as n approaches infinity. This statistic is reasonably satisfactory for moderate or large values of m and n. An exact test can be made using Table C.7 in Appendix C for $n \leqslant 7$ and selected values of m between 3 and 20. This table gives critical values of s for $\alpha = .05$ and $\alpha = .01$.

For the data of Table 3.1.5, Equation 3.1.13 yields $\chi^2 = 4(8-1)(.68) = 19.17$, which, with $n-1=7$ degrees of freedom, is significant beyond the .01 level. We conclude, therefore, that the rankings of the textbooks by the four professors are significantly related. For this example, Table C.7 cannot be used, because it does not include values of n greater than 7.

3.2 BISERIAL CORRELATION

One of the basic assumptions necessary to drawing inferences about the population value, ρ, of the product-moment correlation coefficient is that the joint distribution of X and Y is bivariate normal. In practice, we sometimes encounter variables that have underlying continuous distributions, but which have been artificially dichotomized. For example, in item analysis, during test construction and refinement, we often desire to obtain the correlation between an item score and the total test score. In most achievement testing situations we assume that the achievement tested by a particular item is continuously and normally distributed; however, the item is usually scored simply correct or incorrect, which represents an artificial dichotomization of such an underlying distribution. A statistic that may be used to express the relationship between an artificially dichotomous variable such as item performance and a continuous variable such as total test score, is the *biserial correlation coefficient*. We denote this coefficient by r_{bis}.

Let us assume that we have two continuous normally distributed variables, one of which has been artificially dichotomized into "pass" and "fail" categories. Referring to Figure 3.2.1, we let p stand for the proportion

FIGURE 3.2.1 An Artificially Dichotomized Normal Distribution of Values of Variable Y

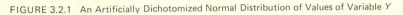

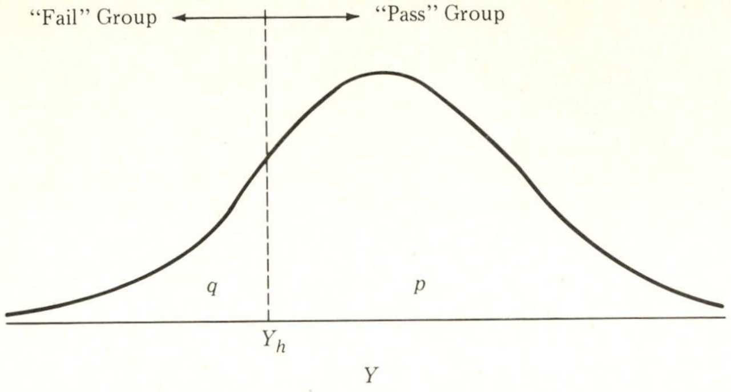

passing and q for the proportion failing. We now wish to find the correlation between the continuous variable, X, and the artificially dichotomized variable, Y.

Each of the groups formed by artificially dichotomizing Y has a mean on variable X. We denote these means by $\overline{X}_p$, the mean on X of the "pass" group, and $\overline{X}_q$, the mean on X of the "fail" group. The size of the biserial correlation coefficient, r_{bis}, depends on the size of the difference $\overline{X}_p - \overline{X}_q$. A large difference tends to indicate a strong relationship, while a difference of zero indicates no relationship at all. A formula for r_{bis}, based on this difference, is

$$r_{bis} = \frac{\overline{X}_p - \overline{X}_q}{s_X}\left(\frac{pq}{h}\right), \tag{3.2.1}$$

where s_X is the standard deviation of X for the total sample and h is the height of the ordinate of the *standard* normal distribution at the point z_h, corresponding to Y_h in Figure 3.2.1. Note that when $\overline{X}_p$ is greater than $\overline{X}_q$, then r_{bis} is positive; when $\overline{X}_p$ is less than $\overline{X}_q$, then r_{bis} is negative. Thus, the interpretation of a positive or negative sign depends on how the "pass" and "fail" groups have been defined. A detailed explanation of the derivation of Equation 3.2.1 is given in Kendall and Stuart (1973, pp. 321–3).

To demonstrate the application of the biserial correlation coefficient, let us consider the data of Table 3.2.1, in which $p = .66$ and $q = .34$. The means are given for the "fail" group ($\overline{X}_q = 76.4118$), the "pass" group ($\overline{X}_p = 90.9394$) and the total group ($\overline{X} = 86.0000$). The respective standard deviations are $s_{X_q} = 11.1304$, $s_{X_p} = 13.9095$, and $s_X = 14.7024$.

The height of the ordinate, h, at the point z_h, which divides the standard normal distribution into two parts with areas $p = .66$ and $q = .34$, may be determined from Table C.1 in Appendix C. Because z_h is the point to the right of which we have a proportion, $p = .66$, of the area, and to the left of which we have a proportion, $q = .34$, of the area, we look under

TABLE 3.2.1 FREQUENCY DISTRIBUTIONS ON TEST *X* FOR GROUPS FAILING AND PASSING ITEM *Y* ON TEST *X* ($n = 100$)

Total Score (X)	*Group Failing Item Y*	*Group Passing Item Y*	*Total Group*
125–129		1	1
120–124		2	2
115–119		2	2
110–114		0	0
105–109		4	4
100–104	1	6	7
95–99	0	10	10
90–94	2	12	14
85–89	4	8	12
80–84	6	7	13
75–79	9	6	15
70–74	5	4	9
65–69	3	3	6
60–64	0	1	1
55–59	2		2
50–54	2		2
Sum (n)	34	66	100
$\bar{X}$	76.4118	90.9394	86.0000
s	11.1304	13.9095	14.7024

the column headed "Area" for the value .34. The value closest to .34 in the "Area" column is .3409. Corresponding to this value, we read under the column headed "h Ordinate" the value .3668. Linear interpolation yields the value $h = .3664$, which we use in the following computations.

Substitution in Equation 3.2.1 produces

$$r_{bis} = \frac{90.9394 - 76.4118}{14.7024} \cdot \frac{(.66)(.34)}{.3664} = .6052.$$

Thus, we conclude that a moderately large positive relationship exists between performance on item Y and total score, X.

To test the null hypothesis, H_0: $\rho_{bis} = 0$, that is, that the population value of the biserial correlation equals zero, we may use the statistic

$$z = \frac{hr_{bis}}{\sqrt{\frac{pq}{n}}}, \qquad (3.2.2)$$

which, under H_0, has approximately the standard normal distribution. For the data of Table 3.2.1,

$$z = \frac{(.3664)(.6052)}{\sqrt{\frac{(.34)(.66)}{100}}} = 4.68.$$

Because the significance probability of this result is much smaller than .05, we conclude that the population value of the biserial coefficient is greater than zero.

Note that the biserial correlation coefficient cannot be interpreted in the same way as the Pearson product-moment correlation, r, nor should r_{bis} be compared with r. The possible range of values of r_{bis} is from minus infinity to plus infinity; thus, occasionally, one computes an r_{bis} which is greater than 1.00 or less than -1.00.

3.3 POINT BISERIAL CORRELATION

The biserial correlation coefficient discussed in the preceding section is appropriate when one wishes to express the relationship between a continuous variable and an artificially dichotomized variable. For some variables, however, we cannot reasonably assume an underlying normal distribution that has been *artificially* dichotomized. An example is sex, which is a *true* dichotomy. Other examples are eye color (say, blue or not blue), employed–unemployed, pilot–not pilot (among airline employees), senator–representative (among members of the U.S. Congress), etc. In each of these examples there is a true dichotomy, not an underlying normal distribution that has been artificially dichotomized. To estimate relationships between variables such as these and a continuous variable, an appropriate statistic is the *point biserial* coefficient of correlation.

Unlike the biserial coefficient, the point biserial is a product-moment correlation coefficient. Each individual or object included in the sample has a score on a continuous variable, X, and a score, such as 0 or 1, that indicates membership in one of the categories of the dichotomous variable, Y. For example, if the dichotomous variable were Sex (Y), a value of 0 might be assigned to all males and a 1 to all females. The point biserial correlation, usually denoted r_{pb}, is then simply the product-moment correlation between X and Y, where Y takes on only the two values 0 and 1. Since r_{pb} is a product-moment correlation, it ranges from -1 to $+1$ and can be interpreted in the same way as a Pearson product-moment r. It should be noted, however, that unless the dichotomous variable has an ordinal scale (so that being in one category is "better" than, or indicates "more" of something, than being in the other), the sign of r_{pb} is meaningless and the relationship is generally expressed in terms of the positive coefficient.

Although r_{pb} may be computed using Equation 2.2.2 or any equivalent equation for the Pearson product-moment r, a special formula may be used because the Y variable is always a dichotomy, consisting of two distinct, nonoverlapping categories that we shall denote by 0 and 1. One form of such a formula is

$$r_{pb} = \frac{\overline{X}_p - \overline{X}_q}{s_X}\sqrt{pq}\sqrt{\frac{n}{n-1}}\,, \tag{3.3.1}$$

where $\overline{X}_p$ is the mean of X for individuals in category 1 on Y, $\overline{X}_q$ is the mean of X for individuals in category 0 on Y, p is the proportion of the total sample in category 1 on Y, q is the proportion of the total sample in category 0 on Y, and s_X is the standard deviation of X for the total sample of n individuals. Note that $p=n_1/n$ and $q=n_0/n$, in which n_0 and n_1 are the numbers in categories 0 and 1, respectively, and $n=n_0+n_1$.

To derive Equation 3.3.1, we first state the formula for the Pearson product-moment correlation as follows:

$$r=\frac{n\Sigma XY-\Sigma X\Sigma Y}{\sqrt{\left[n\Sigma X^2-(\Sigma X)^2\right]\left[n\Sigma Y^2-(\Sigma Y)^2\right]}}. \tag{3.3.2}$$

Now, let us assign the value 0 on variable Y to all members of category 0 and the value 1 to all members of category 1. Then $\Sigma Y=n_1=np$, the number of individuals in category 1. Since Y has only the values 0 and 1, ΣY^2 also equals np. Furthermore, the product XY is zero for all individuals in category 0. Hence, ΣXY is merely the sum of X values for all individuals in category 1, that is, $\Sigma XY=n_1\bar{X}_p$. Substituting ΣXY, ΣY and ΣY^2 in Equation 3.3.2, we have

$$r_{pb}=\frac{n\left(n_1\bar{X}_p\right)-(\Sigma X)(np)}{\sqrt{\left(n\Sigma X^2-(\Sigma X)^2\right)\left(n(np)-(np)^2\right)}}.$$

Dividing both numerator and denominator by n^2 and noting that $\Sigma X=n_1\bar{X}_p+n_0\bar{X}_q$ and that $1-q=p$, we can write

$$r_{pb}=\frac{\dfrac{n_1\bar{X}_p}{n}-p\left(\dfrac{n_1\bar{X}_p}{n}+\dfrac{n_0\bar{X}_q}{n}\right)}{\sqrt{\dfrac{\left(n\Sigma X^2-(\Sigma X)^2\right)}{n}\cdot\dfrac{p(1-p)}{n}}}$$

$$=\frac{p\bar{X}_p-p^2\bar{X}_p-pq\bar{X}_q}{s_X\sqrt{\dfrac{n-1}{n}}\sqrt{pq}}=\frac{p\left[\bar{X}_p(1-p)-q\bar{X}_q\right]}{s_X\sqrt{\dfrac{n-1}{n}}\sqrt{pq}}$$

$$=\frac{\bar{X}_p-\bar{X}_q}{s_X}\sqrt{pq}\sqrt{\frac{n}{n-1}}.$$

For large n, the factor $\sqrt{\dfrac{n}{n-1}}$ is nearly equal to one, and can be dropped. Other equivalent formulas for r_{pb} can easily be derived.

To illustrate the computation of r_{pb} and to see how its value compares with that of r_{bis}, we again use the data of Table 3.2.1. We emphasize, however, that for those data r_{bis}, not r_{pb}, is the appropriate statistic, since we assume that the item score is an artificial dichotomy. Substitution in Equation 3.3.1 yields

$$r_{pb}=\frac{90.9394-76.4118}{14.7024}\sqrt{(.66)(.34)}\sqrt{\frac{100}{99}}=.47.$$

Note that for these data, the value of r_{pb} is considerably less than that of r_{bis}. A comparison of Equations 3.3.1 and 3.2.1 shows that, for $n>2$, $r_{pb}<r_{bis}$. Their ratio is

$$\frac{r_{pb}}{r_{bis}}=\frac{h}{\sqrt{pq}}\sqrt{\frac{n}{n-1}}\,. \tag{3.3.3}$$

When $p=q=.5$, the difference between r_{bis} and r_{pb} is a minimum; as the difference between p and q increases, the difference between r_{bis} and r_{pb} also increases. For large n, that is, $n/(n-1)\cong 1$, we find, for example, that: when $p=q$, $r_{pb}\cong .80\ r_{bis}$; when $p=.3$, $r_{pb}\cong .76\ r_{bis}$; and when $p=.1$, $r_{pb}\cong .59\ r_{bis}$.

A test of significance of r_{pb} can be carried out using the familiar t test for the difference between means of two different populations. Using the notation of this section, the statistic is

$$t=\frac{\overline{X}_p-\overline{X}_q}{\sqrt{s_c^2\left(\frac{1}{np}+\frac{1}{nq}\right)}}, \tag{3.3.4}$$

in which

$$s_c^2=\frac{(np-1)s_{X_p}^2+(nq-1)s_{X_q}^2}{np+nq-2}=\frac{(n-1)s_X^2-npq\left(\overline{X}_p-\overline{X}_q\right)^2}{n-2}.$$

Here s_c^2 is the combined variance estimate obtained from $s_{X_p}^2$ and $s_{X_q}^2$, the variances of X within the two groups defined by the categories 0 and 1, on Y. The t statistic in Equation 3.3.4 has $np+nq-2=n-2$ degrees of freedom.

For the data of Table 3.2.1, $s_c^2=170.0409$ and

$$t=\frac{90.9394-76.4118}{\sqrt{(170.0409)\left(\frac{1}{66}+\frac{1}{34}\right)}}=5.28.$$

We conclude that the population value of the point biserial correlation coefficient is greater than zero, since $t=5.28$ is significant beyond the .0005 level.

3.4 TETRACHORIC CORRELATION

In the behavioral sciences there are many occasions when it is necessary to determine the relationship between two dichotomized variables. For example, in personality or interest inventories, the respondent is often asked to give a yes or no answer to an item, or to check it or omit it. When one is developing and refining these inventories, it is quite helpful to compute correlation coefficients between all possible pairs of items. It is reasonable to assume that the underlying trait measured by a particular item is continuous, but the variable has been artificially dichotomized. Thus, each pair of items requires a correlation between two artificially dichotomous variables. An appropriate measure of relationship in this case is the *tetrachoric* correlation coefficient, which we denote by r_{tet}.

The calculation of r_{tet} is based upon a fourfold table of observed frequencies. For example, suppose that a group of 280 persons responded to two dichotomous items in an opinion poll. The items were:

1. Should public nursery school education be extended to children less than 4 years old?
2. Should the school tax rate be increased to provide for increased support of an extended nursery school program?

The responses to these items were as shown in Table 3.4.1. We might reasonably ask what correlation between the underlying continuously measured traits (e.g., attitudes or opinions) would be most likely to produce these results when the variables were artificially dichotomized. In estimating this correlation, that is, in computing r_{tet}, we assume that the underlying variables have a normal bivariate distribution.

The basic computational formula for r_{tet}, developed by Karl Pearson (1901a), involves an infinite series and, in the general case, requires the solution of at least a quartic equation. It can be found in Lord and Novick (1968, p. 346), as well as in Pearson's original paper. Several alternative formulas are available, however. Let us denote the frequencies in a fourfold table by the letters a, b, c, and d, and their total by $n=a+b+c+d$, as shown in Table 3.4.2. When $a+b=c+d$ and $a+c=b+d$, then an exact formula for r_{tet} is

$$r_{tet}=\sin\left[90°\left(\frac{b+c-a-d}{n}\right)\right]. \qquad (3.4.1)$$

When both variables are split *approximately* at the median, Equation 3.4.1 provides a good approximation. However, when the split is not near the median on one or both of the vairables, as is the case for the 280 responses to the two items given in Table 3.4.1, then a different formula yields a better approximation:

$$r_{tet}=\cos\left(\frac{180°}{1+\sqrt{bc/ad}}\right). \qquad (3.4.2)$$

TABLE 3.4.1 RESPONSES BY 280 PERSONS TO TWO DICHOTOMOUS ITEMS IN AN OPINION POLL

		Item 1		
		No	Yes	Total
Item 2	Yes	27	168	195
	No	45	40	85
	Total	72	208	280

TABLE 3.4.2 REPRESENTATION OF FREQUENCIES IN A FOURFOLD TABLE

		Variable 1		
		0	1	Total
Variable 2	1	a	b	$a+b$
	0	c	d	$c+d$
	Total	$a+c$	$b+d$	n

When either of these formulas is used, b and c must represent the frequencies of individuals who have a value of 1 on both variables or a 0 on both variables, and a and d must represent the frequencies of individuals who have a value of 1 on one variable and 0 on the other. When a fourfold table is set up in this way, the sign of r_{tet} will correctly indicate whether the relationship between the variables is positive or negative. Since r_{tet} is an estimate of the correlation, ρ, in a normal bivariate population, its values may range from -1 to $+1$. Note that since r_{tet} is not a product-moment correlation, it cannot be regarded as an estimate of the product-moment correlation in a non-normal population.

When a large number of r_{tet} values must be calculated, computing diagrams, tables, and other shortcut methods are desirable. An effective method for the rapid calculation of r_{tet} is the technique developed by Davidoff and Goheen (1953) and improved by Jenkins (1955). The tables required for using these procedures are Tables C.8a, C.8b, and C.8c in Appendix C. The method involves the following steps:

1. Denote the frequencies in the fourfold table by one of the following four arrangements, so that $a<d$ and $ad>bc$. The four arrangements are:

	0	1			0	1			0	1			0	1
1	c	d		1	b	a		1	a	b		1	d	c
0	a	b		0	d	c		0	c	d		0	b	a
+				+				–				–		

The sign of r_{tet} is indicated below each arrangement. Note that a, b, c, and d do not necessarily represent the combinations of values they were required to represent for use of Equations 3.4.1 and 3.4.2.

2. Calculate ad/bc and refer to Table C.8a in Appendix C to obtain $|r_{tet}|$ uncorrected for marginal splits.
3. Compute the two marginal proportions, $(a+b)/n$ and $(a+c)/n$. Refer to Table C.8b in Appendix C and find the base correction at the intersection of the two proportions.

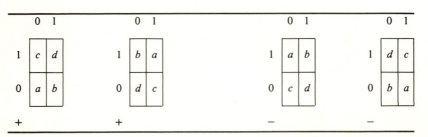

4. Find the multiplier in Table C.8c in Appendix C for the proportions computed in Step 3, and multiply it by the base correction to obtain the final correction.
5. Subtract the final correction from the uncorrected $|r_{tet}|$ obtained in Step 2 to obtain the corrected $|r_{tet}|$.
6. Attach the correct sign, as indicated below the arrangement selected in Step 1, to the corrected $|r_{tet}|$ to obtain r_{tet}.

Using the data of the preceding example, we compute r_{tet} as follows:

1. Selecting the arrangement that makes $a<d$ and $ad>bc$, we have:

	0	1
1	c	d
0	a	b

+

2. $ad/bc=(45)(168)/(40)(27)=7$. Therefore, the uncorrected $|r_{tet}|$ equals .651.
3.

$$(a+b)/n=(45+40)/280=.30.$$

$$(a+c)/n=(45+27)/280=.26.$$

Therefore, the base correction is .022.

4. The multiplier is 1.02 and the final correction is thus $(1.02)(.022)=$ .022.
5. Then the corrected $|r_{tet}|=.651-.022=.629$.
6. Because the sign in Step 1 is +, $r_{tet}=.629$.

Using Equation 3.4.2 instead of Tables C.8, we obtain

$$r_{tet}=\cos\left(\frac{180°}{1+\sqrt{7}}\right)=\cos 49°22'=\cos 49.37°=.651.$$

Note that this value is equal to that for the uncorrected $|r_{tet}|$ obtained by using Table C.8a in Appendix C. Actually, the uncorrected values in Table C.8a are obtained by using Equation 3.4.2.

When $\rho=0$ in the normal bivariate population, which we assume in this section, the sampling distribution of r_{tet} is approximately normal for moderately large n (say 30 or more). The standard error of r_{tet} may be approximated by

$$s_{r_{tet}}=\frac{\sqrt{p_1q_1p_2q_2}}{h_1h_2\sqrt{n}} \tag{3.4.3}$$

in which the p's and q's are proportions based on the cell frequencies shown in Table 3.4.1; that is,

$$p_1=\frac{b+d}{n},\quad q_1=\frac{a+c}{n},\quad p_2=\frac{a+b}{n},\quad \text{and}\quad q_2=\frac{c+d}{n}.$$

The values h_1 and h_2 are the heights of the ordinates of the standard normal

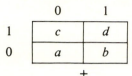

TABLE 3.4.3 PROPORTIONS BASED ON THE DATA OF TABLE 3.4.1

		Item 1 No	*Item 1* Yes	
Item 2	Yes	.10	.60	$p_2=.70$
	No	.16	.14	$q_2=.30$
		$q_1=.26$	$p_1=.74$	

curve that cut off areas p_1 and p_2, respectively. A test of the hypothesis H_0: $\rho=0$ can be carried out using

$$z=\frac{r_{tet}}{s_{r_{tet}}}, \tag{3.4.4}$$

which has approximately the standard normal distribution.

To apply Equations 3.4.3 and 3.4.4 to the data of Table 3.4.1, we first obtain the proportions shown in Table 3.4.3.

Then

$$s_{r_{tet}}=\frac{\sqrt{(.74)(.26)(.70)(.30)}}{(.3225)(.3496)\sqrt{280}}=.10654,$$

and

$$z=\frac{r_{tet}}{s_{r_{tet}}}=\frac{.629}{.10654}=5.90.$$

Because $P(z>5.90)$ is very small, we would reject the null hypothesis that $\rho=0$.

Note that, in general, the sampling variability of r_{tet} is much greater than the sampling variability of r, the product-moment correlation. Therefore, one should be more cautious in the interpretation of r_{tet}, particularly with small samples. We do not recommend the use of r_{tet} unless the sample size is quite large, say at least twice as large as would be considered satisfactory for the use of r. Furthermore, one should not calculate r_{tet} as a substitute for r, after dichotomizing continuous distributions. Such a procedure is tantamount to throwing away at least half of the available data.

3.5 MEASURES OF ASSOCIATION

The correlation techniques we have discussed in the preceding sections all yield measures from which we may infer the degree of relationship between two variables. When we are dealing with discrete variables, or categorized attributes, it is not always appropriate to use the concept of correlation in describing the relationship. If the discrete variables constitute an ordered series, the concept of correlation (including both positive and negative values) *is* appropriate. On the other hand, if the discrete variables consist of unordered categories, it is more appropriate to think in terms of association rather than correlation. The truly discrete sets of categories are said to be "associated" with each other, and the statistic that measures the degree of association is referred to as a "coefficient of association." Such coefficients range only from 0 to 1 instead of from -1 to $+1$, because when categories are unordered, the concepts of negative and positive relationships are not appropriate.

3.5.1 The Phi Coefficient

The phi coefficient, ϕ, may be used to express the degree of association between two *truly* dichotomous variables. Thus, it is analogous to the tetrachoric correlation coefficient, which involves two *artificially* dichotomized variables. However, unlike r_{tet}, ϕ is a product-moment correlation, obtained by scaling each dichotomous variable 0 or 1. Phi may be considered a measure of either correlation or association, depending on whether the categories of the variables involved are ordered or unordered.

The Phi coefficient may be calculated by means of the formula

$$\phi = \frac{bc - ad}{\sqrt{(a+b)(a+c)(b+d)(c+d)}}, \tag{3.5.1}$$

in which the letters a, b, c, and d, again refer to frequencies in the cells of a fourfold table, (Table 3.5.1), also called a *contingency table*.

To derive Formula 3.5.1, we refer to Equation 3.3.2 for the Pearson product-moment correlation coefficient. Because both X and Y are dichotomous variables, the quantities ΣX, ΣY, ΣX^2, ΣY^2, and ΣXY, can be expressed in terms of the cell frequencies, a, b, c, and d. Thus, for the dichotomous variables X and Y, $\Sigma X = b + d$, $\Sigma X^2 = b + d$, $\Sigma Y = a + b$, $\Sigma Y^2 = a + b$, and $\Sigma XY = b$. Substitution in Equation 3.3.2 yields

$$\phi = \frac{nb - (b+d)(a+b)}{\sqrt{\left[n(b+d) - (b+d)^2\right]\left[n(a+b) - (a+b)^2\right]}}.$$

TABLE 3.5.1 REPRESENTATION OF FREQUENCIES IN A FOURFOLD TABLE

		X		
		0	1	
Y	1	a	b	$a+b$
	0	c	d	$c+d$
		$a+c$	$b+d$	n

$n = a + b + c + d$, and so we have

$$\phi = \frac{(a+b+c+d)b - ab - b^2 - ad - bd}{\sqrt{(b+d)(a+c)(a+b)(c+d)}}$$

$$= \frac{bc - ad}{\sqrt{(a+b)(a+c)(b+d)(c+d)}},$$

which is identical to Equation 3.5.1.

As an illustration of the use of ϕ, suppose we wished to test the hypothesis that color of hair in infants is associated with color of eyes. We select a random sample of 151 infants and observe the hair and eye color of each, with results as shown in Table 3.5.2. The application of Equation 3.5.1 yields:

$$\phi = \frac{(60)(54) - (16)(21)}{\sqrt{(76)(70)(81)(75)}} = .51,$$

which suggests that there is an association between hair color and eye color of these infants. The data show that blond hair and blue eyes tend to be associated.

For large n the standard error of ϕ, when H_0 is true, is approximately $1/\sqrt{n}$. Hence, s_ϕ for the data in the fourfold table above is approximately .08.

A formal test of significance of ϕ, that is, a test of the null hypothesis that the population value is zero, can be carried out using a chi-square statistic,

$$\chi^2 = n\phi^2, \tag{3.5.2}$$

which has, under H_0, a chi-square distribution with one degree of freedom. For the data in Table 3.5.1, $\chi^2 = 39.4$, which is significant beyond the .01 level. From this result, and an inspection of Table 3.5.2, we may confirm that blue eyes tend to be associated with blond hair. Note that the same value of χ^2 would have been obtained by use of the more familiar formula for χ^2, not involving ϕ, namely,

$$\chi^2 = \sum_{i=1}^{I} \frac{(o_i - e_i)^2}{e_i}, \tag{3.5.3}$$

TABLE 3.5.2 HAIR AND EYE COLOR OF 151 INFANTS

		Hair Color: Not Blond	Hair Color: Blond	Total
Eye Color	Blue	16	60	76
	Not Blue	54	21	75
	Total	70	81	151

in which o_i = the observed frequency in cell i,
e_i = the expected frequency in cell i, and
I = the number of cells in the table.

When the X and Y categories are ordered, the values of ϕ range from -1.00 to $+1.00$. However, when the X and Y categories are unordered, the sign attached to the coefficient merely reflects the way the categories were defined. For example, it makes no sense to say that there is a negative (or positive) association between hair color and eye color. Because ϕ is a measure of degree of association when the categories are unordered, it can vary only from no association at all (zero) to perfect association ($+1.00$). The *kind* of association present may be described after inspecting the data in the fourfold table.

In most research applications of ϕ, its upper limit is substantially less than $+1.00$. The maximum Phi coefficient, denoted ϕ_{max}, for any given fourfold contingency table is dependent upon the combination of the marginal proportions for that table. Consider Table 3.5.3.

TABLE 3.5.3 CONTINGENCY TABLE WITH MARGINAL PROPORTIONS EXPRESSED IN TERMS OF CELL FREQUENCIES

		Variable 1		
		0	1	
Variable 2	1	a	b	$p_2=(a+b)/n$
	0	c	d	$q_2=(c+d)/n$
		$q_1=(a+c)/n$	$p_1=(b+d)/n$	

TABLE 3.5.4 EXAMPLE OF 2×2 CONTINGENCY TABLES WITH VARIOUS VALUES OF ϕ_{max}

A

0	50	50
50	0	50
50	50	100

$\phi=1.00$

$\phi_{max}=1.00$

B

5	45	50
20	30	50
25	75	100

$\phi=.35$

$\phi_{max}=.58$

C

15	35	50
15	35	50
30	70	100

$\phi=0$

$\phi_{max}=.65$

D

10	30	40
20	10	30
30	40	70

$\phi=.42$

$\phi_{max}=1.00$

E

15	25	40
25	5	30
40	30	70

$\phi=.46$

$\phi_{max}=.75$

The maximum value of ϕ is a function of the marginal proportions, p_1, q_1, p_2, and q_2. It may be computed by:

$$\phi_{\max}=\sqrt{\frac{p_i q_j}{q_i p_j}} \quad ; \quad (p_j \geqslant p_i, i \neq j). \tag{3.5.4}$$

To illustrate, suppose we have the contingency tables shown in Table 3.5.4. Note that when $p_1=p_2$, the maximum value of ϕ is 1.00, as in examples A and D. As the discrepancy between p_1 and p_2 increases, the value of $\phi_{\max}$ decreases, as can be seen from examples B, C, and E. When interpreting an observed value of ϕ, one should take account of the value of $\phi_{\max}$ for the particular contingency table involved.

3.5.2 The Contingency Coefficient

The contingency coefficient, C, measures the degree of association between two discrete variables, each of which may have more than two categories. C is defined as

$$C=\sqrt{\frac{\chi^2}{n+\chi^2}}\,, \tag{3.5.5}$$

in which χ^2 may be computed using Equation 3.5.3.

To illustrate the use of C, we refer to the problem of determining whether there is any association between hair color and color of eyes in infants. Instead of defining just two categories of each variable, let us assume that we can classify the 151 infants into the following categories: Hair—Blond, Brunette, Black, and Auburn; Eyes—Blue, Brown, and Hazel. We shall consider the data of Table 3.5.5. The value of χ^2 for the data of Table 3.5.5, computed from Equation 3.5.3, is 53.66. Substitution of this value in Equation 3.5.5 yields

$$C=\sqrt{\frac{53.66}{151+53.66}}=.51.$$

The χ^2 value of 53.66 with 6 degrees of freedom is significant beyond the .001 level, indicating that *there is* an association between hair color and eye color among these 151 infants. The value of C indicates the *strength* of association between the two characteristics.

TABLE 3.5.5 INFANTS CLASSIFIED ACCORDING TO HAIR COLOR AND EYE COLOR ($n=151$)

		Hair Color				
		Blond	Brunette	Black	Auburn	Total
Eye Color	Blue	60	4	4	8	76
	Brown	11	14	5	5	35
	Hazel	10	8	15	7	40
	Total	81	26	24	20	151

A disadvantage of C is that its upper limit depends on the number of categories involved in the analysis. For a square table in which each variable has K categories, the upper limit of C is $\sqrt{(K-1)/K}$. For any rectangular contingency table, the upper limit may be calculated from this expression with K equal to the smaller number of categories. For the above example, the upper limit of C is, therefore, .82. As the number of categories increases, the upper limit of C increases. However, C cannot attain an upper limit of 1.00 unless the number of categories for both variables is infinite. One should always take account of the maximum value for C when interpreting observed values of the contingency coefficient.

3.5.3 Cramér's Phi Coefficient

A measure of association for $K \times L$ tables, the maximum value of which does not depend upon the number of categories of either variable, is Cramér's Phi, denoted by ϕ'. This nonparametric statistic is a generalization of the phi coefficient discussed above for the fourfold contingency table. Its value always lies between 0 and $+1.00$, regardless of the values of K and L. If we define K and L so that $K \leqslant L$, the formula is

$$\phi' = \sqrt{\frac{\chi^2}{n(K-1)}} \,. \qquad (3.5.6)$$

Note that when $K=L=2$, ϕ' and ϕ are identical.* This can be seen by solving Equation 3.5.2 for ϕ. Thus, ϕ is a special case of ϕ'.

For the data of Table 3.5.5, Cramér's Phi is

$$\phi' = \sqrt{\frac{53.66}{(151)(2)}} = .42.$$

The fact that the maximum value of ϕ' is always 1.00, regardless of the values of K and L, makes this statistic somewhat easier to interpret than the contingency coefficient, C.

3.6 INTRACLASS CORRELATION

Although perhaps better known for its application in tests of hypotheses about population means, as in the analysis of variance, the concept of intraclass correlation is useful in dealing with a particular type of problem involving the relationship between variables. A discussion of it is relevant here also because it is essentially a product-moment correlation, although it represents a somewhat different application of that statistic than we encountered in Chapter 2.

*Except, possibly, for sign, because ϕ can be either positive or negative, as we indicated previously.

As the name implies, the intraclass correlation coefficient is a measure of the relationship within classes, that is, among the members of groups, on some variable of interest. Suppose, for example, that we are interested in determining the correlation between family members on a measure of Liberalism, say X. Note that we are *not* interested in the correlation between husbands' liberalism, say X_H, and wives' liberalism, say X_W, *or* in the correlation between mothers' liberalism, say X_M, and eldest daughters' liberalism, say X_D. Rather, we are interested in the correlation between family members *in general*, on the measure of liberalism.

To obtain the correlation desired, we first identify a population of families that we wish to study, take a random sample of them, and obtain the measure X, for all members of each family selected. Within each family, we identify all of the possible different pairs of family members. Each such pair provides two pairs of observations, the first in one order, and the second in the reverse order. The sample size, then, is *not* the number of families *or* the total number of family members, but the total number of pairs of observations thus generated.

To illustrate the procedure, we consider Table 3.6.1, which shows values of X for a sample of four families. We then list the pairs of observations obtained by following the procedure we have just described. The first value of X in each pair will be denoted by X_1 and the second by X_2. The pairs are shown in Table 3.6.2.

TABLE 3.6.1 VALUES OF *X* (LIBERALISM) FOR MEMBERS OF FOUR FAMILIES

Family	*Number of Family Members*	*Values of X*
1	3	25, 29, 34
2	4	18, 21, 21, 26
3	2	31, 35
4	3	30, 30, 32

TABLE 3.6.2 ALL POSSIBLE ORDERED PAIRS OF OBSERVATIONS OF *X* WITHIN FOUR FAMILIES

Family							
1		2		3		4	
X_1	X_2	X_1	X_2	X_1	X_2	X_1	X_2
25	29	18	21	31	35	30	30
29	25	21	18	35	31	30	30
25	34	18	21			30	32
34	25	21	18			32	30
29	34	18	26			30	32
34	29	26	18			32	30
		21	21				
		21	21				
		21	26				
		26	21				
		21	26				
		26	21				

The total number of pairs is related to the number of families and the numbers of family members by the formula

$$N = \sum_{i=1}^{G} m_i(m_i - 1), \tag{3.6.1}$$

in which G is the number of families in the sample and m_i is the number of members of the ith family. In our example, the number of pairs is $N = 3(2) + 4(3) + 2(1) + 3(2) = 26$.

The intraclass correlation coefficient, r_I, for these data is the product-moment correlation coefficient between X_1 and X_2, computed according to Equation 2.2.2. The value is $r_I = .62$. This value indicates that there is a moderately high positive relationship among family members in terms of the measure of Liberalism, X. In general, a positive value of r_I indicates that members of classes tend to be more homogeneous than would individuals selected at random from a population of individuals; a negative value indicates less homogeneity within classes. Thus, the value $r_I = .62$ seems reasonable, because we would expect that individuals in the *same* family would be more alike in terms of Liberalism than would individuals from *different* families.

For the general case of G classes, with m_i the number of members of the ith class, a more efficient method is available for computing r_I, one that does not require listing all the pairs of observations as in Table 3.6.2. The formula is

$$r_I = \frac{\sum_{i=1}^{G} m_i^2(\bar{X}_i - \bar{X})^2 - \sum_{i=1}^{G}\sum_{j=1}^{m_i}(X_{ij} - \bar{X})^2}{\sum_{i=1}^{G}\left[(m_i - 1)\sum_{j=1}^{m_i}(X_{ij} - \bar{X})^2\right]}, \tag{3.6.2}$$

in which X_{ij} is the value of the jth individual in the ith class, $\bar{X}_i$ is the mean of the ith class, and

$$\bar{X} = \frac{1}{N}\sum_{i=1}^{G}\left[(m_i - 1)\sum_{j=1}^{m_i} X_{ij}\right].$$

Note that $\bar{X}$ is *not* simply $\Sigma\Sigma X_{ij}/\Sigma m_i$, the mean of X for all individuals in all classes combined. Rather, it is the mean, $\bar{X} = \bar{X}_1 = \bar{X}_2$, of each of the variables, X_1 and X_2, in the list of pairs of observations given in Table 3.6.2.

For the data of our example, the preliminary computations are listed in Table 3.6.3. Then, from Equation 3.6.2, we compute

$$r_I = \frac{802.3669 - 350.8284}{729.5385} = .62,$$

which agrees with the result obtained by applying Equation 2.2.2 to the list of pairs of observations given above.

Although r_I is a product-moment correlation, it differs from r, in that it clearly is not based on pairs of observations, all of which are independent.

TABLE 3.6.3 PRELIMINARY COMPUTATIONS REQUIRED FOR COMPUTING THE INTRACLASS CORRELATION FROM EQUATION 3.6.2

Family	m_i	$\bar{X}_i$	$\sum_j X_{ij}$	$\sum_j X_{ij}^2$
1	3	29.3333	88	2622
2	4	21.5000	86	1882
3	2	33.0000	66	2186
4	3	30.6667	92	2824
			$\Sigma\Sigma X_{ij} = 332$	$\Sigma\Sigma X_{ij}^2 = 9514$

$$\bar{X} = 26.3077$$
$$\sum_i m_i^2(\bar{X}_i - \bar{X})^2 = 802.3669$$
$$\sum_i \sum_j (X_{ij} - \bar{X})^2 = 350.8284$$
$$\sum_i [(m_i - 1)\sum_j (X_{ij} - \bar{X})^2] = 729.5385$$

One of the consequences is that its range of values is restricted, compared with r. Although r_I can be as large as $+1.00$, its lower limit depends on the class sizes. In the special case in which $m = m_1 = m_2 = \cdots = m_G$, r_I cannot be less than $-1/m-1$. When all classes are not equal in size, the lower limit depends on the particular distribution of class sizes, but will usually be greater than -1.00. This restriction should be taken account of when interpreting an observed value of r_I.

A further consequence of the lack of independence of the paired observations on which r_I is based is that the inferential procedures discussed in Chapter 2 for the ordinary product-moment correlation are not applicable. Discussions of estimation and hypothesis testing procedures for the intraclass correlation coefficient may be found in Winer (1971, p. 244–47).

3.7 EXERCISES

3.1. The president of a charitable organization wished to determine whether attendance of members at meetings was associated with their promptness in making monthly contributions to the organization's assistance fund. A sample of 135 members was selected and each member was classified as regular or nonregular in attendance, and also as being prompt or not prompt in making monthly contributions. The results were:

		Promptness		
		Prompt	Not Prompt	Total
Attendance	Regular	60	35	95
	Nonregular	20	20	40
	Total	80	55	135

(a) Calculate an appropriate measure of association between attendance and promptness of contribution.
(b) Test the significance of the statistic obtained in part (a). Use $\alpha = .05$.
(c) What should the president of the organization conclude? Explain.

3.2. The following data were obtained when two items of a college aptitude test were administered to 326 high school juniors:

		Item 1		
		Correct	Incorrect	Total
Item 2	Correct	212	43	255
	Incorrect	39	32	71
	Total	251	75	326

(a) What assumptions should be met for the tetrachoric correlation, r_{tet}, to be considered appropriate for expressing the relationship between performance on these items? In this case do those assumptions seem to be reasonably well satisfied?
(b) Compute r_{tet} using the cosine approximation of Equation 3.4.2.
(c) Compute r_{tet} using the appropriate tables in Appendix C. Compare the result with that obtained in (b).

3.3. Some professors believe that students who choose to take their courses in the late afternoon do not perform as well in their academic subjects as do those students who select courses earlier in the day. To test this belief data were collected on 400 students in a liberal arts program. The data are given in the following four-fold contingency table.

		Performance		
		Below Avg.	Above Avg.	Total
Time of Day	Early	38	62	100
	Late	149	151	300
	Total	187	213	400

(a) Compute an appropriate measure of association between time of day and performance.
(b) Test the significance of the statistic computed in (a). Use $\alpha = .05$.
(c) Evaluate and interpret the results of (a) and (b).

3.4. Refer to Exercise 3.3, regarding the association between academic performance of students and hour of the day in which courses are taken. If the two variables were measured on scales having several, rather than just two, values, the results might appear as in the following table.

	Grade					
Time	A	B	C	D	F	Total
AM						
8–10	18	74	47	12	4	155
10–12	21	11	27	14	6	79
PM						
12–2	32	26	14	11	13	96
2–4	10	21	14	10	15	70
Total	81	132	102	47	38	400

(a) Calculate a measure of association based on these data.

(b) Test the significance of the statistic computed in part (a). Use $\alpha = .05$.

(c) Does your conclusion differ from that in Exercise 3.3? Explain.

3.5. In the following table are given the ranks assigned to 10 items on an inventory measuring attitude toward work conditions, by a group of employers and a group of employees.

	Rankings	
Item	*Employers*	*Employees*
Interesting work	2.0	4.5
Generous retirement plans	3.5	2.0
Credit for work done	1.0	4.5
Fair play	5.0	8.5
Noncontributory hospital insurance	7.0	6.0
Counseling on personal problems	6.0	7.0
Promotion on merit	3.5	1.0
Good working climate	9.0	10.0
Job security	10.0	3.0
Understanding and appreciation	8.0	8.5

(a) To what extent do employers agree with employees? Compute an appropriate statistic for expressing the extent of agreement between the two groups.

(b) Use the appropriate table in Appendix C to test the significance of the result in (a). Use $\alpha = .05$.

4 / Multiple Regression and Correlation

In Chapter 1 we considered the case in which we desired to estimate the conditional mean, $\mu_{y \cdot x}$ of a random variable Y, the dependent variable, from a single nonrandom variable X, the independent variable, using a bivariate linear regression model. In Chapter 2 we described the use of the product-moment correlation coefficient as a measure of the linear relationship between two random variables, X and Y. In this chapter we extend the discussion of regression and correlation analysis to cases in which more than two variables are involved. *Multiple regression analysis* deals with the estimation of the conditional mean of a random variable, Y, from *several* X variables, rather than from a single X. The *multiple correlation coefficient* expresses the relationship between Y and a linear combination of *several* Xs. We shall first discuss the mathematical models appropriate for such cases and show how estimates of parameters may be obtained. We shall then show the formulation of the regression equation, again based on least squares methods, and outline computational procedures using both ordinary algebraic methods and matrix methods. Finally, we shall illustrate the computations using data from an applied regression problem.

Although matrix algebra was introduced and used to a limited extent in Chapter 1, the reader may wish to review basic matrix operations before continuing with the remaining chapters. An introduction to matrix algebra is provided in Appendix A.

4.1 MATHEMATICAL MODELS IN MULTIPLE REGRESSION ANALYSIS

Statisticians have studied a number of different mathematical models for multiple regression analysis. These models differ from one another primarily in the assumptions about the variables they include. The basic equation relating these variables may be written

$$Y_i = \beta_0 + \beta_1 X_{i1} + \cdots + \beta_p X_{ip} + e_i. \tag{4.1.1}$$

This equation states that the Y value for the ith individual is a function of: $p+1$ constants, the βs; the values for the ith individual on p independent variables, the Xs; and an error, e_i. Equation 4.1.1 is the analogue in multiple regression of Equation 1.5.1 in bivariate regression. Except for the use of β_0 in place of α, Equation 4.1.1 is a simple extension of Equation 1.5.1 to the case in which there are p Xs instead of just one.

In this chapter we shall consider the Y_is to be values of a random variable, Y, which is usually referred to as *dependent* because its values are predicted on the basis of the known values of the Xs. The Xs are called *independent* variables. They can be regarded as (1) fixed constants, (2) preselected values of random variables, or (3) random variables whose values in a given problem are determined through random sampling. Variables whose values are fixed or preselected are called *deterministic*, whereas those whose values may vary randomly from sample to sample are called *stochastic*. A variable whose value tends to remain the same in repeated experiments is considered deterministic. For example, the temperature at which pure water boils at sea level is considered deterministic, while the SAT score of an eleventh grade student selected at random is stochastic. In the latter case the observed score would depend on which student happened to be selected.

Provided certain assumptions are made about the e_is and the βs, procedures for drawing inferences about the βs are the same whether the Xs are deterministic or stochastic. For this reason it is convenient in practice to consider the Xs as either fixed constants or as preselected values of random variables, which are not subject to sampling variation and are free of measurement error. For discussion of models in which the Xs are assumed to contain measurement error, see Kendall and Stuart (1974), Volume 2.

The necessary assumptions can be stated somewhat more simply if we first express the model in matrix form (see Appendix A). Let

$$\mathbf{Y}=\begin{bmatrix} Y_1 \\ \vdots \\ Y_n \end{bmatrix}, \mathbf{X}=\begin{bmatrix} 1 & X_{11} & X_{12} & \cdots & X_{1p} \\ \vdots & \vdots & \vdots & & \vdots \\ 1 & X_{n1} & X_{n2} & \cdots & X_{np} \end{bmatrix},$$

$$\mathbf{e}=\begin{bmatrix} e_1 \\ \vdots \\ e_n \end{bmatrix}, \text{ and } \boldsymbol{\beta}=\begin{bmatrix} \beta_0 \\ \vdots \\ \beta_p \end{bmatrix},$$

in which n is the number of individuals in the sample and p is the number of Xs. Then the model may be written

$$\mathbf{Y}=\mathbf{X}\boldsymbol{\beta}+\mathbf{e}. \tag{4.1.2}$$

The necessary assumptions are:

1. $\mathbf{e}$ has a multivariate normal distribution (see Section 4.1.1) with mean vector zero ($E(\mathbf{e})=\mathbf{0}$) and variance-covariance matrix $\boldsymbol{\Sigma}=\sigma^2\mathbf{I}_n$, in which $\mathbf{I}_n$ is the identity matrix of order n. This is equivalent to assuming

that e_i has a normal distribution with $E(e_i)=0$ and $\text{Var}(e_i)=\sigma^2$, and that e_i and $e_j (i \neq j)$ are independent.

2. $\boldsymbol{\beta}$ is deterministic and unknown. If $\mathbf{X}$ consists of known constants, then the fact that $\boldsymbol{\beta}$ is deterministic implies that $E(\mathbf{Y})=\mathbf{X}\boldsymbol{\beta}$ and $\text{Var}(\mathbf{Y})=\text{Var}(\mathbf{e})=\boldsymbol{\Sigma}$. If $\mathbf{X}$ consists of preselected values of random variables, then $E(\mathbf{Y}|\mathbf{X})=\mathbf{X}\boldsymbol{\beta}$ and $\text{Var}(\mathbf{Y}|\mathbf{X})=\text{Var}(\mathbf{e}|\mathbf{X})=\boldsymbol{\Sigma}$.

3. If $\mathbf{X}$ is assumed to be stochastic, it is also assumed to be independent of $\mathbf{e}$, and hence of $\boldsymbol{\Sigma}$, and its distribution depends upon neither $\boldsymbol{\beta}$ nor $\boldsymbol{\Sigma}$. Under this assumption, the procedures for drawing inferences about $\boldsymbol{\beta}$ are the same as those used when $\mathbf{X}$ is nonstochastic. Again, $E(\mathbf{Y}|\mathbf{X})=\mathbf{X}\boldsymbol{\beta}$ and $\text{Var}(\mathbf{Y}|\mathbf{X})=\boldsymbol{\Sigma}$.

Note that these assumptions reduce to those made in Chapter 1 in the special case in which there is a single X, that is, when $p=1$. We have expressed them here in matrix notation both for generality and because most of the discussion in this and succeeding chapters will be presented in matrix form.

4.1.1 The Multivariate Normal Distribution

In stating the assumptions of the multiple linear regression model, we referred to the multivariate normal distribution. Because this distribution is assumed in inferential procedures discussed in the remaining chapters, we digress at this point to describe it and to indicate its properties.

The multivariate normal distribution may be viewed as the extension of the bivariate normal distribution to cases in which there are more than two random variables. One may also view the bivariate normal as a special case of the multivariate normal. Either view is correct, but we choose the former to facilitate our discussion here.

We shall first rewrite Equation 2.4.1 with X_1 and X_2 substituted for X and Y:

$$f(X_1,X_2)=\frac{1}{2\pi\sigma_1\sigma_2\sqrt{1-\rho^2}}\exp\left\{\frac{1}{2(1-\rho^2)}\left[\frac{(X_1-\mu_1)^2}{\sigma_1^2}+\frac{(X_2-\mu_2)^2}{\sigma_2^2}-2\rho\frac{(X_1-\mu_1)(X_2-\mu_2)}{\sigma_1\sigma_2}\right]\right\}, \tag{4.1.3}$$

in which the subscripts 1 and 2 refer to X_1 and X_2, respectively. The variance-covariance matrix, denoted by $\boldsymbol{\Sigma}$ above, is defined as

$$\boldsymbol{\Sigma}=\begin{bmatrix}\sigma_1^2 & \rho\sigma_1\sigma_2\\ \rho\sigma_1\sigma_2 & \sigma_2^2\end{bmatrix},$$

which has determinant

$$|\boldsymbol{\Sigma}|=\sigma_1^2\sigma_2^2-\rho^2\sigma_1^2\sigma_2^2=\sigma_1^2\sigma_2^2(1-\rho^2). \tag{4.1.4}$$

The inverse of the variance-covariance matrix is then

$$\boldsymbol{\Sigma}^{-1}=\begin{bmatrix} \dfrac{\sigma_2^2}{\sigma_1^2\sigma_2^2(1-\rho^2)} & \dfrac{-\rho\sigma_1\sigma_2}{\sigma_1^2\sigma_2^2(1-\rho^2)} \\ \dfrac{-\rho\sigma_1\sigma_2}{\sigma_1^2\sigma_2^2(1-\rho^2)} & \dfrac{\sigma_1^2}{\sigma_1^2\sigma_2^2(1-\rho^2)} \end{bmatrix}$$

$$=\begin{bmatrix} \dfrac{1}{\sigma_1^2} & \dfrac{-\rho}{\sigma_1\sigma_2} \\ \dfrac{-\rho}{\sigma_1\sigma_2} & \dfrac{1}{\sigma_2^2} \end{bmatrix}\cdot\frac{1}{1-\rho^2}.$$

We now define a vector of differences between the Xs and their corresponding population means:

$$\mathbf{x}=\begin{bmatrix} X_1-\mu_1 \\ X_2-\mu_2 \end{bmatrix}.$$

Then, taking the product

$$\mathbf{x}'\boldsymbol{\Sigma}^{-1}=\frac{1}{1-\rho^2}\left[\frac{X_1-\mu_1}{\sigma_1^2}-\frac{\rho(X_2-\mu_2)}{\sigma_1\sigma_2}\quad \frac{-\rho(X_1-\mu_1)}{\sigma_1\sigma_2}+\frac{X_2-\mu_2}{\sigma_2^2}\right],$$

and postmultiplying by $\mathbf{x}$, we have

$$\mathbf{x}'\boldsymbol{\Sigma}^{-1}\mathbf{x}=\frac{1}{1-\rho^2}\left[\frac{(X_1-\mu_1)^2}{\sigma_1^2}-2\frac{\rho(X_1-\mu_1)(X_2-\mu_2)}{\sigma_1\sigma_2}+\frac{(X_2-\mu_2)^2}{\sigma_2^2}\right],$$

which, when multiplied by $-1/2$ is seen to be equivalent to the exponent in Equation 4.1.3. For reasons to be stated later, we denote this expression by

$$\chi^2=\mathbf{x}'\boldsymbol{\Sigma}^{-1}\mathbf{x}. \tag{4.1.5}$$

Now, using Equations 4.1.4 and 4.1.5, we can express Equation 4.1.3 in matrix notation as

$$f(X_1,X_2)=(2\pi)^{-1}|\boldsymbol{\Sigma}|^{-\frac{1}{2}}\exp\left(-\tfrac{1}{2}\chi^2\right). \tag{4.1.6}$$

In order to generalize this expression to the case of p Xs, it is reasonable to redefine $\mathbf{x}$ and $\boldsymbol{\Sigma}$ more generally as

$$\mathbf{x}=\begin{bmatrix} X_1-\mu_1 \\ \vdots \\ X_p-\mu_p \end{bmatrix} \tag{4.1.7}$$

and

$$\boldsymbol{\Sigma}=\begin{bmatrix} \sigma_1^2 & \sigma_{12} & \cdots & \sigma_{1p} \\ \sigma_{21} & \sigma_2^2 & \cdots & \sigma_{2p} \\ \vdots & \vdots & \ddots & \vdots \\ \sigma_{p1} & \sigma_{p2} & \cdots & \sigma_p^2 \end{bmatrix}, \tag{4.1.8}$$

in which $\sigma_{ij} = \rho_{ij}\sigma_i\sigma_j$ $(i,j = 1,\ldots,p; i \neq j)$. It can be shown (see, for example, Anderson, 1958, p. 17) that the exponent of 2π should be $-p/2$ and that the exponent of $|\boldsymbol{\Sigma}|$ remains $-1/2$ regardless of the number of variables. Thus, the density function for $N(\boldsymbol{\mu}, \boldsymbol{\Sigma})$, the multivariate normal distribution with centroid $\boldsymbol{\mu}$, and variance-covariance matrix $\boldsymbol{\Sigma}$, is

$$f(X_1,\ldots,X_p) = (2\pi)^{-\frac{p}{2}}|\boldsymbol{\Sigma}|^{-\frac{1}{2}}\exp\left(-\tfrac{1}{2}\chi^2\right), \qquad (4.1.9)$$

in which $\mathbf{x}$ and $\boldsymbol{\Sigma}$ in $\chi^2 = \mathbf{x}'\boldsymbol{\Sigma}^{-1}\mathbf{x}$ are defined by Equations 4.1.7 and 4.1.8, respectively.

For the multivariate normal distribution of the error vector, $\mathbf{e}$, in the model of Equation 4.1.2, it is particularly easy to show why $\mathbf{x}'\boldsymbol{\Sigma}^{-1}\mathbf{x}$ was denoted by χ^2 in Equation 4.1.5. Recall that the variance-covariance matrix of $\mathbf{e}$ was

$$\boldsymbol{\Sigma} = \sigma^2\mathbf{I}_n,$$

so that the e_is ($p = n$ of them, in this case) were mutually independent. Then $\boldsymbol{\Sigma}$ is a diagonal matrix (see Appendix A), the elements of which are the variances, $\sigma_1^2, \sigma_2^2, \ldots, \sigma_n^2$, and

$$\mathbf{x}'\boldsymbol{\Sigma}^{-1}\mathbf{x} = \sum_{i=1}^{n} \frac{[e_i - E(e_i)]^2}{\sigma_i^2},$$

which is the sum of n expressions of the form $[e_i - E(e_i)]^2/\sigma_i^2$. It is well known that such an expression has a chi-square distribution with one degree of freedom; furthermore, the sum of n such expressions, all of them independent, has a chi-square distribution with n degrees of freedom (see, for example, Hays, 1973).

In the more general case of p Xs, it can be shown that $\chi^2 = \mathbf{x}'\boldsymbol{\Sigma}^{-1}\mathbf{x}$ has a chi-square distribution with p degrees of freedom even when the Xs are not independent. For a proof see Tatsuoka (1971, Chapter 5).

4.1.2 Properties of the Multivariate Normal Distribution

A number of properties of the multivariate normal distribution are important in correlation and regression analysis, as well as in other methods discussed in the succeeding chapters:

1. Each of the Xs has a univariate normal marginal distribution. However, even if all Xs have univariate normal marginal distributions, it is not necessarily true that they have a multivariate normal joint distribution.

2. Each conditional distribution of a particular X, say X_j, identified by specifying the values of all of the remaining Xs, that is, the X_ks $(k = 1,\ldots,p; k \neq j)$, is univariate normal with mean $\mu_{j \cdot k}$ and variance $\sigma_{j \cdot k}^2$. The means of all such conditional distributions of a particular X_j lie in the same hyperplane (see Section 4.2.1) and all have the same variance.

3. A linear combination of any of the Xs in a multivariate normal distribution also has a normal distribution. This is important in the context

of multiple correlation, because the bivariate product-moment correlation between X_j and a linear combination of the remaining Xs is a multiple correlation.

4. Recall from Figure 2.4.3 that, in the bivariate normal distribution, the locus of all points having the same value of the density function, $f(X, Y)$, was an ellipse with center at the centroid, (μ_x, μ_y). In the case of p variables having a multivariate normal distribution, the locus of all points having the same value of $f(X_1, \ldots, X_p)$ is an *ellipsoid* with center at the centroid, $\boldsymbol{\mu}$.

5. Any subset of a set of Xs having a multivariate normal distribution also has a multivariate normal distribution. This is particularly important in Chapter 5 where we deal with canonical correlation.

6. The joint conditional distribution of any two (or more) of the Xs, given specified values of some (or all) of the remaining Xs, is also multivariate normal. The means of all such distributions lie in a hyperplane and have the same variance.

4.2 ESTIMATES OF PARAMETERS IN MULTIPLE REGRESSION

The number of parameters associated with the multiple linear regression model depends on the number of independent variables (Xs) included. For illustrative purposes, we first consider the case in which there are two such variables, using ordinary algebraic methods. We then employ matrix notation to generalize procedures to cases in which more than two Xs are included.

The parameters of greatest interest in terms of estimation are the regression coefficients (βs), although the model may also be used to estimate μ_y. Here again, as in Chapter 1, we must distinguish between μ_Y and μ_y. Having specified a particular population for study, the value of μ_Y is a fixed parameter of that population. On the other hand, the value of μ_y and, hence, estimates of it, depend on which Xs one decides to include and how many observations one takes at each combination of values of the Xs. The interpretation of an estimate of μ_y, therefore, depends heavily on the design of the regression study.

Using the method of least squares, as in Chapter 1, we again wish to minimize

$$\sum_{i=1}^{n} (Y_i - \hat{Y}_i)^2,$$

the sum of squared discrepancies between the actual and predicted values of Y in the sample. Let b_0, b_1, and b_2 be sample estimates of the parameters β_0, β_1, and β_2, respectively. Then, for the case of two independent variables, X_1 and X_2,

$$\hat{Y}_i = b_0 + b_1 X_{i1} + b_2 X_{i2}. \tag{4.2.1}$$

Substituting in $\Sigma(Y_i - \hat{Y}_i)^2$ for $\hat{Y}_i$, we have

$$\begin{aligned} G &= \Sigma(Y_i - \hat{Y}_i)^2 = \Sigma(Y_i - b_0 - b_1X_{i1} - b_2X_{i2})^2 \\ &= \Sigma Y_i^2 + nb_0^2 + b_1^2\Sigma X_{i1}^2 + b_2^2\Sigma X_{i2}^2 - 2b_0\Sigma Y_i - 2b_1\Sigma X_{i1}Y_i \\ &\quad - 2b_2\Sigma X_{i2}Y_i + 2b_0b_1\Sigma X_{i1} + 2b_0b_2\Sigma X_{i2} + 2b_1b_2\Sigma X_{i1}X_{i2}. \end{aligned} \tag{4.2.2}$$

To obtain the normal equations, we use partial differentiation of 4.2.2 (see Appendix B) with respect to b_0, b_1, and b_2.

This gives

$$\frac{\partial G}{\partial b_0} = 2nb_0 - 2\Sigma Y_i + 2b_1\Sigma X_{i1} + 2b_2\Sigma X_{i2},$$

$$\frac{\partial G}{\partial b_1} = 2b_1\Sigma X_{i1}^2 - 2\Sigma X_{i1}Y_i + 2b_0\Sigma X_{i1} + 2b_2\Sigma X_{i1}X_{i2},$$

and

$$\frac{\partial G}{\partial b_2} = 2b_2\Sigma X_{i2}^2 - 2\Sigma X_{i2}Y_i + 2b_0\Sigma X_{i2} + 2b_1\Sigma X_{i1}X_{i2}.$$

Setting each of these expressions equal to zero and rearranging, we have:

$$\left.\begin{aligned} nb_0 + (\Sigma X_{i1})b_1 + (\Sigma X_{i2})b_2 &= \Sigma Y_i \\ (\Sigma X_{i1})b_0 + (\Sigma X_{i1}^2)b_1 + (\Sigma X_{i1}X_{i2})b_2 &= \Sigma X_{i1}Y_i \\ (\Sigma X_{i2})b_0 + (\Sigma X_{i1}X_{i2})b_1 + (\Sigma X_{i2}^2)b_2 &= \Sigma X_{i2}Y_i. \end{aligned}\right\} \tag{4.2.3}$$

These are the *normal equations* and they may be solved for the three regression coefficients, b_0, b_1, and b_2. Because the direct algebraic solutions in the specific case of two concomitant variables are of little general interest, we shall not give them here. Instead, we shall use the result in Equations 4.2.3 to develop a more general formulation of the problem and its solution in matrix terms.

4.2.1 Geometric Representation of the Multiple Regression Problem

It is useful at this point to distinguish between the geometric representations of regression problems involving one predictor, X, and those involving two predictors, X_1 and X_2. In Chapter 1 we pointed out that the regression equation, $\hat{Y} = a + bX$, is the equation of a straight line and that the least squares criterion resulted in placement of the line to minimize the sum of squared discrepancies between actual and predicted Y values. The geometric representation of the one-predictor case is given in Figure 4.2.1(a), where an individual's actual and predicted Y values, as well as the discrepancy between them, are shown.

In the two-predictor case, shown in Figure 4.2.1(b), Equation 4.2.1 is represented by a plane rather than a line. The least squares criterion in this case leads to location of the plane to minimize $\Sigma(Y_i - \hat{Y}_i)^2$, just as in the one-predictor case. All predictions, $\hat{Y}_i$, of actual Y values are located in the

FIGURE 4.2.1 Geometric Representation of Regression Equations in the One-Predictor and Two-Predictor Cases

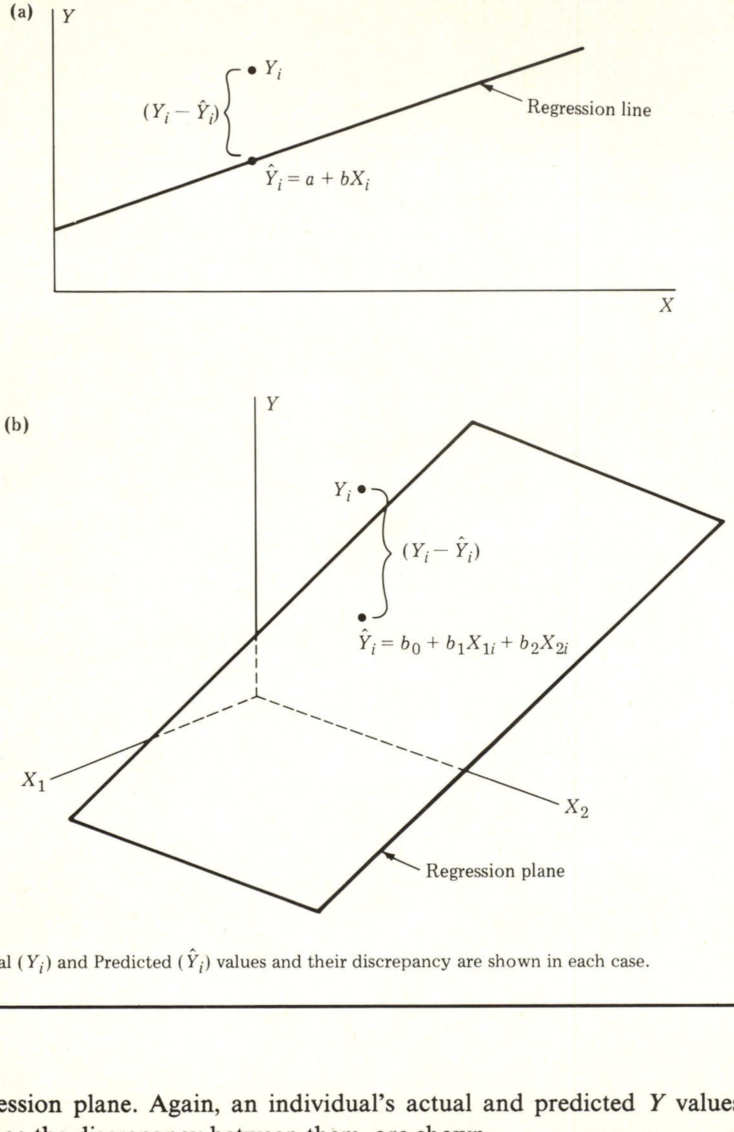

Actual (Y_i) and Predicted ($\hat{Y}_i$) values and their discrepancy are shown in each case.

regression plane. Again, an individual's actual and predicted Y values, as well as the discrepancy between them, are shown.

When there are more than two predictors, it is not possible to represent the problem as in Figure 4.2.1. The regression equation then is the equation of a *hyperplane*, a "plane" of more than two dimensions. Analogously, however, the individual's predicted Y value is located in the regression hyperplane, which again is placed to minimize the sum of squared discrepancies between actual and predicted values on the dependent variable.

4.2.2 The Normal Equations in Matrix Form

The matrix formulation of the multiple regression model was given as Equation 4.1.2. For two independent variables,

$$\boldsymbol{\beta}=\begin{bmatrix}\beta_0\\ \beta_1\\ \beta_2\end{bmatrix} \quad \text{and} \quad \mathbf{X}=\begin{bmatrix}1 & X_{11} & X_{12}\\ \vdots & \vdots & \vdots\\ 1 & X_{n1} & X_{n2}\end{bmatrix}.$$

The sample estimate of $\boldsymbol{\beta}$ may be written

$$\hat{\boldsymbol{\beta}}=\begin{bmatrix}b_0\\ b_1\\ b_2\end{bmatrix}.$$

If $\mathbf{X}$ is premultiplied by its transpose, the product will be

$$\mathbf{X}'\mathbf{X}=\begin{bmatrix}1 & \cdots & 1\\ X_{11} & \cdots & X_{n1}\\ X_{12} & \cdots & X_{n2}\end{bmatrix}\cdot\begin{bmatrix}1 & X_{11} & X_{12}\\ \vdots & \vdots & \vdots\\ 1 & X_{n1} & X_{n2}\end{bmatrix}$$

$$=\begin{bmatrix}n & \Sigma X_{i1} & \Sigma X_{i2}\\ \Sigma X_{i1} & \Sigma X_{i1}^2 & \Sigma X_{i1}X_{i2}\\ \Sigma X_{i2} & \Sigma X_{i1}X_{i2} & \Sigma X_{i2}^2\end{bmatrix}.$$

Postmultiplying by $\hat{\boldsymbol{\beta}}$ gives

$$\mathbf{X}'\mathbf{X}\hat{\boldsymbol{\beta}}=\begin{bmatrix}nb_0+(\Sigma X_{i1})b_1+(\Sigma X_{i2})b_2\\ (\Sigma X_{i1})b_0+(\Sigma X_{i1}^2)b_1+(\Sigma X_{i1}X_{i2})b_2\\ (\Sigma X_{i2})b_0+(\Sigma X_{i1}X_{i2})b_1+(\Sigma X_{i2}^2)b_2\end{bmatrix}.$$

If $\mathbf{Y}$ is premultiplied by $\mathbf{X}'$, the result is

$$\mathbf{X}'\mathbf{Y}=\begin{bmatrix}1 & \cdots & 1\\ X_{11} & \cdots & X_{n1}\\ X_{12} & \cdots & X_{n2}\end{bmatrix}\cdot\begin{bmatrix}Y_1\\ \vdots\\ Y_n\end{bmatrix}=\begin{bmatrix}\Sigma Y_i\\ \Sigma X_{i1}Y_i\\ \Sigma X_{i2}Y_i\end{bmatrix}.$$

Comparing $\mathbf{X}'\mathbf{X}\hat{\boldsymbol{\beta}}$ and $\mathbf{X}'\mathbf{Y}$ with the left and right sides, respectively, of Equations 4.2.3 suggests that the normal equations can be expressed in matrix form as

$$\mathbf{X}'\mathbf{X}\hat{\boldsymbol{\beta}}=\mathbf{X}'\mathbf{Y}. \tag{4.2.4}$$

That Equation 4.2.4 expresses the normal equations, regardless of the number of Xs in the model, can be verified through differentiating the sum of squares of the residuals, expressed in matrix form, with respect to $\boldsymbol{\beta}$. Thus,

$$\mathbf{e}'\mathbf{e}=(\mathbf{Y}-\mathbf{X}\boldsymbol{\beta})'(\mathbf{Y}-\mathbf{X}\boldsymbol{\beta})=\mathbf{Y}'\mathbf{Y}-2\boldsymbol{\beta}'\mathbf{X}'\mathbf{Y}+\boldsymbol{\beta}'\mathbf{X}'\mathbf{X}\boldsymbol{\beta},$$

and differentiation produces (see Appendix B)

$$-2\mathbf{X}'(\mathbf{Y}-\mathbf{X}\boldsymbol{\beta})=\mathbf{0},$$

or, on substitution of $\hat{\boldsymbol{\beta}}$ for $\boldsymbol{\beta}$,

$$\mathbf{X}'\mathbf{X}\hat{\boldsymbol{\beta}}=\mathbf{X}'\mathbf{Y}.$$

Solving Equation 4.2.4 for $\hat{\boldsymbol{\beta}}$ involves computing the inverse $\mathbf{X}'\mathbf{X}$. If the number of independent variables, p, is less than $n-1$, and if the rank of $\mathbf{X}'\mathbf{X}$ is $p+1$, then $\mathbf{X}'\mathbf{X}$ is nonsingular and it has an inverse. In that case the solution for $\hat{\boldsymbol{\beta}}$ is

$$\hat{\boldsymbol{\beta}}=(\mathbf{X}'\mathbf{X})^{-1}\mathbf{X}'\mathbf{Y}. \qquad (4.2.5)$$

However, if there are more parameters than items in the sample ($p>n$) or if some of the Xs are perfectly correlated with one another (for example, if one X is a multiple of another X), then $|\mathbf{X}'\mathbf{X}|=0$ and an inverse does not exist. In that case there is no unique solution to Equation 4.2.5 and one must reduce the number of Xs in the problem by appropriate deletions.

4.2.3 Estimation of Parameters When Bivariate Correlations are Available

Quite often in practice one has available not only the raw data on Y and the Xs for a sample of n individuals, but also the bivariate correlations between all pairs of such variables. We should emphasize that *these correlations must all be based on the same n individuals*. When the correlational data are available, it is convenient to use them in the estimation of the regression coefficients, $\beta_0,\ldots,\beta_p$. This section deals with the formulation of the multiple regression problem when the data are in the form of bivariate correlation coefficients. We consider first the case of two independent variables, X_1 and X_2, and then generalize the procedures to the case in which more than two Xs are involved.

If we solve the first of the normal equations (see Equations 4.2.3) for b_0, the result is

$$b_0=\bar{Y}-b_1\bar{X}_1-b_2\bar{X}_2. \qquad (4.2.6)$$

Substitution for b_0 in Equation 4.2.1 yields

$$\hat{Y}_i=\bar{Y}+b_1(X_{i1}-\bar{X}_1)+b_2(X_{i2}-\bar{X}_2).$$

Using this expression, we may now substitute for $\hat{Y}_i$ in $\Sigma(Y_i-\hat{Y}_i)^2$ to obtain

$$\begin{aligned}G=\Sigma(Y_i-\hat{Y}_i)^2&=\Sigma\left[(Y_i-\bar{Y})-b_1(X_{i1}-\bar{X}_1)-b_2(X_{i2}-\bar{X}_2)\right]^2\\&=S_{yy}+b_1^2S_{11}+b_2^2S_{22}-2b_1S_{1y}-2b_2S_{2y}+2b_1b_2S_{12},\end{aligned}$$

in which

$$S_{yy}=\Sigma(Y_i-\bar{Y})^2,\ S_{11}=\Sigma(X_{i1}-\bar{X}_1)^2,\ S_{22}=\Sigma(X_{i2}-\bar{X}_2)^2,$$

$$S_{1y}=\Sigma(X_{i1}-\bar{X}_1)(Y_i-\bar{Y}),\ S_{2y}=\Sigma(X_{i2}-\bar{X}_2)(Y_i-\bar{Y}),$$

and

$$S_{12}=\Sigma(X_{i1}-\bar{X}_1)(X_{i2}-\bar{X}_2).$$

Again, we obtain the normal equations by differentiating G with respect to b_1 and b_2.

The result is

$$\frac{\partial G}{\partial b_1}=2S_{11}b_1+2S_{12}b_2-2S_{1y}$$

and

$$\frac{\partial G}{\partial b_2}=2S_{22}b_2+2S_{12}b_1-2S_{2y}.$$

Setting each of these expressions equal to zero and rearranging, we have

$$\left.\begin{aligned} S_{11}b_1+S_{12}b_2&=S_{1y}\\ S_{12}b_1+S_{22}b_2&=S_{2y}.\end{aligned}\right\} \tag{4.2.7}$$

These equations can be solved for b_1 and b_2 to yield estimates of β_1 and β_2 exactly equal to those obtained from Equation 4.2.5. The solutions are $b_1=(S_{1y}S_{22}-S_{2y}S_{12})/(S_{11}S_{22}-S_{12}^2)$ and $b_2=(S_{2y}S_{11}-S_{1y}S_{12})/(S_{11}S_{22}-S_{12}^2)$.

Equations 4.2.7 can also be written in matrix form. If we let

$$\mathbf{S}=\begin{bmatrix} S_{11} & S_{12}\\ S_{12} & S_{22}\end{bmatrix},\ \mathbf{b}=\begin{bmatrix} b_1\\ b_2\end{bmatrix},\ \text{and } \mathbf{h}=\begin{bmatrix} S_{1y}\\ S_{2y}\end{bmatrix},$$

then $\mathbf{Sb}=\mathbf{h}$ and, assuming that $\mathbf{S}$ is nonsingular,

$$\mathbf{b}=\mathbf{S}^{-1}\mathbf{h}. \tag{4.2.8}$$

Note that this formulation holds in the general case where the number of Xs equals p. Then $\mathbf{S}$ would be of order p and, $\mathbf{b}$ and $\mathbf{h}$ would be column vectors consisting of p elements. For example, in the case of three independent variables, X_1, X_2, and X_3,

$$\mathbf{S}=\begin{bmatrix} S_{11} & S_{12} & S_{13}\\ S_{12} & S_{22} & S_{23}\\ S_{13} & S_{23} & S_{33}\end{bmatrix},\ \mathbf{b}=\begin{bmatrix} b_1\\ b_2\\ b_3\end{bmatrix},\ \text{and } \mathbf{h}=\begin{bmatrix} S_{1y}\\ S_{2y}\\ S_{3y}\end{bmatrix}.$$

A simple transformation applied to the equation $\mathbf{Sb}=\mathbf{h}$, yields a solution for $\mathbf{b}$ in terms of the bivariate correlations between the Xs and Y. In the case of two Xs ($p=2$), we premultiply both sides of the equation $\mathbf{Sb}=\mathbf{h}$ by the diagonal matrix

$$\mathbf{K}=\begin{bmatrix} \dfrac{1}{\sqrt{S_{11}S_{yy}}} & 0\\ 0 & \dfrac{1}{\sqrt{S_{22}S_{yy}}}\end{bmatrix}.$$

The result on the left side is

$$\mathbf{KSb}=\begin{bmatrix} \sqrt{\dfrac{S_{11}}{S_{yy}}} & \dfrac{S_{12}}{\sqrt{S_{11}S_{yy}}} \\ \dfrac{S_{12}}{\sqrt{S_{22}S_{yy}}} & \sqrt{\dfrac{S_{22}}{S_{yy}}} \end{bmatrix} \cdot \begin{bmatrix} b_1 \\ b_2 \end{bmatrix},$$

and on the right,

$$\mathbf{Kh}=\begin{bmatrix} \dfrac{1}{\sqrt{S_{11}S_{yy}}} & 0 \\ 0 & \dfrac{1}{\sqrt{S_{22}S_{yy}}} \end{bmatrix} \cdot \begin{bmatrix} S_{1y} \\ S_{2y} \end{bmatrix} = \begin{bmatrix} r_{1y} \\ r_{2y} \end{bmatrix}.$$

Because $r_{12}=S_{12}/\sqrt{S_{11}S_{22}}$, $\sqrt{S_{11}/S_{yy}}=s_1/s_y$, and $\sqrt{S_{22}/S_{yy}}=s_2/s_y$, the product **KSb** may be written

$$\mathbf{KSb}=\begin{bmatrix} \dfrac{s_1}{s_y} & r_{12}\dfrac{s_2}{s_y} \\ r_{12}\dfrac{s_1}{s_y} & \dfrac{s_2}{s_y} \end{bmatrix} \cdot \begin{bmatrix} b_1 \\ b_2 \end{bmatrix}.$$

We now let

$$\mathbf{R}=\begin{bmatrix} 1 & r_{12} \\ r_{12} & 1 \end{bmatrix}, \ \mathbf{D}=\begin{bmatrix} \dfrac{s_1}{s_y} & 0 \\ 0 & \dfrac{s_2}{s_y} \end{bmatrix}, \text{ and } \mathbf{c}=\begin{bmatrix} r_{1y} \\ r_{2y} \end{bmatrix}.$$

Then the equation $\mathbf{KSb}=\mathbf{Kh}$ may be written

$$\mathbf{RDb}=\mathbf{c}, \tag{4.2.9}$$

and the solution for $\mathbf{b}$ is

$$\mathbf{b}=(\mathbf{RD})^{-1}\mathbf{c}. \tag{4.2.10}$$

Again, by proper definition of **R**, **D**, **b**, and **c**, Equation 4.2.10 can be applied to the general case of p independent variables.

A slightly modified form of Equation 4.2.10 results if the sample regression equation is expressed in standard score form, that is, $\hat{z}_{iy}=B_1z_{i1}+B_2z_{i2}+\cdots+B_pz_{ip}$, in which $\hat{z}_{iy}$ is a prediction of the standardized value of Y for the ith individual, $z_{ij}=(X_{ij}-\bar{X}_j)/s_j$, and $B_j=(s_j/s_y)b_j$. Then

$$\mathbf{B}=\mathbf{Db}=\begin{bmatrix} B_1 \\ \vdots \\ B_p \end{bmatrix},$$

and Equation 4.2.10 becomes

$$\mathbf{B}=\mathbf{R}^{-1}\mathbf{c}. \tag{4.2.11}$$

The raw score form of the regression coefficients can be obtained from **B** by means of

$$\mathbf{b}=\mathbf{D}^{-1}\mathbf{B}. \tag{4.2.12}$$

The constant b_0 can then be obtained from

$$b_0=\bar{Y}-b_1\bar{X}_1-b_2\bar{X}_2-\cdots-b_p\bar{X}_p. \tag{4.2.13}$$

4.2.4 A Numerical Example

In the preceding sections we have given several different methods for obtaining estimates of $\boldsymbol{\beta}$ in the multiple regression model. All of these produce the same numerical results, but they differ in formulation and computation. For both illustrative and comparative purposes, we show below the computations of the sample regression coefficients by each of the methods described, using the same data set. The illustrative data for two independent variables, X_1 and X_2, are given in Table 4.2.1.

First, using Equations 4.2.3, we have:

$$7b_0+55b_1+77b_2=28$$

$$55b_0+557b_1+657b_2=262$$

$$77b_0+657b_1+959b_2=342$$

TABLE 4.2.1 DATA AND BASIC STATISTICS FOR $N=7$ INDIVIDUALS ON VARIABLES Y, X_1, AND X_2

Individual number	Y	X_1	X_2	*Basic statistics*
1	6	14	17	$n=7$
2	7	12	13	$\Sigma Y=28$
3	3	10	5	$\Sigma X_1=55$
4	5	8	15	$\Sigma X_2=77$
5	1	6	9	$\Sigma Y^2=140$
6	4	4	7	$\Sigma X_1^2=557$
7	2	1	11	$\Sigma X_2^2=959$
$\bar{Y}=4.00000$				$\Sigma X_1Y=262$
$\bar{X}_1=7.85714$				$\Sigma X_2Y=342$
$\bar{X}_2=11.00000$				$\Sigma X_1X_2=657$
$s_y^2=4.66667$	$s_y=2.16025$			
$s_1^2=20.80952$	$s_1=4.56175$			
$s_2^2=18.66667$	$s_2=4.32049$			
	$r_{y1}=.71034$			
	$r_{y2}=.60714$			
	$r_{12}=.43973$			

Dividing each equation by the coefficient of b_0 gives:

$$(1)\quad b_0+7.85714b_1+11.00000b_2=4.00000$$
$$(2)\quad b_0+10.12727b_1+11.94545b_2=4.76364$$
$$(3)\quad b_0+8.53247b_1+12.45455b_2=4.44156$$

Now, if we subtract Equation (1) from Equations (2) and (3), we obtain:

$$2.27013b_1+.94545b_2=.76364$$
$$.67533b_1+1.45455b_2=.44156.$$

Dividing each of these equations by the coefficient of b_1 yields:

$$b_1+.41647b_2=.33639$$
$$b_1+2.15384b_2=.65384.$$

When the first of these equations is subtracted from the second, the result is $1.73737b_2=.31745$, which may be solved to yield $b_2=.18272$. The values of b_1 and b_0 may be obtained by appropriate substitution in the equations above. We find that $b_1=.26029$ and $b_0=-.05507$. Thus, the regression equation for predicting Y from the Xs is

$$\hat{Y}=-.05507+.26029X_1+.18272X_2.$$

If the normal equations given in Equation 4.2.3 are expressed in matrix form, then the same numerical results can be obtained by solution of Equation 4.2.5. For the data of Table 4.2.1, the **X** and **Y** matrices are:

$$\mathbf{X}=\begin{bmatrix}1 & 14 & 17\\ 1 & 12 & 13\\ 1 & 10 & 5\\ 1 & 8 & 15\\ 1 & 6 & 9\\ 1 & 4 & 7\\ 1 & 1 & 11\end{bmatrix} \quad\text{and}\quad \mathbf{Y}=\begin{bmatrix}6\\7\\3\\5\\1\\4\\2\end{bmatrix}.$$

Then

$$\mathbf{X'X}=\begin{bmatrix}7 & 55 & 77\\ 55 & 557 & 657\\ 77 & 657 & 959\end{bmatrix} \quad\text{and}\quad \mathbf{X'Y}=\begin{bmatrix}28\\262\\342\end{bmatrix}.$$

To solve Equation 4.2.5, we must find $(\mathbf{X'X})^{-1}$. The steps are as follows (see Appendix A):

1. $|\mathbf{X'X}|=78960$.
2. The cofactors of the elements of **X′X** are:

$$\begin{bmatrix}102514 & -2156 & -6754\\ -2156 & 784 & -364\\ -6754 & -364 & 874\end{bmatrix}.$$

3. Divide each element in the matrix of cofactors by $|\mathbf{X'X}|$ to obtain $(\mathbf{X'X})^{-1}$.

$$(\mathbf{X'X})^{-1}=\begin{bmatrix}1.29830294 & -.02730496 & -.08553698\\ -.02730496 & .00992908 & -.00460993\\ -.08553698 & -.00460993 & .01106890\end{bmatrix}.$$

4. Check by computing $(\mathbf{X}'\mathbf{X})^{-1}(\mathbf{X}'\mathbf{X})=\mathbf{I}$.
   After obtaining $(\mathbf{X}'\mathbf{X})^{-1}$, we compute

$$\hat{\beta}=(\mathbf{X}'\mathbf{X})^{-1}\mathbf{X}'\mathbf{Y}=\begin{bmatrix} -.05506 \\ .26028 \\ .18273 \end{bmatrix},$$

which agrees with the results obtained using Equations 4.2.3.

Neither of the above methods is very suitable for hand computation, although they may be used when the number of independent variables is less than three. For larger values of p, it is better to use methods based on the correlation coefficients. We now illustrate the use of Equations 4.2.10 and 4.2.11 and introduce a method of solving the normal equations that is appropriate for larger matrices when the work is done without a computer. Of course, large multiple regression problems are now typically done by computer. Furthermore, when several successive problems are to be done regardless of their size, a computer is ordinarily used. However, hand methods are practicable for single problems involving fewer than ten variables.

To use Equation 4.2.10 we must obtain the matrix, **R**, of bivariate correlations between the Xs and also the vector, **c**, of correlation coefficients between the Xs and Y. In the example of Table 4.2.1, these are:

$$\mathbf{R}=\begin{bmatrix} 1.00000 & .43973 \\ .43973 & 1.00000 \end{bmatrix} \quad \text{and} \quad \mathbf{c}=\begin{bmatrix} .71034 \\ .60714 \end{bmatrix}.$$

The matrix **D**, a diagonal matrix whose elements s_j/s_y are the ratios of the standard deviations of the Xs to those of Y, is

$$\mathbf{D}=\begin{bmatrix} 2.11168 & 0 \\ 0 & 2.00000 \end{bmatrix}.$$

Then

$$\mathbf{RD}=\begin{bmatrix} 2.11168 & .87946 \\ .92857 & 2.00000 \end{bmatrix}$$

and

$$(\mathbf{RD})^{-1}=\begin{bmatrix} .58707 & -.25815 \\ -.27257 & .61986 \end{bmatrix}.$$

Therefore,

$$\mathbf{b}=(\mathbf{RD})^{-1}\mathbf{c}=\begin{bmatrix} .58707 & -.25815 \\ -.27257 & .61986 \end{bmatrix}\cdot\begin{bmatrix} .71034 \\ .60714 \end{bmatrix}=\begin{bmatrix} .26029 \\ .18272 \end{bmatrix},$$

and from Equation 4.2.6, $b_0=4.00000-(.26029)(7.85714)-(.18272)\cdot(11.00000)=-.05505$. Note that these results agree, within rounding error, with those obtained from Equations 4.2.3 and 4.2.5.

Equation 4.2.11 produces the standardized form of the regression coefficients, appropriate when the regression equation is to be used in standard-score form. We must first find the inverse of **R**, the matrix of correlations between the Xs. Because $|\mathbf{R}|=.80664$, and **R** is a symmetric matrix of order two, its inverse is easily found to be

$$\mathbf{R}^{-1}=\begin{bmatrix} 1.23971 & -.54514 \\ -.54514 & 1.23971 \end{bmatrix}.$$

Then

$$\mathbf{B}=\mathbf{R}^{-1}\mathbf{c}=\begin{bmatrix} 1.23971 & -.54514 \\ -.54514 & 1.23971 \end{bmatrix}\cdot\begin{bmatrix} .71034 \\ .60714 \end{bmatrix}=\begin{bmatrix} .54964 \\ .36545 \end{bmatrix}.$$

Transforming to raw score form,

$$\mathbf{b}=\mathbf{D}^{-1}\mathbf{B}=\begin{bmatrix} .47356 & 0 \\ 0 & .50000 \end{bmatrix}\cdot\begin{bmatrix} .54964 \\ .36545 \end{bmatrix}=\begin{bmatrix} .26029 \\ .18272 \end{bmatrix},$$

which agrees with the results from Equation 4.2.10. Note that premultiplying $\mathbf{B}$ by $\mathbf{D}^{-1}$ is equivalent to multiplying each element in $\mathbf{B}$ by s_y/s_j. Thus, $b_j = B_j(s_y/s_j)$.

When there are more than three independent variables (Xs), a more suitable hand computational method is that suggested by Crout (1941). The procedure for carrying out the Crout method is described in Appendix D. Applying the Crout method to the data of Table 4.2.1, the Original Matrix is:

Row	*Column*			*Row sums*
	C_1	C_2	C_3	
R_1	1.00000	.43973	.71034	2.15007
R_2	.43973	1.00000	.60714	2.04687

The Auxiliary Matrix is:

Row	*Column*			*Check*	$\Sigma R_i + 1$
	C_1	C_2	C_3		
R_1	1.00000	.43973	.71034	2.15007	2.15007
R_2	.43973	.80664	.36544	1.36544	1.36544

From this matrix we get $B_2 = .36544$ and $B_1 = .71034 - (.43973)(.36544) = .54964$. These values agree with those obtained from Equation 4.2.11.

4.3 THE MULTIPLE CORRELATION COEFFICIENT

Under the assumption of a multivariate normal distribution of the variables $Y, X_1, \ldots, X_p$, the multiple correlation, $\rho_{y.1\ldots p}$, is the maximum linear correlation between Y and a linear combination of $X_1, X_2, \ldots, X_p$. The weights in the linear combination $\beta_1, \beta_2, \ldots, \beta_p$ that maximize $\rho_{y.1\ldots p}$ are the elements of the vector of regression coefficients in the multivariate normal distribution. As defined in Section 4.1, these are fixed constants in the model for multiple linear regression. They are also fixed constants in the model for multiple linear correlation. In the latter model, however, the Xs, as well as Y, are assumed to be random variables having a multivariate normal distribution. Thus, the characteristics and properties of the multivariate normal distribution, as described in Section 4.1.1, form the basis for the following discussion of the multiple correlation coefficient.

A sample estimate of $\rho_{y.1\ldots p}$, which we denote by $R_{y.1\ldots p}$, may be computed using the matrices $\mathbf{R}$ and $\mathbf{c}$ introduced in Equation 4.2.9. For p independent variables, the estimate is

$$R_{y.1\ldots p} = \sqrt{\mathbf{c}'\mathbf{R}^{-1}\mathbf{c}} \,. \tag{4.3.1}$$

Equivalent expressions for $R_{y.1\ldots p}$ are

$$R_{y.1\ldots p} = \sqrt{\mathbf{c}'\mathbf{B}} = \sqrt{\mathbf{c}'\mathbf{Db}} \,. \tag{4.3.2}$$

Thus, the sample multiple correlation coefficient can be computed either from the sample regression coefficients or from the bivariate correlations in $\mathbf{R}$ and $\mathbf{c}$.

For the data of Table 4.2.1, the result using Equation 4.3.2 is

$$R^2_{y.12} = \mathbf{c}'\mathbf{B} = [.71034 \;\; .60714] \cdot \begin{bmatrix} .54964 \\ .36544 \end{bmatrix}$$

$$= .61230.$$

Then

$$R_{y.12} = \sqrt{.61230} = .78250.$$

The coefficient $R_{y.1\ldots p}$ has some properties that are useful in interpreting the results of multiple regression and correlation analyses:

1. $R_{y.1\ldots p}$ is the product-moment correlation coefficient between the variable Y and the linear combination of the Xs: $\hat{Y}_i = b_0 + b_1X_{i1} + \cdots + b_pX_{ip}$. However, as its definition ensures, it ranges in value between 0 and +1, instead of between −1 and +1 as in the case of a bivariate product-moment correlation coefficient. Furthermore, when some Xs are correlated positively, and some negatively, with Y, it is not appropriate to consider directionality in expressing the relationship between Y and a linear combination of the Xs.

2. The above property implies that the square of $R_{y.1\ldots p}$ expresses the proportion of the total Y variance accounted for by the linear relationship between Y and the Xs. Thus, for the data of Table 4.2.1, the linear combination $\hat{Y}_i = b_0 + b_1X_{i1} + b_2X_{i2}$ explains or accounts for approximately 61% of the variance in Y.

3. The value of $R_{y.1\ldots p}$ is always at least as large as the largest (in absolute value) bivariate correlation in $\mathbf{c}$, as guaranteed by the definition of the multiple correlation coefficient. Therefore, increasing the number of Xs in a regression problem cannot reduce the value of the multiple correlation

between Y and the Xs. There is no guarantee, however, that the multiple correlation coefficient will be significantly increased by including additional Xs. This point is discussed further in a later section of this chapter.

4.4 TESTS OF SIGNIFICANCE IN MULTIPLE REGRESSION AND CORRELATION

The second property of $R_{y.1\ldots p}$ listed in the preceding section implies that the total variance of Y can be divided into two parts, one due to the relationship between Y and the linear combination of the Xs, and the other to unexplained or error variance. We may describe the breakdown of the total Y sum of squares, S_{yy}, in the form of an analysis-of-variance table, as shown in Table 4.4.1.

Under the assumption of a multivariate normal distribution of Y and the Xs, the above table provides a test of significance of the multiple correlation coefficient. If the null hypothesis, H_0: $\rho_{y.1\ldots p}=0$, is true, the ratio of the mean square for regression to the mean square for error has the F distribution with p and $n-p-1$ degrees of freedom. Thus, the value of the test statistic,

$$F=\frac{(n-p-1)R^2_{y.1\ldots p}}{p\left(1-R^2_{y.1\ldots p}\right)}, \tag{4.4.1}$$

may be compared with the critical value, $F_{1-\alpha;p,n-p-1}$, to test the null hypothesis at the α level that the population multiple correlation coefficient equals zero. This statistic also provides a test of the null hypothesis H_0: $\beta_1=\cdots=\beta_p=0$, that is, that the βs associated with the Xs in the model are all equal to zero.

4.4.1 Inferences About Parameters in the Multiple Regression Model

Before we consider procedures for estimation and tests of significance in the multiple regression model, we should emphasize again the distinction we made in Chapter 1 between μ_y and μ_Y. This distinction is also necessary in this chapter, because of the difference between the models for multiple

TABLE 4.4.1 BREAKDOWN OF TOTAL SUM OF SQUARES IN MULTIPLE REGRESSION ANALYSIS

Source of variation	*Sum of squares*	*Degrees of freedom*	*Mean square*
Regression	$S_{yy}R^2_{y.1\ldots p}$	p	$S_{yy}R^2_{y.1\ldots p}/p$
Error	$S_{yy}(1-R^2_{y.1\ldots p})$	$n-p-1$	$\dfrac{S_{yy}\left(1-R^2_{y.1\ldots p}\right)}{n-p-1}$
Total	S_{yy}	$n-1$	

correlation analysis and multiple regression analysis. In multiple correlation we assume a multivariate normal population from which X and Y observations in the sample have been drawn *at random*. It is the mean of the marginal distribution of Y in this multivariate normal population that we have denoted by μ_Y. The sample mean, $\bar{Y}$, provides an unbiased estimate of μ_Y if the X and Y observations have been drawn at random from this population.

In the model for multiple regression analysis, we consider the X_js to be fixed constants. We may select (nonrandomly) any set of values of a particular X_j that we wish. Furthermore, for any combination of values of the X_js, we may specify any desired number of Y observations and randomly select that specified number. Therefore, the parameters of the resulting population of Y values depend heavily, not only on the particular values of the X_js selected, but also on the number of Y observations taken at each combination of values of the X_js. It is the mean of this "conditional" population of Y values that we have denoted by μ_y. In the regression model, the sample mean, $\bar{Y}$, is an unbiased estimate of μ_y.

Because μ_Y and μ_y may differ markedly from each other, as the bivariate example in Section 1.5 (Chapter 1) suggests, we must consider which of these population means we really wish to estimate. Usually, the focus is on μ_Y rather than μ_y, because the mean of the *entire* population of Y values is typically more interesting and important in research than is the mean of a subpopulation defined by the design of a particular multiple regression study. In fact, estimating μ_y may make little sense, because a multiple regression study is ordinarily designed to produce accurate predictions of individual scores and of conditional means, rather than to provide an accurate estimate of the population mean, μ_Y.

In practical research settings, however, it is often difficult to support the assumption of a multivariate normal population and to obtain truly random samples. Therefore, in the discussion of estimation procedures that follows, we have decided to denote the population mean by μ_y instead of by μ_Y. We strongly urge that the distinction between these parameters be considered when results of estimation procedures are interpreted.

The error mean square in Table 4.4.1 provides an estimate of the conditional variance of $Y.\sigma^2_{y.1\ldots p}$. Let

$$s^2_{y.1\ldots p} = \frac{S_{yy}\left(1 - R^2_{y.1\ldots p}\right)}{n - p - 1}.$$

A confidence interval for μ_y can be obtained from the following probability statement:

$$P\left[\bar{Y} + t_{\frac{\alpha}{2};n-p-1}\frac{s_{y.1\ldots p}}{\sqrt{n}} < \mu_y < \bar{Y} + t_{1-\frac{\alpha}{2};n-p-1}\frac{s_{y.1\ldots p}}{\sqrt{n}}\right] = 1 - \alpha. \quad (4.4.2)$$

Note that this statement is analogous to that for one concomitant variable as given in Chapter 1.

Confidence or prediction intervals can also be constructed for the mean of Y for a given set of X values, and for the Y value of an individual having a given set of X values. Recall from Chapter 1 that $\mu_{y \cdot x}$ is the mean of the Y values in the population for a given value of X, while $Y_{i \cdot x}$ is the value of Y for an individual member of the population having a given value of X. The analogues here are $\mu_{y \cdot \mathbf{X}}$, the mean of Y for all individuals having a particular vector, $\mathbf{X}$, of values of the Xs; and $Y_{i \cdot \mathbf{X}}$, the value of Y for an individual having X values $\mathbf{X}_i$. As in Chapter 1, $\hat{Y}_i = b_0 + b_1 X_{i1} + \cdots + b_p X_{ip}$ is an estimator of both $\mu_{y \cdot \mathbf{X}}$ and $Y_{i \cdot \mathbf{X}}$. The standard error differs, however, depending on whether we are estimating $\mu_{y \cdot \mathbf{X}}$ or $Y_{i \cdot \mathbf{X}}$. The standard errors, which we denote by $s_{\hat{\mu}_{y \cdot \mathbf{X}}}$ and $s_{\hat{Y}_{i \cdot \mathbf{X}}}$ are:

$$s_{\hat{\mu}_{y \cdot \mathbf{X}}} = s_{y.1 \ldots p} \sqrt{\frac{1}{n} + \mathrm{tr}(\mathbf{S}^{-1}\mathbf{x}\mathbf{x}')} \tag{4.4.3}$$

and

$$s_{\hat{Y}_{i \cdot \mathbf{X}}} = s_{y.1 \ldots p} \sqrt{1 + \frac{1}{n} + \mathrm{tr}(\mathbf{S}^{-1}\mathbf{x}\mathbf{x}')} \;, \tag{4.4.4}$$

in which $\mathrm{tr}(\mathbf{S}^{-1}\mathbf{x}\mathbf{x}')$ is the *trace* of the matrix product, $\mathbf{S}^{-1}\mathbf{x}\mathbf{x}'$ (see Appendix A for a definition of the trace of a matrix). The matrix $\mathbf{S}$ is the generalization to p Xs of $\mathbf{S}$, as used in Equation 4.2.8, and, also for p Xs,

$$\mathbf{x} = \begin{bmatrix} X_1 - \bar{X}_1 \\ X_2 - \bar{X}_2 \\ \vdots \\ X_p - \bar{X}_p \end{bmatrix}. \tag{4.4.5}$$

Note that $\mathbf{x}$ denotes a somewhat different vector here than it does in Equation 4.1.7. Because $\mathbf{x}$ is widely used in the statistical literature to denote both vectors, we use it here also for both. Which meaning is intended will be clear from the context. The confidence or prediction intervals are found as usual. The probability statements are:

$$P\left[\hat{Y}_i + t_{\frac{\alpha}{2};n-p-1} s_{\hat{\mu}_{y \cdot \mathbf{X}}} < \mu_{y \cdot \mathbf{X}} < \hat{Y}_i + t_{1-\frac{\alpha}{2};n-p-1} s_{\hat{\mu}_{y \cdot \mathbf{X}}}\right] = 1 - \alpha \tag{4.4.6}$$

and

$$P\left[\hat{Y}_i + t_{\frac{\alpha}{2};n-p-1} s_{\hat{Y}_{i \cdot \mathbf{X}}} < Y_{i \cdot \mathbf{X}} < \hat{Y}_i + t_{1-\frac{\alpha}{2};n-p-1} s_{\hat{Y}_{i \cdot \mathbf{X}}}\right] = 1 - \alpha. \tag{4.4.7}$$

An example of the application of these statements is given later in this chapter.

One may also wish to test hypotheses or construct confidence intervals for the β_js ($j = 1, \ldots, p$). To do this it is necessary to compute the standard errors, s_{b_j}, of the estimators, b_j. These are

$$s_{b_j} = s_{y.1 \ldots p} \sqrt{c_{jj}} \;, \tag{4.4.8}$$

where c_{jj} is the element in the jth column and jth row of the matrix $\mathbf{S}^{-1}$, defined as in Equation 4.4.3. To test the null hypothesis H_0: $\beta_j = 0 (j=1,\ldots,p)$, an appropriate statistic is

$$t = \frac{b_j}{s_{b_j}}, \qquad (4.4.9)$$

which, under H_0, has the t distribution with $n-p-1$ degrees of freedom. A confidence interval for $\beta_j (j=1,\ldots,p)$ may be obtained from

$$P\left[b_j + t_{\frac{\alpha}{2};n-p-1} s_{b_j} < \beta_j < b_j + t_{1-\frac{\alpha}{2};n-p-1} s_{b_j}\right] = 1-\alpha. \qquad (4.4.10)$$

Examples of the use of this equation are given later in this chapter.

4.4.2 Assessing the Contribution of an Added Predictor

In addition to carrying out one or more of the inferential procedures described in the preceding sections, one may also be interested in determining whether a significant improvement in prediction could be made by adding one or more, say m, concomitant variables to the original p such variables in the regression equation. Suppose, for example, that one had used X_1 and X_2 and wished to determine whether the addition of X_3 would significantly increase the accuracy with which Y could be predicted. The test might then be carried out on the basis of Table 4.4.2.

The significance of the additional variable, X_3, may be tested by means of the statistic

$$F = \frac{(n-4)\left(R^2_{y.123} - R^2_{y.12}\right)}{1 - R^2_{y.123}}, \qquad (4.4.11)$$

which, under the hypothesis of no additional contribution from X_3, has the F distribution with one and $n-4$ degrees of freedom. In general, when there are p original variables and m additional variables whose contribution is to be assessed, the statistic is

$$F = \frac{(n-p-m-1)\left(R^2_{y.1\ldots p+m} - R^2_{y.1\ldots p}\right)}{m\left(1 - R^2_{y.1\ldots p+m}\right)}, \qquad (4.4.12)$$

TABLE 4.4.2 ANALYSIS OF VARIANCE TO INDICATE CONTRIBUTION OF ADDED PREDICTOR

Source of variation	*Sum of squares*	*Degrees of freedom*	*Mean square*
Total regression	$SS_r = S_{yy} R^2_{y.123}$	$p+m=3$	$SS_r/p+m$
(i) X_1 and X_2	$SS_p = S_{yy} R^2_{y.12}$	$p=2$	SS_p/p
(ii) X_3	$SS_m = S_{yy}(R^2_{y.123} - R^2_{y.12})$	$m=1$	SS_m/m
Error	$SS_e = S_{yy}(1 - R^2_{y.123})$	$n-p-m-1=n-4$	$SS_e/n-p-m-1$
Total	$SS_t = S_{yy}$	$n-1$	

which has the F distribution with m and $n-p-m-1$ degrees of freedom when the null hypothesis is true.

This procedure may also be used to determine whether certain variables already in the regression equation are making a significant contribution to the prediction of Y. In this case, Equation 4.4.12 may be used with $p+m$ as the number of original variables, m as the number whose contribution is to be tested, and p as the number remaining after m are removed.

The test based on Table 4.4.2 is not the only criterion to be utilized in evaluating the contribution of an additional predictor variable. One should consider two other indicators, namely, the size of the multiple correlation coefficient and the standard error of estimate. As we pointed out earlier in this chapter, the size of the multiple correlation coefficient cannot decrease as additional independent variables enter the regression equation. The experienced researcher can usually decide, however, whether the magnitude of the increase in $R_{y.1\ldots p}$, as a predictor is added, is important in terms of the accuracy with which the criterion variable is estimated. It is important to keep in mind that the size of the multiple correlation coefficient can be arbitrarily increased as predictors are added and, theoretically, can even reach 1.00 if the number of predictors is large enough. Thus, an increase in $R_{y.1\ldots p}$ as predictors are added is not always an indication that accuracy of prediction is being increased.

Another indicator of the contribution of an added predictor is the standard error of estimate, $s_{y.1\ldots p}$. As in the bivariate case, this statistic provides a measure of the discrepancy between predicted and actual values of the criterion variable. It may be computed by means of the equation

$$s_{y.1\ldots p}=\sqrt{\frac{S_{yy}(1-R^2_{y.1\ldots p})}{n-p-1}}\ . \tag{4.4.13}$$

As additional predictors are introduced, the size of $s_{y.1\ldots p}$ is usually reduced because $R^2_{y.1\ldots p}$ increases. Note, however, that if p is relatively large compared to n (but not larger), an increase in p accompanied by a very small increase in $R^2_{y.1\ldots p}$, as a variable is added, may actually lead to an increase in $s_{y.1\ldots p}$. Such an increase would indicate that the additional variable was making no positive contribution and was actually decreasing the accuracy of prediction. For an example of such a case, see Draper and Smith (1966, p. 118).

Before concluding this discussion of the contributions made by added predictor variables, we should point out that the bivariate correlation between the criterion (Y) and a predictor (X_j) is sometimes a poor indicator of the contribution the predictor might make in the regression equation. There are cases in which a variable having a relatively large bivariate correlation with the criterion contributes very little to predictive accuracy, given that other predictors are already present in the equation. The presence of the other predictors may cause the information contained in the added one to be almost totally redundant. There also are cases in which a potential predictor variable has nearly a zero bivariate correlation with the criterion, but is correlated in such a way with other predictors that it contributes significantly when added to the regression equation. Such a variable is called a *suppressor variable* because it suppresses irrelevant information in other predictors, thus increasing overall accuracy of estimation. An excellent discussion of suppressor variables may be found in McNemar (1969).

An illustration of the use of a suppressor variable is given by Sorenson (1966) in a study designed to assess the utility of various measures for prediction of job effectiveness in a skilled mechanical repair job. The criterion measure, *C*, was based on evaluations given by first-level supervisors through direct observation of mechanics' job performances. From a total of 34 predictors considered, 3 were selected for operational use. These were: Survey of Mechanical Insight (*SMI*), Test of Mechanical Comprehension (*TMC*), and Background Survey Questionnaire (*BSQ*). The product-moment correlations among *C*, *SMI*, *TMC*, and *BSQ*, based on 63 subjects, were:

Variable	*SMI*	*TMC*	*BSQ*
Criterion(*C*)	.22	−.04	.30
SMI		.71	.09
TMC			−.02

Note that *TMC* has a near-zero correlation with *C* and with *BSQ*, but has a correlation of .71 with *SMI*. These correlations suggest that *TMC* is acting as a suppressor variable. Logical support for that conclusion comes from a consideration of the nature of the *SMI* and *TMC* instruments. As Sorenson states: "The Survey of Mechanical Insight is almost exclusively a 'nuts and bolts' type of test; that is, all of its 35 items are composed of drawings of noncomplex mechanical apparatus. ... By contrast, the Test of Mechanical Comprehension, Form BB, is not limited to simple mechanical apparatus types of items. ... It would ... seem to be more of an achievement test of elementary physics than a test of mechanical knowledge and experience. ... In this study, the ability to achieve a high score on the more 'nuts and bolts' oriented Survey of Mechanical Insight without benefit of a high score on the more school-achievement-oriented Test of Mechanical Comprehension was associated with success on the job of industrial mechanic." One might conclude, then, that the *TMC* acted to suppress that part of the *SMI* which was more school-achievement oriented, thus increasing the predictive effectiveness of the combination of the three tests.

4.5 PARTIAL CORRELATION

Sometimes it is necessary to eliminate the effect of a third variable in order to assess the relationship between a given pair of variables. For example, among young children, we would expect to find a positive correlation between size of vocabulary and size of shoe. It is doubtful, however, that there is really a cause-effect relationship between these variables. If the influence of age could be eliminated, we would expect the correlation between them to be practically zero.

The partial correlation coefficient expresses the relationship between two variables with the influence of one or more other variables eliminated. The rationale for the method is basically simple. Suppose we have three variables, Y, X_1, and X_2, and we wish to compute the partial correlation between Y and X_1 with X_2 partialed out (or "held constant," as it is sometimes described). We first compute the regression equations for estimating Y and X_1 from X_2, that is, $\hat{Y}=a_{Y2}+b_{Y2}X_2$ and $\hat{X}_1=a_{12}+b_{12}X_2$. We then subtract from Y and X_1, respectively, their predicted values, based on X_2. This step yields two derived variables, $Y-\hat{Y}$ and $X_1-\hat{X}_1$. These two variables represent the portions of Y and X_1, respectively, that are not predictable on the basis of X_2. In other words, they represent Y and X_1, with the "effect" of X_2 removed. The partial correlation between Y and X_1 with X_2 partialed out is merely the product-moment correlation between the derived variables, $Y-\hat{Y}$ and $X_1-\hat{X}_1$. It is usually denoted by $r_{y1.2}$.

The partial correlation coefficient is not ordinarily computed by using the derived variables defined above. It can be shown to be a function of the bivariate product-moment correlations between Y, X_1, and X_2:

$$r_{y1.2}=\frac{r_{Y1}-r_{Y2}r_{12}}{\sqrt{(1-r_{Y2}^2)(1-r_{12}^2)}}. \tag{4.5.1}$$

If one wishes to partial out two variables, say X_2 and X_3, the formula may be written

$$r_{y1.23}=\frac{r_{y1.2}-r_{y3.2}r_{13.2}}{\sqrt{(1-r_{y3.2}^2)(1-r_{13.2}^2)}}. \tag{4.5.2}$$

In general, one may compute partial correlations of any order from those of the next lower order. The term "order" refers to the number of variables held constant. Thus, r_{Y1} is a zero-order correlation coefficient, whereas $r_{y1.2}$ is a first-order partial correlation coefficient.

The null hypothesis, H_0: $\rho_{y1.2\ldots q}=0$, that the population partial correlation coefficient equals zero, may be tested using either a t test or Fisher's Z transformation. The statistics are:

$$t=\frac{r_{y1.2\ldots q}\sqrt{n-P-2}}{\sqrt{1-r_{y1.2\ldots q}^2}} \tag{4.5.3}$$

and

$$z=Z\sqrt{n-3-P}\ , \tag{4.5.4}$$

in which

P is the number of variables partialed out,

q equals $P+1$,

Z is Fisher's Z transformation of $r_{y1.2\ldots q}$, and

n is the number in the sample on which $r_{y1.2\ldots q}$ is based.

The statistic t has $n-P-2$ degrees of freedom and z has approximately the standard normal distribution. An example of the application of partial correlation may be found later in this chapter.

4.6 SELECTING A SATISFACTORY SUBSET OF PREDICTORS

It sometimes happens in multivariate research that one has available a relatively large set of possible predictor variables and wishes to select a subset that predicts a given criterion with an accuracy nearly equal to that which can be obtained using the entire set. Such a reduction in the number of predictors is often necessary because data on certain potentially useful predictors may be very difficult or time-consuming or expensive to obtain. If nearly the same predictive accuracy can be obtained with fewer variables, a research enterprise of seemingly doubtful feasibility may sometimes be executed with relative ease.

Several methods have been used for selecting a good subset of predictors. Only two of these will be discussed here and these will be outlined briefly rather than described in detail. For a more complete discussion with numerical examples, see Draper and Smith (1966, Chapter 6).

The two methods treated here are usually called the *backward elimination* and the *stepwise regression* methods. These methods were selected for illustrative purposes because they differ basically in their approach to the problem, each having its own advantages over the other. Although both procedures may yield the same result, that is, the same set of predictors, there are circumstances in which one may be preferable to the other. In describing the methods, we shall assume that we start with a set of K potential predictors and wish to select a good subset.

The backward elimination procedure involves the following steps:

1. Compute the regression equation based on *all* of the K potential predictors.

2. Test the significance of the contribution of *each* of the K predictors as if it were the last to enter the regression equation. The statistic given as Equation 4.4.12 may be used for these tests.

3. Pick out the smallest of the Fs among the total of K Fs and call it F_s. If F_s is not statistically significant, that is, if it is smaller than a preselected critical value, F_0, remove X_s, the variable associated with F_s, from the regression equation and recompute the regression equation based on the

remaining $K-1$ variables. Carry out Steps 2 and 3 again, based on the reduced set of potential predictors.

4. When a subset of predictors is obtained for which F_s is statistically significant, adopt the regression equation as computed on the basis of that subset.

The backward elimination procedure is preferred by some researchers because it permits inspection of the contributions of all of the potential predictors. However, it is often somewhat slower, requires larger computer storage, and is usually more expensive than certain other methods, particularly if K is large. It may also be difficult to carry out if the initial $K \times K$ correlation matrix is nearly singular, that is, if the value of its determinant is close to zero.

The steps in the stepwise regression procedure are as follows:

1. Select the X having the largest (in absolute value) bivariate correlation with the criterion, Y, and call this X_1. Compute the regression equation, $\hat{Y}=b_0+b_1X_1$. Test the significance of the regression of X_1 on Y. If it is significant, proceed to Step 2; if it is not significant, there is no subset of the original set of K predictors that will provide a better estimate of Y than will the sample mean, $\bar{Y}$.

2. Compute the squares of the partial correlations between Y and the remaining Xs, with the X selected in Step 1 partialed out. Select the X, say X_2, corresponding to the largest of these squared partial correlations and compute the regression equation, including X_1 and X_2.

3. Test the significance of the contribution of X_1 and of X_2 as if each were the *last* to enter the equation. (a) If both are significant, go to Step 4. (b) If the contribution of X_2 is not significant, the process is terminated and the regression equation includes X_1 only. (c) If the contribution of X_2 is significant, but that of X_1 is not, remove X_1 from the regression equation, return it to the pool of possible predictors, and go to Step 4.

4. Compute the squared partial correlations between Y and each of the Xs not in the equation, with the set of Xs already in the equation partialed out. Select the X having the largest squared partial correlation with Y and test the significance of its contribution. If it is *not* significant, the variable should not be included in the equation and the process is terminated. If its contribution *is* significant, retain it in the equation. Then test the significance of each other X in the equation as if it were the last to enter. If all are significant, retain them in the equation. If any are not significant, remove them from the equation and return them to the pool of possible predictors.

5. Repeat Step 4 until the X having the largest squared partial correlation with Y does not make a significant contribution to the prediction of Y. Then the process is terminated. Of course, the process is also terminated if *all* variables in the original pool make significant contributions to the prediction of Y.

The stepwise method is usually preferable to others because the contribution of each variable is examined at every step. It sometimes happens that a variable included in the regression equation at an early stage is rejected later when the presence of other variables causes the information provided by the earlier one to be redundant. Thus, the method is not only more efficient than backward elimination, but is somewhat more likely to select a smaller satisfactory subset of the original K variables.

We should emphasize again that in using any of the procedures discussed in this and previous chapters it is important to use large samples. Only then can one be reasonably sure that results obtained can be replicated on the basis of different samples drawn from the same population. While we hesitate to attempt to give specific sample size recommendations that would be appropriate in all situations, a useful rule to observe is that *the sample size in multiple regression problems should be at least* 100 *or at least* 20 *times the number of variables, whichever is larger*. In some cases, the application of this rule will yield sample sizes larger than necessary for the levels of precision required in the estimates of population parameters. However, if one is to err in choosing a sample size in these problems, it is much better to be on the high rather than on the low side.

4.7 ASSESSING THE RELATIVE CONTRIBUTIONS OF PREDICTORS

After carrying out a study involving multiple regression analysis, one sometimes wishes to determine which independent variables contributed the most and which the least to the prediction of the dependent variable, Y. When predictors (Xs) are mutually uncorrelated, the determination of their relative contributions may easily be expressed in percentage terms by means of the equation

$$R_{y.1\ldots p}^2 = r_{Y1}^2 + r_{Y2}^2 + \cdots + r_{Yp}^2. \qquad (4.7.1)$$

Thus, the percentage contribution of any predictor, X_j, when the X_js are all uncorrelated, is simply

$$\text{Percentage Contribution} = 100 \frac{r_{Yj}^2}{R_{y.1\ldots p}^2}. \qquad (4.7.2)$$

The percentage of the *total* Y variance accounted for by X_j is r_{Yj}^2.

When the X_js are *not* uncorrelated with each other, as is almost invariably the case in practice, the above equations no longer hold because a portion of the contribution of any predictor is a joint contribution with one or more other predictors with which it is correlated. In this case, a rather rough ordering of the X variables in terms of their relative contributions can be obtained by ranking them on the basis of their *standardized* regression coefficients, the B_js. Such an ordering is only approximate, however, because correlations between predictors also affect the magnitudes of the B_js.

A better procedure to follow when the X_js are correlated makes use of the correlation between the dependent variable, Y, and the residual variable, $X_j' = (X_j - \hat{X}_j)$, in which $\hat{X}_j$ is the value of X_j predicted on the basis of the $p-1$ independent X variables remaining when X_j is excluded, that is, $X_1, X_2, \ldots, X_{j-1}, X_{j+1}, \ldots, X_p$. This correlation, $r_{YX_j'}$, was called by Dunlap and Cureton (1930) a *semipartial correlation* and by McNemar (1969) a *part correlation*. Kerlinger and Pedhazur (1973) also call it a semipartial correlation, the term we shall use here. Note that, because X_j' represents the part of X_j that is not predictable from the $p-1$ remaining Xs, X_j' is uncorrelated with the other predictors. We shall denote the semipartial correlation between Y and X_j' by

$$r_{Y(X_j \cdot 1,2,\ldots,j-1,j+1,\ldots,p)}.$$

Of course, one may wish to compute a semipartial correlation by partialing out only one or two of the $p-1$ predictors remaining when X_j is excluded. For example, if there are a total of ten predictors and $X_j = X_3$, the symbol

$$r_{y(3.12)}$$

would represent the semipartial correlation between Y and $X_3' = X_3 - \hat{X}_3$, where $\hat{X}_3$ is the regression estimate of X_3 based only on X_1 and X_2.

Now, utilizing semipartial correlations and, for simplicity, setting $p=4$, Equation 4.7.1 can be restated as

$$R^2_{y.1234} = r^2_{Y1} + r^2_{Y(2.1)} + r^2_{Y(3.12)} + r^2_{Y(4.123)}. \tag{4.7.3}$$

For other values of p, the reader can easily deduce the correct modification of this formula. Equation 4.7.3 can be used to compute the percentage contributions of X_1, X_2, X_3 and X_4, *but only if X_1 enters the regression first, X_2 second, X_3 third, and X_4 fourth*. For example, the contribution of X_3 in that case is $100(r^2_{Y(3.12)}/R^2_{y.1234})$. If X_3 enters the equation first, instead of third, its percentage contribution is likely to be larger. Thus, the percentage contribution of any of the X_js depends on the position at which it enters the equation. If the order were X_4 first, X_3 second, X_2 third, and X_1 fourth, for example, Equation 4.7.3 would become

$$R^2_{y.1234} = r^2_{Y4} + r^2_{Y(3.4)} + r^2_{Y(2.34)} + r^2_{Y(1.234)},$$

and the percentage contribution of X_3 would be $100(r^2_{Y(3.4)}/R^2_{y.1234})$, a value likely to be different from $100(r^2_{Y(3.12)}/R^2_{y.1234})$.

Because of the effect of order described above, we suggest computing the percentage contribution of each X_j for each of the $p!$ orderings of the p independent variables. The mean of these $p!$ values for each X_j then indicates the average percentage contribution of each variable to the prediction of Y. Although this method may not be feasible if p is larger than 5 or 6, it is often possible to identify the subset of 5 or 6 most important variables in the total set (using, for example, the stepwise regression method discussed earlier). One can then apply the method described here to determine the relative percentage contribution of each variable in the subset. An example of the application of this method will be given later in this chapter.

4.8 APPLICATIONS

We may best illustrate the application of multiple regression analysis through a numerical example.

4.8.1 Application of Multiple Regression Analysis

The U.S. Navy maintains a program of psychological testing of recruits for the purposes of classification and selection. The battery is composed essentially of four tests: The General Classification Test (*GCT*); The Arithmetic Reasoning Test (*ARI*); The Mechanical Aptitude Test (*MECH*); and the Clerical Aptitude Tests (*CLER*). Scores on these tests are reported as ordinary standard scores, known as Navy Standard Scores (*NSS*). The standard scale for *NSS* has a mean of 50 and a standard deviation of 10. Among the selection uses of these tests is that of selecting those recruits who will attend one of the Class "A" Schools that exist to train apprentices in one of the nearly six dozen U.S. Navy occupational specialties (ratings).

The data that follow are taken from a study conducted by the Bureau of Naval Personnel to validate the Test Battery for selecting recruits to attend the Class "A" School for Electrician's Mates. The criterion, Y, is the final grade in the course (*GRADE*). Values of Y range from 0 to 100; the passing score is 63. Means and standard deviations of Y and the Xs, as well as bivariate product-moment correlation coefficients, are given in Table 4.8.1 for a sample of 1181 individuals. We shall use this information to illustrate a number of the procedures discussed earlier in this chapter.

Computation of Regression Coefficients. Two of the important reasons for collecting the data of this example were: (1) to determine how well the criterion, Y, could be predicted on the basis of the test battery and (2) to develop a regression equation that could aid in assigning recruits to training programs. To accomplish these goals it was necessary to compute the regression coefficients and determine how strong the relationship was between the Xs and Y.

The standardized regression coefficients were first computed using Equation 4.2.11 and the Crout method (see Appendix D). The Original and Auxiliary Matrices are shown in Tables 4.8.2 and 4.8.3.

TABLE 4.8.1 MEANS, STANDARD DEVIATIONS, AND CORRELATIONS FOR FIVE VARIABLES, BASED ON A SAMPLE OF 1181 INDIVIDUALS

Variable	X_2	X_3	X_4	Y	*Mean*	*Standard Deviation*
GCT (X_1)	.205	.248	.047	.216	56.72	5.81
ARI (X_2)		.072	.195	.240	55.44	5.39
MECH (X_3)			.040	.401	54.33	7.54
CLER (X_4)				.106	50.56	7.28
GRADE (Y)					80.94	7.39

TABLE 4.8.2 ORIGINAL MATRIX OF THE CROUT METHOD APPLIED TO TABLE 4.8.1

Variable	*GCT*	*ARI*	*MECH*	*CLER*	*GRADE*	*Row sum*
GCT	1.000	.205	.248	.047	.216	1.716
ARI	.205	1.000	.072	.195	.240	1.712
MECH	.248	.072	1.000	.040	.401	1.761
CLER	.047	.195	.040	1.000	.106	1.388

TABLE 4.8.3 AUXILIARY MATRIX OF THE CROUT METHOD APPLIED TO THE DATA OF TABLE 4.8.1

	GCT	*ARI*	*MECH*	*CLER*	*GRADE*	*CHECK*	$\Sigma R_i + 1$
GCT	1.00000	.20500	.24800	.04700	.21600	1.71600	1.71600
ARI	.20500	.95798	.02209	.19349	.20430	1.41988	1.41988
MECH	.24800	.02116	.93803	.02585	.36578	1.39163	1.39163
CLER	.04700	.18536	.02425	.96130	.05109	1.05109	1.05109

The regression coefficients were found to be:

$$B_1 = .08501$$

$$B_2 = .18636$$

$$B_3 = .36446$$

$$B_4 = .05109.$$

From these we may compute the multiple correlation coefficient using Equation 4.3.2. The result is

$$R^2_{y.1234} = \mathbf{c}'\mathbf{B} = [.216 \quad .240 \quad .401 \quad .106] \cdot \begin{bmatrix} .08501 \\ .18636 \\ .36446 \\ .05109 \end{bmatrix} = .21465.$$

Then $R_{y.1234} = \sqrt{.21465} = .46331$.

A test of significance of the multiple correlation coefficient may be carried out using Equation 4.4.1. The value of the test statistic is

$$F = \frac{1176(.21465)}{4(.78535)} = 80.36,$$

which is significant beyond the .001 level. This result clearly indicates that there is a statistically significant relationship between Y and the linear combination of the Xs. While the relationship is only moderately strong, as indicated by the size of the multiple correlation coefficient, one would probably conclude that the test battery would serve as a useful predictor of Y.

In standard score form the multiple regression equation is

$$\hat{z}_y = .08501 z_1 + .18636 z_2 + .36446 z_3 + .05109 z_4,$$

in which z_1, z_2, z_3, and z_4 are the standard scores on X_1, X_2, X_3, and X_4, respectively. To express the equation in raw score form, it is necessary to obtain the values of the b_js, as well as that of b_0. Using Equations 4.2.6 and 4.2.12, we have:

$$b_1 = (7.39/5.81)(.08501) = .10813$$

$$b_2 = (7.39/5.39)(.18636) = .25551$$

$$b_3 = (7.39/7.54)(.36446) = .35721$$

$$b_4 = (7.39/7.28)(.05109) = .05186$$

$$b_0 = 80.94 - .10813(56.72) - .25551(55.44) - .35721(54.33)$$
$$- .05186(50.56) = 38.61.$$

Thus, the regression equation in raw score form is

$$\hat{Y}_i = 38.61 + .10813X_{i1} + .25551X_{i2} + .35721X_{i3} + .05186X_{i4}.$$

To illustrate the use of this equation, suppose that we consider a given recruit with scores on the four tests in the battery as follows:

X_1: $GCT = 55$
X_2: $ARI = 47$
X_3: $MECH = 52$
X_4: $CLER = 42$.

Substituting these scores in the regression equation gives:

$$\hat{Y} = 38.61 + .10813(55) + .25551(47) + .35721(52)$$
$$+ .05186(42)$$
$$= 77.32.$$

Our best available estimate, therefore, of the final course grade for this recruit is 77.32.

Test of $H_0 : \beta_j = 0$. The data of this example may also be used to illustrate the test of $H_0 : \beta_j = 0$, that is, the null hypotheses that each of the population regression coefficients is equal to zero. The fact that the test of $H_0 : \rho_{y \cdot 1234}$ resulted in rejection, has shown that not all of the β_js are zero. We might then wish to test hypotheses about the individual β_js to identify the one(s) different from zero.

To carry out this test, we must compute $s_{y \cdot 1234}$ and $\mathbf{S}^{-1}$ for use in Equation 4.4.9. The matrix $\mathbf{S}$ of sums of squares and cross-products of the Xs, is

$$\begin{array}{cccc} GCT & ARI & MECH & CLER \end{array}$$

$$\mathbf{S} = \begin{bmatrix} 39841.035 & 7579.658 & 12844.556 & 2341.923 \\ 7579.658 & 34269.077 & 3430.805 & 9048.822 \\ 12844.556 & 3430.805 & 67106.782 & 2567.494 \\ 2341.923 & 9048.822 & 2567.494 & 62618.982 \end{bmatrix},$$

and the inverse is

$$\mathbf{S}^{-1} = \begin{bmatrix} .000027796 & -.000005639 & -.000005031 & -.000000018 \\ -.000005639 & .000031612 & -.000000370 & -.000004342 \\ -.000005031 & -.000000370 & .000015899 & -.000000410 \\ -.000000018 & -.000004342 & -.000000410 & .000016614 \end{bmatrix}.$$

The value of $s_{y.1234}$ may be obtained using Equation 4.4.13, written in a somewhat different form as

$$s_{y.1234} = s_y \sqrt{\frac{n-1}{n-p-1}\left(1 - R^2_{y.1234}\right)} \ .$$

Substituting for s_y from Table 4.8.1 and for $R^2_{y.1234}$, we have

$$s_{\hat{\mu}_{y.\mathbf{x}}} = 6.560\sqrt{\frac{1}{1181} + .00277583} = .39483$$

Then the tests of $H_0\!:\beta_j = 0$ may be carried out by computing s_{b_j} and $t = b_j / s_{b_j}$. The results are shown in Table 4.8.4.

On the basis of these results we conclude that *GCT* (X_1), *ARI* (X_2), and *MECH* (X_3) are better predictors of final course grade, Y, than is *CLER* (X_4). In fact, we cannot even reject the null hypothesis that $\beta_4 = 0$. Therefore, we might consider removing X_4 from the regression equation for predicting Y.

Confidence Intervals for the β_js. The values s_{b_j} obtained above can be used in constructing confidence intervals for the β_js. The 95% intervals are shown in Table 4.8.5.

Confidence Intervals for μ_y, $\mu_{y\cdot X}$, and $Y_{i\cdot X}$. We may also wish to construct confidence intervals for the population mean of Y, for the mean of one or more of the conditional distributions of Y, or for the Y value of an individual having a particular vector of X values. The latter may be of particular interest in making an assignment decision about an individual.

A 95% confidence interval for μ_y may be obtained by substitution in Equation 4.4.2. The limits are

$$\text{Lower: } 80.94 - 1.96\frac{6.560}{\sqrt{1181}} = 80.57$$

$$\text{Upper: } 80.94 + 1.96\frac{6.560}{\sqrt{1181}} = 81.31.$$

TABLE 4.8.4 TESTS OF THE HYPOTHESIS $H_0\!:\beta_j = 0$

Hypothesis	s_{b_j}	t
$\beta_1 = 0$	$6.560\sqrt{.000027796} = .0346$	3.13 ($p < .005$)
$\beta_2 = 0$	$6.560\sqrt{.000031612} = .0369$	6.92 ($p < .001$)
$\beta_3 = 0$	$6.560\sqrt{.000015899} = .0262$	13.65 ($p < .001$)
$\beta_4 = 0$	$6.560\sqrt{.000016614} = .0267$	1.94 ($p < .100$)

TABLE 4.8.5 95% CONFIDENCE INTERVALS FOR THE β_j s

β_j	*Lower limit*	*Upper limit*
β_1	$.10813 - 1.96(.0346) = .040$	$.10813 + 1.96(.0346) = .176$
β_2	$.25551 - 1.96(.0369) = .183$	$.25551 + 1.96(.0369) = .328$
β_3	$.35721 - 1.96(.0262) = .306$	$.35721 + 1.96(.0262) = .408$
β_4	$.05186 - 1.96(.0267) = -.001$	$.05186 + 1.96(.0267) = .104$

The extremely narrow interval is due primarily to the large sample size, $n=1181$.

To obtain 95% confidence intervals for $\mu_{y \cdot \mathbf{x}}$ and $Y_{i \cdot \mathbf{x}}$, we first need to specify $\mathbf{X}$. For illustration we shall use $\mathbf{X}'=[55 \quad 47 \quad 52 \quad 42]$. Then the vector of deviations (Equation 4.4.5) is

$$\mathbf{x}=\begin{bmatrix} -1.72 \\ -8.44 \\ -2.33 \\ -8.56 \end{bmatrix},$$

from which we compute

$$\mathbf{xx}'=\begin{bmatrix} 2.9584 & 14.5168 & 4.0076 & 14.7232 \\ 14.5168 & 71.2336 & 19.6652 & 72.2464 \\ 4.0076 & 19.6652 & 5.4289 & 19.9448 \\ 14.7232 & 72.2464 & 19.9448 & 72.2736 \end{bmatrix}.$$

From Equations 4.4.3 and 4.4.4 we find:

$$s_{\hat{\mu}_{y.\mathbf{x}}}=6.560\sqrt{\frac{1}{1181}+.00277583}=.39483$$

and

$$s_{\hat{Y}_{i.\mathbf{x}}}=6.560\sqrt{1+\frac{1}{1181}+.00277583}=6.57187.$$

Then the 95% confidence interval for $\mu_{y.\mathbf{x}}$ is, from Equation 4.4.6:

Lower: $77.32-1.96(.39483)=76.55$

Upper: $77.32+1.96(.39483)=78.09$.

Using Equation 4.4.7, we find the limits of the 95% prediction interval for $Y_{i.\mathbf{x}}$ to be:

Lower: $77.32-1.96(6.57187)=64.15$

Upper: $77.32+1.96(6.57187)=90.49$.

As expected, the latter interval is much wider than the former because of the much larger standard error of $\hat{Y}_{i.\mathbf{x}}$.

Assessing the Relative Contributions of Predictors. To illustrate the method we described earlier for assessing the relative contributions of predictors, we shall again use the Navy data involving four predictors. Because we have shown that the contribution of X_4 was not statistically significant, we shall not include it in the illustration, but shall compute the average percentage contributions based on all orderings of X_1, X_2, and X_3. The orderings are:

$$X_1X_2X_3$$
$$X_1X_3X_2$$
$$X_2X_1X_3$$
$$X_2X_3X_1$$
$$X_3X_1X_2$$
$$X_3X_2X_1.$$

We describe in detail the computations for the first order above, namely,

$X_1X_2X_3$. The formula for $R^2_{y.123}$ is then

$$R^2_{y.123} = r^2_{Y1} + r^2_{Y(2.1)} + r^2_{Y(3.12)}.$$

We must compute each of the following:

$$(1) \qquad r_{Y1} = .216; r^2_{Y1} = .0467.$$

$$(2) \quad r_{Y(2.1)} = \frac{r_{Y2} - r_{Y1}r_{12}}{\sqrt{1 - r^2_{12}}} = \frac{.240 - (.216)(.205)}{\sqrt{1 - (.205)^2}}$$

$$= .200.$$

$$r^2_{Y(2.1)} = .0400.$$

$$(3) \quad r_{Y(3.1,2)} = \frac{r_{Y(3.1)} - r_{Y(2.1)}r_{3(2.1)}}{\sqrt{1 - r^2_{3(2.1)}}}.$$

$$r_{Y(3.1)} = \frac{r_{Y3} - r_{Y1}r_{13}}{\sqrt{1 - r^2_{13}}} = \frac{.401 - (.216)(.248)}{\sqrt{1 - (.248)^2}} = .3586.$$

$$r_{3(2.1)} = \frac{r_{23} - r_{13}r_{12}}{\sqrt{1 - r^2_{12}}} = \frac{.072 - (.248)(.205)}{\sqrt{1 - (.205)^2}} = .0216.$$

$$r_{Y(3.1,2)} = \frac{.3586 - (.2000)(.0216)}{\sqrt{1 - (.0216)^2}} = .3544.$$

$$r^2_{Y(3.1,2)} = .1256.$$

The multiple correlation coefficient is then

$$R_{y.123} = \sqrt{.0467 + .0400 + .1256} = \sqrt{.2123}$$

$$= .4608.$$

The relative percentage contributions are:

$$X_1: \ 100\frac{.0467}{.2123} = 22.0\%$$

$$X_2: \ 100\frac{.0400}{.2123} = 18.8\%$$

$$X_3: \ 100\frac{.1256}{.2123} = 59.2\%.$$

Using the same procedure, we obtain the relative percentage contributions of each variable for the remaining five orderings of X_1, X_2, and X_3. The contributions for all six orderings are shown in Table 4.8.6.

Thus, X_3 is by far the best predictor, while X_1 and X_2 are distant seconds, X_2 being somewhat better than X_1.

It is clear from the above results that when predictors are correlated, their relative contributions to the predictability of the criterion depend heavily on the order of their appearance in the regression equation. A variable entered first usually accounts for a much larger proportion of variance in the criterion than it would if it were entered second, third, or later. The average percentage contribution, considering all possible permutations of the X_js, seems to provide a reasonably accurate picture of a

TABLE 4.8.6 RELATIVE PERCENTAGE CONTRIBUTIONS OF X_1, X_2, AND X_3 FOR EACH OF SIX ORDERINGS AND MEAN PERCENTAGE CONTRIBUTION IN A MULTIPLE REGRESSION ANALYSIS

Order	*Variable*		
	X_1	X_2	X_3
$X_1X_2X_3$	22.0%	18.8%	59.2%
$X_1X_3X_2$	22.0	17.4	60.6
$X_2X_1X_3$	13.7	27.1	59.2
$X_2X_3X_1$	3.2	27.1	69.7
$X_3X_1X_2$	6.8	17.5	75.7
$X_3X_2X_1$	3.2	21.1	75.7
Mean %	11.8	21.5	66.7

given predictor's contribution *as a member of the set of predictors being considered.* Of course, as a member of a different set, the contribution of any predictor may be quite different.

4.8.2 Application of Partial Correlation

Using Equation 4.5.1 and the data of Table 4.8.1, the partial correlation coefficient between Y and X_1 with X_2 partialed out is

$$r_{y1.2} = \frac{.216 - (.240)(.205)}{\sqrt{1-(.240)^2}\sqrt{1-(.205)^2}} = .17555.$$

Note that the correlation between scores on the *GCT* and *GRADE* reduces from .216 to .175 when *ARI*, which is related to both, is partialed out.

If both X_2 and X_3 are partialed out, the result, using Equation 4.5.2, is

$$r_{y1.23} = \frac{.175 - (.396)(.239)}{\sqrt{1-(.396)^2}\sqrt{1-(.239)^2}} = .09070.$$

Thus, we see that the correlation between *GRADE* and *GCT* reduces even further when both *ARI* and *MECH* are partialed out.

4.9 MULTIVARIATE CORRECTION FOR RESTRICTION IN RANGE

Research workers in the behavioral sciences, especially in education and psychology, frequently use restricted samples in their studies. In such samples, subjects have been preselected on one or more of the variables included in the analysis. For example, in a study intended to establish the validity (see, for example, Mehrens and Lehmann, 1973, Chapter 5) of a particular set of predictors, the sample subjects might constitute a relatively homogeneous group on one or more of the variables in the predictor set, that is, they may have been selected, directly or indirectly, on one or more of the independent variables. In such a case, the validity coefficient (usually a multiple correlation coefficient) would be attenuated, and the estimated test validity would be markedly lower than in an unrestricted sample.

In order to understand the rationale and procedures involved in making corrections for multivariate selection, we must first distinguish between *explicit* selection and *incidental* selection. When there is direct selection of cases on the basis of at least one variable, the selection is referred to as explicit. When there is an indirect selection effect upon a variable or variables, caused by the explicit selection, it is called incidental selection. For example, in the study of the validity of the Navy test battery for selecting naval recruits as students in the Electricians Mates Class "A" School, let us assume that the 1181 subjects had been preselected on the basis of two or more of the tests prior to being sent to the School. This would be a clear example of *explicit* selection. *Incidental* selection occurs on other variables in the predictor set if they are positively and significantly correlated with those on which the explicit selection took place.

When either explicit or incidental selection, or both, occur, the students in the sample tend to have higher scores on certain of the classification tests than the general, unrestricted, population of Naval recruits. Thus, they tend to earn higher grades in School than would an unrestricted sample. In other words, because the distributions of the predictor variables are negatively skewed for the sample in which explicit selection occurs, the distribution of the criterion scores will also tend to be negatively skewed. Variability in both predictor and criterion score distributions would tend to be larger in an unrestricted sample than in one subjected to incidental selection. Since the magnitude of the correlation coefficient, multiple as well as bivariate, is reduced by a restriction in variability, a correction for this restriction is required in order to provide the most accurate possible estimate of the true relationship between criterion and predictors.

Thorough understanding of the theory underlying the multivariate correction for restriction in range requires more than the minimal level of understanding of matrix algebra that we assume in our treatment here. A full exposition of the theory may be found in Gulliksen (1950, Chapter 13). Our purpose here is to indicate when the correction may be needed and to show its effect on the regression equation and on the magnitude of the sample estimate of the multiple correlation coefficient. A step-by-step procedure for correcting a multiple correlation coefficient, when there is explicit selection on two of four predictor variables, is given in Appendix E. The data used in that development are again those of Table 4.8.1, with explicit selection on the basis of $GCT(X_1)$ and $ARI(X_2)$.

The procedure outlined in Appendix E yields results analogous to those in Table 4.8.1, but corrected for explicit selection on X_1 and X_2. These results are given in Table 4.9.1. This new matrix of corrected correlations yields a multiple correlation coefficient equal to .67, which is considerably

TABLE 4.9.1 MEANS, STANDARD DEVIATIONS, AND CORRELATIONS, CORRECTED FOR RESTRICTION ON *GCT* AND *ARI*, FOR THE EXAMPLE OF TABLE 4.8.1

Variable	X_2	X_3	X_4	Y	*Mean*	*Standard deviation*
$GCT(X_1)$	.205	.435	.226	.486	56.72	10.54
$ARI(X_2)$		.320	.314	.490	55.44	8.76
$MECH(X_3)$			.125	.499	54.33	8.11
$CLER(X_4)$				.206	50.56	7.52
$GRADE(Y)$					80.94	8.31

larger than the value .46, based on Table 4.8.1. The standardized regression coefficients are $B_1 = .30433$, $B_2 = .34682$, $B_3 = .25610$, and $B_4 = -.00369$.

When the relative contributions of the predictor variables are recomputed, based on the corrected correlations, there is a marked change from those based on the uncorrected values. The average percentage contributions based on the corrected correlations are: *GCT*, 32.4%; *ARI*, 36.5%; and *MECH*, 31.1%. These results, when compared with those in Section 4.8.1, suggest that the correction for restriction in range may affect markedly the estimates of relative contributions of predictor variables, as well as their multiple correlation with the criterion variable. We recommend, therefore, that the correction for restriction in range be applied whenever it is deemed appropriate and whenever the necessary population parameters are known.

4.10 THE USE OF DUMMY VARIABLES IN REGRESSION ANALYSIS

So far in this chapter we have considered the independent variables (the Xs) in regression problems to be measured on a continuous scale. Although the actual measurements on such variables as age, weight, achievement, aptitude, and attitude are nearly always discrete (for example, weight to the nearest pound, age at last birthday, number of correct answers on a test), the trait or characteristic measured is considered continuous. That is, in theory, an infinite number of values can occur in the population being measured. There are, however, quite a number of characteristics of persons or objects that are not continuous. Attributes such as sex, eye color, political party membership, religious affiliation, and professorial rank, are examples. We call such variables *categorical* to distinguish them from continuous. Some consist of a set of *ordered* categories, that is, categories that may be arranged in order according to the amount of something represented by the categories. Professorial rank is an example of an ordered categorical variable. On the other hand, sex, eye color, and religious affiliation are *unordered* categorical variables. The categories are simply different from one another; they cannot be arranged hierarchically on the basis of a common attribute possessed in different amounts by members of different categories.

The use of ordered categorical variables as predictors in multiple regression analysis is very similar to that of continuous variables. Ordinarily a numerical value is assigned to each category of the variable, with the numbers assigned being ordered from small to large to correspond to the ordering represented in the categories themselves. For example, if the categories of response to an attitude item were Strongly Agree, Agree, Undecided, Disagree, and Strongly Disagree, we might assign the values 5, 4, 3, 2, and 1, respectively, to the five responses, so that the value 5 would represent the most agreement and 1 the least. Alternatively, of course, we might use $+2$, $+1$, 0, -1, and -2 to reflect agreement versus disagreement. However, the predictive utility of the variable would be the same in either case. In the assignment of such numbers, one should attempt to ensure that the difference between consecutive numbers reflects as accurately as possible the difference between the categories they represent in terms of the amount of the characteristic measured by the variable. Our example, for instance, implies that the difference in strength of agreement between Strongly Agree and Agree is equal to the difference between Agree and Undecided. The assumptions of the regression model do not require such equality, because the Xs are regarded as fixed constants. However, the matter of correspondence between the numbers and the categories they represent is of concern from a measurement viewpoint. Thus, the numbers assigned might be 1, 3, 4, 6, and 7, if one believed that such a set better represented amounts of the measured characteristic corresponding to the various categories.

When one desires to use *unordered* categorical variables in multiple regression problems, a different procedure is usually required. If the variable is dichotomous, it may be treated as an ordered categorical variable by assigning a different number to each category. For example, the variable *Sex* has two values, Male and Female, to which we might assign the numbers 0 and 1 (or any other pair of different values), respectively. Such a variable, whose values are 0 and 1, is called a "dummy" variable because the values merely indicate group membership. If a 1 is assigned to Female, then the variable measures "femaleness." If the assignment were reversed, the variable would measure "maleness." In either case, the results of analysis would be interpretable, because we could always determine whether "femaleness" or "maleness" were related in some way to the variable being predicted.

Suppose, however, that the variable to be used in the regression equation were "eye color," consisting of the categories, Blue, Brown, and Other. It would make no sense to assign the values 1, 2, and 3, for example, to these categories because any such assignment (of the possible $3! = 6$ different assignments of the three numbers to the three categories) would imply differences in magnitude between the categories which do not really exist. In this case, we can use two dummy variables, say X_1 and X_2, which take on values to represent the three categories as follows:

	X_1	X_2
Blue	1	0
Brown	0	1
Other	−1	−1

The choice of these values for X_1 and X_2 is arbitrary; other values could be used. We selected the above values because of their utility in a particular application of dummy variables to be described later. In general, however, if the number of categories of an unordered categorical variable is r, we can use $r-1$ dummy variables to denote the categories. For example, if $r=4$, we might use:

Category	X_1	X_2	X_3
1	1	0	0
2	0	1	0
3	0	0	1
4	−1	−1	−1

Note in these examples that each X takes on only three different values, namely 1, 0, and −1. It is the combination of these values for the several Xs that denotes the category.

We shall not provide an illustrative numerical example of the application of dummy variables at this point, in part because we feel that the procedure for their use is straightforward, but principally because we prefer to illustrate their application in a well-known and important kind of statistical analysis. We refer to the analysis of variance, which many researchers view as quite distinct from regression analysis. We shall see that there is a close correspondence between the two methods.

4.11 ANALYSIS OF VARIANCE AS A MULTIPLE REGRESSION PROBLEM

We consider in this section the relationship between analysis of variance and multiple regression analysis. Although all fixed-effects analysis-of-variance problems can be shown to be special cases of multiple regression analysis, we limit our discussion here to the one- and two-way fixed-effects designs. It is assumed that the reader is already familiar with simple analysis of variance techniques.

In the one-way ANOVA, the model may be written $Y_{ij} = \mu + \alpha_j + e_{ij}$, in which $\alpha_j = \mu_j - \mu$ denotes the effect of population j ($j = 1,\ldots,J$), and e_{ij} is the error associated with the score of the ith individual in population j. Thus, the model "explains" the value Y_{ij} in terms of the grand mean over all populations, μ; an effect of being in population j, α_j; and an error, e_{ij}, associated with the ith individual in population j. Note that if we define $\Sigma_j \mu_j / J = \mu$, then $\Sigma_j \alpha_j = 0$. Therefore, the jth α is determined if the remaining $J-1$ α's are known.

Since any Y_{ij} is either a member or is not a member of population j, dummy variables may be employed to express the model in multiple regression terms as

$$Y_{ij} = \mu + \alpha_1 X_1 + \alpha_2 X_2 + \cdots + \alpha_{J-1} X_{J-1} + e_{ij} \tag{4.11.1}$$

or as

$$E(Y) = \mu + \alpha_1 X_1 + \alpha_2 X_2 + \cdots + \alpha_{J-1} X_{J-1}. \tag{4.11.2}$$

Thus, the Xs indicate membership in one of the J populations. For example, with only three populations ($J=3$), we would require only two dummy variables, X_1 and X_2, as follows:

Population	X_1	X_2
1	1	0
2	0	1
3	−1	−1

For $J=4$, we would need three dummy variables, X_1, X_2, and X_3, and the values for members of each population would be:

Population	X_1	X_2	X_3
1	1	0	0
2	0	1	0
3	0	0	1
4	−1	−1	−1

The model of Equation 4.11.2 can be expressed in matrix terms as

$$E(\mathbf{Y}) = \mathbf{X}\boldsymbol{\beta}. \tag{4.11.3}$$

To show what **Y**, **X**, and $\boldsymbol{\beta}$ are like, it will be helpful to consider a simple example. Suppose we have samples from three treatment populations representing three different methods of teaching statistics, and wish to test the hypothesis H_0: $\mu_1 = \mu_2 = \mu_3$. We obtain the following data on a common measure of statistics achievement, Y:

Method 1	*Method* 2	*Method* 3
24	36	19
33	29	21
21	34	23
27	39	27
	37	31
	40	

Then we may write

$$\mathbf{Y}=\begin{bmatrix}24\\33\\21\\27\\36\\29\\34\\39\\37\\40\\19\\21\\23\\27\\31\end{bmatrix},\quad \mathbf{X}=\begin{bmatrix}1&1&0\\1&1&0\\1&1&0\\1&1&0\\1&0&1\\1&0&1\\1&0&1\\1&0&1\\1&0&1\\1&0&1\\1&-1&-1\\1&-1&-1\\1&-1&-1\\1&-1&-1\\1&-1&-1\end{bmatrix},\quad \text{and } \hat{\beta}=\begin{bmatrix}\hat{\mu}\\\hat{\alpha}_1\\\hat{\alpha}_2\end{bmatrix}.$$

Note that the vector **Y** consists simply of the Y values of members of the three groups, with the Method 1 values listed first, the Method 2 values next, followed by the Method 3 values. The matrix **X** consists of a column of ones to represent the parameter μ in the model (Equation 4.11.2) followed by two columns representing the dummy variables, X_1 and X_2, that denote group membership. Thus, for members of Group 1, $X_1=1$ and $X_2=0$; for Group 2, $X_1=0$ and $X_2=1$; and for Group 3, $X_1=-1$ and $X_2=-1$. The elements of the vector $\hat{\boldsymbol{\beta}}$ are the sample estimates of μ, α_1, and α_2, respectively, for which we must solve by the methods of this chapter.

Having specified **Y**, **X**, and $\hat{\boldsymbol{\beta}}$, we may proceed as in multiple regression analysis to obtain $\hat{\boldsymbol{\beta}}$. It may be computed as suggested by Equation 4.2.5, that is,

$$\hat{\beta}=(\mathbf{X}'\mathbf{X})^{-1}\mathbf{X}'\mathbf{Y}.$$

For our example,

$$\mathbf{X'X} = \begin{bmatrix} 15 & -1 & 1 \\ -1 & 9 & 5 \\ 1 & 5 & 11 \end{bmatrix},$$

$$(\mathbf{X'X})^{-1} = \begin{bmatrix} .0685185 & .0148148 & -.0129630 \\ .0148148 & .1518519 & -.0703704 \\ -.0129630 & -.0703704 & .1240741 \end{bmatrix},$$

and

$$\mathbf{X'Y} = \begin{bmatrix} 441 \\ -16 \\ 94 \end{bmatrix}.$$

Then

$$\hat{\mathbf{B}} = (\mathbf{X'X})^{-1}\mathbf{X'Y} = \begin{bmatrix} 28.7610997 \\ -2.5111212 \\ 7.0722064 \end{bmatrix}.$$

Having obtained $\hat{\boldsymbol{\beta}}$, the error sum of squares, usually denoted by SS_W in the one-way analysis of variance, can be computed from

$$SS_W = \mathbf{Y'Y} - \hat{\boldsymbol{\beta}}'\mathbf{X'Y}. \tag{4.11.4}$$

The product $\mathbf{Y'Y}$ is simply the sum of squares of the Y values, namely, $\Sigma_{j=1}^{J}\Sigma_{i=1}^{n_j} Y_{ij}^2$. The elements of $\hat{\boldsymbol{\beta}}$, as indicated above, are estimates of μ, α_1, and α_2. Thus, $\hat{\mu}$ is the average of the separate sample means, which, for $J=3$, is equal to $\bar{Y}' = (\bar{Y}_1 + \bar{Y}_2 + \bar{Y}_3)/3$. We use $\bar{Y}'$ to denote this mean, to distinguish it from $\bar{Y} = \Sigma_{j=1}^{J}\Sigma_{i=1}^{n_j} Y_{ij}/n$, in which $n = n_1 + n_2 + n_3$. When the sample sizes are all equal, that is, $n_1 = n_2 = n_3$, then $\bar{Y}' = \bar{Y}$. For the present example, $\bar{Y} = 29.40$ and $\bar{Y}' = 28.76$. The estimates of α_1 and α_2 are $\hat{\alpha}_1 = \bar{Y}_1 - \bar{Y}' = -2.51$ and $\hat{\alpha}_2 = \bar{Y}_2 - \bar{Y}' = 7.07$. Becasue $\Sigma_j \alpha_j = 0$, we find that $\hat{\alpha}_3 = -\hat{\alpha}_1 - \hat{\alpha}_2 = -4.56$.

An inspection of $\mathbf{X}'$ and $\mathbf{Y}$ shows that the elements of the product $\mathbf{X'Y}$ are, for $J=3$,

$$\mathbf{X'Y} = \begin{bmatrix} \sum_j \sum_i Y_{ij} \\ \sum_i Y_{i1} - \sum_i Y_{i3} \\ \sum_i Y_{i2} - \sum_i Y_{i3} \end{bmatrix}.$$

Then the product $\hat{\boldsymbol{\beta}}'\mathbf{X}'\mathbf{Y}$ is

$$\begin{bmatrix} \bar{Y}' & \bar{Y}_1-\bar{Y}' & \bar{Y}_2-\bar{Y}' \end{bmatrix} \cdot \begin{bmatrix} \sum_j \sum_i Y_{ij} \\ \sum_i Y_{i1}-\sum_i Y_{i3} \\ \sum_i Y_{i2}-\sum_i Y_{i3} \end{bmatrix}$$

$$\begin{aligned}
&= \bar{Y}'\Sigma\Sigma Y + (\bar{Y}_1-\bar{Y}')(\Sigma Y_1-\Sigma Y_3) + (\bar{Y}_2-\bar{Y}')(\Sigma Y_2-\Sigma Y_3) \\
&= \bar{Y}'\Sigma\Sigma Y + \bar{Y}_1\Sigma Y_1 - \bar{Y}_1\Sigma Y_3 - \bar{Y}'\Sigma Y_1 + \bar{Y}'\Sigma Y_3 + \bar{Y}_2\Sigma Y_2 \\
&\quad - \bar{Y}_2\Sigma Y_3 - \bar{Y}'\Sigma Y_2 + \bar{Y}'\Sigma Y_3 \\
&= \bar{Y}'(\Sigma\Sigma Y - \Sigma Y_1 + \Sigma Y_3 - \Sigma Y_2 + \Sigma Y_3) + n_1\bar{Y}_1^2 + n_2\bar{Y}_2^2 - (\bar{Y}_1+\bar{Y}_2)\Sigma Y_3 \\
&= n_1\bar{Y}_1^2 + n_2\bar{Y}_2^2 + \bar{Y}'(3\Sigma Y_3) - (3\bar{Y}'-\bar{Y}_3)\Sigma Y_3 \\
&= n_1\bar{Y}_1^2 + n_2\bar{Y}_2^2 + 3\bar{Y}'\Sigma Y_3 - 3\bar{Y}'\Sigma Y_3 + n_3\bar{Y}_3^2 \\
&= \sum_{j=1}^{3} n_j\bar{Y}_j^2.
\end{aligned}$$

(Note that summation limits and many i and j subscripts have been omitted to simplify notation.) Now, subtracting from $\mathbf{Y}'\mathbf{Y}=\Sigma\Sigma Y$, we have, for $J=3$,

$$\begin{aligned}
\mathbf{Y}'\mathbf{Y}-\hat{\boldsymbol{\beta}}'\mathbf{X}'\mathbf{Y} &= \sum_{j=1}^{3}\sum_{i=1}^{n_j} Y_{ij}^2 - \sum_{j=1}^{3} n_j\bar{Y}_j^2 \\
&= \sum_{j=1}^{3}\sum_{i=1}^{n_j} (Y_{ij}-\bar{Y}_j)^2 \\
&= SS_W,
\end{aligned}$$

which shows for the case $J=3$ that Equation 4.11.4 yields the correct value for SS_W. Of course, with somewhat more effort, it can be shown for the general case that

$$SS_W = \mathbf{Y}'\mathbf{Y}-\hat{\boldsymbol{\beta}}'\mathbf{X}'\mathbf{Y} = \sum_{j=1}^{J}\sum_{i=1}^{n_j} (Y_{ij}-\bar{Y}_j)^2. \qquad (4.11.5)$$

For our example,

$$\begin{aligned}
SS_W &= 13639 - \begin{bmatrix} 28.7610997 & -2.5111212 & 7.0722064 \end{bmatrix} \cdot \begin{bmatrix} 441 \\ -16 \\ 94 \end{bmatrix} \\
&= 250.38969.
\end{aligned}$$

As indicated in the regression model of Equation 4.11.1 or Equation 4.11.2, Y_{ij} is a function of both μ and the X_js, the latter indicating group membership. This implies that we are using the X_js as predictors in the same sense that we used Xs as predictors earlier in the chapter. Therefore, we can measure the contribution the X_js are making by removing them from the regression model and computing the error sum of squares for the reduced model. We shall denote this sum of squares for the reduced model by $SS_{W(\alpha)}$. The increase (if any) in SS_W when the X_js are removed will then be a measure of their contribution to predicting Y. A large increase would indicate a large contribution, a small increase, a small contribution. It should be noted that removing the X_js from the model is equivalent to setting $\alpha_1 = \alpha_2 = \cdots = \alpha_J = 0$, which, in turn, is equivalent to the usual null hypothesis in the one-way ANOVA, H_0: $\mu_1 = \mu_2 = \cdots = \mu_J$. Thus, the increase in SS_W when the X_js are removed is evidence concerning the truth of the null hypothesis. A small increase would suggest accepting H_0, a large increase would suggest rejection.

When the X_js are removed, so that the reduced model is $Y_{ij} = \mu + e_{ij}$, then $\mathbf{X}' = [1 \quad 1 \quad 1 \quad \ldots \quad 1]$, that is, a row of n 1's, so that $\mathbf{X}'\mathbf{X} = n$, $(\mathbf{X}'\mathbf{X})^{-1} = 1/n$, and $\mathbf{X}'\mathbf{Y} = \Sigma_j \Sigma_i Y_{ij}$. Then $\hat{\boldsymbol{\beta}} = (\mathbf{X}'\mathbf{X})^{-1}\mathbf{X}'\mathbf{Y} = \Sigma\Sigma Y/n = \bar{Y}$, and

$$SS_{W(\alpha)} = \mathbf{Y}'\mathbf{Y} - \hat{\boldsymbol{\beta}}'\mathbf{X}'\mathbf{Y} = \Sigma\Sigma Y^2 - \bar{Y}\Sigma\Sigma Y$$

$$= \Sigma\Sigma Y^2 - n\bar{Y}^2 = \Sigma\Sigma\left(Y_{ij} - \bar{Y}\right)^2,$$

which is usually called, in one-way ANOVA, the total sum of squares, SS_T. It is also clear that

$$SS_{W(\alpha)} - SS_W = \Sigma\Sigma\left(Y_{ij} - \bar{Y}\right)^2 - \Sigma\Sigma\left(Y_{ij} - \bar{Y}_j\right)^2$$

$$= \sum_j n_j\left(\bar{Y}_j - \bar{Y}\right)^2,$$

which is the between groups sum of squares, SS_B. For our example, $\mathbf{X}'\mathbf{X} = 15$, $(\mathbf{X}'\mathbf{X})^{-1} = 1/15$, $\mathbf{X}'\mathbf{Y} = 441$, and $\mathbf{Y}'\mathbf{Y} = 13639$. Therefore,

$$\hat{\boldsymbol{\beta}} = (1/15)(441) = 29.40,$$

$$SS_{W(\alpha)} = 13639 - (29.40)(441) = 673.60,$$

and

$$SS_B = SS_{W(\alpha)} - SS_W = 423.21.$$

The analysis of variance may then be summarized as in Table 4.11.1.

Our purpose in discussing the correspondence between regression analysis and analysis of variance was to provide a different view of the latter method than typically results from the more traditional or "classical"

TABLE 4.11.1 ANALYSIS OF VARIANCE SUMMARY OF THE EXAMPLE IN SECTION 4.11

Source	*Sum of squares*	*Degrees of freedom*	*Mean square*	*F*
Between	$SS_B = 423.21$	2	211.61	10.14
Within	$SS_W = 250.39$	12	20.87	
Total	$SS_T = 673.60$	14		

approach usually presented in introductory statistics texts, and to emphasize the importance of the model that underlies analysis of variance, but is sometimes not given sufficient emphasis in textbook treatments of the topic. Clearly, the computing methods given in traditional presentations are more economical than those given here, particularly for hand or desk calculator computation. Therefore, in the one-way case, we do not suggest that ANOVA necessarily *should* be carried out using a multiple regression procedure, but only that it *can* be. In more complex designs, however, the regression approach may sometimes be the *only* method by which computations can be carried out. In such cases, the methods discussed here are of much more than academic interest.

4.11.1 Regression Analysis Applied to Two-Way ANOVA Problems

The methods presented in the preceding section can easily be extended to analyze two-way and more complex fixed-effects designs and to many types of mixed models. The basic procedures are very similar to those for the one-way analysis. Suppose, for example, that we wished to analyze the following 2×3 design with 3 observations for each treatment combination:

		Teaching Method 1	2	3
		Y_{111}	Y_{112}	Y_{113}
	1	Y_{211}	Y_{212}	Y_{213}
Aptitude		Y_{311}	Y_{312}	Y_{313}
Level		Y_{121}	Y_{122}	Y_{123}
	2	Y_{221}	Y_{222}	Y_{223}
		Y_{321}	Y_{322}	Y_{323}

in which Y_{ijk} represents the ith observation ($i = 1,2,3$) in the jth aptitude level ($j = 1,2$) and the kth teaching method ($k = 1,2,3$). The model for this case may be written

$$Y_{ijk} = \mu + \alpha_j + \beta_k + (\alpha\beta)_{jk} + e_{ijk}, \qquad (4.11.5)$$

in which α_j is the effect of being at the jth level of aptitude, β_k is the effect of receiving the kth teaching method, and $(\alpha\beta)_{jk}$ is the interaction effect associated with treatment combination jk. Again, it is convenient to define the effects so that $\Sigma_j \alpha_j = 0$, $\Sigma_k \beta_k = 0$, $\Sigma_j(\alpha\beta)_{jk} = 0$ ($k = 1,2,3$), and $\Sigma_k(\alpha\beta)_{jk} = 0$ ($j = 1,2$). If we again use dummy variables to denote group membership, the model may be written as

$$Y_{ijk} = \mu + \alpha_1 X_1 + \beta_1 X_2 + \beta_2 X_3 + (\alpha\beta)_{11} X_4 + (\alpha\beta)_{12} X_5 + e_{ijk} \quad (4.11.6)$$

or

$$E(Y) = \mu + \alpha_1 X_1 + \beta_1 X_2 + \beta_2 X_3 + (\alpha\beta)_{11} X_4 + (\alpha\beta)_{12} X_5. \quad (4.11.7)$$

The tests for the main effects and the interaction may be carried out in a manner similar to that employed in the one-way case. Suppose we define

$$\mathbf{Y}=\begin{bmatrix} Y_{111}\\ Y_{211}\\ Y_{311}\\ Y_{112}\\ Y_{212}\\ Y_{312}\\ Y_{113}\\ Y_{213}\\ Y_{313}\\ Y_{121}\\ Y_{221}\\ Y_{321}\\ Y_{122}\\ Y_{222}\\ Y_{322}\\ Y_{123}\\ Y_{223}\\ Y_{323} \end{bmatrix}, \quad \mathbf{X}=\begin{bmatrix} 1 & 1 & 1 & 0 & 1 & 0\\ 1 & 1 & 1 & 0 & 1 & 0\\ 1 & 1 & 1 & 0 & 1 & 0\\ 1 & 1 & 0 & 1 & 0 & 1\\ 1 & 1 & 0 & 1 & 0 & 1\\ 1 & 1 & 0 & 1 & 0 & 1\\ 1 & 1 & -1 & -1 & -1 & -1\\ 1 & 1 & -1 & -1 & -1 & -1\\ 1 & 1 & -1 & -1 & -1 & -1\\ 1 & -1 & 1 & 0 & -1 & 0\\ 1 & -1 & 1 & 0 & -1 & 0\\ 1 & -1 & 1 & 0 & -1 & 0\\ 1 & -1 & 0 & 1 & 0 & -1\\ 1 & -1 & 0 & 1 & 0 & -1\\ 1 & -1 & 0 & 1 & 0 & -1\\ 1 & -1 & -1 & -1 & 1 & 1\\ 1 & -1 & -1 & -1 & 1 & 1\\ 1 & -1 & -1 & -1 & 1 & 1 \end{bmatrix}, \quad \text{and } \hat{\boldsymbol{\beta}}=\begin{bmatrix} \hat{\mu}\\ \hat{\alpha}_1\\ \hat{\beta}_1\\ \hat{\beta}_2\\ \widehat{(\alpha\beta)}_{11}\\ \widehat{(\alpha\beta)}_{12} \end{bmatrix}.$$

Then $\hat{\boldsymbol{\beta}}$ can be obtained using Equation 4.2.5, and SS_W can be computed from Equation 4.11.4. Tests of H_0: $\alpha_1=\alpha_2=0$, H_0: $\beta_1=\beta_2=\beta_3=0$, and H_0: $(\alpha\beta)_{jk}=0$ ($j=1,2$, and $k=1,2,3$), can be carried out by computing the sum of squares within groups for the following reduced models:

1. H_0: $\alpha_1=\alpha_2=0$ (Test of main effect of Aptitude Level):

$$Y_{ijk}=\mu+\beta_1X_2+\beta_2X_3+(\alpha\beta)_{11}X_4+(\alpha\beta)_{12}X_5+e_{ijk}. \quad (4.11.8)$$

2. H_0: $\beta_1=\beta_2=\beta_3=0$ (Test of main effect of Teaching Method):

$$Y_{ijk}=\mu+\alpha_1X_1+(\alpha\beta)_{11}X_4+(\alpha\beta)_{12}X_5+e_{ijk}. \quad (4.11.9)$$

3. H_0: $(\alpha\beta)_{11}=\cdots=(\alpha\beta)_{23}=0$ (Test of method by level interaction):

$$Y_{ijk}=\mu+\alpha_1X_1+\beta_1X_2+\beta_2X_3+e_{ijk}. \quad (4.11.10)$$

As in the one-way analysis, the sum of squares for each of these effects is obtained by subtracting SS_W for the full model from that for the reduced model. Thus, $SS_\alpha=SS_{W\alpha}-SS_W$, $SS_\beta=SS_{W\beta}-SS_W$, and $SS_{\alpha\beta}=SS_{W\alpha\beta}-SS_W$, in which $SS_{W\alpha}$, $SS_{W\beta}$, and $SS_{W\alpha\beta}$ are the sums of squares within groups based on the reduced models in Equations 4.11.8, 4.11.9, and 4.11.10, respectively. Then the ANOVA summary table can be set up in the usual way.

We illustrate this approach with the following data:

		Teaching Method		
		1	2	3
Aptitude Level	1	16	20	12
		9	22	14
		12	27	19
	2	23	17	30
		27	18	34
		30	24	36

For this example we have

$$\mathbf{Y}' = [16\ 9\ 12\ 20\ 22\ 27\ 12\ 14\ 19\ 23\ 27\ 30\ 17\ 18\ 24\ 30\ 34\ 36].$$

The matrix **X** sometimes called the *design* matrix, is the same as that given at the beginning of this section. Then,

$$\mathbf{X}'\mathbf{X} = \begin{bmatrix} 18 & 0 & 0 & 0 & 0 & 0 \\ 0 & 18 & 0 & 0 & 0 & 0 \\ 0 & 0 & 12 & 6 & 0 & 0 \\ 0 & 0 & 6 & 12 & 0 & 0 \\ 0 & 0 & 0 & 0 & 12 & 6 \\ 0 & 0 & 0 & 0 & 6 & 12 \end{bmatrix},$$

$$(\mathbf{X}'\mathbf{X})^{-1} = \begin{bmatrix} \frac{1}{18} & 0 & 0 & 0 & 0 & 0 \\ 0 & \frac{1}{18} & 0 & 0 & 0 & 0 \\ 0 & 0 & \frac{12}{108} & \frac{-6}{108} & 0 & 0 \\ 0 & 0 & \frac{-6}{108} & \frac{12}{108} & 0 & 0 \\ 0 & 0 & 0 & 0 & \frac{12}{108} & \frac{-6}{108} \\ 0 & 0 & 0 & 0 & \frac{-6}{108} & \frac{12}{108} \end{bmatrix},$$

and

$$\mathbf{X}'\mathbf{Y} = \begin{bmatrix} 390 \\ -88 \\ -28 \\ -17 \\ 12 \\ 65 \end{bmatrix}.$$

Therefore,

$$\hat{\beta}=\begin{bmatrix} 21.666667 \\ -4.888889 \\ -2.166667 \\ -.333333 \\ -2.277778 \\ 6.555556 \end{bmatrix}$$

and

$$SS_W = \mathbf{Y'Y} - \boldsymbol{\beta}'\mathbf{X'Y} = 148.666667.$$

Now, to find the sum of squares for Level, SS_α, we remove from $\mathbf{X}$ the column corresponding to α_1 and compute the within-groups sum of squares based on the reduced model. Then,

$$(\mathbf{X'X})^{-1}=\begin{bmatrix} \frac{1}{18} & 0 & 0 & 0 & 0 \\ 0 & \frac{12}{108} & \frac{-6}{108} & 0 & 0 \\ 0 & \frac{-6}{108} & \frac{12}{108} & 0 & 0 \\ 0 & 0 & 0 & \frac{12}{108} & \frac{-6}{108} \\ 0 & 0 & 0 & \frac{-6}{108} & \frac{12}{108} \end{bmatrix} \quad \text{and } \mathbf{X'Y}=\begin{bmatrix} 390 \\ -28 \\ -17 \\ 12 \\ 65 \end{bmatrix},$$

so that

$$SS_{W\beta}=9494 - [21.666667 \quad -2.166667 \quad -.333333 \quad -2.277778 \quad 6.555556]\cdot\begin{bmatrix} 390 \\ -28 \\ -17 \\ 12 \\ 65 \end{bmatrix}$$

$$=578.888889.$$

Therefore, $SS_\alpha = SS_{W\alpha} - SS_W = 578.888889 - 148.666667 = 430.222222$. Similarly, we find that

$$SS_{W\beta}=9494 - [21.666667 \quad -4.888889 \quad -2.277778 \quad 6.555556]\cdot\begin{bmatrix} 390 \\ -88 \\ 12 \\ 65 \end{bmatrix}$$

$$=214.999999,$$

so that

$$SS_\beta = SS_{W\beta} - SS_W = 66.333333$$

and

$$SS_{W\alpha\beta}=9494 - [21.666667 \quad -4.888889 \quad -2.166667 \quad -.333333]\cdot\begin{bmatrix} 390 \\ -88 \\ -28 \\ -17 \end{bmatrix}$$

$$=547.444444,$$

TABLE 4.11.2 ANALYSIS OF VARIANCE OF APTITUDE LEVEL AND TEACHING METHOD

Source	Sum of squares	Degrees of freedom	Mean square	F
Aptitude level (α)	$SS_\alpha = 430.222$	1	430.222	34.73*
Teaching method (β)	$SS_\beta = 66.333$	2	33.167	2.68†
Interaction ($\alpha\beta$)	$SS_{\alpha\beta} = 398.778$	2	199.389	16.09‡
Within cells	$SS_W = 148.667$	12	12.389	
Total	$SS_T = 1044.000$	17		

*$P < .001$
†$P > .10$
‡$P < .005$

which yields $SS_{\alpha\beta} = SS_{W\alpha\beta} - SS_W = 398.777778$. The total sum of squares, SS_T, may be obtained by expressing the model as $Y_{ijk} = \mu + e_{ijk}$, in which case $\mathbf{X}' = [1 \quad 1 \quad 1 \quad \ldots \quad 1]$, that is, a row matrix of 1's equal in number to the number in all groups combined. Then $SS_T = 9494 - (390)(21.666667) = 1044$, and the analysis of variance may be summarized as in Table 4.11.2.

These results are identical to those that would be obtained using the traditional computational formulas for the two-way ANOVA.

The procedure used in this example can easily be adapted for use in analyzing higher-order balanced fixed-effects designs, as well as in carrying out computations for unbalanced fixed effects and certain mixed and random designs. For a more complete discussion of such applications, see, for example, Mendenhall (1968), Ward and Jennings (1973), or Searle (1971).

4.12 EXERCISES

1. Given the following data on a criterion variable, Y, and two independent variables, X_1 and X_2, for a sample of $n = 25$ students in high school:

Student	X_1	X_2	Y	Student	X_1	X_2	Y
1	46	53	283				
2	49	59	277	14	59	57	281
3	45	55	283	15	49	57	283
4	63	63	288	16	46	53	279
5	55	53	287	17	54	56	279
6	62	57	291	18	49	59	294
7	57	57	293	19	59	60	279
8	59	56	291	20	65	68	297
9	60	57	284	21	48	56	283
10	60	63	294	22	63	62	305
11	67	65	294	23	51	60	281
12	49	49	277	24	62	60	281
13	62	65	291	25	58	57	292

(a) Compute the necessary sums of squares and cross-products and set up the normal equations as shown in Equation 4.2.3.
(b) Use Equation 4.2.5 to compute the vector of regression coefficients, $\hat{\boldsymbol{\beta}}$.
(c) Compute the bivariate product-moment correlations, r_{Y1}, r_{Y2}, and r_{12}, between Y, X_1, and X_2. Use these results and Equations 4.2.6 and 4.2.10 to compute the vector of regression coefficient, $\hat{\boldsymbol{\beta}}$. Compare with the result obtained in (b).
(d) Formulate the regression equation for estimating Y based on X_1 and X_2. Use it to obtain the estimate, $\hat{Y}_i$, for each of the 25 students in the sample.
(e) Using the results in (d), compute the product-moment correlation, $r_{Y\hat{Y}}$, between the actual Y values and the estimated Y values. Show that this result is actually the multiple correlation coefficient by independently computing $R_{y.12}$ using Equation 4.3.1 or 4.3.2.
(f) Again using the results in (d), obtain for each of the 25 students in the sample the discrepancy, $Y-\hat{Y}$, between actual and estimated Y values. Use this set of discrepancies to compute

$$s_{y.12}=\sqrt{\frac{\Sigma(Y-\hat{Y})^2}{n-3}}\ .$$

Show that this result is equal to the standard error of estimate obtained by using Equation 4.4.13.
(g) Test the null hypothesis H_0: $\rho_{y.12}=0$, using Equation 4.4.1. Set $\alpha=.05$.
(h) Suppose that a third independent variable, X_3, were being considered as a predictor. The data on X_3 for the sample of 25 students are as follows:

Student	X_3
1	46
2	30
3	28
4	39
5	49
6	37
7	32
8	33
9	39
10	45
11	36
12	38
13	36

Student	X_3
14	44
15	39
16	38
17	34
18	33
19	42
20	47
21	33
22	44
23	27
24	39
25	22

Use Equation 4.4.11 to determine whether X_3 would make a significant contribution to the estimation of Y. Use $\alpha=.05$.
(i) What percent of the variance in Y is explained or accounted for by the information contained in X_1 and X_2? by X_1, X_2, and X_3? On the basis

of this comparison, does X_3 make a useful contribution to the accuracy of estimation of Y? Explain.

(j) Compute the standard error of estimate when X_1, X_2, and X_3 are used as predictors. How does it differ from the result in (f)? What is the meaning of this difference in terms of the contribution of X_3 to the estimation of Y?

2. The Semantic Differential (Osgood et al., 1957) can be scored to yield three scales representing attitude toward mental disorders. These three scales are Evaluation (E), Understandability (U), and Adjustment (A). The data of the following table present bivariate correlations for 128 student nurses who were administered the Semantic Differential upon admission to Nursing School and whose grade-point averages were computed at the end of their third year.

	E	*U*	*A*	*GPA*
E	1.000	.593	.568	.296
U		1.000	.412	.327
A			1.000	.234
GPA				1.000

(a) Calculate the partial correlation between the U scale and GPA, holding E constant.

(b) Calculate the partial correlation between U and GPA, holding both E and A constant.

(c) How do you account for the large reduction in correlation between U and GPA when E is held constant?

(d) Why do you think that also removing the effect of A reduces the correlation between U and GPA even further?

3. The bivariate correlational data for four psychological tests (Xs) used in predicting Y, the final grade point average (GPA) of 122 first-year college students are given below:

	X_2	X_3	X_4	$GPA(Y)$	*Mean*	*S.D.*
X_1	.024	−.055	−.069	−.099	19.467	5.349
X_2		.610	.586	.297	61.426	6.795
X_3			.436	.330	39.106	7.404
X_4				.247	45.600	7.681
$GPA(Y)$					2.780	0.338

(a) Calculate $R_{y.1234}$ using the Crout Method (Appendix D).

(b) Write the multiple regression equation for predicting GPA from scores on the four psychological tests.

(c) Suppose a student had the following set of scores on the four predictors (Xs):

$$X_1 = 30,\ X_2 = 65,\ X_3 = 24,\ X_4 = 49$$

What would be his predicted score on Y? Compute a 95% confidence interval for the Y value of a student with these scores on the four predictors (Xs).

4. A sample of 174 student nurses in a hospital nursing school were administered a 25-item Semantic Differential word list and a Nursing Aptitude Test (NAT). The Semantic Differential yielded three subscale scores: Evaluation (E), Understanding (U), and Adjustment (A). These four measures were used to predict final grade-point average (GPA) at the end of the second year of training. The correlation matrix was:

	NAT	*E*	*U*	*A*	*GPA*
NAT	1.000				
E	−.134	1.000			
U	.067	.419	1.000		
A	−.018	.435	.307	1.000	
GPA	−.017	.164	.094	−.065	1.000
Mean	19.540	62.226	41.626	49.690	2.520
S.D.	5.232	5.504	7.159	7.228	0.458

(a) Calculate* the multiple correlation between GPA and the four predictors and test its significance.
(b) Write the multiple regression equation for predicting GPA from the four predictors.
(c) Calculate the relative contributions of each of the three Semantic Differential subscales (Omit the NAT.).
(d) Comment on the effectiveness of the predictor set in predicting GPA.

5. A battery of six tests was administered to 310 high school students in order to determine whether they would be effective in predicting final grades (Y) in a course in mathematics. The tests were: General Intelligence (X_1), Verbal Fluency (X_2), Quantitative Ability (X_3), Vocabulary (X_4), Logical Reasoning (X_5), and Spatial Relations Ability (X_6). The correlation matrix was:

	X_1	X_2	X_3	X_4	X_5	X_6	Y	*Mean*	*S.D.*
X_1	1.000							38.697	8.470
X_2	.425	1.000						47.842	16.867
X_3	.407	.401	1.000					29.500	13.382
X_4	.242	.497	.277	1.000				10.584	4.190
X_5	.255	.267	.283	.289	1.000			12.403	8.485
X_6	.332	.359	.342	.259	.151	1.000		11.363	3.236
Y	.263	.292	.243	.248	.218	.253	1.000	5.639	1.254

(a) Calculate (using a computer program) the multiple correlation coefficient between the six predictor tests and final grades in mathematics.

*First calculate using the Crout method with a hand or desk calculator. Then check your results using a computer program.

(b) Test the statistical significance of the multiple correlation coefficient obtained in (a).
(c) Calculate 95% confidence intervals for (1) μ_y, (2) $\mu_{y \cdot \mathbf{X}}$ in which $\mathbf{X}' = [40 \quad 50 \quad 30 \quad 15 \quad 20 \quad 15]$, and (3) $Y_{i \cdot \mathbf{X}}$ in which $\mathbf{X}'$ is the same as in (2). Comment on the relative widths of these intervals.

5/Canonical Correlation

In the preceding chapter we saw that if we wish to determine the degree of linear relationship between a single variable, Y, and a linear combination of the variables in a given set of Xs, say $X_1, X_2, \ldots, X_p$, then the multiple correlation coefficient may be an appropriate statistic to use. Let us suppose, however, that we have instead, *two sets*, that is, a set of Ys, say $Y_1, Y_2, \ldots, Y_q$, as well as a set of Xs, and that we wish to determine the degree of linear relationship between linear combinations of variables in the two sets. Then the models discussed in Chapter 4 do not apply. An appropriate statistical procedure in this case is known as *canonical* correlation analysis.

Canonical correlation analysis was developed more than four decades ago by Hotelling (1935, 1936). One may view the method as an extension of multiple correlation analysis; however, it might be more appropriate to view multiple correlation as a special case of canonical correlation. The method has many applications in the behavioral sciences, especially in psychology. For example, suppose one is interested in determining the degree of linear relationship between a set of aptitude variables and a set of achievement variables, or between a set of variables measuring psychological attributes and a set measuring physiological attributes. The method of canonical correlation analysis provides a measure of such relationships in the form of a correlation coefficient. Many applications also occur in the field of psychological testing. If one wishes, for example, to determine the reliability of a test battery or of a multifactor test profile, canonical correlation analysis will assist in providing such information.

5.1 RATIONALE UNDERLYING CANONICAL CORRELATION

In psychological and educational studies we usually find that almost any set of variables we identify measures several different dimensions (traits, attributes, or characteristics). In fact, more than one dimension might be measured by a single variable in a set. It would be extremely rare to find several tests in a battery measuring *exactly* the same dimension; it would be equally rare to find that a test battery was measuring a number of dimensions equal to the number of tests it comprised. Thus, in practice we find that the number of dimensions measured by a set of variables is usually greater than one, but less than the number of variables in the set. A method for determining the number and kinds of dimensions measured by a set of variables is discussed in Chapter 8.

The different dimensions measured by a set of tests are not usually emphasized equally. For example, a set of achievement tests may measure verbal better than quantitative achievement. A set of mathematics tests may measure computational skills better than arithmetic reasoning ability. A set of psychological variables might measure perceptual speed better than it measures associative memory. Thus, the several dimensions measured by any set of variables can usually be ordered at least roughly on the basis of the emphasis accorded them in the set.

Canonical correlation analysis may be viewed conceptually as a stepwise procedure that first selects two dimensions, one from each set of variables, that have a stronger linear relationship than any other such pair. The method then identifies the next most highly correlated pair of dimensions, then the next, etc. The maximum number of such pairs of dimensions identified by the method is equal to the number of variables in the smaller set. Thus, if two sets consisted of two and three variables, respectively, the maximum number of dimensions identified by the method in each set would be two.

Before proceeding to discuss the mathematics of canonical correlation, we should indicate how a dimension is characterized in the method. Again, it is useful to refer to multiple correlation and regression. The regression coefficients in a multiple regression equation are chosen so as to maximize the correlation between Y and a linear combination of the Xs. In general, these regression coefficients differ from one another in magnitude, with the result that some Xs receive more weight than others in the linear combination. Thus, the linear combination of the Xs that best predicts Y can be thought of as defining a particular dimension of the Xs. A different set of weights would produce a different linear combination, defining a different dimension. For example, if the Xs were tests in an aptitude battery, a different dimension of the battery would be emphasized if Y were a verbal achievement test than if Y were a quantitative achievement test. Thus, the weights applied to the Xs, that is, the regression coefficients, define the dimension of the set that is the best predictor of Y.

In canonical correlation, a linear combination of the variables in a set defines *each* dimension measured by the set. Each of these linear combinations is called a *canonical variate*. They differ from one another in the weights they assign to the variables in the set. *Canonical correlations* are product-moment correlations between pairs of canonical variates, each pair consisting of one canonical variate from each set. Thus, each canonical correlation is a measure of the degree of linear relationship between two dimensions, one measured by each set of variables. The maximum number of canonical correlations that may be defined for a given problem is equal to the number of variables in the smaller set.

Not all of the canonical correlations associated with a particular problem are necessarily statistically significant, nor are all of them of practical or theoretical significance. Typically, one finds that only the largest ones are meaningful, the number depending on the sizes and compositions of the sets. Furthermore, the largest canonical correlation may not be between the two dimensions that are measured best by the variables in their respective sets. It may happen that the strongest relationship between sets is between dimensions measured only moderately well, or even poorly, in either or both sets.

5.2 THE MATHEMATICS OF CANONICAL CORRELATION

In the preceding section we indicated that a canonical correlation coefficient is the product-moment correlation between two canonical variates, each of which is a linear combination of the variables in its set. In this section, we first show how this correlation can be formulated, both in terms of matrices of sums of squares and cross-products, and in terms of matrices of bivariate correlations between the variables in each set. We then derive an equation from which the canonical correlation coefficient and the appropriate weights for the canonical variates can be determined. The weights to be calculated are those that maximize the correlation between the canonical variates in a pair. This maximum correlation is the canonical correlation between them.

Let us consider two variables, Z_x and Z_y, the first a linear combination of p and the other of q variables. That is,

$$Z_x = u_1 X_1 + \cdots + u_p X_p$$

and

$$Z_y = v_1 Y_1 + \cdots + v_q Y_q,$$

in which the Xs are the p variables in one set and the Ys are the q variables in the other. The u's and v's are weights in the linear combinations, the values of which are to be determined by the procedures described below. Now, considering the first of these variables, Z_x, we can easily show that its mean, $\bar{Z}_x$, is equal to

$$\bar{Z}_x = u_1 \bar{X}_1 + \cdots + u_p \bar{X}_p.$$

We now define

$$\begin{aligned} z_x = Z_x - \bar{Z}_x &= u_1 X_1 + \cdots + u_p X_p - u_1 \bar{X}_1 - \cdots - u_p \bar{X}_p \\ &= u_1(X_1 - \bar{X}_1) + \cdots + u_p(X_p - \bar{X}_p). \end{aligned}$$

Squaring, and summing over n observations, gives

$$\begin{aligned}\Sigma z_x^2 &= u_1^2\Sigma(X_1-\bar{X}_1)^2+\cdots+u_p^2\Sigma(X_p-\bar{X}_p)^2\\ &\quad+2u_1u_2\Sigma(X_1-\bar{X}_1)(X_2-\bar{X}_2)+\cdots+2u_{p-1}u_p\Sigma(X_{p-1}-\bar{X}_{p-1})(X_p-\bar{X}_p)\\ &= u_1^2S_{11}+\cdots+u_p^2S_{pp}+2u_1u_2S_{12}+\cdots+2u_{p-1}u_pS_{p-1,p},\end{aligned}$$

in which $S_{jk}=\Sigma_{i=1}^n(X_{ij}-\bar{X}_j)(X_{ik}-\bar{X}_k)$; $j,k=1,\ldots,p$. Now let

$$\mathbf{u}=\begin{bmatrix}u_1\\ \vdots\\ u_p\end{bmatrix}\quad\text{and}\quad \mathbf{S}_{xx}=\begin{bmatrix}S_{11} & S_{12} & \cdots & S_{1p}\\ \vdots & \vdots & & \vdots\\ S_{p1} & S_{p2} & \cdots & S_{pp}\end{bmatrix}.$$

Then the expression for Σz_x^2 may be written

$$\Sigma z_x^2=\mathbf{u}'\mathbf{S}_{xx}\mathbf{u}. \tag{5.2.1}$$

In a similar manner, we can show that

$$\Sigma z_y^2=\mathbf{v}'\mathbf{S}_{yy}\mathbf{v} \tag{5.2.2}$$

and that

$$\Sigma z_x z_y=\mathbf{u}'\mathbf{S}_{xy}\mathbf{v}, \tag{5.2.3}$$

where

$$\mathbf{v}=\begin{bmatrix}v_1\\ \vdots\\ v_q\end{bmatrix},\qquad \mathbf{S}_{yy}=\begin{bmatrix}S_{11} & S_{12} & \cdots & S_{1q}\\ \vdots & \vdots & & \vdots\\ S_{q1} & S_{q2} & \cdots & S_{qq}\end{bmatrix},$$

and

$$\mathbf{S}_{xy}=\begin{bmatrix}S_{X_1Y_1} & S_{X_1Y_2} & \cdots & S_{X_1Y_q}\\ \vdots & \vdots & & \vdots\\ S_{X_pY_1} & S_{X_pY_2} & \cdots & S_{X_pY_q}\end{bmatrix}.$$

In the matrix $\mathbf{S}_{yy}$, $S_{jk}=\Sigma_{i=1}^n(Y_{ij}-\bar{Y}_j)(Y_{ik}-\bar{Y}_k)$; $j,k=1,\ldots,q$. In $\mathbf{S}_{xy}$, $S_{X_jY_k}=\Sigma_{i=1}^n(X_{ij}-\bar{X}_j)(Y_{ik}-\bar{Y}_k)$; $j=1,\ldots,p$ and $k=1,\ldots,q$. Now, using Equations 5.2.1, 5.2.2, and 5.2.3, the correlation coefficient $r_{Z_xZ_y}$ can be formulated as

$$r_{Z_xZ_y}=\frac{\Sigma z_xz_y}{\sqrt{\Sigma z_x^2\Sigma z_y^2}}=\frac{\mathbf{u}'\mathbf{S}_{xy}\mathbf{v}}{\sqrt{(\mathbf{u}'\mathbf{S}_{xx}\mathbf{u})(\mathbf{v}'\mathbf{S}_{yy}\mathbf{v})}}.$$

Because we shall usually find it more convenient to work with correlation matrices than with the matrices $\mathbf{S}_{xx}$, $\mathbf{S}_{yy}$, and $\mathbf{S}_{xy}$, consider the following transformation for the case $p=2$. The correlation coefficient between X_1 and X_2 may be written

$$r_{12}=\frac{S_{12}}{\sqrt{S_{11}S_{22}}}.$$

Furthermore, we may write

$$r_{11} = \frac{S_{11}}{\sqrt{S_{11}S_{11}}} = r_{22} = \frac{S_{22}}{\sqrt{S_{22}S_{22}}} = 1.$$

We now define a transformation matrix,

$$\mathbf{D} = \begin{bmatrix} \sqrt{S_{11}} & 0 \\ 0 & \sqrt{S_{22}} \end{bmatrix}.$$

Then it can easily be shown that

$$\mathbf{DRD} = \mathbf{S},$$

where

$$\mathbf{R} = \begin{bmatrix} 1 & r_{12} \\ r_{21} & 1 \end{bmatrix} \quad \text{and} \quad \mathbf{S} = \begin{bmatrix} S_{11} & S_{12} \\ S_{21} & S_{22} \end{bmatrix}.$$

Now, considering the p variables in the X set and the q variables in the Y set, we may define

$$\mathbf{D}_x = \begin{bmatrix} \sqrt{S_{X_1X_1}} & & 0 \\ & \ddots & \\ 0 & & \sqrt{S_{X_pX_p}} \end{bmatrix},$$

$$\mathbf{D}_y = \begin{bmatrix} \sqrt{S_{Y_1Y_1}} & & 0 \\ & \ddots & \\ 0 & & \sqrt{S_{Y_qY_q}} \end{bmatrix},$$

$$\mathbf{R}_{xx} = \begin{bmatrix} 1 & r_{X_1X_2} & \cdots & r_{X_1X_p} \\ r_{X_2X_1} & 1 & \cdots & r_{X_2X_p} \\ \vdots & \vdots & \ddots & \vdots \\ r_{X_pX_1} & r_{X_pX_2} & \cdots & 1 \end{bmatrix},$$

$$\mathbf{R}_{yy} = \begin{bmatrix} 1 & r_{Y_1Y_2} & \cdots & r_{Y_1Y_q} \\ r_{Y_2Y_1} & 1 & \cdots & r_{Y_2Y_q} \\ \vdots & \vdots & \ddots & \vdots \\ r_{Y_qY_1} & r_{Y_qY_2} & \cdots & 1 \end{bmatrix},$$

and

$$\mathbf{R}_{xy} = \begin{bmatrix} r_{X_1Y_1} & r_{X_1Y_2} & \cdots & r_{X_1Y_q} \\ r_{X_2Y_1} & r_{X_2Y_2} & \cdots & r_{X_2Y_q} \\ \vdots & \vdots & \ddots & \vdots \\ r_{X_pY_1} & r_{X_pY_2} & \cdots & r_{X_pY_q} \end{bmatrix}.$$

Then we can rewrite Equations 5.2.1, 5.2.2, and 5.2.3 as

$$\Sigma z_x^2 = \mathbf{u}'\mathbf{D}_x\mathbf{R}_{xx}\mathbf{D}_x\mathbf{u}, \tag{5.2.4}$$

$$\Sigma z_y^2 = \mathbf{v}'\mathbf{D}_y\mathbf{R}_{yy}\mathbf{D}_y\mathbf{v}, \tag{5.2.5}$$

and

$$\Sigma z_x z_y = \mathbf{u}'\mathbf{D}_x\mathbf{R}_{xy}\mathbf{D}_y\mathbf{v}. \tag{5.2.6}$$

To simplify the notation, we set

$$\left.\begin{array}{c}\mathbf{u}'\mathbf{D}_x = \mathbf{c}' \\ \text{and} \\ \mathbf{v}'\mathbf{D}_y = \mathbf{d}'.\end{array}\right\} \tag{5.2.7}$$

Then $\mathbf{D}_x\mathbf{u} = \mathbf{c}$ and $\mathbf{D}_y\mathbf{v} = \mathbf{d}$. Equations 5.2.4, 5.2.5, and 5.2.6 then become

$$\Sigma z_x^2 = \mathbf{c}'\mathbf{R}_{xx}\mathbf{c}, \tag{5.2.8}$$

$$\Sigma z_y^2 = \mathbf{d}'\mathbf{R}_{yy}\mathbf{d}, \tag{5.2.9}$$

and

$$\Sigma z_x z_y = \mathbf{c}'\mathbf{R}_{xy}\mathbf{d}. \tag{5.2.10}$$

Making use of Equations 5.2.8, 5.2.9, and 5.2.10, we may express the product-moment correlation between Z_x and Z_y as

$$r_{Z_xZ_y} = \frac{\Sigma z_x z_y}{\sqrt{\Sigma z_x^2 \Sigma z_y^2}} = \frac{\mathbf{c}'\mathbf{R}_{xy}\mathbf{d}}{\sqrt{(\mathbf{c}'\mathbf{R}_{xx}\mathbf{c})(\mathbf{d}'\mathbf{R}_{yy}\mathbf{d})}}.$$

The problem now is to choose the weights **c** and **d** that maximize this expression.

Because multiplication of Z_x and Z_y by any arbitrary constants would not change the value of the correlation coefficient between them, it is necessary to determine **c** and **d** only up to proportionality constants. Therefore, it will be convenient to choose **c** and **d** under the condition

$$\mathbf{c}'\mathbf{R}_{xx}\mathbf{c} = \mathbf{d}'\mathbf{R}_{yy}\mathbf{d} = 1. \tag{5.2.11}$$

Then the function, F, to be maximized is

$$F = \mathbf{c}'\mathbf{R}_{xy}\mathbf{d} - \frac{\sqrt{\lambda}}{2}\mathbf{c}'\mathbf{R}_{xx}\mathbf{c} - \frac{\sqrt{\gamma}}{2}\mathbf{d}'\mathbf{R}_{yy}\mathbf{d},$$

where $\sqrt{\lambda}/2$ and $\sqrt{\gamma}/2$ are LaGrange multipliers (see Appendix B). Taking partial derivatives of F with respect to **c** and **d**′, and setting each of these equal to zero, we have

$$\mathbf{R}_{xy}\mathbf{d} - \sqrt{\lambda}\,\mathbf{R}_{xx}\mathbf{c} = \mathbf{0} \tag{5.2.12}$$

and

$$\mathbf{c}'\mathbf{R}_{xy} - \sqrt{\gamma}\,\mathbf{d}'\mathbf{R}_{yy} = \mathbf{0}'. \tag{5.2.13}$$

Premultiplying Equation 5.2.12 by **c**′ and postmultiplying Equation 5.2.13 by **d** gives

$$\mathbf{c}'\mathbf{R}_{xy}\mathbf{d} - \sqrt{\lambda}\,\mathbf{c}'\mathbf{R}_{xx}\mathbf{c} = 0$$

and

$$\mathbf{c}'\mathbf{R}_{xy}\mathbf{d} - \sqrt{\gamma}\ \mathbf{d}'\mathbf{R}_{yy}\mathbf{d} = 0,$$

which, under the condition 5.2.11, yields

$$\mathbf{c}'\mathbf{R}_{xy}\mathbf{d} = \sqrt{\lambda} = \sqrt{\gamma}\ .$$

Since this result shows that $\sqrt{\gamma} = \sqrt{\lambda}$, we may replace $\sqrt{\gamma}$ by $\sqrt{\lambda}$ in Equation 5.2.13. Then, because $(\mathbf{c}'\mathbf{R}_{xy})' = \mathbf{R}_{yx}\mathbf{c}$, where $\mathbf{R}_{yx}$ denotes $\mathbf{R}'_{xy}$, and because $(\mathbf{d}'\mathbf{R}_{yy})' = \mathbf{R}_{yy}\mathbf{d}$, we can write

$$\mathbf{R}_{yx}\mathbf{c} = \sqrt{\lambda}\ \mathbf{R}_{yy}\mathbf{d}. \tag{5.2.14}$$

Solving Equation 5.2.12 for **c**, we have

$$\mathbf{c} = \frac{1}{\sqrt{\lambda}}\mathbf{R}_{xx}^{-1}\mathbf{R}_{xy}\mathbf{d}. \tag{5.2.15}$$

Substituting for **c** in Equation 5.2.14 gives

$$\mathbf{R}_{yx}\frac{1}{\sqrt{\lambda}}\mathbf{R}_{xx}^{-1}\mathbf{R}_{xy}\mathbf{d} = \sqrt{\lambda}\ \mathbf{R}_{yy}\mathbf{d},$$

which, when premultiplied by $\sqrt{\lambda}\ \mathbf{R}_{yy}^{-1}$, becomes

$$\sqrt{\lambda}\ \mathbf{R}_{yy}^{-1}\mathbf{R}_{yx}\frac{1}{\sqrt{\lambda}}\mathbf{R}_{xx}^{-1}\mathbf{R}_{xy}\mathbf{d} = \lambda\mathbf{d}.$$

On rearranging, we have

$$(\mathbf{R}_{yy}^{-1}\mathbf{R}_{yx}\mathbf{R}_{xx}^{-1}\mathbf{R}_{xy} - \lambda\mathbf{I})\mathbf{d} = \mathbf{0}. \tag{5.2.16}$$

The number of non-zero roots (values of λ) obtained by solving Equation 5.2.16 is equal to p if $p \leqq q$ or to q if $q \leqq p$. In other words, the number of values of λ is equal to the smaller of p and q. The *largest* value of λ, which we shall denote by λ_1, is the square of the maximum correlation between the variables Z_x and Z_y, defined at the beginning of this section. We could denote this correlation by $r_{Z_xZ_y}$; instead, however, we shall use the symbol R_{c_1}, to distinguish it from the infinite number of other values of $r_{Z_xZ_y}$ that might be obtained by choosing the weights on the Xs and Ys differently. Thus, $R_{c_1} = \sqrt{\lambda_1}$ is the canonical correlation between that pair of canonical variates, one from each set, which, among all possible pairs, has the strongest linear relationship. We denote these two canonical variates by Z_{x1} and Z_{y1}.

We denote the other values of λ obtained from solving Equation 5.2.16, in order of descending magnitude, by $\lambda_2, \lambda_3, \ldots, \lambda_r$, where r is the number of non-zero values obtained. The corresponding canonical correlations are $R_{c_2}, \ldots, R_{c_r}$, and the corresponding pairs of canonical variates are $(Z_{x2}, Z_{y2}), \ldots, (Z_{xr}, Z_{yr})$. As we stated above, the canonical correlation, R_{c_1}, between Z_{x1} and Z_{y1}, is the largest correlation between any pair of canonical variates that might be based on the sample data. Among all possible remaining linear combinations, there is a set consisting of those that are not correlated with either Z_{x1} or Z_{y1}. The pair Z_{x2} and Z_{y2} have the maximum correlation, R_{c_2}, among pairs in that set. In general, R_{c_j} is the correlation between the pair Z_{xj} and Z_{yj} that are most highly correlated with each other, but are uncorrelated with Z_{xk} and Z_{yk} $(j,k=1,\ldots,r;k\neq j)$. Thus, the Z_xs are uncorrelated with each other and with all but one of the Z_ys, while the Z_ys are uncorrelated with each other and with all but one of the Z_xs.

Each pair of canonical variates, Z_{xj} and Z_{yj}, has associated with it vectors $\mathbf{c}_j$ and $\mathbf{d}_j$ of weights on the variables in the X and Y sets, respectively. Following the computation of λ_j, the vector $\mathbf{d}_j$ may be computed from Equation 5.2.16. Then $\mathbf{c}_j$ may be computed by substitution for $\mathbf{d}_j$ in Equation 5.2.15. These computational procedures are illustrated in the next section. It is important to note that the elements of $\mathbf{c}_j$ and $\mathbf{d}_j$ are standardized, as we can see from Equations 5.2.7, whereas those in $\mathbf{u}$ and $\mathbf{v}$, defined earlier, are not. Therefore, the elements of $\mathbf{u}$ and $\mathbf{v}$ for the jth pair of canonical variates, which can be obtained from $\mathbf{u}_j=\mathbf{D}_x^{-1}\mathbf{c}_j$ and $\mathbf{v}_j=\mathbf{D}_y^{-1}\mathbf{d}_j$, should be used with *raw* X and Y scores in the expressions, $Z_{xj}=u_{j1}X_1+\cdots+u_{jp}X_p$ and $Z_{yj}=v_{j1}Y_1+\cdots+v_{jq}Y_q$. The corresponding expressions in standard form for the jth pair of canonical variates are $x_j=c_{j1}z_{X1}+\cdots+c_{jp}z_{X_p}$ and $y_j=d_{j1}z_{Y_1}+\cdots+d_{jq}z_{Y_q}$, in which x_j and y_j denote the standardized forms of Z_{xj} and Z_{yj}, respectively, and the z_{X_y}s and z_{Y_j}s are the standardized forms of the Xs and Ys. In the following sections we shall use the standaradized vectors, $\mathbf{c}_j$ and $\mathbf{d}_j$, because they are more useful than $\mathbf{u}_j$ and $\mathbf{v}_j$ for interpreting the results of canonical correlation analyses.

5.3 COMPUTATION OF CANONICAL CORRELATION COEFFICIENTS

As we indicated in Section 5.2, the computation of R_{c_j}, $\mathbf{c}_j$, and $\mathbf{d}_j$, involves the solution of Equation 5.2.16 for λ_j. To simplify some of the matrix expressions that follow, we first define the matrix

$$\mathbf{M}=\mathbf{R}_{yy}^{-1}\mathbf{R}_{yx}\mathbf{R}_{xx}^{-1}\mathbf{R}_{xy}.$$

Then Equation 5.2.16 becomes

$$(\mathbf{M}-\lambda\mathbf{I})\mathbf{d}=\mathbf{0}. \tag{5.3.1}$$

The solution of Equation 5.3.1 for λ may be carried out using the method discussed in Appendix A for the solution of equations of the general form $(\mathbf{A}-\lambda\mathbf{I})\mathbf{a}=\mathbf{0}$. We first state the *characteristic equation* of matrix $\mathbf{M}$ as

$$|\mathbf{M}-\lambda\mathbf{I}|=0. \tag{5.3.2}$$

The roots, λ, of this equation are called the *characteristic roots* or *eigenvalues* of $\mathbf{M}$.

The solution of an equation of the form of 5.3.2 can be shown most easily by means of an example. Consider a problem in which the set of Xs and the set of Ys each consist of only two variables, that is, $p=q=2$. Suppose that, based on a sample of $n=100$ cases, the matrices $\mathbf{R}_{xx}$, $\mathbf{R}_{yy}$, $\mathbf{R}_{xy}$, and $\mathbf{R}_{yx}$, were computed and found to be

$$\left[\begin{array}{c|c} \mathbf{R}_{xx} & \mathbf{R}_{xy} \\ \hline \mathbf{R}_{yx} & \mathbf{R}_{yy} \end{array}\right] = \left[\begin{array}{cc|cc} 1.0 & .3 & .5 & .4 \\ .3 & 1.0 & .5 & .6 \\ \hline .5 & .5 & 1.0 & .5 \\ .4 & .6 & .5 & 1.0 \end{array}\right].$$

Computing $\mathbf{M}=\mathbf{R}_{yy}^{-1}\mathbf{R}_{yx}\mathbf{R}_{xx}^{-1}\mathbf{R}_{xy}$, we find

$$\mathbf{M}=\begin{bmatrix} .256410 & .237363 \\ .256410 & .294506 \end{bmatrix}.$$

Then

$$|\mathbf{M}-\lambda\mathbf{I}| = \begin{vmatrix} .256410-\lambda & .237363 \\ .256410 & .294506-\lambda \end{vmatrix}$$
$$=(.256410-\lambda)(.294506-\lambda)-(.256410)(.237363).$$

Multiplying and setting the result equal to zero, we have the quadratic equation

$$\lambda^2-.550916\lambda+.014652=0,$$

which, when solved for λ yields $\lambda_1=.522895$ and $\lambda_2=.028021$. Because $R_{c_j}=\sqrt{\lambda_j}$, we find that the two values for the canonical correlation are $R_{c_1}=.723$ and $R_{c_2}=.167$. For the same reason discussed earlier in connection with the multiple correlation coefficient, $R_{y.1\ldots p}$, we always take the positive square root of λ_j. Thus, the range of values of R_{c_j} is from 0 to +1.

Because $p=q=2$ in this example, the determinantal equation, $|\mathbf{M}-\lambda\mathbf{I}|=0$, is quadratic. If the smaller of p and q had been 3 instead, then this equation would have been cubic, yielding three values of λ. As we indicated, the number of non-zero roots (values of λ) will be p if $p\leqq q$, and q if $q\leqq p$.

Because the data of this example are fictitious and used here only for illustrating computations, we cannot sensibly interpret the results. However, some useful observations can be made. One is that $R_{c_1}=.723$ is larger than any of the correlations in $\mathbf{R}_{xy}$. It will always be true that R_{c_1} is at least as large as the largest (in absolute value) bivariate correlation in $\mathbf{R}_{xy}$. A second observation is that R_{c_1} is considerably larger than R_{c_2}. In fact, as we will show below, only R_{c_1} is statistically significant. Apparently, each variable set is measuring only one dimension that is meaningfully related to a dimension of the other set. This is quite likely to be the case when the sets are very small. Finally, we should mention that $\lambda_1=R_{c_1}^2$ and $\lambda_2=R_{c_2}^2$ can be interpreted just as the square of any product-moment correlation, that is, as indicating the proportion of variance of either variable accounted for by the other. The value $\lambda_1=.522895$ means that about 52% of the variance of Z_{x1} is explained by Z_{y1}, and vice versa. The value $\lambda_2=.028021$ can be interpreted similarly for Z_{x2} and Z_{y2}, the other canonical variates of the sets X and Y, respectively.

5.4 SIGNIFICANCE TESTS IN CANONICAL CORRELATION

A test for the statistical significance of canonical correlations can be carried out using Wilks' (1932) Λ criterion, originally proposed for testing hypotheses about the centroids of several multivariate normal populations. For a discussion of the rationale for the use of the Λ criterion in such tests, see Anderson (1958), Chapter 8.

It can be shown (see, for example, Tatsuoka, 1971, Chapter 6) that Λ may be expressed in terms of λ_j as

$$\Lambda = \prod_{j=1}^{r} (1-\lambda_j), \tag{5.4.1}$$

in which r is the number of nonzero values of λ obtained by solving Equation 5.3.2. Ordinarily, of course, r will equal the smaller of p and q.

To test the significance of the maximum canonical correlation, R_{c_1}, we first compute Λ (which we now denote Λ_1) using Equation 5.4.1. We then substitute for Λ_1 in

$$V_1 = -\left[n-1-\tfrac{1}{2}(p+q+1)\right]\ln\Lambda_1. \tag{5.4.2}$$

The statistic of Equation 5.4.2 actually provides a test of the null hypothesis that all r of the population canonical correlations are zero, H_0: $\rho_{c_1} = \cdots = \rho_{c_r} = 0$. Bartlett (1947) showed that under the assumption of a multivariate normal distribution and when H_0 is true, V_1 has approximately a chi-square distribution with pq degrees of freedom for large n. Because we recommend a sample size of $n > 100$ or n greater than 20 times the number of variables for any correlational study (see Chapter 4), this statistic will give a good approximation if our recommendation is followed.

If the test based on Bartlett's V_1 results in rejection of the above null hypothesis, we conclude that $\rho_{c_1} > 0$, since this is the population parameter corresponding to the largest root obtained by solving Equation 5.3.2, λ_1, and, therefore, to the largest sample canonical correlation, R_{c_1}. To see how we may test the remaining canonical correlations for significance, let us express Equation 5.4.2 as

$$V_1 = -\left[n-1-\frac{1}{2}(p+q+1)\right]\sum_{j=1}^{r}\ln(1-\lambda_j).$$

V_1 is thus seen to be the sum of r terms, all of them, by definition, mutually uncorrelated. In fact, each of these functions can be shown to have an approximate chi-square distribution. It follows that we may remove any of these functions from the sum and that the remaining sum will have an approximate chi-square distribution. For example, to test the significance of R_{c_2}, we may compute

$$\Lambda_2 = \prod_{j=2}^{r} (1-\lambda_j),$$

and

$$V_2 = -\left[n-1-\tfrac{1}{2}(p+q+1)\right]\ln\Lambda_2,$$

which, under H_0: $\rho_{c_2} = \cdots = \rho_{c_r} = 0$, has an approximate chi-square distribution with $(p-1)(q-1)$ degrees of freedom. In general, to test the significance of the kth canonical correlation ($k = 1, \ldots, r$), one may compute

$$\Lambda_k = \prod_{j=k}^{r}(1-\lambda_j) \tag{5.4.3}$$

and

$$V_k = -\left[n-1-\tfrac{1}{2}(p+q+1)\right]\ln\Lambda_k, \tag{5.4.4}$$

which, under H_0: $\rho_{c_k} = \rho_{c_{k+1}} = \cdots = \rho_{c_r} = 0$, has an approximate chi-square distribution with degrees of freedom equal to $(p-k+1)(q-k+1)$.

For the illustrative problem of the preceding section we found that $\lambda_1 = .522895$. Then

$$\Lambda_1 = \prod_{j=1}^{2}(1-\lambda_j) = (1-.522895)(1-.028021) = .463736,$$

and

$$V_1 = -\left[99-\tfrac{1}{2}(5)\right](-.76844) = 74.15.$$

With $pq = 4$ degrees of freedom, this value of chi-square is significant well beyond the .001 level.

To test the significance of $\lambda_2 = .028021$, we compute $\Lambda_2 = 1 - .028021 = .971979$. Then $V_2 = -[99 - .5(5)](-.02842) = 2.74$. With $(p-1)(q-1) = 1$ degree of freedom, this result is not significant at the .05 level.

5.5 DETERMINATION OF THE WEIGHTS c AND d

After computing the values of λ and testing their significance, the next step in a canonical correlation analysis is to compute the vectors of weights, **c** and **d**, which define pairs of canonical variates. As we indicated in Section 5.1, the weights on the variables in a set depend on the emphasis to be accorded them in defining a particular dimension of the set. The computational steps that follow result in vectors of weights **c** and **d**, on the X and Y sets, respectively, that produce the correlations $R_{c_1}, R_{c_2}, \ldots, R_{c_r}$ between pairs of canonical variates in the two sets.

In Equation 5.3.1, $(\mathbf{M}-\lambda\mathbf{I})\mathbf{d} = \mathbf{0}$, each characteristic root or eigenvalue, λ_j, has associated with it a *characteristic vector*, $\mathbf{d}_j$, called an *eigenvector*. The numerical values that make up this eigenvector are the weights on the set of Y variables that define the canonical variate Z_{yj}. To compute the eigenvector $\mathbf{d}_j$ corresponding to λ_j, we carry out the following steps:

1. In the expression $\mathbf{M}-\lambda_j\mathbf{I}$, substitute the value of λ_j obtained by the method given in Section 5.3.

2. Compute the cofactors (see Appendix A) of the elements in *any* row of $\mathbf{M}-\lambda_j\mathbf{I}$. Denote the vector of these cofactors by $\mathbf{f}$.
3. Compute $\theta=\sqrt{\mathbf{f}'\mathbf{R}_{yy}\mathbf{f}}$. This step is necessary to obtain values of $\mathbf{d}_j$ (and eventually $\mathbf{c}_j$) scaled so that the canonical variates Z_{xj} and Z_{yj} will have unit variance.
4. Compute $\mathbf{d}_j=(1/\theta)\mathbf{f}$. Once $\mathbf{d}_j$ has been computed, $\mathbf{c}_j$ can be obtained by substitution in Equation 5.2.15.

For the example of Section 5.3, the results for $\lambda_1=.522895$ are:

1. $\mathbf{M}-.522895\mathbf{I}=\begin{bmatrix} -.266485 & .237363 \\ .256410 & -.228389 \end{bmatrix}.$
2. The cofactors of the first row of $\mathbf{M}-\lambda_1\mathbf{I}$ are $\mathbf{f}'=[\,-.228389 \quad -.256410\,]$.
3. $\theta=\sqrt{\mathbf{f}'\mathbf{R}_{yy}\mathbf{f}}=\sqrt{.176469}=.420082.$
4. $\mathbf{d}_1=\dfrac{1}{\theta}\mathbf{f}=\begin{bmatrix} -.228389/.420082 \\ -.256410/.420082 \end{bmatrix}=\begin{bmatrix} -.543677 \\ -.610381 \end{bmatrix}.$

$$\mathbf{c}_1=\frac{1}{\sqrt{\lambda}}\mathbf{R}_{xx}^{-1}\mathbf{R}_{xy}\mathbf{d}=\begin{bmatrix} -.493243 \\ -.734414 \end{bmatrix}.$$

If we had used the cofactors of the second row instead of the first, we would have found, for Step 2,

2. $\mathbf{f}'=[\,-.237363 \quad -.266485\,]$.

Then

3. $\theta=\sqrt{\mathbf{f}'\mathbf{R}_{yy}\mathbf{f}}=\sqrt{.190609}=.436588.$
4. $\mathbf{d}_1=\dfrac{1}{\theta}\mathbf{f}=\begin{bmatrix} -.543677 \\ -.610381 \end{bmatrix},$

which is exactly the same as the values obtained above, using the first row of $\mathbf{M}-\lambda_1\mathbf{I}$.

As computed here the signs of all elements of $\mathbf{c}_1$ and $\mathbf{d}_1$ are negative. One can, however, reverse all the signs of *both* $\mathbf{c}$ and $\mathbf{d}$ if desired. For example, let us suppose that the trait measured by the first canonical variate of the set of Xs in the above problem were dishonesty, and by that of the Ys untruthfulness. If the signs of both $\mathbf{c}$ and $\mathbf{d}$ were reversed, these variates would be measures of honesty and truthfulness, respectively. Thus, a change of signs might facilitate discussion and interpretation of results.

We will find it useful in later sections to have available the values $\mathbf{c}_2$ and $\mathbf{d}_2$, corresponding to $R_{c_2}=.167$, the canonical correlation between the second pair of canonical variates, Z_{x2} and Z_{y2}. Using the same computational procedure we described above, we obtain

$$\mathbf{c}_2=\begin{bmatrix} .924993 \\ -.748022 \end{bmatrix} \quad \text{and} \quad \mathbf{d}_2=\begin{bmatrix} 1.018699 \\ -.980188 \end{bmatrix}.$$

Note that these differ noticeably from $\mathbf{c}_1$ and $\mathbf{d}_1$, indicating that the Xs and Ys receive different emphases in defining the two different pairs of canonical variates involved in this problem. In the next section we shall discuss the use of these weights in defining the canonical variates and in interpreting the results of canonical correlation analysis.

5.6 INTERPRETING RESULTS OF CANONICAL CORRELATION ANALYSIS

As we have shown above, canonical correlation analysis leads to a set of correlation coefficients and a set of standardized weights, **c** and **d**, on the X and Y variables corresponding to each coefficient. Each of the canonical correlations, R_{c_j}, may be interpreted in essentially the same way as any product-moment correlation. A minor exception to this statement is that R_{c_j} is always positive, by definition; furthermore, it would not usually make sense to speak of the directionality of the relationship when some Xs and Ys are correlated positively, and some negatively, with their respective canonical variates. $R_{c_j}^2$ may still be interpreted as an expression of the proportion of variance of Z_{xj} explained by Z_{yj}, and vice versa. Note that $R_{c_j}^2$ indicates *only* the proportion of variance of a particular canonical variate explained by the corresponding canonical variate in the other set.

The sets of weights $\mathbf{c}_j$ and $\mathbf{d}_j$, corresponding to the jth canonical correlation, may be interpreted in roughly the same way as the standardized regression coefficients in multiple regression analysis. Their relative sizes indicate the emphasis accorded each variable in a set compared with other variables in the set. A particular variable may recieve a larger weight in defining one canonical variate, say Z_{x1}, than in defining another, say Z_{x2} or Z_{x3}. The relative sizes of the weights, as well as their signs, are the basis for defining the canonical variates, that is, for determining what each is measuring. For example, if a measure of quantitative achievement received a much larger weight than a measure of verbal achievement in the vector **c**, defining a particular canonical variate, then we would conclude that the variate was measuring primarily quantitative achievement.

In the example introduced in Section 5.3, we found that the vectors of weights, $\mathbf{c}_1, \mathbf{d}_1$ and $\mathbf{c}_2, \mathbf{d}_2$, differed noticeably in terms of the relative magnitudes of weights on the Xs and Ys. Considering just

$$\mathbf{c}_1 = \begin{bmatrix} -.493243 \\ -.734414 \end{bmatrix} \quad \text{and} \quad \mathbf{c}_2 = \begin{bmatrix} .924993 \\ -.748022 \end{bmatrix},$$

for example, we note first that the elements of $\mathbf{c}_1$ have the same sign, while those of $\mathbf{c}_2$ have different signs. Thus, X_1 and X_2 are both negatively (or positively) related to the canonical variate, Z_{x1}, the sign depending on how Z_{x1} is defined. On the other hand, if X_1 is positively related to Z_{x2}, then X_2 is negatively related to Z_{x2}, and vice versa, again depending on how Z_{x2} is defined. Furthermore, in $\mathbf{c}_1$ the ratio of the weights on X_1 and X_2 is about 1 : 1.5, whereas in $\mathbf{c}_2$ it is about 1 : .8. Thus, X_2 would play a more important role than X_1 in defining Z_{x1}, but a somewhat less important role in defining Z_{x2}.

5.6.1 The Structure Vectors

Although the canonical correlations and their associated weights, **c** and **d**, are of primary interest, certain other statistics may be helpful in interpreting the results of canonical correlation analyses. One of these is the vector of correlations between each variable of a set and any one of the canonical variates of the set. Such a vector of correlations is referred to as a *structure* vector, a term borrowed from factor analysis (see Chapter 8), in which the structure matrix is the matrix of correlations between the variables and the factors. We may write expressions for the ith individual's values on the jth canonical variates of the sets of Xs and Ys as

$$x_{ij} = c_{j1} z_{X_1 i} + \cdots + c_{jp} z_{X_p i}$$

and

$$y_{ij} = d_{j1} z_{Y_1 i} + \cdots + d_{jq} z_{Y_q i},$$

in which x_{ij} and y_{ij} denote the standardized forms of Z_{xij} and Z_{yij}, respectively, and the $z_{X_j i}$s and $z_{Y_j i}$s are the standardized forms of the individual's values on the Xs and Ys, respectively. If we now define $\mathbf{s}_{Xj}$ as the vector of correlations between the Xs and x_j, that is, the structure vector for the set of Xs, then we may write

$$\mathbf{s}_{Xj} = \frac{1}{n} \sum_{i=1}^{n} \mathbf{z}_{Xi} x_{ij}, \qquad (5.6.1)$$

in which $\mathbf{z}_{Xi}$ is the vector of standardized X values for individual i and x_{ij} is the value of the jth canonical variate for individual i. Using matrix notation, the above expressions for x_{ij} and y_{ij} may be written

$$x_{ij} = \mathbf{c}_j' \mathbf{z}_{Xi}$$

and

$$y_{ij} = \mathbf{d}_j' \mathbf{z}_{Yi},$$

in which $\mathbf{z}_{Yi}$ is the vector of standardized Y values for individual i. Now, substituting for x_{ij} in Equation 5.6.1, we have

$$\mathbf{s}_{Xj} = \frac{1}{n} \sum_{i=1}^{n} \mathbf{z}_{Xi} \mathbf{c}_j' \mathbf{z}_{Xi}.$$

Because $\mathbf{c}_j' \mathbf{z}_{Xi} = \mathbf{z}_{Xj}' \mathbf{c}_j$, substitution yields

$$\mathbf{s}_{Xj} = \frac{1}{n} \sum_{i=1}^{n} \mathbf{z}_{Xi} \mathbf{z}_{Xi}' \mathbf{c}_j = \mathbf{R}_{xx} \mathbf{c}_j. \qquad (5.6.2)$$

Similarly, we may show that the structure vector for the set of Y variables is

$$\mathbf{s}_{Yj} = \mathbf{R}_{yy} \mathbf{d}_j. \qquad (5.6.3)$$

The structure vectors $\mathbf{s}_{Xj}$ and $\mathbf{s}_{Yj}$ are useful in interpreting results of canonical correlations because they show which variables in a set are most highly correlated with a given canonical variate of the set and which least. Furthermore, the squares of the elements of these vectors indicate the proportion of the variance of each X or Y variable explained or accounted for by the canonical variate, x_j or y_j. As we shall see, this information is useful in judging the importance of x_j and y_j in their respective sets of variables. It also aids in determining what is actually being measured by the canonical variates of both sets, and thus helps to clarify the nature of the relationships between the sets.

5.6.2 Redundancy

Another useful concept in connection with the interpretation of canonical correlation analysis is that of redundancy. *Redundancy analysis*, as it has become known, was proposed independently by Stewart and Love (1968) and by Miller (1969). The basic concept of redundancy in the context of canonical correlation analysis is quite simple. Suppose one wishes to answer the question, "To what extent does information contained in the set of X variables duplicate information in the set of Y variables, assuming that the set Y is already available?" In other words, to what extent is the information in X redundant, given that the information in Y is available? The question could also be put in the other direction, that is, "To what extent is the set Y redundant, given X?" That the answers to these questions are not necessarily the same can be seen by considering the case in which the set Y is a subset of the set X. Then Y is totally redundant, given X, but X is obviously *not* totally redundant, given Y.

The answers to the questions in the preceding paragraph provide information somewhat different than that provided by examining the values of $R_{c_j}^2$. The value of $R_{c_1}^2$, for example, indicates the proportion of variance in x_1, the first canonical variate of the set X, that is explained or accounted for by y_1, the first canonical variate of the set Y. This is useful information, but it tells us something about only x_1 and y_1 and the relationship between the dimensions they measure. Thus, the interpretation of $R_{c_1}^2$ depends on the importance of x_1 and y_1 in their respective sets. If both are of major importance, accounting for large proportions of the variances of the X and Y variables, then $R_{c_1}^2$ says a great deal more about the strength of the relationship between the sets than if either x_1 or y_1 or both are of only minor importance within their respective sets. Both Miller and the Stewart and Love team proposed to quantify the concept of redundancy by combining the measures of explained variance, $R_{c_j}^2$, with measures of the importance of the x_j and the y_j in explaining variance within their respective sets. The result of their efforts is the interpretation of redundancy in the following quantitative terms.

Let us consider the first pair of canonical variates, x_1 and y_1, whose correlation is R_{c_1}. Taking x_1 first, we may express its degree of importance in its own set in terms of the proportion of the total variance of the Xs for which it accounts. In standard form, each X has variance equal to one, so the total variance of the Xs is p. The proportion to be determined, then, is equal to the sum of the squared correlations between the Xs and x_1, divided by p. Because these correlations are exactly those contained in the structure vector $\mathbf{s}_{X1}$, defined in Section 5.6.1, the proportion desired can be written $\mathbf{s}'_{X1}\mathbf{s}_{X1}/p$. The squared correlations in the numerator of this expression range in value from 0 to +1, and so their sum, when divided by p, also ranges from 0 to +1. The analogous expression for y_1 is $\mathbf{s}'_{Y1}\mathbf{s}_{Y1}/q$.

Stewart and Love's measure of the redundancy of x_1, given that the information in y_1 is available, is simply the product

$$R_{dX_1} = \frac{\mathbf{s}'_{X1}\mathbf{s}_{X1}}{p} R^2_{c_1}. \tag{5.6.4}$$

The factor $\mathbf{s}'_{X1}\mathbf{s}_{X1}/p$ thus adjusts $R^2_{c_1}$ according to the relative importance of x_1 in explaining the variance in the set of Xs. A similar development leads to the corresponding measure,

$$R_{dY_1} = \frac{\mathbf{s}'_{Y1}\mathbf{s}_{Y1}}{q} R^2_{c_1}, \tag{5.6.5}$$

of the redundancy of y_1, given that x_1 is available. Of course, analogous measures can be formulated for other pairs of canonical variates.

The total redundancy of the set X, given that the entire set of canonical variates based on Y is available, is obtained by computing R_{dX_j} and R_{dY_j} for each pair of canonical variates, x_j and y_j, and summing them. Thus,

$$R_{dX} = \sum_{j=1}^{r} R_{dX_j} \tag{5.6.6}$$

and

$$R_{dY} = \sum_{j=1}^{r} R_{dY_j}, \tag{5.6.7}$$

where r is the number of nonzero values of λ_j.

Nicewander and Wood (1974, 1975) have objected to this redundancy index on the grounds that it has no theoretical or practical use and that its mathematical basis is extremely weak. Miller (1975) argued, in defense of the index, that Nicewander and Wood had misquoted Stewart and Love and misinterpreted certain of the statements in their 1968 publication of it. Gleason (1976) presented the mathematical basis for a generalization of the index to forms of canonical analysis other than canonical correlation. Gleason concludes by stating, "Thus, one must take exception to the objections of Nicewander and Wood. For although Stewart and Love may not have provided the relevant derivation, their redundancy index rests on a sound mathematical foundation. Moreover, the theoretical and practical significance of the index as a measure of the extent to which one matrix can be reconstructed from another seems indisputable."

We believe that redundancy is an interesting and useful concept in the proper interpretation of the relationships between sets of variables. The acceptability of the Stewart-Love-Miller quantifications of the concept depends primarily on their utility in affording more precise interpretation of the results of canonical correlation analysis. Equations 5.6.4 and 5.6.5 show that R_{dX} and R_{dY} range from 0, when there is no redundancy, to +1, when there is total redundancy. The interpretation of the intermediate range of values of the index is less clear. However, we suggest that if the index, as defined by Stewart and Love, is useful to the researcher in clarifying relationships between sets of variables, then it should be used.

To illustrate the computation of the structure vectors and the redundancy measures, we return to the data of the example presented earlier in this chapter. The structure vectors for the first canonical variates, x_1 and y_1, are:

$$\mathbf{s}_{X1}=\mathbf{R}_{xx}\mathbf{c}_1=\begin{bmatrix} 1.0 & .3 \\ .3 & 1.0 \end{bmatrix}\cdot\begin{bmatrix} .493243 \\ .734414 \end{bmatrix}=\begin{bmatrix} .713567 \\ .882387 \end{bmatrix}$$

and

$$\mathbf{s}_{Y1}=\mathbf{R}_{yy}\mathbf{d}_1=\begin{bmatrix} 1.0 & .5 \\ .5 & 1.0 \end{bmatrix}\cdot\begin{bmatrix} .543677 \\ .610381 \end{bmatrix}=\begin{bmatrix} .848868 \\ .882220 \end{bmatrix}.$$

Note that the signs of both $\mathbf{c}_1$ and $\mathbf{d}_1$ have been changed so that the correlations in $\mathbf{s}_{X1}$ and $\mathbf{s}_{Y1}$ will be positive. Also note that the elements of $\mathbf{s}_{X1}$ and $\mathbf{s}_{Y1}$ are quite large, a result which suggests that, on the average, x_1 and y_1 account for relatively large proportions of the variances of the variables in the X and Y sets, respectively.

For the second canonical variates in this problem, the corresponding values are:

$$\mathbf{s}_{X2}=\mathbf{R}_{xx}\mathbf{c}_2=\begin{bmatrix} .700587 \\ -.470524 \end{bmatrix}$$

and

$$\mathbf{s}_{Y2}=\mathbf{R}_{yy}\mathbf{d}_2=\begin{bmatrix} .528605 \\ -.470838 \end{bmatrix}.$$

Thus, on the average, the second canonical variates x_2 and y_2 account for smaller proportions of the variances of the variables in their respective sets compared to the first canonical variates, x_1 and x_2.

The redundancy indices for the data of this problem are:

First canonical variates:

$$R_{dX_1}=\frac{\mathbf{s}'_{X1}\mathbf{s}_{X1}}{p}R^2_{c_1}=(.643892)(.522895)=.336688$$

$$R_{dY_1}=\frac{\mathbf{s}'_{Y1}\mathbf{s}_{Y1}}{q}R^2_{c_1}=(.749444)(.522895)=.391881$$

Second canonical variates:

$$R_{dX_2} = \frac{\mathbf{s}'_{X2}\mathbf{s}_{X2}}{p} R_{c_2}^2 = (.356108)(.028021) = .009979$$

$$R_{dY_2} = \frac{\mathbf{s}'_{Y2}\mathbf{s}_{Y2}}{q} R_{c_2}^2 = (.250556)(.028021) = .007021$$

Total:

$$R_{dX} = .336688 + .009979 = .346667, \text{ or about } .35$$

$$R_{dY} = .391881 + .007021 = .398902, \text{ or about } .40.$$

Thus, one would conclude that the information in the set Y is slightly more redundant, given the availability of X, than is the information in X, given Y. As expressed by this index, there seems to be only a modest amount of overlap between the two sets of variables.

5.7 PRECAUTIONS WHEN USING CANONICAL CORRELATION ANALYSIS

Weiss (1972) and Thorndike and Weiss (1973) have called attention to the fact that canonical correlation is subject to sampling error, just as are other statistics, but is, perhaps, more seriously affected. Therefore, we suggest that the following precautions be taken in the use of the method in order to avoid drawing erroneous conclusions:

1. Cross-validation of both the canonical correlations and the canonical weights is highly desirable, especially if the sample size is small compared to the number of variables in the X and Y sets. Weiss reinforces our earlier suggestion that there be a ratio of 20:1 between the number in the sample and the number of variables.

2. Cross-validation also is necessary if individual prediction rather than merely statistical description of relationships is an objective of the canonical correlation analysis. This may be accomplished by using the initial sample canonical coefficients to compute the values of the canonical variates, x_j and y_j (or Z_{xj} and Z_{yj}) for an independent sample, and then calculating the Pearson product-moment correlation coefficients between each pair of canonical variates. These may then be compared with the initial sample canonical correlations.

3. Sample *size* is not the only problem with which we have to contend in canonical correlation analysis; the manner in which the sample is selected may also bias the results. As in multiple regression analysis, especially in the behavioral sciences, restriction in range is likely to exist in some of the variables in each set. Hence, corrections such as those described in the previous chapter should generally be considered. In the case of canonical correlation analysis, the correction for restriction in range has to be made for the matrix formed by joining together $\mathbf{R}_{xx}$, $\mathbf{R}_{xy}$, $\mathbf{R}_{yx}$, and $\mathbf{R}_{yy}$.

There still remain a number of unanswered questions and controversies regarding canonical correlation analysis and its applications, particularly to the solution of problems in the behavioral sciences. These questions deal primarily with the nature of hypotheses that may be properly tested by the method and, as we indicated above, the degree of overlap or redundancy between the two sets. Darlington, Weinberg, and Wahlberg (1973) point out a number of items of controversy surrounding the method. They also propose ways in which some of these may be resolved and suggest other available multivariate techniques to substitute for canonical correlation analysis when it appears to be inappropriate for the problem being studied.

5.8 EXERCISES

5.1. Suppose that the matrices $\mathbf{R}_{xx}$, $\mathbf{R}_{yy}$, $\mathbf{R}_{xy}$, and $\mathbf{R}_{yx}$, based on a sample of 100 cases, were computed and found to be:

$$\left[\begin{array}{c|c} \mathbf{R}_{xx} & \mathbf{R}_{xy} \\ \hline \mathbf{R}_{yx} & \mathbf{R}_{yy} \end{array}\right] =$$

	X_1	X_2	X_3	Y_1	Y_2
	1.0	.1	.6	.7	.2
	.1	1.0	.4	.3	.8
	.6	.4	1.0	.5	.3
	.7	.3	.5	1.0	.4
	.2	.8	.3	.4	1.0

(a) Using a desk or hand calculator, compute the canonical correlations between the set of Xs and the set of Ys. Also compute the vectors of weights associated with each of these canonical correlations.

(b) Test the statistical significance of each of the two canonical correlation coefficients obtained in (a). Use $\alpha = .05$.

(c) Analyze the redundancy of the set of Xs given the set of Ys, and of the set of Ys given the set of Xs.

(d) Describe the relationships between the two sets of variables, using the results in (a), (b), and (c).

5.2. Repeat Exercise 5.1, using a computer program. Compare the results with those obtained by use of a desk or hand calculator.

5.3. The Rhode Island Pupil Identification Scale (RIPIS) is an observation scale used by teachers in kindergarten and the early grades to detect learning problems in young children. It is composed of two parts. Part I consists of 5 subscales and Part II of 4 subscales. The test-retest reliability of the RIPIS was determined on the basis of the relationship between scores obtained for 603 pupils in December and again in May. The correlation matrices for Parts I and II are given in Tables 5.8.1 and 5.8.2. The subscales

TABLE 5.8.1 CORRELATION MATRIX FOR RIPIS, PART I, ADMINISTERED TO 603 PUPILS

	X_1	X_2	X_3	X_4	X_5	Y_1	Y_2	Y_3	Y_4	Y_5
X_1	1.00000									
X_2	.43013	1.00000								
X_3	.57512	.28988	1.00000							
X_4	.50919	.33185	.53610	1.00000						
X_5	.54519	.42877	.67154	.63187	1.00000					
Y_1	.62168	.20598	.45628	.34250	.45615	1.00000				
Y_2	.39493	.74916	.22285	.26732	.36937	.45007	1.00000			
Y_3	.51915	.25124	.76110	.44242	.59628	.60384	.31351	1.00000		
Y_4	.46165	.28554	.49557	.73449	.54307	.44212	.26021	.55656	1.00000	
Y_5	.44748	.36011	.54767	.52022	.79361	.49928	.39886	.68499	.63246	1.00000
Mean	3.7645	10.5207	7.2653	11.0100	11.7231	3.6733	10.1592	7.1294	11.0017	11.5191
S.D.	1.5415	4.1034	3.0872	3.9900	4.4544	1.4744	3.7533	3.0105	3.9150	4.4663

TABLE 5.8.2 CORRELATION MATRIX FOR RIPIS, PART II, ADMINISTERED TO 603 PUPILS

	X_1	X_2	X_3	X_4	Y_1	Y_2	Y_3	Y_4
X_1	1.00000							
X_2	.57447	1.00000						
X_3	.61714	.65074	1.00000					
X_4	.63733	.53111	.60909	1.00000				
Y_1	.79213	.43989	.47764	.51847	1.00000			
Y_2	.48528	.70152	.56788	.46162	.53419	1.00000		
Y_3	.57949	.56828	.82444	.60372	.58321	.68181	1.00000	
Y_4	.51072	.42477	.53844	.75123	.56789	.54025	.64007	1.00000
Mean	13.1003	5.3388	9.3569	10.5164	12.4934	4.8980	8.9128	10.2993
S.D.	4.6627	2.3423	3.8148	3.7989	4.7494	1.9842	3.6127	4.0352

TABLE 5.8.3 MATRIX OF CORRELATIONS BETWEEN MRT AND RIPIS, PART I, ADMINISTERED TO 51 KINDERGARTEN PUPILS

WM	*Listening*	*Matching*	*Alphabet*	*Numbers*	*Copying*	*BP*	*SMC*	*A*	*S-C*	*ME*
1. 1.00000										
2. .34539	1.00000									
3. .32570	.38585	1.00000								
4. .47825	.55623	.40519	1.00000							
5. .60644	.57915	.65033	.71697	1.00000						
6. .43631	.36365	.56612	.47717	.50355	1.00000					
7. −.04640	−.20198	−.38847	−.07993	−.21398	−.38315	1.00000				
8. −.11607	−.33598	−.46717	−.19884	−.34587	−.46767	.71683	1.00000			
9. −.28869	−.24520	−.42263	−.27644	−.37321	−.49663	.79669	.58170	1.00000		
10. −.18533	−.11246	−.42384	−.31400	−35663	−.43082	.68333	.57285	.66036	1.00000	
11. −.31726	−.14390	−.45359	−.30805	−.43178	−.45989	.73061	.67157	.73610	.84714	1.00000

1 = MRT: Word Meaning
2 = MRT: Listening
3 = MRT: Matching
4 = MRT: Alphabet
5 = MRT: Numbers
6 = MRT: Copying
7 = RIPIS: Body Perception
8 = RIPIS: Sensory-Motor Coordination
9 = RIPIS: Attention
10 = RIPIS: Self-Concept
11 = RIPIS: Memory for Events

TABLE 5.8.4 MATRIX OF CORRELATIONS BETWEEN MRT AND RIPIS, PART II, ADMINISTERED TO 150 FIRST GRADE PUPILS

	WM	*Listening*	*Matching*	*Alphabet*	*Numbers*	*Copying*	*MRS*	*DPC*	*SSA*	*MSCO*
1. WM	1.00000									
2. L	.27129	1.00000								
3. M	.22588	.16754	1.00000							
4. A	.34660	.35019	.31365	1.00000						
5. N	.40836	.40995	.42236	.61906	1.00000					
6. C	.14113	.19958	.35357	.32843	.39996	1.00000				
7. MRS	−.17120	−.20349	−.43404	−.40463	−.35868	−.31528	1.00000			
8. DPC	−.07640	−.05624	−.21678	−.33999	−.29251	−.01089	.54734	1.00000		
9. SSA	−.02367	.05233	−.12005	−.23671	−.19147	−.06169	.35896	.51418	1.00000	
10. MSCO	−.17751	−.17801	−.34535	−.44033	−.44989	−.17135	.45716	.38150	.61283	1.00000

1 = MRT: Word Meaning
2 = MRT: Listening
3 = MRT: Matching
4 = MRT: Alphabet
5 = MRT: Numbers
6 = MRT: Copying
7 = RIPIS: Memory for Reproduction of Symbols
8 = RIPIS: Directional or Positional Constancy
9 = RIPIS: Spatial and Sequential Arrangement
10 = RIPIS: Memory for Symbols for Cognitive Operations

administered in December are denoted by X and those administered in May are denoted by Y.

(a) Perform a canonical correlation analysis for each of the two parts.

(b) How many of the canonical correlations are statistically significant ($\alpha = .05$) for Part I? for Part II?

(c) In a test-retest reliability study, how many canonical correlations would you expect to be statistically significant?

(d) Interpret the results of these analyses for each part.

5.4. Evidence of the validity of the RIPIS (see Exercise 5.3) was obtained through its correlation with the Metropolitan Readiness Test (MRT). The MRT was administered to 51 kindergarten children during May. At the end of May, the classroom teachers rated the pupils on Part I of the RIPIS. The MRT was also administered to 150 first grade pupils prior to the beginning of school in September. At the end of October, the classroom teachers rated the pupils on Part II of the RIPIS. The correlations obtained in these two administrations are given in Tables 5.8.3 and 5.8.4. (Note: High scores on the MRT indicate a high degree of readiness to do school work, whereas high scores on the RIPIS indicate the presence of problems that impede learning.)

(a) For each part of the RIPIS calculate the canonical correlations with the MRT and test their significance using $\alpha = .05$.

(b) Interpret the results obtained in (a). What conclusions should be drawn concerning the validity of the RIPIS, based on these data?

5.5. Further evidence of validity of Part II of the RIPIS was obtained through its correlations with the Cognitive Aptitude Test (CAT), based on a sample of 355 fourth-grade pupils. The results are given in Table 5.8.5.

(a) Perform a canonical correlation analysis on these data.

(b) Test the significance of the canonical correlations obtained in (a).

TABLE 5.8.5 CORRELATION MATRIX FOR FOUR RIPIS VARIABLES AND THREE CAT SCALES ($N = 355$)

	X_1	X_2	X_3	X_4	Y_1	Y_2	Y_3	*Mean*	*S.D.*
X_1	1.00000							12.0930	4.4948
X_2	.47877	1.00000						4.6366	1.8732
X_3	.50335	.63512	1.00000					8.2310	3.3879
X_4	.50190	.50438	.66494	1.00000				9.5775	3.8166
Y_1	−.06871	−.04104	−.13281	−.14791	1.00000			94.5493	29.5459
Y_2	−.03413	.07229	−.20314	−.09255	.38313	1.00000		60.5521	53.5368
Y_3	−.08812	.00840	−.08376	−.12725	.89643	.32813	1.00000	95.4930	30.4558

RIPIS
X_1 = Memory for Reproduction of Symbols
X_2 = Directional or Positional Constancy
X_3 = Spatial and Sequential Arrangement
X_4 = Memory for Symbols for Cognitive Operations

CAT
Y_1 = Verbal
Y_2 = Quantitative
Y_3 = Non-verbal

How many are statistically significant ($\alpha = .05$)?

(c) What conclusions should be drawn on the basis of the analyses in (a) and (b)?

5.6. Refer to the study by Merenda, Novack, and Bonaventura (1972) described in Exercise 8.4. Use the data given in that exercise, and carry out a canonical correlation analysis to determine the nature and degree of relationship between the CTMM and the SAT, for the sample of first-grade children. Compare your results and conclusions with those of the authors.

6/ Discriminant Analysis

Research in the behavioral sciences involving prediction models is frequently concerned with one or the other of two major issues: *classification* and *selection*. We have dealt with selection in Chapters 1 and 4, the more general model being the multiple linear regression model discussed in Chapter 4. In multiple regression analysis a linear combination of predictors (Xs) is used to estimate an individual's value on a continuous criterion measure, Y. In many research applications of the multiple regression models, the estimate of the criterion value is used in making a selection decision; that is, if the estimate is sufficiently large (or small, as the case may be), the individual is selected. For example, in selecting individuals for advanced training, a large estimated value on a measure of achievement in the training program might lead to selection.

The issue of classification involves answering the question, To which group does one best belong? The objective of the analysis is to determine the degree to which an individual's profile of scores on a set of X measures (the independent variables) corresponds to or resembles the typical profiles of each of a given set of discrete classes. In this situation we have a set of X variables that (as in the multiple regression problem) are assumed to be continuous and each pair of which are linearly related. However, in this case, the values of the dependent variable, Y, represent truly discrete categories. In the simplest case the discrete dependent variable is a dichotomy, that is, it consists of only two categories. The problem, therefore, reduces to one of analyzing the independent set (Xs) in such a way as to test the hypothesis that a particular profile based on the X measures resembles that of the members of Category A more closely than that of Category B. The statistical method applied to this problem is known as *discriminant analysis*.

Discriminant analysis involves the answers to two important questions. First, Can the groups be distinguished on the basis of the set of X measures on which data are presumed available? In other words, Can it be shown that the groups differ significantly in terms of their means on a linear combination of the X variables? If the answer to this question is yes, then it makes sense to try to answer the second, namely, How shall a particular individual be classified in terms of group membership? Which one of perhaps several groups does he or she most closely resemble?

In this chapter we first deal with these questions in the special case of two groups. We then treat the general case of G groups but limit our discussion somewhat in the interest of maintaining a moderate level of mathematical sophistication. We do, however, suggest references to which the interested reader may turn for further discussions of multigroup classification problems.

6.1 THE FISHER LINEAR DISCRIMINANT FUNCTION

In the mid 1930s Sir Ronald A. Fisher developed a method for the solution of the two-group case known as *linear discriminant analysis* (Fisher, 1936). Linear discriminant analysis is analogous to a multiple regression analysis in which the dependent variable, Y, assumes only two values, each indicating membership in one or the other of the two groups. It is also similar to methods of multivariate analysis developed by Hotelling (1931) and by Mahalonobis (1936). Its relationships to these methods will be discussed later in this chapter.

6.1.1 Computation of the Discriminant Equation

Recall that in Chapter 5 we defined a linear combination of p independent variables (Xs) as

$$Z = u_1X_1 + u_2X_2 + \cdots + u_pX_p. \tag{6.1.1}$$

In the context of discriminant analysis, this expression is called a *discriminant equation*. Because we are now dealing with two groups, this expression can be written

$$Z_{i1} = u_1X_{i11} + u_2X_{i21} + \cdots + u_pX_{ip1}$$

to represent the value of Z for the ith individual in Group 1; here, $i = 1, 2, \ldots, n_1$. Similarly, for Group 2,

$$Z_{i2} = u_1X_{i12} + u_2X_{i22} + \cdots + u_pX_{ip2}.$$

The values of the u's in the discriminant equation are chosen so as to maximize the separation between the two groups.

In Fisher's discriminant function analysis, separation between groups is expressed in terms of the difference between the means of the discriminant functions for the two groups. Let us define the group means, $\bar{Z}_1$ and $\bar{Z}_2$, as

$$\bar{Z}_1 = \frac{\sum_{i=1}^{n_1} Z_{i1}}{n_1} = \frac{\sum_{i=1}^{n_1} (u_1X_{i11} + u_2X_{i21} + \cdots + u_pX_{ip1})}{n_1}$$

$$= u_1\bar{X}_{11} + u_2\bar{X}_{21} + \cdots + u_p\bar{X}_{p1} \tag{6.1.2}$$

and

$$\bar{Z}_2 = \frac{\sum_{i=1}^{n_2} Z_{i2}}{n_2} = \frac{\sum_{i=1}^{n_2} (u_1 X_{i12} + u_2 X_{i22} + \cdots + u_p X_{ip2})}{n_2}$$

$$= u_1 \bar{X}_{12} + u_2 \bar{X}_{22} + \cdots + u_p \bar{X}_{p2}. \tag{6.1.3}$$

If we let $d_j = \bar{X}_{j1} - \bar{X}_{j2}$ denote the difference between group means on variable X_j, and if $D = \bar{Z}_1 - \bar{Z}_2$, then

$$D = u_1 d_1 + u_2 d_2 + \cdots + u_p d_p \tag{6.1.4}$$

is the quantity that must be maximized.

Now D is a random variable and its value is influenced by the variability of Z_{i1} and Z_{i2} within the two groups. Accordingly, Fisher proposed that the criterion for maximum separation be the maximization of the ratio

$$\frac{D^2}{SS_W}, \tag{6.1.5}$$

in which SS_W is the within-groups sum of squares of the variable Z, defined in Equation 6.1.1 To simplify the notation in Sections 6.1.1 through 6.1.3, we shall denote this sum of squares by V. Thus, $V = SS_W$.

To understand what V is, and to see how it may be calculated, recall from Equation 5.2.1 that for a single group, say Group 1,

$$\sum_{i=1}^{n_1} z_{i1}^2 = \mathbf{u}'\mathbf{S}_{xx}^{(1)}\mathbf{u},$$

where $z_{i1} = Z_{i1} - \bar{Z}_1$ and $\mathbf{u}', \mathbf{S}_{xx}^{(1)}$, and $\mathbf{u}$ are defined as in Chapter 5, except for the addition of the superscript (1) to indicate the group number. Similarly, we can write

$$\sum_{i=1}^{n_2} z_{i2}^2 = \mathbf{u}'\mathbf{S}_{xx}^{(2)}\mathbf{u}$$

for Group 2. If we now let

$$\mathbf{W} = \mathbf{S}_{xx}^{(1)} + S_{xx}^{(2)}, \tag{6.1.6}$$

we can write the sum $\Sigma_{i=1}^{n_1} z_{i1}^2 + \Sigma_{i=1}^{n_2} z_{i2}^2$, as $\mathbf{u}'\mathbf{S}_{xx}^{(1)}\mathbf{u} + \mathbf{u}'\mathbf{S}_{xx}^{(2)}\mathbf{u}$. Then V may be expressed as

$$V = \mathbf{u}'(\mathbf{S}_{xx}^{(1)} + S_{xx}^{(2)})\mathbf{u} = \mathbf{u}'\mathbf{W}\mathbf{u}. \tag{6.1.7}$$

Now, writing D in matrix notation as $D = \mathbf{u}'\mathbf{d}$, where $\mathbf{d}' = [d_1 d_2 \ldots d_p]$, we can express the ratio D^2/V, as

$$\lambda = \frac{D^2}{V} = \frac{(\mathbf{u}'\mathbf{d})^2}{\mathbf{u}'\mathbf{W}\mathbf{u}}. \tag{6.1.8}$$

In this expression λ is called the *discriminant criterion*. (Note that d here is not equivalent to the vector of weights denoted by d in Chapter 5.)

To determine the value of **u** that maximizes the discriminant criterion, λ,

we take the first derivative (see Appendix B) of Equation 6.1.8 with respect to **u** and set it equal to zero. The result is

$$\frac{\partial\lambda}{\partial\mathbf{u}}=\frac{V\left(\frac{\partial D^2}{\partial\mathbf{u}}\right)-D^2\frac{\partial V}{\partial\mathbf{u}}}{V^2}=\frac{\frac{\partial D^2}{\partial\mathbf{u}}}{V}-\left(\frac{D}{V}\right)^2\frac{\partial V}{\partial\mathbf{u}}=0.$$

Then $(1/v)\partial D^2/\partial\mathbf{u}=(D/V)^2\partial V/\partial\mathbf{u}$. Rewriting the derivative on the left side, we have $(1/V)(2D)\partial D/\partial\mathbf{u}=(D^2/V^2)\partial V/\partial\mathbf{u}$. Because $\partial D/\partial\mathbf{u}=\partial(\mathbf{u}'\mathbf{d})/\partial\mathbf{u}=\mathbf{d}$ and $\partial V/\partial\mathbf{u}=\partial(\mathbf{u}'\mathbf{W}\mathbf{u})/\partial\mathbf{u}=2\mathbf{W}\mathbf{u}$, (see Appendix B), we have $2(D/V)\mathbf{d}=(D/V)^2(2\mathbf{W}\mathbf{u})$. Dividing by 2 and multiplying by $(V/D)^2$, we have

$$\left(\frac{V}{D}\right)\mathbf{d}=\mathbf{W}\mathbf{u}. \tag{6.1.9}$$

Assuming that $\mathbf{W}^{-1}$ exists (see Appendix A), we can solve for **u** by premultiplying both sides by $\mathbf{W}^{-1}$. Then

$$\mathbf{u}=\left(\frac{V}{D}\right)\mathbf{W}^{-1}\mathbf{d}. \tag{6.1.10}$$

Since the multiplier, V/D, will not affect the proportionality among the elements of **u**, it will be convenient to set it equal to one. There is a certain advantage to this which we shall see later. Thus, with $V/D=1$, we may state

$$\mathbf{u}=\mathbf{W}^{-1}\mathbf{d}. \tag{6.1.11}$$

The values u_1 through u_p obtained by means of Equation 6.1.11 may be substituted in Equation 6.1.1 to yield the discriminant equation.

6.1.2 An Illustration of the Computations

To illustrate the computation of the discriminant equation, $Z=u_1X_1+\cdots+u_pX_p$, we shall use the following fictitious example. Suppose we have values of X_1 and X_2 available for two groups as shown in Table 6.1.1. We wish to determine the weights, **u**, that will result in the maximum separation of the groups according to Fisher's criterion. According to Equation 6.1.11, we must first compute $\mathbf{W}^{-1}$ and **d**. The elements of **W** are obtained by computing $\mathbf{S}_{xx}^{(1)}$ and $S_{xx}^{(2)}$ and adding them. For Group 1, $S_{11}=46.1$, $S_{22}=32.1$, and $S_{21}=S_{12}=35.1$. For group 2, $S_{11}=77.6$, $S_{22}=78.1$, and $S_{21}=S_{12}=53.4$. Then

$$\mathbf{W}=\begin{bmatrix}46.1 & 35.1\\ 35.1 & 32.1\end{bmatrix}+\begin{bmatrix}77.6 & 53.4\\ 53.4 & 78.1\end{bmatrix}=\begin{bmatrix}123.7 & 88.5\\ 88.5 & 110.2\end{bmatrix},$$

and

$$\mathbf{W}^{-1}=\begin{bmatrix}.01900167 & -.01525996\\ -.01525996 & .02132946\end{bmatrix}.$$

TABLE 6.1.1 VALUES OF X_1 AND X_2 FOR TWO GROUPS

Group 1 ($n_1=10$)			*Group* 2 ($n_2=10$)		
No.	X_1	X_2	*No.*	X_1	X_2
1	1	10	1	4	13
2	5	12	2	9	17
3	4	11	3	9	19
4	2	8	4	6	11
5	7	13	5	12	20
6	5	11	6	11	14
7	3	10	7	8	15
8	2	9	8	7	13
9	1	9	9	8	17
10	7	14	10	14	18

To compute **d** we first find the means of X_1 and X_2 for both groups. For Group 1, $\overline{X}_1=3.7$ and $\overline{X}_2=10.7$. For group 2, $\overline{X}_1=8.8$ and $\overline{X}_2=15.7$. Thus,

$$\mathbf{d}=\begin{bmatrix} 3.7-8.8 \\ 10.7-15.7 \end{bmatrix}=\begin{bmatrix} -5.1 \\ -5.0 \end{bmatrix}.$$

Now, using Equation 6.1.11, we compute

$$\mathbf{u}=\mathbf{W}^{-1}\mathbf{d}=\begin{bmatrix} -.02060872 \\ -.02882150 \end{bmatrix}.$$

The discriminant equation is then

$$Z=-.02060872X_1-.02882150X_2.$$

By substituting the means, $\overline{X}_1$ and $\overline{X}_2$, for each of the two groups in this equation, we obtain, for Group 1,

$$\overline{Z}_1=-.02060872(3.7)-.02882150(10.7)=-.384642314,$$

and, for Group 2,

$$\overline{Z}_2=-.02060872(8.8)-.02882150(15.7)=-.633854286,$$

so that $D=\overline{Z}_1-\overline{Z}_2=.249211972$. We now consider a method for determining whether this difference is statistically significant.

6.1.3 Testing the Significance of *D*

To test the significance of D we shall use a method analogous to the univariate analysis of variance test of the hypothesis, H_0: $\mu_1-\mu_2=0$. The between-groups sum of squares in the univariate analysis of variance is $\Sigma_{g=1}^{G} n_g(\overline{X}_g-\overline{X})^2$, which, for only two groups may be written $SS_B=n_1(\overline{X}_1-\overline{X})^2+n_2(\overline{X}_2-\overline{X})^2$. In this expression $\overline{X}$ is the mean of the entire (combined) group. In the present case we are dealing with the variable $Z=u_1X_1+\cdots+u_pX_p$, instead of with a single X. Then the above expression for SS_B may be written $SS_B=n_1(\overline{Z}_1-\overline{Z})^2+n_2(\overline{Z}_2-\overline{Z})^2$, where, again, $\overline{Z}$ is the mean of all n_1+n_2 individuals in both groups, that is, $\bar{Z}=(n_1\bar{Z}_1+n_2\bar{Z}_2)/(n_1+n_2)$. We now show that SS_B can be expressed in terms of $D=\bar{Z}_1-\bar{Z}_2$. Substituting for $\bar{Z}$ in the above expression for SS_B produces

$$SS_B=n_1\left[\bar{Z}_1-\frac{n_1\bar{Z}_1+n_2\bar{Z}_2}{n_1+n_2}\right]^2+n_2\left[\bar{Z}_2-\frac{n_1\bar{Z}_1+n_2\bar{Z}_2}{n_1+n_2}\right]^2$$

$$=n_1\left[\frac{n_2(\bar{Z}_1-\bar{Z}_2)}{n_1+n_2}\right]^2+n_2\left[\frac{-n_1(\bar{Z}_1-\bar{Z}_2)}{n_1+n_2}\right]^2$$

$$=n_1\left[\frac{n_2D}{n_1+n_2}\right]^2+n_2\left[\frac{-n_1D}{n_1+n_2}\right]^2=\frac{n_1n_2}{n_1+n_2}D^2.$$

An expression for the within-groups sum of squares is available from Equation 6.1.7. In general, SS_W may be obtained by computing $V=\mathbf{u}'\mathbf{W}\mathbf{u}$. However, in the special case where the multiplier, V/D, has been set equal to one, it is clear that $V=D$. Then $SS_W=D$ and Table 6.1.2 is the analysis of variance table. Note that SS_B here has p degrees of freedom and that SS_w has n_1+n_2-p-1 degrees of freedom. Therefore, in the univariate test of H_0: $\mu_1-\mu_2=0$, $p=1$ and SS_B and SS_w have 1 and n_1+n_2-2 degrees of freedom, respectively.

For the data of our example, the results are shown in Table 6.1.3. The F value obtained here has probability less than .005 with 2 and 17 degrees of freedom. We would thus conclude that the difference between means of Z for the populations from which these samples were drawn is not equal to zero.

TABLE 6.1.2 ANALYSIS OF VARIANCE TABLE FOR TESTING THE SIGNIFICANCE OF *D*

Source	*Sum of Squares*	*Degrees of Freedom*	*Mean Square*	*F*
Between Groups	$SS_B=\dfrac{n_1n_2}{n_1+n_2}D^2$	p	$MS_B=\dfrac{SS_B}{p}$	$\dfrac{MS_B}{MS_W}$
Within Groups	$SS_W=D$	n_1+n_2-p-1	$MS_W=\dfrac{D}{n_1+n_2-p-1}$	

TABLE 6.1.3 SUMMARY OF ANALYSIS OF VARIANCE BASED ON TABLE 6.1.1

Source	*Sum of Squares*	*Degrees of Freedom*	*Mean Square*	*F*
Between Groups	$SS_B=.31053$	2	$MS_B=.15527$	10.59
Within Groups	$SS_W=.24921$	17	$MS_W=.01466$	

6.1.4 The Relationship Between Discriminant Analysis and Multiple Regression Analysis

Earlier in this chapter we stated that the two-group Fisher linear discriminant analysis is analogous to multiple regression analysis. We now show the relationship between these two methods.

The general form of the multiple regression equation may be stated for p independent variables (Xs) as

$$\hat{Y}_i = b_0 + b_1 X_{i1} + \cdots + b_p X_{ip},$$

where $b_0 = \bar{Y} - b_1 \bar{X}_1 - \cdots - b_p \bar{X}_p$. Substituting for b_0, we obtain

$$\hat{Y}_i = \bar{Y} + b_1(X_{i1} - \bar{X}_1) + \cdots + b_p(X_{ip} - \bar{X}_p).$$

Now suppose we define a dummy variable, Y, that takes on only two values, namely, $n_2/(n_1+n_2)$ for each member of Group 1 and $-n_1/(n_1+n_2)$ for each member of Group 2. Since n_1 and n_2 are the sizes of Groups 1 and 2, respectively, the mean of Y for the combined groups is

$$\bar{Y} = \frac{n_1\left(\dfrac{n_2}{n_1+n_2}\right) + n_2\left(\dfrac{-n_1}{n_1+n_2}\right)}{n_1+n_2} = 0.$$

Thus, the regression equation for estimating Y becomes

$$\hat{Y}_i = b_1(X_{i1} - \bar{X}_1) + \cdots + b_p(X_{ip} - \bar{X}_p),$$

and the normal equations can be expressed in matrix notation as

$$\mathbf{h} = \mathbf{S}\mathbf{b}, \tag{6.1.12}$$

where **h**, **S**, and **b** are generalized to p independent variables from their definition for two independent variables in Equation 4.2.8. Then

$$\mathbf{h} = \begin{bmatrix} S_{1y} \\ S_{2y} \\ \vdots \\ S_{py} \end{bmatrix}, \quad S = \begin{bmatrix} S_{11} & S_{12} & \cdots & S_{1p} \\ S_{21} & S_{22} & \cdots & S_{2p} \\ \vdots & \vdots & \ddots & \vdots \\ S_{p1} & S_{p2} & \cdots & S_{pp} \end{bmatrix}, \text{ and } \mathbf{b} = \begin{bmatrix} b_1 \\ b_2 \\ \vdots \\ b_p \end{bmatrix}.$$

In the matrix **S**, the jkth element, S_{jk} $(j,k = 1,\ldots,p)$, is the sum of squares or cross-products of deviations of X_j and X_k from their respective sample means. This notation is the same as that used in previous chapters.

Now let us consider the nature of the elements of **h** when Y is defined as a dummy variable indicating group membership. Any element of **h**, say h_{jy}, may be written, for the jth X, as

$$h_{jy} = \sum_{i=1}^{n_1} Y_{i1}(X_{ij1} - \bar{X}_j) + \sum_{i=1}^{n_2} Y_{i2}(X_{ij2} - \bar{X}_j)$$

$$= \frac{n_2}{n_1+n_2} \sum_{i=1}^{n_1} (X_{ij1} - \bar{X}_j) + \frac{-n_1}{n_1+n_2} \sum_{i=1}^{n_2} (X_{ij2} - \bar{X}_j),$$

because $\bar{Y}=0$ and $Y=n_2/(n_1+n_2)$ for Group 1 and $Y=-n_1/(n_1+n_2)$ for Group 2. Multiplying the first term on the right by n_1/n_1 and the second by n_2/n_2, we have

$$\begin{aligned} h_{jy} &= \frac{n_1 n_2}{n_1+n_2}\left(\bar{X}_{j1}-\bar{X}_j\right) - \frac{n_1 n_2}{n_1+n_2}\left(\bar{X}_{j2}-\bar{X}_j\right) \\ &= \frac{n_1 n_2}{n_1+n_2}\left(\bar{X}_{j1}-\bar{X}_{j2}\right) = \frac{n_1 n_2}{n_1+n_2} d_j, \end{aligned}$$

where d_j is again defined as in Equation 6.1.4. Thus,

$$\mathbf{h} = \frac{n_1 n_2}{n_1+n_2}\mathbf{d}, \tag{6.1.13}$$

in which $\mathbf{d}'=[d_1 \ldots d_j \ldots d_p]$ as before. Substituting for $\mathbf{h}$ in Equation 6.1.12, we have

$$\frac{n_1 n_2}{n_1+n_2}\mathbf{d} = \mathbf{Sb}.$$

If we let $c=(n_1+n_2)/n_1 n_2$, then this equation is

$$\mathbf{d} = c\mathbf{Sb}. \tag{6.1.14}$$

Now we may refer to $\mathbf{S}$ as the *total* sums of squares and cross-products matrix of the Xs. Because we are dealing here with two groups, $\mathbf{S}$ may be broken down into two parts, the sum of squares and cross-products *between* groups and that *within* groups. We already have an expression for the within-groups sum of squares and cross-products matrix, namely, $\mathbf{W}$, as defined in Equation 6.1.6. Therefore, we need only to determine the nature of the between-groups sum of squares and cross-products matrix.

We shall denote the latter matrix by $\mathbf{B}$. A general element of this matrix is

$$b_{jk} = n_1\left(\bar{X}_{j1}-\bar{X}_j\right)\left(\bar{X}_{k1}-\bar{X}_k\right) + n_2\left(\bar{X}_{j2}-\bar{X}_j\right)\left(\bar{X}_{k2}-\bar{X}_k\right),$$

where the subscripts j and k refer to X_j and X_k, respectively, and the subscripts 1 and 2 refer to Groups 1 and 2, respectively. The means, $\bar{X}_j$ and $\bar{X}_k$, for the combined groups may be expressed as

$$\bar{X}_j = \frac{n_1\bar{X}_{j1}+n_2\bar{X}_{j2}}{n_1+n_2} \text{ and } \bar{X}_k = \frac{n_1\bar{X}_{k1}+n_2\bar{X}_{k2}}{n_1+n_2}.$$

Substituting in the expression for b_{jk} we have

$$\begin{aligned} b_{jk} = n_1 &\left[\bar{X}_{j1} - \frac{\left(n_1\bar{X}_{j1}+n_2\bar{X}_{j2}\right)}{n_1+n_2}\right]\left[\bar{X}_{k1} - \frac{\left(n_1\bar{X}_{k1}+n_2\bar{X}_{k2}\right)}{n_1+n_2}\right] \\ + n_2 &\left[\bar{X}_{j2} - \frac{\left(n_1\bar{X}_{j1}+n_2\bar{X}_{j2}\right)}{n_1+n_2}\right]\left[\bar{X}_{k2} - \frac{\left(n_1\bar{X}_{k1}+n_2\bar{X}_{k2}\right)}{n_1+n_2}\right]. \end{aligned}$$

Now, within the first pair of brackets, the expression reduces to

$$\bar{X}_{j1} - \frac{(n_1\bar{X}_{j1} + n_2\bar{X}_{j2})}{n_1+n_2} = \frac{n_1\bar{X}_{j1} + n_2\bar{X}_{j1} - n_1\bar{X}_{j1} - n_2\bar{X}_{j2}}{n_1+n_2}$$

$$= \frac{n_2}{n_1+n_2}(\bar{X}_{j1} - \bar{X}_{j2}) = \frac{n_2}{n_1+n_2} d_j.$$

Similarly, we find that the other three expressions in brackets reduce to $(n_2/(n_1+n_2))\,d_k$, $(-n_1/(n_1+n_2))d_j$, and $(-n_1/(n_1+n_2))d_k$, respectively. Substituting in the above expression for b_{jk}, the result is

$$b_{jk} = \frac{n_1 n_2^2}{(n_1+n_2)^2} d_j d_k + \frac{n_1^2 n_2}{(n_1+n_2)^2} d_j d_k$$

$$= \frac{n_1 n_2}{n_1+n_2} d_j d_k.$$

Thus, the between-groups sum of squares and cross-products matrix, **B**, may be expressed as

$$\mathbf{B} = \frac{n_1 n_2}{n_1+n_2}\mathbf{dd}' = \left(\frac{1}{c}\right)\mathbf{dd}'. \qquad (6.1.15)$$

Recall now from Equation 6.1.9, with $V/D=1$, that $\mathbf{d}=\mathbf{Wu}$. Adding $\mathbf{Bu}$ to both sides of this equation gives

$$\mathbf{d}+\mathbf{Bu}=\mathbf{Wu}+\mathbf{Bu}=(\mathbf{W}+\mathbf{B})\mathbf{u}=\mathbf{Su},$$

because the sum of the within- and between-groups sum of squares and cross-products matrices is equal to the total. Now, substituting for **B** from Equation 6.1.15 yields

$$\mathbf{d}+\frac{1}{c}\mathbf{dd}'\mathbf{u}=\mathbf{Su}.$$

Hence, since $D=\mathbf{d}'\mathbf{u}$ (see Equation 6.1.4),

$$\mathbf{d}\left(1+\frac{D}{c}\right)=\mathbf{Su},$$

or

$$\mathbf{d}=\left(\frac{c}{c+D}\right)\mathbf{Su}, \qquad (6.1.16)$$

where the expression in parentheses is a scalar.

Equations 6.1.14 and 6.1.16 give us two expressions for **d**, one in terms of **u**, the discriminant weights, and one in terms of **b**, the regression coefficients. Equating these two expressions, we have

$$\left(\frac{c}{c+D}\right)\mathbf{Su}=c\mathbf{Sb},$$

or

$$\mathbf{u}=(c+D)\mathbf{b}. \qquad (6.1.17)$$

From this result we can see that the discriminant weights, **u**, are proportional to the regression coefficients, **b**, when the dependent variable, Y, takes on just two values indicating group membership. We should emphasize that the relationship shown here between the two methods of analysis holds *only* in the two-group case.

We now illustrate the above relationship using the data of the example given for discriminant analysis. The vector of regression coefficients may be obtained from Equation 4.2.8, $\mathbf{b}=\mathbf{S}^{-1}\mathbf{h}$. Here,

$$\mathbf{S}=\begin{bmatrix} S_{11} & S_{12} \\ S_{12} & S_{22} \end{bmatrix}=\begin{bmatrix} 253.75 & 216.00 \\ 216.00 & 235.20 \end{bmatrix},$$

$$\mathbf{S}^{-1}=\begin{bmatrix} .0180562 & -.0165822 \\ -.0165822 & .0194803 \end{bmatrix}, \quad \text{and } \mathbf{h}=\begin{bmatrix} -25.5 \\ -25.0 \end{bmatrix}.$$

Then

$$\mathbf{b}=\mathbf{S}^{-1}\mathbf{h}=\begin{bmatrix} -.0458781 \\ -.0641614 \end{bmatrix}.$$

To see that these values are proportional to the values of $\mathbf{u}$ obtained in the discriminant analysis, it is convenient to compute b_1/b_2 and u_1/u_2. We find

$$b_1/b_2=\frac{-.0458781}{-.0641614}=.7150,$$

and

$$u_1/u_2=\frac{-.0206087}{-.0288215}=.7150.$$

Note that using Equation 6.1.17, with $c+D=.449212$, we may obtain

$$\mathbf{u}=.449212\begin{bmatrix} -.0458781 \\ -.0641614 \end{bmatrix}=\begin{bmatrix} -.0206090 \\ -.0288221 \end{bmatrix},$$

which agrees, allowing for rounding error, with the values previously obtained using Equation 6.1.11.

6.1.6 Other Methods of Discrimination Between Two Groups

Our discussion of discriminant analysis in the two-group case would not be complete without treating, at least in moderate detail, the work of Hotelling (1931) and Mahalanobis (1936). Both of these statisticians dealt with the problem in ways very closely related to the Fisher procedure. In this section we shall discuss both of these methods of analysis and show how they are related to each other as well as to Fisher's method.

Hotelling (1931) showed how the univariate t test of the null hypotheses $H_0: \mu=\mu_0$ and $H_0: \mu_1=\mu_2$ could be generalized to the multivariate case, that is, the case in which the corresponding null hypotheses are:

(1) $H_0: \boldsymbol{\mu}=\boldsymbol{\mu}_0$, where $\boldsymbol{\mu}$ is a vector of population means on several variables, $X_1,\ldots,X_p$.

(2) $H_0: \boldsymbol{\mu}_1=\boldsymbol{\mu}_2$, where $\boldsymbol{\mu}_1$ and $\boldsymbol{\mu}_2$ are vectors of population means on several variables, $X_1,\ldots,X_p$.

The test of the first of these hypotheses, $H_0: \mu = \mu_0$, does not properly belong in a chapter dealing with discrimination between two groups. However, it will be instructive to discuss this test because it represents the simplest case dealing with the extension of a univariate to a multivariate test of population means. The second test, that is, the test of $H_0: \mu_1 = \mu_2$, is directly relevant to the topic in this cahpter, that of distinguishing between two populations on the basis of a number of independent (X) variables.

Considering the first of these hypotheses, Hotelling showed that the statistic

$$T_1^2 = n(n-1)(\overline{\mathbf{X}} - \mu_0)'\mathbf{S}^{-1}(\overline{\mathbf{X}} - \mu_0), \tag{6.1.18}$$

when multiplied by the constant $(n-p)/(n-1)p$, has the F distribution with p and $n-p$ degrees of freedom. The subscript, 1, on T_1^2 indicates that the statistic is appropriate for use in the one-sample case. In Equation 6.1.18,

$$\overline{\mathbf{X}} = \begin{bmatrix} \overline{X}_1 \\ \overline{X}_2 \\ \vdots \\ \overline{X}_p \end{bmatrix}, \mu_0 = \begin{bmatrix} \mu_1 \\ \mu_2 \\ \vdots \\ \mu_p \end{bmatrix},$$

and $\mathbf{S}$ is the same as $\mathbf{S}$ in Equation 6.1.12.

It is easy to show that T_1^2 is a simple extension of the square of the familiar univariate statistic, $t = (\overline{X} - \mu_0)/(s/\sqrt{n})$, appropriate for testing $H_0: \mu = \mu_0$. The square of this univariate statistic is $t^2 = (\overline{X} - \mu_0)^2/(s^2/n)$, which, under $H_0: \mu = \mu_0$, has an F distribution with 1 and $n-1$ degrees of freedom. Now suppose that in Equation 6.1.18 there were only one X, say X_1. Then

$$T_1^2 = \frac{n(n-1)(\overline{X} - \mu_0)^2}{S_{11}} = \frac{(\overline{X} - \mu_0)^2}{s^2/n},$$

which is seen to be equivalent to the t^2 statistic above. Thus, T_1^2 may be viewed as a direct extension of the univariate one-sample t statistic to the multivariate case.

As an illustration of the application of the test of $H_0: \mu = \mu_0$, we consider the following example. Two measures, X_1 and X_2, of reading comprehension were obtained from a class of 25 children following the use of a new method of teaching reading comprehension in the third grade. Past experience with a more traditional method resulted in mean performances on the test of $\mu_1 = 55$ and $\mu_2 = 47$. In the sample of 25 children, the means were $\overline{X}_1 = 58$ and $\overline{X}_2 = 51$. The sums of squares and cross-products matrix, $\mathbf{S}$, was $\mathbf{S} = \begin{bmatrix} 946 & 452 \\ 452 & 721 \end{bmatrix}$. With $\alpha = .01$, we wish to test the hypothesis that the new method results in improved reading comprehension for this population of third grade children. The null hypothesis is H_0: $\boldsymbol{\mu} = \begin{bmatrix} 55 \\ 47 \end{bmatrix}$. Therefore

$$\overline{\mathbf{X}} - \boldsymbol{\mu}_0 = \begin{bmatrix} 58-55 \\ 51-47 \end{bmatrix} = \begin{bmatrix} 3 \\ 4 \end{bmatrix}$$

$$\text{and} \quad \mathbf{S}^{-1} = \begin{bmatrix} .001509 & -.000946 \\ -.000946 & .001980 \end{bmatrix}.$$

Then, from Equation 6.1.18, we find that $T_1^2 = 13.5342$ and $((n-p)/(n-1)p)T_1^2 = 6.4851$, which is distributed as F with $p=2$ and $n-p=23$ degrees of freedom. Since this value exceeds the 99th percentile, 5.66, in the F distribution with 2 and 23 degrees of freedom, we should reject H_0 and conclude that the new method results in improved reading comprehension.

Now considering the test of the hypothesis H_0: $\boldsymbol{\mu}_1 = \boldsymbol{\mu}_2$, Hotelling showed that the statistic

$$T_2^2 = \frac{n_1 n_2 (n_1 + n_2 - 2)}{n_1 + n_2} \mathbf{d}'\mathbf{W}^{-1}\mathbf{d}, \tag{6.1.19}$$

when multiplied by the constant $(n_1 + n_2 - d - 1)/(n_1 + n_2 - 2)p$, has an F distribution with p and $n_1 + n_2 - p - 1$ degrees of freedom. Here, the subscript, 2, on T_2^2 indicates that the statistic is appropriate for the two-sample test. In Equation 6.1.19, $\mathbf{d} = \overline{\mathbf{X}}_1 - \overline{\mathbf{X}}_2$ is the vector of differences between group means of the Xs and $\mathbf{W}$ is the matrix of pooled within-groups sums of squares and cross-products already defined in Equation 6.1.6. Again it is easy to show that T_2^2 is a simple extension of the square of the familiar univariate statistic used in testing H_0: $\boldsymbol{\mu}_1 = \boldsymbol{\mu}_2$, that is,

$$t^2 = \frac{(\overline{X}_1 - \overline{X}_2)^2}{s_p^2\left(\dfrac{1}{n_1} + \dfrac{1}{n_2}\right)},$$

in which $s_p^2 (S^{(1)} + S^{(2)})/(n_1 + n_2 - 2)$, the superscripts denoting groups (1) and (2). When there is only one X, say X_1, Equation 6.1.19 becomes

$$T_2^2 = \frac{n_1 n_2 (n_1 + n_2 - 2)}{n_1 + n_2} (\overline{X}_{11} - \overline{X}_{12})^2 (S_{11}^{(1)} + S_{11}^{(2)})^{-1}$$

$$= \frac{(\overline{X}_{11} - \overline{X}_{12})^2}{\dfrac{(S_{11}^{(1)} + S_{11}^{(2)})}{n_1 + n_2 - 2}\left(\dfrac{1}{n_1} + \dfrac{1}{n_2}\right)}.$$

After dropping the subscripts indicating the variable number, this becomes

$$T_2^2 = \frac{(\overline{X}_1 - \overline{X}_2)^2}{s_p^2\left(\dfrac{1}{n_1} + \dfrac{1}{n_2}\right)},$$

which is the same as t^2 above.

As an illustration of the use of T_2^2 to test H_0: $\boldsymbol{\mu}_1=\boldsymbol{\mu}_2$, we return to the example of Section 6.1.2. There we found that, based on samples of $n_1=n_2=10$,

$$\mathbf{d}=\begin{bmatrix}-5.1\\-5.0\end{bmatrix}\quad\text{and}\quad\mathbf{W}^{-1}=\begin{bmatrix}.01900167 & -.01525996\\-.01525996 & .02132946\end{bmatrix}.$$

Substitution in Equation 6.1.19 yields $T_2^2=22.429$ and $((n_1+n_2-p-1)/(n_1+n_2-2)p)T_2^2=10.59$, which has an F distribution with $p=2$ and $n_1+n_2-p-1=17$ degrees of freedom. The result is significant well beyond the .005 level.

It is instructive to note that this value, $F=10.59$, is precisely equal to that obtained in the test of the significance of D in Section 6.1.3. We now show that this equality is not merely a coincidence. Recall from Equation 6.1.11 that the vector of discriminant weights $\mathbf{u}=\mathbf{W}^{-1}\mathbf{d}$. Equation 6.1.4, defining D, may be written in matrix notation as

$$D=\mathbf{d}'\mathbf{u}=\mathbf{d}'\mathbf{W}^{-1}\mathbf{d} \tag{6.1.20}$$

Substituting for $\mathbf{d}'\mathbf{W}^{-1}\mathbf{d}$ in Equation 6.1.19 yields

$$T_2^2=\left(\frac{n_1n_2(n_1+n_2-2)}{n_1+n_2}\right)D, \tag{6.1.21}$$

which shows how T_2^2 and D are related. Now to show the equivalence of the statistic of Section 6.1.3, namely,

$$F=\frac{n_1n_2}{n_1+n_2}\left(\frac{n_1+n_2-p-1}{p}\right)D, \tag{6.1.22}$$

and the statistic of Hotelling (1931), namely,

$$F=\left(\frac{n_1+n_2-p-1}{(n_1+n_2-2)p}\right)T_2^2, \tag{6.1.23}$$

we merely substitute for T_2^2 in the latter on the basis of Equation 6.1.21. The result is

$$\begin{aligned}F&=\left(\frac{n_1+n_2-p-1}{(n_1+n_2-2)p}\right)\left(\frac{n_1n_2(n_1+n_2-2)}{n_1+n_2}\right)D\\&=\frac{n_1n_2}{n_1+n_2}\left(\frac{n_1+n_2-p-1}{p}\right)D,\end{aligned}$$

which is identical to Equation 6.1.22. We may conclude, then, that the tests proposed by Fisher (1936) and by Hotelling (1931) are equivalent.

Before concluding our discussion of the two-group case, we should mention briefly a method due to Mahalanobis (1936), usually referred to as the Mahalanobis D^2 method. The statistic proposed by Mahalanobis is

$$D_M^2=\mathbf{d}'\hat{\boldsymbol{\Sigma}}^{-1}\mathbf{d}, \tag{6.1.24}$$

where the subscript, M, is used to distinguish D_M^2 from D. Note that D_M^2 differs from D of Equation 6.1.20 only because $\mathbf{W}^{-1}$ is replaced by $\hat{\boldsymbol{\Sigma}}^{-1}$, which is the inverse of the pooled sample within-groups variance-covariance matrix. Thus, $\hat{\boldsymbol{\Sigma}}^{-1}=(n_1+n_2-2)\mathbf{W}^{-1}$, and through substitution in Equation 6.1.24 we can express D_M^2 as

$$D_M^2=(n_1+n_2-2)\mathbf{d}'\mathbf{W}^{-1}\mathbf{d}=(n_1+n_2-2)D,$$

or

$$D=\frac{1}{n_1+n_2-2}D_M^2. \qquad (6.1.25)$$

Note also that Hotelling's T_2^2 is related to D_M^2, that is,

$$T_2^2=\frac{n_1n_2}{n_1+n_2}D_M^2. \qquad (6.1.26)$$

Thus, by substitution in Equation 6.1.23, we see that the statistic

$$F=\frac{n_1+n_2-p-1}{(n_1+n_2-2)p}\left(\frac{n_1n_2}{n_1+n_2}\right)D_M^2 \qquad (6.1.27)$$

has an F distribution with p and n_1+n_2-p-1 degrees of freedom.

We may conclude from this discussion that the procedures proposed by Fisher and Mahalanobis, probably based on the earlier work of Hotelling, are essentially equivalent. In fact, when used for testing the hypothesis, H_0: $\boldsymbol{\mu}_1=\boldsymbol{\mu}_2$, the three methods yield exactly the same F statistic.

To illustrate the computations and test based on D_M^2, we shall again use the data of Section 6.1.2. For these data,

$$\mathbf{d}=\begin{bmatrix}-5.1\\-5.0\end{bmatrix},$$

and

$$\hat{\boldsymbol{\Sigma}}^{-1}=(n_1+n_2-2)\mathbf{W}^{-1}=18\begin{bmatrix}.01900167 & -.01525996\\ -.01525996 & .02132946\end{bmatrix}$$

$$=\begin{bmatrix}.34203006 & -.27467928\\ -.27467928 & .38393028\end{bmatrix}.$$

Therefore, from Equation 6.1.24, we find that $D_M^2=\mathbf{d}'\hat{\boldsymbol{\Sigma}}^{-1}\mathbf{d}=4.4858$, and substituting in Equation 6.1.27 yields

$$F=\frac{17(100)}{36(20)}(4.4858)=10.59,$$

which is exactly the same as that obtained in Section 6.1.3.

6.2 THE GENERAL CASE OF *G* GROUPS

So far in this chapter we have dealt with discriminant analysis in the case of only two groups. The reason for the emphasis on that special case is in part historical and in part heuristic. The earliest developments of discriminant analysis, by Hotelling, Fisher, and Mahalanobis, were restricted to the two-group case primarily for practical reasons; without computers the computational labor made application of the more general methods intractable. Furthermore, the mathematical treatment is somewhat easier to describe in the two-group case than in the more general case of G groups. Thus, we chose to develop the basic concepts of discriminant analysis in the context of Fisher's two-group method, which was described in some detail in Section 6.1.1.

Although many practical research problems in the social sciences involve discrimination between only two groups, there are obviously many others for which more general methods are required. The early workers in the development of methods of discriminant analysis were aware of the potential need for generalization to more than two groups. Fisher (1938), for example, indicated that in the case of G groups there would be $G-1$ discriminant functions. However, it was not until the late 1940s that notable progress was made in generalizing the earlier methods. Rao (1948) illustrated the use of two discriminant functions in classifying individuals into one of three Indian castes. Rao and Slater (1949) used discriminant analysis to determine whether patients classified in five neurotic groups could be distinguished on the basis of a number of personality variables. Lubin (1950), in England, and Bryan (1950), in the United States, proposed very similar generalized methods of discriminant analysis. Thus, three different statisticians, working independently and in widely dispersed geographic settings, published generalizations at about the same time.

Since the methods proposed by Lubin and Bryan are quite closely related to the Fisher method which was treated in detail in Section 6.1.1, much of the basic mathematics required for extension to the general case of G groups is already available to us. Therefore, in this section, we show how the Fisher method can be extended to yield discriminant equations when there are more than two groups involved; how a discriminant criterion is defined; and how it may be tested for significance.

6.2.1 Computation of a Discriminant Equation in the Case of *G* Groups

A discriminant equation for the case of G groups is identical to that given in Equation 6.1.1. Thus, the score on a linear combination of p Xs for the ith member $(i=1,2,\ldots,n_g)$ of group g $(g=1,2,\ldots,G)$ may be written

$$Z_{ig}=u_iX_{i1g}+u_2X_{i2g}+\cdots+u_pX_{ipg}. \tag{6.2.1}$$

The mean of the random variable Z, that is, the mean of this linear combination, for the gth group, may be denoted by $\bar{Z}_g$. In the case of two groups, separation between them was expressed in terms of $D=\bar{Z}_1-\bar{Z}_2$. When three or more groups are involved, separation between groups is expressed in terms of the variability among group means on the variable Z. More specifically, such variability is expressed, as in the univariate analysis of variance, by the sum of squares among group means; that is, $SS_A=\Sigma n_g(\bar{Z}_g-\bar{Z})^2$, where $\bar{Z}$ is the grand mean based on the $n=n_1+n_2+\cdots+n_G$

individuals in all groups combined. Because variability among means is due in part to variability among individuals, the discriminant criterion λ, in the case of G groups, is defined as the ratio of variability among group means, SS_A, to that within groups, SS_W. Thus,

$$\lambda = \frac{SS_A}{SS_W}. \tag{6.2.2}$$

This expression is analogous to Equation 6.1.8 in the two-group case. Again the values of the u's in Equation 6.2.1 are chosen to maximize λ.

In order to determine values of the u's that maximize λ, SS_A and SS_W must be expressed in terms of the u's. We have already shown that, for two groups, the within-groups sum of squares may be expressed as (see Equation 6.1.7)

$$SS_W = \mathbf{u}'\mathbf{W}\mathbf{u}. \tag{6.2.3}$$

If we define $\mathbf{W}$ more generally as

$$\mathbf{W} = \mathbf{S}_{xx}^{(1)} + \mathbf{S}_{xx}^{(2)} + \cdots + \mathbf{S}_{xx}^{(G)}, \tag{6.2.4}$$

the within-groups sum of squares is still expressed accurately by Equation 6.2.3. However, we must derive an equation that also expresses the sum of squares among groups, SS_A, in terms of the u's.

As indicated above, the sum of squares among group means for the variable, Z, may be written

$$SS_A = \sum_{g=1}^{G} n_g(\bar{Z}_g - \bar{Z})^2, \tag{6.2.5}$$

in which $\bar{Z}_g = u_1\bar{X}_{1g} + \cdots + u_p\bar{X}_{pg}$ is the mean of Z for group g and $\bar{Z} = u_1\bar{X}_1 + \ldots + u_p\bar{X}_p$ is the grand mean of Z for all groups combined. In these expressions the set $(\bar{X}_{1g}, \bar{X}_{2g}, \ldots, \bar{X}_{pg})$ is called the *centroid* of group g and the set $(\bar{X}_1, \bar{X}_2, \ldots, \bar{X}_p)$ is called the centroid of the total sample.

Let us define the vectors

$$\mathbf{u} = \begin{bmatrix} u_1 \\ \vdots \\ u_p \end{bmatrix}, \quad \bar{\mathbf{X}}_g = \begin{bmatrix} \bar{X}_{1g} \\ \vdots \\ \bar{X}_{pg} \end{bmatrix}, \quad \text{and} \quad \bar{\mathbf{X}} = \begin{bmatrix} \bar{X}_1 \\ \vdots \\ \bar{X}_p \end{bmatrix},$$

in which $\bar{\mathbf{X}}_g$ and $\bar{\mathbf{X}}$ express the sample centroids in vector form. Then $\bar{Z}_g = \mathbf{u}'\bar{\mathbf{X}}_g$ and $\bar{Z} = \mathbf{u}'\bar{\mathbf{X}}$, so that

$$\begin{aligned} (\bar{Z}_g - \bar{Z})^2 &= (\mathbf{u}'\bar{\mathbf{X}}_g - \mathbf{u}'\bar{\mathbf{X}})^2 \\ &= (\mathbf{u}'(\bar{\mathbf{X}}_g - \bar{\mathbf{X}}))^2 \\ &= \mathbf{u}'\left[(\bar{\mathbf{X}}_g - \bar{\mathbf{X}})(\bar{\mathbf{X}}_g - \bar{\mathbf{X}})'\right]\mathbf{u}. \end{aligned} \tag{6.2.6}$$

(Note: In going from the second to the third step here, we have made use of a property of the multiplication of vectors, namely that the two scalar products of two vectors are equal, that is, that $\mathbf{a}'\mathbf{b}=\mathbf{b}'\mathbf{a}$, where $\mathbf{a}$ and $\mathbf{b}$ are both $p \times 1$ vectors. Thus, $(\mathbf{a}'\mathbf{b})^2=\mathbf{a}'\mathbf{b}\mathbf{b}'\mathbf{a}$.)

Since $\mathbf{u}$ does not change in value from group to group, we may substitute for $(\bar{Z}_g-\bar{Z})^2$ in Equation 6.2.5 on the basis of Equation 6.2.6 to obtain

$$SS_A=\mathbf{u}'\left[\sum_{g=1}^{G} n_g(\bar{\mathbf{X}}_g-\bar{\mathbf{X}})(\bar{\mathbf{X}}_g-\bar{\mathbf{X}})'\right]\mathbf{u}. \tag{6.2.7}$$

In this equation $\bar{\mathbf{X}}_g-\bar{\mathbf{X}}$ is a $p \times 1$ vector so the product, $(\bar{\mathbf{X}}_g-\bar{\mathbf{X}})(\bar{\mathbf{X}}_g-\bar{\mathbf{X}})'$ is a $p \times p$ matrix of the squares and cross-products of the deviations of the X means for group g from the grand means of the corresponding Xs. When summed over the G groups, the expression in brackets is the sums of squares and cross-products matrix among groups, which we shall denote by $\mathbf{A}$, that is,

$$\mathbf{A}=\sum_{g=1}^{G} n_g(\bar{\mathbf{X}}_g-\bar{\mathbf{X}})(\bar{\mathbf{X}}_g-\bar{\mathbf{X}})'. \tag{6.2.8}$$

A typical element of $\mathbf{A}$ may be written

$$a_{jk}=\sum_{g=1}^{G} n_g(\bar{X}_{jg}-\bar{X}_j)(\bar{X}_{kg}-\bar{X}_k),$$

where $j,k=1,2,\ldots,p$. Note that for $G=2$, the matrix $\mathbf{A}$ is identical to the matrix $\mathbf{B}$ defined in Section 6.1.4.

Now substituting in Equation 6.2.7 yields

$$SS_A=\sum_{g=1}^{G} n_g(\bar{Z}_g-\bar{Z})^2=\mathbf{u}'\mathbf{A}\mathbf{u}, \tag{6.2.9}$$

and Equation 6.2.2, the discriminant criterion, may be expressed in terms of the u's as

$$\lambda=\frac{SS_A}{SS_W}=\frac{\mathbf{u}'\mathbf{A}\mathbf{u}}{\mathbf{u}'\mathbf{W}\mathbf{u}}. \tag{6.2.10}$$

To determine the values of the u's that maximize λ, we must first obtain the partial derivative of λ with respect to $\mathbf{u}$ and set this equal to zero. The result is

(see Appendix B)

$$\frac{\partial\lambda}{\partial\mathbf{u}}=\frac{(\mathbf{u}'\mathbf{W}\mathbf{u})\dfrac{\partial(\mathbf{u}'\mathbf{A}\mathbf{u})}{\partial\mathbf{u}}-(\mathbf{u}'\mathbf{A}\mathbf{u})\dfrac{\partial(\mathbf{u}'\mathbf{W}\mathbf{u})}{\partial\mathbf{u}}}{(\mathbf{u}'\mathbf{W}\mathbf{u})^2}$$

$$=\frac{2[(\mathbf{u}'\mathbf{W}\mathbf{u})(\mathbf{A}\mathbf{u})-(\mathbf{u}'\mathbf{A}\mathbf{u})(\mathbf{W}\mathbf{u})]}{(\mathbf{u}'\mathbf{W}\mathbf{u})^2}=\mathbf{0}.$$

Dividing both numerator and denominator by $\mathbf{u}'\mathbf{W}\mathbf{u}$ yields

$$\frac{2(\mathbf{A}\mathbf{u}-\lambda\mathbf{W}\mathbf{u})}{\mathbf{u}'\mathbf{W}\mathbf{u}}=\mathbf{0}$$

because $\mathbf{u'Au}/\mathbf{u'Wu}=\lambda$. Now, multiplying both sides by $\mathbf{u'Wu}/2$, we have $(\mathbf{A}-\lambda\mathbf{W})\mathbf{u}=\mathbf{0}$. Assuming that $\mathbf{W}^{-1}$ exists (see Appendix A), premultiplying by $\mathbf{W}^{-1}$ results in

$$(\mathbf{W}^{-1}\mathbf{A}-\lambda\mathbf{I})\mathbf{u}=\mathbf{0}. \qquad (6.2.11)$$

Note that this equation is identical in form to Equation 5.2.16, obtained in the context of canonical correlation. The general method discussed in Appendix A for the solution of matrix equations having this form can also be used for the solution of Equation 6.2.11. The characteristic equation of the matrix, $\mathbf{W}^{-1}\mathbf{A}$, is

$$|\mathbf{W}^{-1}\mathbf{A}-\lambda\mathbf{I}|=0. \qquad (6.2.12)$$

The number of characteristic roots (values of λ) of this equation is equal to the rank (see Appendix A) of $\mathbf{W}^{-1}\mathbf{A}$. Now the rank of the product of two matrices cannot exceed the rank of the one having the smaller rank. Because we have already assumed, in obtaining $\mathbf{W}^{-1}$, that $\mathbf{W}$ is non-singular, its rank must be equal to p, the number of X's. Because of the fact that $\Sigma_{g=1}^{G} n_g(\bar{Z}_g-\bar{Z})=0$, the rank of $\mathbf{A}$ is $G-1$, which is ordinarily less than p. Therefore, in general, the rank of $\mathbf{W}^{-1}\mathbf{A}$, and, hence, the number of values of λ, is equal to the smaller of p and $G-1$. It is usually equal to $G-1$, however, because the number of groups is usually less than or equal to the number of variables.

As we shall see, each value of λ obtained by solving Equation 6.2.12 has associated with it an eigenvector, $\mathbf{u}$, the elements of which are the weights applied to the Xs in determining the corresponding discriminant function, Z. The first such function, say Z_1, defines a dimension on which the groups differ maximally. The second, say Z_2, defines a dimension *uncorrelated with the first*, on which group differences are second in magnitude. All succeeding functions are uncorrelated with the first two and with each other. Thus, the discriminant analysis produces a number (the smaller of p and $G-1$) of discriminant functions, all mutually uncorrelated and ordered from greatest to least in terms of the extent to which they discriminate between the groups.

6.2.2 An Illustration of the Computations

To illustrate the computation of the characteristic roots (eigenvalues) of $\mathbf{W}^{-1}\mathbf{A}$, we shall use the following fictitious data for three groups ($G=3$) and three variables ($p=3$). These data are taken from Rulon, et al. (1967), and consist of scores on three scales of an Activity Preference Inventory (API), namely, Outdoor (X_1), Convivial (X_2), and Conservative (X_3), for three groups of airline employees, Passenger Agents ($n=10$), Mechanics ($n=13$), and Operations Control Agents ($n=14$). A high score on each scale indicates high preference for that type of activity. The data are given in Table 6.2.1.

TABLE 6.2.1 SCORES OF THREE GROUPS OF AIRLINE EMPLOYEES ON THREE SCALES OF AN ACTIVITY PREFERENCE INVENTORY

Group		X_1	X_2	X_3
Passenger agents ($n = 10$)		10	22	13
		20	25	12
		10	24	5
		13	21	11
		11	22	11
		8	29	14
		22	22	6
		15	21	4
		11	23	5
		12	26	9
	Mean	13.20000000	23.50000000	9.0000000000
	St. Dev.	4.5411697	2.5495098	3.7118429
Mechanics ($n = 13$)		18	26	10
		12	16	10
		17	24	5
		15	22	13
		17	19	12
		20	19	11
		17	24	11
		16	19	8
		14	24	7
		16	22	5
		24	14	7
		11	25	12
		17	19	11
	Mean	16.46153846	21.00000000	9.384615385
	St. Dev.	3.3069468	3.6514837	2.6937725
Operations control agents ($n = 14$)		4	12	11
		13	20	16
		13	15	18
		13	16	7
		17	15	10
		11	12	19
		15	16	14
		15	18	14
		4	10	15
		10	12	9
		17	18	9
		15	18	14
		20	13	19
		18	11	19
	Mean	13.21428571	14.71428571	13.85714286
	St. Dev.	4.742258	3.099096	4.111095
	Total Mean	14.35135135	19.2972973	10.97297297

We should emphasize that these data are used here only to illustrate the computation and interpretation of results of a discriminant analysis. As we noted earlier, we recommend that the number of individuals in the sample be at least 20 times the number of variables, but not less than 100. The example used here falls far short of that requirement, but will serve nevertheless to show how the analysis should be carried out and interpreted.

We first need to compute the sums of squares and cross-products matrices within and among groups, **W** and **A**, respectively. These are:

$$\mathbf{W}=\begin{bmatrix} 609.187912 & -10.142857 & -13.879121 \\ -10.142857 & 343.357143 & -1.571429 \\ -13.879121 & -1.571429 & 430.791209 \end{bmatrix}$$

and

$$\mathbf{A}=\begin{bmatrix} 89.244520 & 71.277993 & -66.769528 \\ 71.277993 & 508.372587 & -303.131274 \\ -66.769528 & -303.131274 & 188.181764 \end{bmatrix}.$$

Then

$$\mathbf{W}^{-1}=\begin{bmatrix} .001643552 & .000048794 & .000053130 \\ .000048794 & .002913917 & .000012201 \\ .000053130 & .000012201 & .002323066 \end{bmatrix}$$

and

$$\mathbf{W}^{-1}\mathbf{A}-\lambda\mathbf{I}=\begin{bmatrix} .146608-\lambda & .125849 & -.114532 \\ .211238 & 1.481135-\lambda & -.884261 \\ -.149499 & -.694204 & .429913-\lambda \end{bmatrix}.$$

To obtain the roots (values of λ) of $\mathbf{W}^{-1}\mathbf{A}$, we need to solve the determinantal equation $|\mathbf{W}^{-1}\mathbf{A}-\lambda\mathbf{I}|=0$. We find that

$$|\mathbf{W}^{-1}\mathbf{A}-\lambda\mathbf{I}|=\lambda^3-2.057656\lambda^2+.259370\lambda=0.$$

Dividing through by λ yields

$$\lambda^2-2.057656\lambda+.259370=0,$$

which can be solved using the quadratic formula to yield $\lambda_1=1.922762$ and $\lambda_2=.134894$. A convenient check on the computation of the roots can be made, based on the fact that the sum of the roots must be equal to the trace of $\mathbf{W}^{-1}\mathbf{A}$ (see Appendix A). Here, we find that the trace of $\mathbf{W}^{-1}\mathbf{A}$ is

$$\text{tr}(\mathbf{A})=.146608+1.481135+.429913=2.057656,$$

and

$$\lambda_1+\lambda_2=1.922762+.134894=2.057656.$$

In the next section we shall describe a method for determining whether the roots (eigenvalues) of $\mathbf{W}^{-1}\mathbf{A}$ are statistically significant.

6.2.3 Testing the Significance of the Discriminant Criterion, λ

As indicated at the beginning of this chapter, discriminant analysis deals with the answers to two basic questions, namely, Can the groups be distinguished on the basis of the set of X measures, and How can individuals be classified as members of one of the groups on the basis of the Xs? By itself, computation of the λs answers neither of these questions. It does, however, provide a set of statistics that are essential in answering the first question.

Before describing the appropriate test, we should indicate what hypothesis is being tested. Clearly, the groups, or populations, represented in a discriminant analysis cannot be distinguished on the basis of the X measures if their means on any linear combination of the Xs are all equal. Therefore, the null hypothesis is that the population parameters corresponding to the sample estimates, $\bar{Z}_g$, are all equal. From Equations 6.2.9 and 6.2.10, we can see that there is a direct relationship between λ and $SS_A = \Sigma_{g=1}^{G} n_g(\bar{Z}_g - \bar{Z})^2$. If all $\bar{Z}_g$ are equal, then $\lambda = 0$. Thus, an alternative statement of the null hypothesis is that the population parameter corresponding to λ is equal to zero. In other words, the test of significance of λ answers the question, Is λ sufficiently different from zero to permit the conclusion that the groups differ significantly on the basis of the linear combination, Z, associated with λ?

A test of significance of the λs may be carried out using Wilk's (1932) likelihood ratio criterion, introduced in Chapter 5. We first define

$$\frac{1}{\Lambda_1} = (1+\lambda_1)(1+\lambda_2)\ldots(1+\lambda_r)$$

$$= \prod_{i=1}^{r} (1+\lambda_i), \tag{6.2.13}$$

in which r is the number of nonzero values of λ obtained by solving Equation 6.2.12. Bartlett's (1947) V statistic can now be expressed as

$$V_1 = -\left[n - 1 - \frac{(p+G)}{2}\right] \ln \Lambda_1$$

$$= \left[n - 1 - \frac{(p+G)}{2}\right] \ln\left(\frac{1}{\Lambda_1}\right), \tag{6.2.14}$$

The statistic V_1 provides a test of the null hypothesis that the population parameters corresponding to λ_1 through $\lambda_{hr.}$ are equal to zero. When the null hypothesis is true and the p variables have a multivariate normal distribution, V_1 has approximately a chi-square distribution with $p(G-1)$ degrees of freedom for large n. If this test results in rejection of the null hypothesis, we conclude that at least one of the parameters corresponding to the λs is greater than zero. Because λ_1 is the maximum root, we consider it to be statistically significant.

If the above test results in rejection, one may remove the first root, λ_1, from Equation 6.2.13 to obtain

$$\frac{1}{\Lambda_2} = (1+\lambda_2)\ldots(1+\lambda_r) = \prod_{i=2}^{r} (1+\lambda_i),$$

and the corresponding Bartlett's V is

$$V_2 = \left[n - 1 - \frac{(p+G)}{2}\right] \ln\left(\frac{1}{\Lambda_2}\right).$$

This statistic provides a test of the null hypothesis that the parameters corresponding to λ_2 through λ_r are all equal to zero. If this hypothesis is rejected, we conclude that λ_2 is significant because it is the maximum among λ_2 through λ_r.

In general, a test of significance of the kth root, λ_k, may be carried out by computing

$$V_k = \left[n - 1 - \frac{(p+G)}{2} \right] \ln\left(\frac{1}{\Lambda_k} \right), \tag{6.2.15}$$

in which

$$\frac{1}{\Lambda_k} = \prod_{i=k}^{r} (1+\lambda_i). \tag{6.2.16}$$

V_k has an approximate chi-square distribution with $(p-k+1)(G-k)$ degrees of freedom, again assuming a multivariate normal distribution of the p Xs. Equations 6.2.15 and 6.2.16 may be used to test successively the significance of the functions remaining after λ_1 through λ_{k-1} have been removed. When such a test is found nonsignificant, we conclude that no values of λ smaller than λ_k will be significant and no further tests need to be carried out.

For the data of the above example, we found that $\lambda_1 = 1.922762$ and $\lambda_2 = .134894$. Then, from Equation 6.2.13,

$$\frac{1}{\Lambda_1} \prod_{i=1}^{2} (1+\lambda_i) = (1+1.922762)(1+.134894)$$
$$= 3.317025,$$

and

$$V_1 = \left[37 - 1 - \frac{(3+3)}{2} \right] \ln(3.317025)$$
$$= 33(1.199068) = 39.57.$$

With $p(G-1) = 3(2) = 6$ degrees of freedom, this result is significant well beyond the .0005 level. We would therefore conclude that the three groups of airline employees can be differentiated on the basis of the first discriminant function.

Considering the second root, λ_2, of $\mathbf{W}^{-1}\mathbf{A}$, we have

$$\frac{1}{\Lambda_2} = 1 + .134894 = 1.134894$$

and

$$V_2 = \left[37 - 1 - \frac{(3+3)}{2} \right] (.126540) = 4.18.$$

With $(p-k+1)(G-k) = (2)(1) = 2$ degrees of freedom, this result is not significant. We conclude, therefore, that the three groups of airline employees are significantly different on the basis of only one of the two possible dimensions represented in the variables (Xs) in this example.

Once we have determined that the groups can be differentiated on the basis of one or more dimensions measured by the variables under consideration, it is important to define those dimensions and to describe how the groups differ in terms of them. The dimensions along which groups have been found to differ are defined on the basis of the emphasis (weights) given to the variables in the discriminant equations. Therefore, we must have a procedure for determining the discriminant weights (eigenvectors) associated with each significant discriminant criterion (eigenvalue), λ. In other words, we need to determine the values of the u's in Equation 6.2.1 corresponding to each significant λ.

A procedure for obtaining the eigenvector, $\mathbf{u}$, associated with each eigenvalue, λ, has been suggested in connection with canonical correlation in Chapter 5. The steps are:

1. Substitute the value of λ in the matrix $\mathbf{W}^{-1}\mathbf{A}-\lambda\mathbf{I}$. This step simply involves subtracting the value of λ from each of the diagonal elements of $\mathbf{W}^{-1}\mathbf{A}$. For the above example, we had only one significant value of λ, namely $\lambda_1 = 1.922762$. Subtracting this value from each diagonal element of $\mathbf{W}^{-1}\mathbf{A}$ yields

$$\mathbf{W}^{-1}\mathbf{A}-\lambda_1\mathbf{I} = \begin{bmatrix} -1.776153 & .125849 & -.114532 \\ .211238 & -.441627 & -.884261 \\ -.149499 & -.694204 & -1.492849 \end{bmatrix}.$$

2. Compute the cofactors (see Appendix A) of the elements in any row of $\mathbf{W}^{-1}\mathbf{A}-\lambda\mathbf{I}$. Denote the vector of these cofactors by $\mathbf{f}$. For the above matrix,

$$\mathbf{f}_1 = \begin{bmatrix} .045424 \\ .447543 \\ -.212665 \end{bmatrix}$$

is the vector of cofactors of the elements in the first row of $\mathbf{W}^{-1}\mathbf{A}-\lambda_1\mathbf{I}$.

3. Divide each element of $\mathbf{f}$ by the square root of the sum of squares of the elements of $\mathbf{f}$, a process referred to as *normalizing* the vector. The resulting vector is $\mathbf{u}$. For the above data,

$$\mathbf{u}_1 = \begin{bmatrix} .091290 \\ .899442 \\ -.427400 \end{bmatrix}.$$

The elements of $\mathbf{u}$ found by the above procedure may be substituted in Equation 6.2.1 to yield, for the data of this example, the discriminant equation

$$Z_{ig1} = .091290X_{i1g} + .899442X_{i2g} - .427400X_{i3g}. \qquad (6.2.17)$$

If other discriminant criteria (λs) had been found statistically significant, the corresponding weights, $\mathbf{u}$, would yield different discriminant functions. Each such function is a measurement of some dimension (characteristic) on the basis of which the groups involved have been found to differ. The nature of the differences revealed by discriminant analysis will be discussed in the next section.

6.2.4 Interpreting the Results of Discriminant Analysis

Although the data of the example were fictitious, it is instructive to discuss the results obtained in Sections 6.2.2 and 6.2.3. First, we may wish to determine *how* the groups differ from one another in terms of the dimensions measured by the discriminant functions. To do this it is necessary to compute the mean for each group on each significant discriminant function. For the above data, group means on the first discriminant function can be obtained by substituting in

$$\bar{Z}_{g1} = u_1\bar{X}_{1g} + u_2\bar{X}_{2g} + u_3\bar{X}_{3g}.$$

The results are: $\bar{Z}_{11} = 18.4953$ for Passenger Agents, $\bar{Z}_{21} = 16.3800$ for Mechanics, and $\bar{Z}_{31} = 8.5184$ for Operations Control Agents. The mean for all groups combined is $\bar{Z}_1 = 13.9771$. Thus, it is clear that Passenger Agents and Mechanics are quite similar in terms of their means on the first discriminant function, while Operations Control Agents have a much lower mean.

Having determined how the groups differ on this dimension, we would find it helpful to *define* the characteristic measured by the function Z_1. Because Z_1 is a linear combination of several Xs, a description of what it measures must be based on the relative emphasis accorded the Xs in the linear combination. One might suspect that the elements of $\mathbf{u}_1$, that is, the weights applied to the Xs in the function Z_1, would provide such information. However, because $\mathbf{u}_1$ is based on analysis of the raw X scores, the relative magnitudes of its elements are influenced by differences, both in scaling and in variability, among the Xs, as well as by the emphasis given to them in Z_1. Therefore, to be useful in indicating the relative contributions of the Xs to Z_1, the elements of $\mathbf{u}_1$ must be transformed to standard form.

To obtain the standardized form of $\mathbf{u}$, we define a diagonal transformation matrix, $\mathbf{H}$, the elements of which are the standard deviations of the Xs, computed from the diagonal elements of the matrix, $\mathbf{W}$. Thus, a general element of $\mathbf{H}$ is

$$h_{jj} = \sqrt{\frac{\sum_{g=1}^{G} \sum_{i=1}^{n_g} (X_{ijg} - \bar{X}_{jg})^2}{n - G}}.$$

Note that $h_{jk} = 0$ $(j \neq k)$. Then the vector of standardized weights, which we shall denote by $\mathbf{u}_s$, may be obtained by

$$\mathbf{u}_s = \mathbf{u}'\mathbf{H} = \mathbf{H}\mathbf{u}. \qquad (6.2.18)$$

For the data of our example, the transformation matrix is

$$\mathbf{H} = \begin{bmatrix} 4.232882 & 0 & 0 \\ 0 & 3.177851 & 0 \\ 0 & 0 & 3.559541 \end{bmatrix}.$$

The vector of standardized weights corresponding to $\mathbf{u}_1$ is

$$\mathbf{u}_{s1} = \mathbf{H}\mathbf{u}_1 = \begin{bmatrix} .386420 \\ 2.858292 \\ -1.521348 \end{bmatrix},$$

which shows that the relative contributions of the three measures are roughly in the proportion $1:7:-4$. Because the Xs are measures of preferences for Outdoor (X_1), Convivial (X_2), and Conservative (X_3) activities, one should probably define the function Z_1, based on the first discriminant criterion, λ_1, as a measure of preference for Convivial, Liberal (because the weight on X_3 is negative) activities. The fact that both Passenger Agents and Mechanics had much larger means (18.5 and 16.4, respectively) than Operations Control Agents (8.5) is consistent with such an interpretation of Z_1. We would expect the latter group to prefer fewer convivial and more conservative activities than either of the other groups.

Although the second discriminant criterion in this example, $\lambda_2 = .134894$, was not statistically significant, it will be instructive to compute the corresponding discriminant function, Z_2, and compare it with Z_1. Following the same procedures used to obtain Z_1, we find that

$$\mathbf{u}_2 = \begin{bmatrix} .925482 \\ -.298440 \\ -.233274 \end{bmatrix} \qquad (6.2.19)$$

and $Z_{ig2} = .925482X_{i1g} - .298440X_{i2g} - .233274X_{i3g}$. The means of the three groups are $\bar{Z}_{12} = 3.1035$ for Passenger Agents, $\bar{Z}_{22} = 6.7784$ for Mechanics, and $\bar{Z}_{32} = 4.6057$ for Operations Control Agents. The mean of the combined sample is $\bar{Z}_2 = 4.9631$. The standardized form of $\mathbf{u}_2$ is

$$\mathbf{u}_{s2} = \begin{bmatrix} 3.917456 \\ -.948398 \\ -.830348 \end{bmatrix}.$$

Inspection of the elements of $\mathbf{u}_{s2}$ reveals that they are roughly in the proportion $5:-1:-1$. Therefore, it appears that Z_2 is primarily a measure of preference for Outdoor activity. Although the difference between means on Z_2 is not statistically significant, the fact that the Mechanics Group has a noticeably larger mean than either Passenger Agents or Operations Control Agents is consistent with what we would expect on the basis of the usual work location for the Mechanics Group.

It is clear from an inspection of the rough weightings on Z_1 and Z_2, that is,

Variable	Z_1	Z_2
Outdoor (X_1)	1	5
Convivial (X_2)	7	−1
Conservative (X_3)	−4	−1

that the two discriminant functions, Z_1 and Z_2, are measures of quite different dimensions. Furthermore, as we pointed out earlier, these dimensions are uncorrelated in the total sample and ordered on the basis of the extent to which they discriminate between the groups. In summary, then, the discriminant analysis of the above data permits the following conclusions:

1. It is possible to distinguish among the three groups on the basis of linear combinations of the set of *X*s.

2. The three groups differ significantly on the basis of a measure, Z_1, of preference for Convivial, Liberal activities, with Passenger Agents and Mechanics having much larger means on this dimension than Operations Control Agents.

3. The groups differ, but not significantly (probably because of the small sample size in our illustrative data) on the basis of a second measure, Z_2, of preference for outdoor activities, with Mechanics having a noticeably higher mean than either of the other groups.

6.2.5 The Relationship Between Discriminant Analysis and Canonical Correlation

Before concluding our discussion of discriminant analysis in the general case of *G* groups, we should mention briefly its relationship to canonical correlation, with which we dealt in some detail in Chapter 5. Earlier in this chapter we described the correspondence in the two-group case between discriminant analysis and multiple regression analysis. At one time this correspondence led researchers to the conclusion that discriminant analysis was, *in general*, merely a special case of multiple regression analysis. Actually, if we realize that multiple regression analysis is, itself, a special case of canonical correlation analysis, the correspondence holds. In other words, canonical correlation analysis is the most general analysis, of which multiple regression and discriminant analysis are *both* special cases.

Several special types of canonical correlation analysis are listed in Table 6.2.2, together with a description of the types of variables making up each set. It can be seen that many of the methods discussed in previous chapters may be regarded as special cases of canonical correlation. Discriminant analysis is no exception. If we define each variable in one set as dichotomous, with values of 0 or 1 to indicate membership in each of *G* groups, then discriminant analysis can be seen to be a special case of canonical correlation analysis. The number of variables in the set is then equal to the number of groups, *G*. Tatsuoka (1953) proved mathematically that discriminant analysis and canonical correlation analysis yield identical results. A complete discussion of the correspondence between these methods is given in Tatsuoka (1971).

TABLE 6.2.2 THE NATURE OF THE VARIABLES IN SEVERAL SPECIAL TYPES OF CANONICAL CORRELATION ANALYSIS

	Variables in			
	set Y		*set X*	
Type of analysis	*Number*	*Type*	*Number*	*Type*
Canonical correlation	*p*	Continuous	*q*	Continuous
Multiple regression	1	Continuous	*q*	Continuous
Bivariate regression and correlation	1	Continuous	1	Continuous
Point biserial	1	Dichotomous	1	Continuous
Discriminant analysis (Two groups)	1	Dichotomous	*q*	Continuous
Discriminant analysis (*G* groups)	*G*	Each dichotomous, denoting group membership	*q*	Continuous

6.3 THE CLASSIFICATION PROBLEM: ASSIGNING INDIVIDUALS TO GROUPS

As we indicated at the beginning of this chapter, discriminant analysis involves the answers to two important questions. The first, which we have considered in previous sections, is whether two or more groups can be distinguished from one another on the basis of a set of X measures. Assuming that the answer to this question is yes, we can address the second, namely, How can the discriminant equation be used to determine which one of the groups an individual most closely resembles?

The methods to be discussed here may be applied in a number of settings. For example, they may be used by the armed forces in assigning new personnel to training programs. They also may be used in industry for assigning new employees to particular job categories. In education, they may be found useful as aids in both educational and vocational counseling to help guide students toward educational programs and career choices that seem most appropriate in terms of their interests, attitudes, special aptitudes and general intellectual ability.

The use of these methods requires that data on the Xs be obtained initially on a sufficiently large sample of individuals *who are already members of the groups involved*. Again our recommendation is that the total sample size should be at least $20p$, that is, 20 times the number of variables (Xs) upon which the classification is to be based. The information obtained from this sample may then be used to classify (or assign) *new* applicants (or recruits or students) according to the group they most closely resemble. In practice, of course, it is important that the information obtained from such a sample be carefully validated before it is put to use in classifying new individuals. Methods of cross-validation using a second, independent, sample, such as those discussed in connection with canonical correlation (Chapter 5), should be used to help minimize the probability of misclassification.

Our discussion in this section will not treat all of the methods that have been proposed for dealing with the classification problem. Those we have selected seem to us to be straightforward conceptually and require a level of mathematical and statistical sophistication consistent with that in other chapters. We do mention briefly some other available procedures and provide references to them.

6.3.1 Measuring Group Resemblance

In the context of discriminant analysis, the decision to classify an individual as a member of a particular group is ordinarily based on the concept of group resemblance. The methods to be discussed here generally involve measuring the extent to which an individual resembles each of several groups. The individual is then classified as a member of the group he or she most closely resembles. The question is, How can group resemblance be measured?

Let us consider the simplest case first, namely, the case in which we have only two populations and we wish to classify an individual as a member of one of them on the basis of a single X variable. Intuitively, one might decide that an appropriate measure of resemblance to either of the two populations would be the difference between the individual's X value, X_i, and the mean, μ_g, of that population. Then the measures of resemblance to each of the two populations would be $X_i - \mu_1$ and $X_i - \mu_2$ and the individual would be classified as a member of the population corresponding to the smaller absolute value of these two differences. In this case, the difference $|X_i - \mu_g|$ may be regarded as a measure of *distance* between the individual and the gth population; the greater the distance, the smaller is the resemblance to that population.

A little thought about the measure $X_i - \mu_g$ should convince us that it has one important disadvantage. If the two populations had different standard deviations, that is, if $\sigma_1 \neq \sigma_2$, then the differences $X_i - \mu_1$ and $X_i - \mu_2$ would not be directly comparable. For example, if $\sigma_1 = 2\sigma_2$, then a difference $X_i - \mu_1$ twice as large as $X_i - \mu_2$ would have to be considered equivalent to $X_i - \mu_2$ in terms of standard deviation units. We can easily overcome this problem by expressing the measure of an individual's resemblance to the gth population in terms of a standard score with respect to that population, as

$$z_{ig} = \frac{X_i - \mu_g}{\sigma_g}.$$

We can further improve the measure by using the square of this expression, because all values of the measure will then be positive, and thus directly comparable. The measure of distance between individual i and population g then becomes

$$z_{ig}^2 = \frac{(X_i - \mu_g)^2}{\sigma_g^2}. \tag{6.3.1}$$

Clearly, if this value is relatively large, the individual bears little resemblance to population g; if it is small, say close to zero, the individual strongly resembles population g.

An important observation should be made at this point about the distribution of z_{ig}^2, defined in Equation 6.3.1, in a population of individuals. It is well known (see, for example, Freeman, 1963, p. 206 or Hays, 1973, p. 432) that if the distribution of X is normal with mean μ and variance σ^2, then $z^2 = \frac{(X-\mu)^2}{\sigma^2}$ has a chi-square distribution with one degree of freedom. The importance of this fact in this context is that it permits us to express an individual's resemblance to population g in terms of the *probability* of a distance at least as large as his from μ_g, given that the individual is a member of population g. As we shall see, such a probability forms the basis for a well-known measure of resemblance, called a *centour score*, which we shall discuss later in this chapter.

We now need to generalize Equation 6.3.1 to the case of G groups (populations) and p variables (Xs). Let us first express that equation as

$$z_{ig}^2 = (X_i - \mu_g)^2 (\sigma_g^2)^{-1} = (X_i - \mu_g)(\sigma_g^2)^{-1}(X_i - \mu_g). \tag{6.3.2}$$

Now if there are p Xs instead of just one, the expression $X_i - \mu_g$ becomes a $p \times 1$ vector, which we shall denote by $\mathbf{X}_{ig}$, and σ_g^2 becomes a square symmetrical matrix of order p, whose diagonal elements are the variances of the Xs and whose off-diagonal elements are the covariances of the Xs. We denote this variance-covariance matrix for population g by $\mathbf{\Sigma}_g$. Then, for the gth population, we can generalize Equation 6.3.2 to yield

$$z_{ig}^2 = \mathbf{X}_{ig}' \mathbf{\Sigma}_g^{-1} \mathbf{X}_{ig}, \tag{6.3.3}$$

in which

$$\mathbf{X}_{ig} = \begin{bmatrix} X_{i1} - \mu_{1g} \\ X_{i2} - \mu_{2g} \\ \vdots \\ X_{ip} - \mu_{pg} \end{bmatrix} \text{ and } \mathbf{\Sigma}_g = \begin{bmatrix} \sigma_{1g}^2 & \sigma_{12g} & \cdots & \sigma_{1pg} \\ \sigma_{21g} & \sigma_{2g}^2 & \cdots & \sigma_{2pg} \\ \vdots & \vdots & \ddots & \vdots \\ \sigma_{p1g} & \sigma_{p2g} & \cdots & \sigma_{pg}^2 \end{bmatrix}.$$

This expression is completely analogous to that in Equation 6.3.1. In both equations, z_{ig}^2 is a measure of the distance between individual i and population g.

Now what statement can be made about the distribution of z_{ig}^2 in Equation 6.3.3? We consider first the special case in which the Xs are independent, using the same argument as in Section 4.1.1. Then $\mathbf{\Sigma}_g$ becomes the diagonal matrix

$$\boldsymbol{\Sigma}_g = \begin{bmatrix} \sigma_{1g}^2 & 0 & \cdots & 0 \\ 0 & \sigma_{2g}^2 & \cdots & 0 \\ \vdots & \vdots & \ddots & \vdots \\ 0 & 0 & \cdots & \sigma_{pg}^2 \end{bmatrix},$$

and $z_{ig}^2 = \sum_{j=1}^{p} \frac{(X_{ij} - \mu_{jg})^2}{\sigma_{jg}^2}$. Note that if we assume a normal distribution for each X, then each of the terms in this sum has a chi-square distribution with one degree of freedom. It is well known (see, for example, Freeman, 1963, p. 207 or Hays, 1973, p. 436) that the sum of n independent chi-square variables, each with one degree of freedom, is itself a chi-square variable with n degrees of freedom. Thus, if all the Xs are independent, z_{ig}^2 has a chi-square distribution with p degrees of freedom. Ordinarily, of course, the Xs are not all independent, but it can be shown (see Tatsuoka, 1971, Chapter 5, for a thorough and readable proof) that if we assume a multivariate normal distribution of the Xs, then Equation 6.3.3 has a chi-square distribution with p degrees of freedom even when the Xs are not all independent.

Equation 6.3.3 was expressed in terms of population parameters that, in practice, are not usually known. These parameters may, however, be estimated on the basis of sample data and Equation 6.3.3 must be revised accordingly. There are two cases to consider:

1. If it can be assumed that all population variance-covariance matrices are equal, so that $\boldsymbol{\Sigma}_1 = \boldsymbol{\Sigma}_2 = \ldots = \boldsymbol{\Sigma}_g = \boldsymbol{\Sigma}$, then a single, pooled estimate of $\boldsymbol{\Sigma}$ can be obtained from the G samples. Such an estimate is

$$\mathbf{V} = \left(\frac{1}{n_1 + n_2 + \cdots + n_G - G} \right) \mathbf{W}, \tag{6.3.4}$$

in which $\mathbf{W}$ is the pooled within-groups sums of squares and cross-products matrix defined in Equation 6.2.4. Now, with

$$\mathbf{x}_{ig} = \begin{bmatrix} X_{i1} - \bar{X}_{1g} \\ X_{i2} - \bar{X}_{2g} \\ \vdots \\ X_{ip} - \bar{X}_{pg} \end{bmatrix}$$

as an estimate of $\mathbf{X}_{ig}$, we may rewrite Equation 6.3.3 as

$$z_{ig(P)}^2 = \mathbf{x}_{ig}' \mathbf{V}^{-1} \mathbf{x}_{ig}, \tag{6.3.5}$$

where the subscript (P) indicates the use of the pooled estimate, $\mathbf{V}$, of $\boldsymbol{\Sigma}$.

2. If the assumption of equal population variance-covariance matrices is not tenable, then the estimate of $\boldsymbol{\Sigma}_g$ must be based on the sample data for

the gth group. In that case,

$$\mathbf{V}_g = \mathbf{S}_{xx}^{(g)}\left(\frac{1}{n_g - 1}\right), \tag{6.3.6}$$

in which $\mathbf{S}_{xx}^{(g)}$ is the sums of squares and cross-products matrix for group g, defined as in Chapter 5. Then the measure of the ith individual's resemblance to group g may be written

$$z_{ig(S)}^2 = \mathbf{x}_{ig}' \mathbf{V}_g^{-1} \mathbf{x}_{ig}, \tag{6.3.7}$$

where the subscript (S) means that separate estimates, $\mathbf{V}_g$, of $\mathbf{\Sigma}_g$ are used.

It is clear that, because $\mathbf{V}_g$ is likely to vary from one group to another, the magnitude of $z_{ig(S)}^2$ depends not only on the distance of the individual from group g, but also on the variability within group g. Tatsuoka (1971) therefore suggests using the following modification of Equation 6.3.7:

$$z_{ig(T)}^2 = \mathbf{x}_{ig}' \mathbf{V}_g^{-1} \mathbf{x}_{ig} + \ln|\mathbf{V}_g|, \tag{6.3.8}$$

where the subscript (T) refers to Tatsuoka. The effect of this modification, as well as the use of Equations 6.3.5 and 6.3.7, will be illustrated below. We should point out that Equations 6.3.5, 6.3.7, and 6.3.8 are no longer distributed as chi square with p degrees of freedom.

The classification rule to be used in connection with each of the expressions $z_{ig(P)}^2$, $z_{ig(S)}^2$, and $z_{ig(T)}^2$, is to assign the individual to the group for which the value of the statistic is smallest.

6.3.2 An Example

To illustrate the use of Equations 6.3.5, 6.3.7, and 6.3.8 for classification, we use the data of Section 6.2.1 concerning the Activity Preference Inventory administered to three groups of airline employees. Let us consider the classification of two new employees who were found to have the following scores on X_1 (Outdoor Activities), X_2 (Convivial Activities), and X_3 (Conservative Activities):

Employee	X_1	X_2	X_3
1	11	27	8
2	13	15	14

The question we wish to answer is as follows: Which of the three groups, Passenger Agents, Mechanics, or Operations Control Agents, does each of these employees most closely resemble, in terms of the activity preferences measured by X_1, X_2, and X_3?

The variance-covariance matrices and their determinants for the three groups are:

Passenger Agents

$$\mathbf{V}_1 = \begin{bmatrix} 20.622222 & -3.444444 & -4.111111 \\ -3.444444 & 6.500000 & 3.888889 \\ -4.111111 & 3.888889 & 13.777778 \end{bmatrix}$$

$$|\mathbf{V}_1| = 1371.772154$$

Mechanics

$$\mathbf{V}_2 = \begin{bmatrix} 10.935897 & -4.916667 & -2.025641 \\ -4.916667 & 13.333333 & .166667 \\ -2.025641 & .166667 & 7.256410 \end{bmatrix}$$

$$|\mathbf{V}_2| = 830.964207$$

Operations Control Agents

$$\mathbf{V}_3 = \begin{bmatrix} 22.489011 & 6.142857 & 3.648352 \\ 6.142857 & 9.604396 & -2.967033 \\ 3.648352 & -2.967033 & 16.901099 \end{bmatrix}$$

$$|\mathbf{V}_3| = 2553.960987.$$

It appears from an inspection of these matrices that the assumption of the equality of the population variance-covariance matrices is not supported. We note, for example, that $s_{23} = 3.888889$ for Passenger Agents is about 23 times larger than $s_{23} = .166667$ for Mechanics. In spite of these differences, we shall use the data to illustrate the use of Equation 6.3.5, as well as the application of Equations 6.3.7 and 6.3.8.

The vectors of differences between individual scores and group centroids are shown in Table 6.3.1.

To use Equation 6.3.5, we first obtain the pooled variance-covariance matrix, **V**, by applying Equation 6.3.4. Then

$$\mathbf{V} = \begin{bmatrix} 17.917292 & -0.298319 & -0.408209 \\ -0.298319 & 10.098740 & -0.046218 \\ -0.408209 & -0.046218 & 12.670330 \end{bmatrix},$$

TABLE 6.3.1 VECTORS OF DIFFERENCES BETWEEN INDIVIDUAL SCORES AND GROUP CENTROIDS ON THE ACTIVITY PREFERENCE INVENTORY

Employee 1	*Employee* 2
$\mathbf{x}_{11} = \begin{bmatrix} -2.200000 \\ 3.500000 \\ -1.000000 \end{bmatrix}$	$\mathbf{x}_{21} = \begin{bmatrix} -0.200000 \\ -8.500000 \\ 5.000000 \end{bmatrix}$
$\mathbf{x}_{12} = \begin{bmatrix} -5.461538 \\ 6.000000 \\ -1.384615 \end{bmatrix}$	$\mathbf{x}_{22} = \begin{bmatrix} -3.461538 \\ -6.000000 \\ 4.615385 \end{bmatrix}$
$\mathbf{x}_{13} = \begin{bmatrix} -2.214286 \\ 12.285714 \\ -5.857143 \end{bmatrix}$	$\mathbf{x}_{23} = \begin{bmatrix} -0.214286 \\ 0.285714 \\ 0.142857 \end{bmatrix}$

which has inverse

$$\mathbf{V}^{-1}=\begin{bmatrix} 0.055881 & 0.001659 & 0.001806 \\ 0.001659 & 0.099073 & 0.000415 \\ 0.001806 & 0.000415 & 0.078984 \end{bmatrix}.$$

The values of $z^2_{ig(P)}$ for the two employees are then:

Employee	Passenger agents	Mechanics	Operations control agents
1	*1.5426*	5.2966	17.8345
2	9.1016	5.9069	*0.0120*

The values in italics are the smallest of the three for each employee, and thus indicate the group most closely resembled. We conclude that Employee 1 was most like the Passenger Agent group and Employee 2 the group of Operations Control Agents.

To use Equations 6.3.7 or 6.3.8, we first find the inverses of $\mathbf{V}_1$, $\mathbf{V}_2$, and $\mathbf{V}_3$. These are:

$$\mathbf{V}_1^{-1}=\begin{bmatrix} 0.054260 & 0.022940 & 0.009715 \\ 0.022940 & 0.194804 & -0.048140 \\ 0.009715 & -0.048140 & 0.089067 \end{bmatrix}$$

$$\mathbf{V}_2^{-1}=\begin{bmatrix} 0.116400 & 0.042529 & 0.031517 \\ 0.042529 & 0.090560 & 0.009792 \\ 0.031517 & 0.009792 & 0.146382 \end{bmatrix}$$

$$\mathbf{V}_3^{-1}=\begin{bmatrix} 0.060111 & -0.044889 & -0.020856 \\ -0.044889 & 0.143611 & 0.034901 \\ -0.020856 & 0.034901 & 0.069797 \end{bmatrix}.$$

Then the values of $z^2_{ig(S)}$ and $z^2_{ig(T)}$ for the two employees are shown in Table 6.3.2.

Note first that, although these results are noticeably different from those obtained using Equation 6.3.5, the *orderings* of values for the two employees are the same. Hence, the decisions concerning group resemblance are the same. Note further that the effect of the second term, $|\mathbf{V}_g|$, in Equation 6.3.8, is to sharply reduce the relative discrepancies between the obtained values. For Employee 2, for example, the ratio of the value for Mechanics to that for Operations Control Agents is $7.9903/.0255=313$, using Equation

TABLE 6.3.2 VALUES OF $z^2_{ig(S)}$ AND $z^2_{ig(T)}$ FOR TWO EMPLOYEES

Equation	Employee	Passenger agents	Mechanics	Operations control agents
6.3.7	1	*2.7645*	4.5395	21.2442
	2	20.4539	7.9903	*.0255*
6.3.8	1	*9.9884*	11.2621	29.0896
	2	27.6778	14.7129	*7.8709*

6.3.7, while the corresponding ratio, using Equation 6.3.8, is only 14.7129/7.8709 = 1.9.

We should emphasize that the data of our example are used only for illustrating computations and, because of the small sample sizes, would not provide an adequate basis for making classification decisions in practice. As we stated earlier, we recommend that the size of the total (combined) sample be $20p$, where p is the number of X variables. We also recommend the use of a second, independent sample for cross-validation of the procedures based on Equations 6.3.5, 6.3.7, and 6.3.8.

6.3.3 Centour Scores

The centour score was originally proposed by Tiedeman et al. (1953) and later described in Rulon et al. (1967, p. 167), as a measure of an individual's resemblance to a group. The name "centour" was derived from the words "*cen*tile" and "con*tour*," with reference to the approximate chi-square distribution of the variables in Equations 6.3.5 and 6.3.7. In this context, the term "contour" refers to the *shape* of a chi-square distribution with p degrees of freedom and the term "centile" refers to the *percentile rank* of a particular value of $z^2_{ig(S)}$, as defined by Equation 6.3.7, in that distribution. A centour score is then defined as 100 minus the percentile rank of $z^2_{ig(S)}$ in the distribution of chi-square with p degrees of freedom, that is

$$\text{Centour Score} = 100 - PR_{z^2_{ig(S)};p}. \tag{6.3.9}$$

Thus, the definition of a centour score is based on the assumption of a multivariate normal distribution of the X variables and on the consequent approximate chi-square distribution of the statistic $z^2_{ig(S)}$. A *large* value of the centour score with reference to a particular population thus indicates close resemblance to that population.

Because it is essentially a measure of the *probability* of membership in a particular population, the centour score has an important advantage over the measures of resemblance considered so far in this section. The decision rule based on $z^2_{ig(S)}$ was to assign the individual to the population for which the value of $z^2_{ig(S)}$ was smallest. Such a rule is quite appropriate when the set of populations to which the individual might reasonably be assigned is totally represented in the sample on which the discriminant equation is based. Suppose, however, that a given individual closely resembles a population other than those represented in such a sample and closely resembles *none* of those that are represented. Then the decision rule based on the smallest $z^2_{ig(S)}$ would not be appropriate, because the smallest $z^2_{ig(S)}$ might still suggest a large distance from the reference population. The centour scores, on the other hand, would reflect such a lack of close resemblance to the populations represented in the sample because *all* of the centours would be small. Furthermore, they could be interpreted in a probability sense. Each of them would represent the complement of the probability of a distance from the reference population centroid at least as large as that observed for a particular individual. For example, an individual with centour scores of 30, 35, and 40 with reference to three populations clearly doesn't resemble any of them closely and should probably not be assigned to any of them. On the other hand, a person with centour scores of 30, 55, and 90 resembles the third population closely enough to warrant his or her assignment to it. Further discussion of classification based on the probability of population membership is given in Section 6.3.5.

To illustrate the application of the centour score, we shall use the data of the preceding section concerning the three groups of airline employees. Using Equation 6.3.7, we obtained a value of $z^2_{12(S)} = 4.5395$ as a measure of the resemblance of Employee 1 to the Mechanics group. Because there were three Xs involved in that example, $p = 3$ and we find that the percentile rank of $z^2_{12(S)} = 4.5395$ in the chi-square distribution with three degrees of freedom is approximately 79%. Thus, the centour score for Employee 1 with reference to the Mechanics group is $100 - 79 = 21$. Using the same procedure, we find that the centour scores for the two employees with reference to the three groups are:

Employee	*Passenger agents*	*Mechanics*	*Operations control agents*
1	*43*	21	0
2	0	5	*99*

Because the largest centour score for Employee 1 is associated with the group of Passenger Agents, we conclude that he or she most closely resembles that group. Employee 2 most closely resembles the group of Operations Control Agents.

6.3.4 The Use of Discriminant Scores in Classification

At the beginning of this section we indicated that the discriminant equation, obtained from a large sample of individuals who were already members of the groups of interest, could be used to assign individuals to those groups. Our discussion so far has dealt with measures of resemblance based on the entire set of p individual X variables. These same measures, namely $z^2_{ig(P)}$ and $z^2_{ig(S)}$, can just as easily be based on the r discriminant scores obtained by substituting the X values for an individual in the discriminant equations. In fact, we could argue that such measures would be preferred over those based on the Xs themselves, because r is generally less than p and we therefore can save time and effort in carrying out the analysis. If we further reduced the number of variables by using only those discriminant functions, r' in number, that have been found to be statistically significant, we would gain an additional advantage, in that the classification would then be based on the set of discriminant functions that are least subject to sampling error.

That is, they are least likely to reflect differences between sample groups that may not indicate real differences between the corresponding populations.

The use of the statistically significant discriminant functions in place of the entire set of p individual X variables does not require measures of group resemblance essentially different from those defined in Equations 6.3.5, 6.3.7 and 6.3.8. The only required changes in these equations involve redefining $\mathbf{x}_{ig}$, $\mathbf{V}_g$, and $\mathbf{V}$ in terms of the r' significant discriminant functions. Thus, corresponding to $\mathbf{x}_{ig}$ and $\mathbf{V}_g$, respectively, we define

$$\mathbf{z}_{ig} = \begin{bmatrix} Z_{i1} - \bar{Z}_{1g} \\ Z_{i2} - \bar{Z}_{2g} \\ \vdots \\ Z_{ir'} - \bar{Z}_{r'g} \end{bmatrix}, \tag{6.3.10}$$

$$\mathbf{V}_{(Z)} = \mathbf{U}'\mathbf{V}\mathbf{U}, \tag{6.3.11}$$

and

$$\mathbf{V}_{g(Z)} = \mathbf{U}'\mathbf{V}_g\mathbf{U}, \tag{6.3.12}$$

in which $\mathbf{V}$ and $\mathbf{V}_g$ are defined by Equations 6.3.4 and 6.3.6, respectively, and $\mathbf{U}$ is a $p \times r'$ matrix, the columns of which are the vectors, $\mathbf{u}_1, \mathbf{u}_2, \ldots, \mathbf{u}_{r'}$, of weights on the Xs in the r' significant discriminant functions. Then the analogues of Equations 6.3.5, 6.3.7, and 6.3.8, based on the r' significant discriminant functions, become

$$z'^2_{ig(P)} = \mathbf{z}'_{ig}\mathbf{V}^{-1}_{(Z)}\mathbf{z}_{ig}, \tag{6.3.13}$$

$$z'^2_{ig(S)} = \mathbf{z}'_{ig}\mathbf{V}^{-1}_{g(Z)}\mathbf{z}_{ig}, \tag{6.3.14}$$

and

$$z'^2_{ig(T)} = \mathbf{z}'_{ig}\mathbf{V}^{-1}_{g(Z)}\mathbf{z}_{ig} + \ln|\mathbf{V}_{g(Z)}|. \tag{6.3.15}$$

We now illustrate the use of these equations, again using the data of Section 6.2.2.

According to Equation 6.2.1, the kth discriminant score of the ith individual in group g may be written

$$Z_{igk} = u_{1k}X_{i1g} + u_{2k}X_{i2g} + \cdots + u_{pk}X_{ipg}.$$

Then the mean for group g on this function may be written

$$\bar{Z}_{gk} = u_{1k}\bar{X}_{1g} + u_{2k}\bar{X}_{2g} + \cdots + u_{pk}\bar{X}_{pg}.$$

In the example of Section 6.2.2 only one discriminant function was statistically significant, so r' actually equals one. However, in the interest of better illustrating the more general application of Equations 6.3.13, 6.3.14, and 6.3.15, we shall use both functions, ignoring the fact that the second was not statistically significant. The discriminant equations for the two functions were given by Equations 6.2.17 and 6.2.19, and the means for the three groups on the two functions were as shown in 6.3.3.

TABLE 6.3.3 MEANS OF THREE GROUPS OF EMPLOYEES ON TWO DISCRIMINANT FUNCTIONS

Group	*First function*	*Second function*
Passenger agents	$\bar{Z}_{11} = 18.495304$	$\bar{Z}_{21} = 3.103540$
Mechanics	$\bar{Z}_{12} = 16.380061$	$\bar{Z}_{22} = 6.778414$
Operations control agents	$\bar{Z}_{13} = 8.518427$	$\bar{Z}_{23} = 4.605726$

TABLE 6.3.4 THE VECTORS z_{ig} FOR TWO EMPLOYEES

Group	*Employee* 1	*Employee* 2
Passenger agents	$\mathbf{z}_{11} = \begin{bmatrix} 3.374608 \\ -2.847327 \end{bmatrix}$	$\mathbf{z}_{21} = \begin{bmatrix} -9.800513 \\ 1.185275 \end{bmatrix}$
Mechanics	$\mathbf{z}_{12} = \begin{bmatrix} 5.489850 \\ -6.522201 \end{bmatrix}$	$\mathbf{z}_{22} = \begin{bmatrix} -7.685270 \\ -2.489600 \end{bmatrix}$
Operations control agents	$\mathbf{z}_{13} = \begin{bmatrix} 13.351484 \\ -4.349513 \end{bmatrix}$	$\mathbf{z}_{23} = \begin{bmatrix} 0.176364 \\ -0.316911 \end{bmatrix}$

We find, then, that for the two employees whose scores on X_1, X_2, and X_3 were given at the beginning of Section 6.3.2, the vectors z_{ig}, defined by Equation 6.3.10, are as shown in Table 6.3.4.

Using Equation 6.3.11, we find that

$$\mathbf{V}_{(Z)}^{-1} = \begin{bmatrix} 0.093879 & 0.000000 \\ 0.000000 & 0.057904 \end{bmatrix}.$$

Hence, the application of Equation 6.3.13 yields:

Employee	*Passenger agents*	*Mechanics*	*Operations control agents*
1	*1.5385*	5.2925	17.8304
2	9.0984	5.9037	*.0087*

Again, we conclude that Employee 1 most closely resembles the Passenger Agent group and Employee 2 the group of Operations Control Agents.

TABLE 6.3.5 EQUATIONS 6.3.14 AND 6.3.15 APPLIED TO TWO EMPLOYEES' SCORES

Equation	*Employee*	*Passenger agents*	*Mechanics*	*Operations control agents*
6.3.14	1	*2.7644*	3.9960	18.1720
	2	20.4414	7.9874	*.0122*
6.315	1	*7.4592*	8.9433	23.4607
	2	25.1362	12.9347	*5.3009*

To use Equations 6.3.14 and 6.3.15, we first need to find $\mathbf{V}_{g(Z)}^{-1}$ for each group, using Equation 6.3.12. These matrices, as well as the determinants $|\mathbf{V}_{g(Z)}|$, which are needed for Equation 6.3.15, are:

Passenger agents

$$\mathbf{V}_{1(Z)}^{-1} = \begin{bmatrix} .212209 & .000077 \\ .000077 & .043083 \end{bmatrix}$$

$$|\mathbf{V}_{1(Z)}| = 109.379434$$

Mechanics

$$\mathbf{V}_{2(Z)}^{-1} = \begin{bmatrix} .103436 & .035937 \\ .035937 & .081153 \end{bmatrix}$$

$$|\mathbf{V}_{2(Z)}| = 140.793125$$

Operations control agents

$$\mathbf{V}_{3(Z)}^{-1} = \begin{bmatrix} .079032 & -.023607 \\ -.023607 & .079263 \end{bmatrix}$$

$$|\mathbf{V}_{3(Z)}| = 198.092156$$

Then we find the results listed in Table 6.3.5.

These results are seen to be consistent with our assignment decisions based on Equation 6.3.13.

6.3.5 The Use of Prior Probabilities in Classification

So far in this section we have considered the problem of classification in terms of the extent to which an individual *resembles* each of several populations to which he might be assigned. The measures of resemblance that we have discussed have been based on: (1) the individual's *distance* from the centroid of a reference population and (2) the *probability* of observing a distance at least as large as that observed, given that the individual is indeed a member of that reference population.

Several statisticians, among them Rao, Anderson, Tatsuoka, and Geisser have devised decision rules that incorporate the individual's *prior* probability of membership in a given population, prior in the sense that such a probability is known before measurements of the individual are taken. These probabilities are ordinarily based on the relative sizes of the populations to which the individual might be assigned. For example, suppose we have a classification problem involving three populations with sizes 2,000, 3,000, and 5,000. If we selected a single individual at random from the three populations combined, the three prior probabilities would be .2, .3, and .5. The classification methods that make use of prior probabilities are designed to *increase* the probability of assigning the individual to a population having a relatively *large* prior probability and to *decrease* the probability of assignment to a population having a relatively *small* prior probability. In a vocational counseling situation, for example, accounting for prior probabilities would mean that a person resembling equally each of two occupational groups would have greater probability of assignment to the group having the larger membership.

Before we describe classification procedures that use prior probabilities, we should distinguish between two kinds of conditional probabilities that are centrally involved in such procedures and which are easily confused. The first is the probability that an individual has a particular vector of scores on the X variables, *given that he is a member of population g*. We denote this conditional probability by

$$P(\mathbf{X}_i | H_g), \tag{6.3.16}$$

in which H_g means "member of group g" and $\mathbf{X}_i$ is the vector of scores for individual i on the p Xs. The second conditional probability that we need to define is, in a sense, the *inverse* of that in Equation 6.3.16, namely, the probability that an individual is a member of population g, *given that his vector of scores on the Xs is* $\mathbf{X}_i$. We denote this conditional probability by

$$P(H_g | \mathbf{X}_i). \tag{6.3.17}$$

To see how these two conditional probabilities are related we make use of the multiplication theorem for probabilities, namely, that for two events, A and B,

$$P(A \cap B) = P(B)P(A|B) = P(A)P(B|A), \tag{6.3.18}$$

where $P(A \cap B)$ denotes the probability of the intersection of the events A and B. Regarding H_g and $\mathbf{X}_i$ as the events of interest in the classification problem, we may use Equation 6.3.18 to express the probability that individual i, selected at random, is a member of group g *and* has the score vector $\mathbf{X}_i$. Thus, $P(H_g \cap \mathbf{X}_i) = P(\mathbf{X}_i)P(H_g|\mathbf{X}_i) = P(H_g)P(\mathbf{X}_i|H_g)$, and, dividing by $P(\mathbf{X}_i)$,

$$P(H_g|\mathbf{X}_i) = \frac{P(H_g)P(\mathbf{X}_i|H_g)}{P(\mathbf{X}_i)}. \tag{6.3.19}$$

Now, if the set of G groups involved in the classification problem represents *all* populations to which the individual might conceivably (or at least reasonably) be assigned, then the sum $\Sigma_{g=1}^{G} P(H_g|\mathbf{X}_i)$ must be equal to one. In other words, if an individual cannot conceivably be classified in any other group, then certainly he must be classified as a member of one of the G groups in the set. Therefore, we may write

$$\sum_{g=1}^{G} \frac{P(H_g)P(\mathbf{X}_i|H_g)}{P(\mathbf{X}_i)} = 1,$$

which becomes

$$P(\mathbf{X}_i) = \sum_{g=1}^{G} \left[P(H_g)P(\mathbf{X}_i|H_g) \right],$$

because $P(\mathbf{X}_i)$ is a constant with respect to summation on the index g.

Substitution for $P(\mathbf{X}_i)$ in Equation 6.3.19 yields

$$P(H_g|\mathbf{X}_i) = \frac{P(H_g)P(\mathbf{X}_i|H_g)}{\sum_{g=1}^{G}[P(H_g)P(\mathbf{X}_i|H_g)]}. \tag{6.3.20}$$

This equation represents a particular application of a fundamental theorem in probability due to the English clergyman, Thomas Bayes (1763). The probability $P(H_g)$ is the *prior* probability that an individual selected at random is a member of group g. The probability $P(H_g|\mathbf{X}_i)$ is referred to as the *posterior* probability of membership in group g because it is conditional on having observed the individual's score vector, $\mathbf{X}_i$. The decision rule used in connection with Equation 6.3.20 is to assign the individual to the group for which $P(H_g|\mathbf{X}_i)$ is largest.

To compute $P(H_g|\mathbf{X}_i)$ from Equation 6.3.20, we first need to have an expression for $P(\mathbf{X}_i|H_g)$. Assuming a multivariate normal distribution for the Xs in group g, we see from Chapter 4 that the density function is

$$f(X_1,\ldots,X_p) = (2\pi)^{-p/2}|\boldsymbol{\Sigma}_g|^{-1/2}\exp(-\chi_g^2/2).$$

The probability, $P(\mathbf{X}_i|H_g)$, that we wish to evaluate is the probability that an individual drawn at random from population g will have a score vector located between $\mathbf{X}_i$ and $\mathbf{X}_i + d\mathbf{X}$, in which

$$d\mathbf{X} = \begin{bmatrix} dX_1 \\ dX_2 \\ \vdots \\ dX_p \end{bmatrix}.$$

Using the sample estimate, $\mathbf{V}_g$ of Equation 6.3.6, in place of $\boldsymbol{\Sigma}_g$, we can express this probability as

$$P(\mathbf{X}_i|H_g) = (2\pi)^{-p/2}|\mathbf{V}_g|^{-1/2}\exp\left(-\tfrac{1}{2}z_{ig(S)}^2\right)dX_1,\ldots,dX_p, \tag{6.3.21}$$

in which $z_{ig(S)}^2$ is defined by Equation 6.3.7. If it can be assumed that $\boldsymbol{\Sigma}_1 = \boldsymbol{\Sigma}_2 = \ldots = \boldsymbol{\Sigma}_G = \boldsymbol{\Sigma}$, then this expression becomes

$$P(\mathbf{X}_i|H_g) = (2\pi)^{-p/2}|\mathbf{V}|^{-1/2}\exp\left(-\tfrac{1}{2}z_{ig(P)}^2\right)dX_1,\ldots,dX_p. \tag{6.3.22}$$

Since $\exp(-\frac{1}{2}z_{ig(T)}^2) = |\mathbf{V}_g|^{-1/2}\exp(-\frac{1}{2}z_{ig(S)}^2)$, another expression for $P(\mathbf{X}_i|H_g)$, equivalent to Equation 6.3.21, is

$$P(\mathbf{X}_i|H_g) = (2\pi)^{-p/2}\exp\left(-\tfrac{1}{2}z_{ig(T)}^2\right)dX_1,\ldots,dX_p, \tag{6.3.23}$$

Now, substitution in Equation 6.3.20 yields

$$P(H_g|\mathbf{X}_i) = \frac{P(H_g)\exp\left(-\frac{1}{2}z_{ig(T)}^2\right)}{\sum_{g=1}^{G}P(H_g)\exp\left(-\frac{1}{2}z_{ig(T)}^2\right)}, \tag{6.3.24}$$

because the common factors in the denominator, namely, $(2\pi)^{-p/2}$ and $dX_1, \ldots, dX_p$, cancel the corresponding factors in the numerator. Expressions incorporating $z^2_{ig(P)}$ and $z^2_{ig(S)}$ can be obtained similarly by substitution from Equations 6.3.22 and 6.3.21, respectively. The resulting equations are:

$$P'(H_g|\mathbf{X}_i) = \frac{P(H_g)|\mathbf{V}|^{-1/2}\exp\left(-\frac{1}{2}z^2_{ig(P)}\right)}{\sum_{g=1}^{G} P(H_g)|\mathbf{V}|^{-1/2}\exp\left(-\frac{1}{2}z^2_{ig(P)}\right)} \tag{6.3.25}$$

and

$$P(H_g|\mathbf{X}_i) = \frac{P(H_g)|\mathbf{V}_g|^{-1/2}\exp\left(-\frac{1}{2}z^2_{ig(S)}\right)}{\sum_{g=1}^{G} P(H_g)|\mathbf{V}_g|^{-1/2}\exp\left(-\frac{1}{2}z^2_{ig(S)}\right)}. \tag{6.3.26}$$

As we indicated earlier, the decision rule based on Equation 6.3.20, and hence on Equations 6.3.24, 6.3.25 and 6.3.26, is to classify the individual as a member of the group for which $P(H_g|\mathbf{X}_i)$ or $P'(H_g|\mathbf{X}_i)$ is the largest. If the prior probabilities, $P(H_g)$, are unknown, or if one does not wish to take them into account, then it is convenient to set each equal to $1/G$.

To illustrate the application of Equations 6.3.24, 6.3.25, and 6.3.26, we return to the data for the two airline employees used earlier in this section. Because we already have computed $z^2_{ig(P)}$, $z^2_{ig(S)}$, and $z^2_{ig(T)}$ for these individuals, we can use them to obtain the necessary values for the numerator and denominator of each equation. For illustrative purposes, we shall use as prior probabilities the values, $P(H_1) = .2$, $P(H_2) = .5$, and $P(H_3) = .3$. The results are shown in Table 6.3.6.

Then, the desired probabilities are as shown in Table 6.3.7.

The assignment decisions using Equation 6.3.25 are the same as those obtained earlier for these two employees. However, the effect of using the prior probabilities selected above can be seen in the results for Employee 1 obtained from Equation 6.3.24 (or its equivalent, 6.3.26), with the largest probability now being associated with the Mechanics group. On the basis of the methods discussed earlier, Employee 1 had been assigned to the

TABLE 6.3.6 NUMERATOR AND DENOMINATOR VALUES FOR EQUATIONS 6.3.24, 6.3.25, AND 6.3.26 BASED ON TWO EMPLOYEES' SCORES

		Numerator values			
Equation	*Employee*	*Passenger Agents*	*Mechanics*	*Operations control agents*	*Denominator values (Sum)*
6.3.24	1	.0013554	.0017924	.0000002	.0031480
and 6.3.26	2	.0000002	.0003192	.0058611	.0061805
6.3.25	1	.0019327	.0007395	.0000009	.0026731
	2	.0000441	.0005450	.0062319	.0068210

TABLE 6.3.7 PROBABILITIES OBTAINED FROM EQUATIONS 6.3.24, 6.3.25, AND 6.3.26 FOR TWO EMPLOYEES

Equation	*Employee*	*Passenger agents*	*Mechanics*	*Operations control agents*
6.3.24 and 6.3.26	1	.4306	*.5694*	.0000
	2	.0001	.0516	*.9483*
6.3.25	1	*.7230*	.2767	.0003
	2	.0065	.0799	*.9163*

TABLE 6.3.8 PROBABILITIES OBTAINED FROM EQUATIONS 6.3.24, 6.3.25, AND 6.3.26 WITH PRIOR PROBABILITIES EQUAL TO 1/3

Equation	*Employee*	*Passenger Agents*	*Mechanics*	*Operations control agents*
6.3.24 and 6.3.26	1	*.6540*	.3459	.0000
	2	.0001	.0316	*.9683*
6.3.25	1	*.8670*	.1327	.0003
	2	.0100	.0494	*.9407*

Passenger Agents group, but the larger prior probability, $P(H_2)=.5$, for the Mechanics, compared with $P(H_1)=.2$ for the Passenger Agents, outweighed this employee's somewhat closer resemblance to the Passenger Agents. The effect of the prior probabilities can be seen even more clearly by setting $P(H_1)=P(H_2)=P(H_3)=1/3$. Then we find the results listed in Table 6.3.8.

These results show that when prior probabilities are equal, the assignment decisions for both employees are consistent with those based on the methods discussed earlier.

This example shows clearly that the use of prior probabilities can have a noticeable effect on classification decisions, primarily because the probabilities given by Equations 6.3.24 and 6.3.25 are not measures of group resemblance alone, but instead combine group resemblance and, essentially, population (group) size. We could argue that the use of prior probabilities thus provides a better decision rule because it is more economical, that is, it does mathematically what the user of classification methods would otherwise have to do himself, probably on a more subjective basis. However, we choose not to take such a position. Instead we urge caution in the use of prior probabilities for two reasons:

1. Unless one can be quite certain of their accuracy, the use of prior probabilities can lead to incorrect classification decisions. For example, the decision for Employee 1 above would have been different if the values of $P(H_1)$ and $P(H_2)$ had been reversed. In most practical applications, the values of the prior probabilities are not known with sufficient accuracy to justify their use.

2. It seems better from a counseling point of view, where classification methods have wide potential applicability, first to measure the individual's resemblance to each of several groups and then to make a classification decision *for that individual* that takes into account, perhaps rather subjectively, as much information as is available about the individual and the current or anticipated educational or vocational scene. For example, to discourage Bill from preparing for Career A, whose members he most closely resembles, because few Career A people are needed, is a decision that should not be made solely by application of a mathematical equation. We need to know that Bill resembles group A and that group A is relatively small, but the proper application of these facts in making a classification decision should be determined specifically for Bill.

Computer programs for discriminant analysis such as those found in the SPSS and BMD packages usually provide for the use of prior probabilities. We suggest in using these programs that, in most cases, all prior probabilities be set equal to $1/G$, for the reasons discussed. We also urge that caution be exercised in the interpretation of output from such programs. The SPSS program currently facilitates interpretation by providing (1) the largest and second largest values of $P(H_g|\mathbf{X}_i)$ for individual i and (2) the probability that a member of the group for which individual i's value of $P(H_g|\mathbf{X}_i)$ is largest, would be as far from the centroid as individual i. This latter probability is denoted by $P(X/G)$ on the printout but it is not equivalent to the conditional probability that we have denoted by $P(\mathbf{X}_i|H_g)$. Instead, it is the probability that would be obtained by referring a value of $z_{ig(P)}^{\prime 2}$, as defined by Equation 6.3.13, to an appropriate chi-square distribution. If this probability is quite small, we would probably conclude that individual i belongs to *none* of the G groups included in the classification problem.

6.3.6 Other Classification Methods

As we noted earlier in this section, we do not intend our discussion of classification procedures to be encyclopedic in character. We have tried to present in some detail a basic approach which is consistent in terms of mathematical and statistical level with material in other chapters. However, we believe it is important at least to mention methods proposed by several other statisticians and to indicate where possible how they are related to those we have already discussed.

Mahalanobis' D_M^2. We have mentioned Mahalanobis' work in connection with the two-group test of the hypothesis that two population centroids are equal. The statistic D_M^2, defined by Equation 6.1.24, was used as a measure of the distance between two sample centroids. By replacing $\mathbf{d}$ in that Equation with the vector $\mathbf{x}_{ig}$ of Equation 6.3.5, we see that D_M^2 is identical to the measure of resemblance, $z_{ig(P)}^2$, with $\mathbf{V}^{-1}$ used as a sample estimate of $\mathbf{\Sigma}^{-1}$. Then D_M^2 becomes a measure of the distance between the vector of X values for individual i and the sample centroid of group g.

Cronbach and Gleser (1953) suggested modifying D_M^2 by replacing $\mathbf{V}$, the pooled estimate of Σ, with $\mathbf{V}_g$, the sample estimate of Σ_g for group g. With this modification, the statistic, say $D_M'^2$, becomes equivalent to $z_{ig(S)}^2$, defined by Equation 6.3.7.

Geisser. We mention Geisser's (1964) classification procedure primarily because it is employed in a computer program for classification given by Cooley and Lohnes (1971). In terms of the methods discussed earlier in this chapter, it is most similar to Equation 6.3.25 in that it assumes equal population variance-covariance matrices and employs prior probabilities. The statistic used in the Cooley-Lohnes program is

$$P(H_g|\mathbf{X}_i) = \frac{P_{ig}}{\sum_{k=1}^{G} P_{ik}},$$

in which

$$P_{ig} = P(H_g)\left[\frac{n_g}{n_g+1}\right]^{p/2} \frac{\Gamma\left[\frac{1}{2}(n-G+1)\right]}{\Gamma\left[\frac{1}{2}(n-p-G+1)\right]}$$

$$\cdot\left[1+\frac{n_g z_{ig(P)}^2}{(n_g+1)(n-G)}\right]^{-\frac{1}{2}(n-G+1)}.$$

The classification rule is to assign the individual to the group for which $P(H_g|X_i)$ is largest. For a more complete discussion of the rationale underlying this method and examples of its application see Geisser (1964) and Cooley and Lohnes (1971).

Rao. We referred earlier to the Indian statistician C. R. Rao, in connection with his extension of Mahalanobis' two-group discriminant analysis to the general case of G groups. Rao served as Professor of Statistics and later as Director of the Indian Statistical Institute in Calcutta and was closely associated with Mahalanobis in their work in multivariate analysis. To understand Rao's procedure we shall first consider the numerator of Equation 6.3.25, namely,

$$P(H_g)|\mathbf{V}| - 1/2 \exp\left(-\tfrac{1}{2}z_{ig(P)}^2\right) = P(H_g)|\mathbf{V}|^{-1/2}\exp\left[-\tfrac{1}{2}(\mathbf{x}_{ig}'\mathbf{V}^{-1}\mathbf{x}_{ig})\right].$$

Taking natural logarithms of the expression on the right of the equality sign and denoting the result by S_{ig}^*, we have

$$S_{ig}^* = -\tfrac{1}{2}\ln|\mathbf{V}| - \tfrac{1}{2}(\mathbf{x}_{ig}'\mathbf{V}^{-1}\mathbf{x}_{ig}) + \ln P(H_g).$$

Because we are assuming that $\Sigma_1 = \ldots = \Sigma_G = \Sigma$, the term $-\frac{1}{2}\ln|\mathbf{V}|$ is common to all groups and may be eliminated. Furthermore, because $\mathbf{x}_{ig} = \mathbf{X}_i - \overline{\mathbf{X}}_g$, the expression $-\frac{1}{2}(\mathbf{x}_{ig}'\mathbf{V}^{-1}\mathbf{x}_{ig})$ may be written (see Appendix A)

$$-\tfrac{1}{2}(\mathbf{X}_i - \overline{\mathbf{X}}_g)'\mathbf{V}^{-1}(\mathbf{X}_i - \overline{\mathbf{X}}_g) = -\tfrac{1}{2}(\mathbf{X}_i'\mathbf{V}^{-1}\mathbf{X}_i) + \overline{\mathbf{X}}_g'\mathbf{V}^{-1}\mathbf{X}_i - \tfrac{1}{2}(\overline{\mathbf{X}}_g'\mathbf{V}^{-1}\overline{\mathbf{X}}_g).$$

Now, we may also eliminate the term $-\frac{1}{2}(\mathbf{X}_i'\mathbf{V}^{-1}\mathbf{X}_i)$, because it, too, is common to all groups for the ith individual. Then

$$S_{ig}=\bar{\mathbf{X}}_g'\mathbf{V}^{-1}\mathbf{X}_i-\tfrac{1}{2}\left(\bar{\mathbf{X}}_g'\mathbf{V}^{-1}\bar{\mathbf{X}}_g\right)+\ln P(H_g)$$

is Rao's classification statistic. Since S_{ig} is derived directly from Equation 6.3.25 by eliminating elements that are common to all groups, the values of S_{ig} must be proportional to those of $P'(H_g|\mathbf{X}_i)$. Therefore, the assignment decisions based on Rao's procedure must be identical to those based on the use of Equation 6.3.25. For a more complete discussion and an example, see Rao (1973).

Anderson. The classification procedure proposed by Anderson (1958) is designed to minimize the expected cost of misclassification, under the assumptions that the Xs in the gth group (population) have a multivariate normal distribution with mean vector $\boldsymbol{\mu}_g$ and variance-covariance matrix $\boldsymbol{\Sigma}$, common to all of the G populations. He defines a function

$$u_{gk}(\mathbf{X}_i)=\ln\frac{P(\mathbf{X}_i|H_g)}{P(\mathbf{X}_i|H_k)},\quad k=1,2,\ldots,G;\, k\neq g. \tag{6.3.27}$$

In this expression the conditional probabilities are given by Equation 6.3.21, except that the common population variance-covariance matrix, $\boldsymbol{\Sigma}$, replaces $\mathbf{V}_g$.

Substitution for $P(\mathbf{X}_i|H_g)$ and $P(\mathbf{X}_i|H_k)$ in Equation 6.3.27 yields

$$\begin{aligned}u_{gk}(\mathbf{X}_i)&=\ln\frac{(2\pi)^{-p/2}|\boldsymbol{\Sigma}|^{-1/2}\exp\left[-\tfrac{1}{2}(\mathbf{X}_i-\boldsymbol{\mu}_g)'\boldsymbol{\Sigma}^{-1}(\mathbf{X}_i-\boldsymbol{\mu}_g)\right]}{(2\pi)^{-p/2}|\boldsymbol{\Sigma}|^{-1/2}\exp\left[-\tfrac{1}{2}(\mathbf{X}_i-\boldsymbol{\mu}_k)'\boldsymbol{\Sigma}^{-1}(\mathbf{X}_i-\boldsymbol{\mu}_k)\right]}\\&=\tfrac{1}{2}\left[(\mathbf{X}_i-\boldsymbol{\mu}_k)'\boldsymbol{\Sigma}^{-1}(\mathbf{X}_i-\boldsymbol{\mu}_k)-(\mathbf{X}_i-\boldsymbol{\mu}_g)'\boldsymbol{\Sigma}^{-1}(\mathbf{X}_i-\boldsymbol{\mu}_g)\right]\\&=\left[\mathbf{X}_i-\tfrac{1}{2}(\boldsymbol{\mu}_g+\boldsymbol{\mu}_k)\right]'\boldsymbol{\Sigma}^{-1}(\boldsymbol{\mu}_g-\boldsymbol{\mu}_k).\end{aligned} \tag{6.3.28}$$

Anderson refers to this as a "discriminant function," but this terminology is not consistent with our usage in this chapter. The user of packaged computer programs for discriminant analysis should keep in mind this difference in terminology.

The classification rule proposed by Anderson is to assign the ith individual to the gth group if

$$u_{gk}(\mathbf{X}_i)>\ln\frac{P(H_k)}{P(H_g)},\quad k=1,\ldots,G;\ k\neq g,$$

where, again, $P(H_k)$ and $P(H_g)$ are prior probabilities. If all prior probabilities are equal, that is, if $P(H_k)=1/G$, then

$$\ln\frac{P(H_k)}{P(H_g)}=0,$$

and the rule is to assign the ith individual to group g if $u_{gk}(X_i)\geqslant 0$ $(k=1,\ldots,G;k\neq g)$. When $\boldsymbol{\Sigma}$ and $\boldsymbol{\mu}_k$ are unknown, as will usually be the case, sample estimates are used. For a more complete discussion of this method and a numerical example, we refer the reader to Anderson (1958), Chapter 6.

6.3.7. Estimating the Validity of Classification Procedures

Before concluding our discussion of classification methods, we should mention briefly how the validity of any of the methods discussed above may be assessed. The usual procedure is to obtain a sample of individuals whose group membership is known and compare the predicted group membership with the actual. The measure of validity is then the percentage of individuals correctly classified by the method in question. Such measures should be based on samples sufficiently large to ensure adequate precision. We suggest cross-validation using a different sample whenever feasible.

Cooley and Lohnes have carried out validity studies of several of the methods discussed in this chapter, including those proposed by Anderson, by Tatsuoka, and by Geisser. A discussion of these studies and their results may be found in Cooley and Lohnes (1971), Chapter 10.

6.4. EXERCISES

6.1. A researcher hypothesized that patients who receive electroconvulsive therapy (ECT) differ symptomatically from those who do not receive ECT. To test this hypothesis, a study was conducted using 670 inpatients as subjects. In this group, 427 received ECT and 243 did not. Means for the two groups on the Brief Psychiatric Rating Scale (BPRS), obtained following treatment, were:

BPRS Scale	*Group*	
	ECT	*Non-ECT*
X_1: Thinking Disturbance	4.167	3.021
X_2: Withdrawal Retardation	4.007	2.967
X_3: Hostile Suspiciousness	6.595	6.502
X_4: Anxious Depression	9.044	7.601

The within-groups sums of squares and cross-products matrix was:

$$\mathbf{W}=\begin{bmatrix} 10177.190120 & 1551.676450 & 4533.971740 & -897.862310 \\ 1551.676450 & 8343.603800 & -0.775190 & 417.615020 \\ 4533.971740 & -0.775190 & 9810.290750 & -1332.098690 \\ -897.862310 & 417.615020 & -1332.098690 & 7259.221530 \end{bmatrix}.$$

(a) Carry out an appropriate analysis to determine whether patients who received the ECT treatment were symptomatically different from those who did not.

(b) State a conclusion and defend it on the basis of the results in (a).

6.2. A study was conducted to determine whether graduate students in psychology could be separated into two groups: (1) those who earned the degree sought, and (2) those who did not, on the basis of scores on four measures obtained prior to entry into their graduate programs. The measures were: X_1, Undergraduate grade point average (GPA); X_2, Graduate Record Examination-Q (GRE-Q); X_3, Graduate Record Examination-V (GRE-V); and X_4, Graduate Record Examination-Advanced (GRE-ADV). The means of the two groups, each of size 20, were:

Measure	*Group*	
	Earned degree	*Did not earn degree*
X_1: GPA	3.132	2.666
X_2: GRE-Q	625.000	524.500
X_3: GRE-V	625.000	537.000
X_4: GRE-ADV	616.500	533.500

The within-groups sums of squares and cross-products matrix, $\mathbf{W} = \mathbf{S}_{xx}^{(1)} + \mathbf{S}_{xx}^{(2)}$, was:

$$\mathbf{W} = \begin{bmatrix} 8.166 & -93.195 & 484.630 & 302.160 \\ -93.195 & 550395.000 & 207870.000 & 144935.000 \\ 484.630 & 207870.000 & 484120.000 & 235660.000 \\ 302.160 & 144935.000 & 235660.000 & 231110.000 \end{bmatrix}$$

(a) Calculate Hotelling's T^2 for these data.
(b) Test the statistical significance of T^2, using $\alpha = .05$.
(c) Draw a conclusion regarding the hypothesis tested in this study. Support your conclusion based on the results in (a) and (b).

6.3. A study was conducted to investigate the predictive validity of a set of five predictors of the progress that had been made or was being made in the academic programs of 102 graduate students in psychology. The predictors were: Undergraduate Grade-Point Average (UGPA), the Verbal (V), Quantitative (Q), and Advanced (A) tests of the Graduate Record Examination, and the Miller Analogies Test (MAT). The 102 students were classified into four categories: A = progress without any delays or interruption; B = progress impeded by minor problem(s); C = progress impeded by major problems; and D = outright scholastic failure. A multiple discriminant analysis of the data yielded the following results:

Statistic	*Group*			
	A	*B*	*C*	*D*
n	54	18	22	8
$\bar{X}_{UGPA}$	3.0896	2.8278	2.8659	2.9699
$\bar{X}_V$	636.4814	581.1111	583.1816	540.0000
$\bar{X}_Q$	597.4072	620.3333	529.5454	562.5000
$\bar{X}_A$	588.7036	596.6665	570.0000	526.2500
$\bar{X}_{MAT}$	67.0926	66.6111	62.5000	54.1250
s_{UGPA}	0.4237	0.4699	0.4689	0.5180
s_V	96.9731	91.6454	85.0439	87.0139
s_Q	88.7702	45.5024	84.2612	123.4909
s_A	69.7164	58.2111	64.6603	68.0210
s_{MAT}	13.1512	14.3574	15.4079	11.2813

Three discriminant functions were computed as follows:

Discriminant function	*Eigenvalue*	*Wilks' Λ*	*Bartlett's V*	*Degrees of freedom*
1	0.21737	0.6552	40.798	15
2	0.16760	0.7976	21.817	8
3	0.07373	0.9313	6.865	3

(a) Is there evidence of significant discrimination between the groups? Explain.
(b) Should this prediction system be made operational? Explain.
(c) Suppose that the standardized discriminant function coefficients were computed and found to be:

Variable	*Function* 1	*Function* 2	*Function* 3
UGPA	.362	−.275	.104
V	.808	.096	.234
Q	−.279	−.920	.478
A	−.304	−.235	−.518
MAT	−.214	.114	−.661

How would you describe each of the three functions? What variable(s) seem to be most effective in discriminating between the groups?
(d) The classification table for predicted vs. actual group membership is given below. On the basis of your review of this table would you change any of your answers, particularly to (a) and (b)? Explain.

	Predicted group membership			
Actual group	*A*	*B*	*C*	*D*
A	53.7%	14.8%	24.1%	7.4%
B	11.1	66.7	5.6	16.7
C	13.6	13.6	50.0	22.7
D	12.5	25.0	12.5	50.0

6.4. A study was made to determine whether three groups of tenth-grade students could be distinguished on the basis of five tests from the Iowa Tests of Educational Development (ITED). Group A consisted of 36 students who were achieving at or above grade level (Achievers); Group B consisted of 27 underachievers who had been counseled; and Group C consisted of 23 underachievers who had not received counseling. The data are given in the table below.

(a) For Groups B and C only:

1. Compute the linear discriminant function and test its significance, using $\alpha = .05$.
2. Compute centour scores for all students in these groups.
3. Use the centour scores to determine the percentage of correct classifications.

4. Comment on the effectiveness of these tests in distinguishing between the groups.

(b) For all three groups;
1. Perform a multiple discriminant function analysis using the method in Section 6.2.
2. Perform a classification analysis using the Geisser method (SPSS Program).
3. Evaluate the results, both in terms of the number of statistically significant functions found and in the percentage of hits and misses in predicting actual group membership.
4. Repeat the classification analysis using the Mahalanobis D^2 and the Anderson method of classification (BMD Program). Compare the results with those in 2.

Standard Scores on Iowa Tests of Educational Development

	X_1	X_2	X_3	X_4	X_5
Group	*Social studies*	*Natural sciences*	*Quant. thinking*	*Reading*	*Vocabulary*
Group *A*	20	13	10	11	15
(Achievers)	12	18	13	15	21
($n=36$)	13	15	7	17	21
	24	24	20	24	26
	17	10	9	15	11
	22	8	10	11	8
	12	10	3	14	11
	11	12	20	10	15
	22	27	23	20	22
	13	24	19	19	22
	22	24	21	19	22
	24	22	18	18	18
	25	15	15	17	17
	20	15	20	19	19
	11	16	14	17	20
	15	15	10	17	20
	14	17	10	19	22
	25	5	6	8	6
	9	15	12	18	18
	11	22	12	16	20
	9	5	12	12	13
	18	23	23	24	22
	25	23	9	14	15
	24	6	10	11	1
	20	13	8	9	13
	14	26	27	27	26
	15	15	18	22	20
	19	25	23	27	26
	14	15	9	11	18
	19	9	14	11	16
	17	15	14	18	23
	6	18	23	19	20
	16	24	19	23	26
	19	23	21	24	24
	11	21	28	18	19
	17	16	19	23	25

	X_1	X_2	X_3	X_4	X_5
Group *B*	17	11	13	8	16
(Underachievers)	9	18	14	9	14
(Counseled)	19	14	14	13	14
$n=27$	17	26	19	22	23
	15	17	18	10	16
	17	21	21	16	21
	25	16	13	15	19
	10	16	11	10	16
	17	10	19	12	18
	26	6	13	9	16
	13	11	14	14	18
	20	24	16	15	21
	24	21	13	12	22
	10	15	16	23	18
	9	21	18	15	16
	19	13	15	14	14
	23	21	12	18	17
	17	18	10	12	11
	9	21	11	18	18
	20	10	16	17	16
	20	23	16	23	21
	7	25	33	24	22
	13	16	4	9	18
	14	9	7	9	12
	13	24	22	21	22
	20	16	10	16	19
	21	26	24	27	23
Group *C*	12	9	14	11	9
(Underachievers)	16	20	16	13	18
(Uncounseled)	14	15	8	14	17
$n=23$	11	8	8	12	11
	12	11	9	15	16
	13	15	21	14	14
	15	24	18	17	18
	9	7	10	11	8
	6	6	14	12	2
	17	15	14	13	17
	6	21	12	16	16
	20	21	22	16	19
	16	18	12	15	15
	14	22	16	11	15
	20	13	8	9	2
	15	10	7	13	17
	6	21	14	19	20
	18	24	13	19	18
	18	13	9	17	22
	10	11	6	10	9
	17	16	20	19	18
	23	14	13	13	10
	14	16	9	18	19

7/ Multivariate Analysis of Variance

In this chapter we assume that the reader is already familiar with the basic concepts of univariate analysis of variance (ANOVA) and understands how the procedure may be applied in testing hypotheses about population means. The topics discussed here may be viewed as extensions of ANOVA to the case in which the dependent variable is a vector rather than a scalar. Instead of having only one observation of a given individual in the sample, we have at least two. In some models there may be more than one observation of the *same* variable, perhaps at different times, for each individual. In other models, each individual may be observed on two or more *different* variables. In either case, hypotheses about means, analogous to those in ANOVA, may be tested using a method called *multivariate* analysis of variance (MANOVA).

The MANOVA method is particularly useful when the hypotheses to be investigated in a research study are concerned with traits or characteristics that are too complex or too broad in scope to be measured satisfactorily by a single test. For example, consider the concept of vocational maturity, or the extent to which the individual has acquired or developed the attitudes and knowledge considered essential to make appropriate career decisions. There is considerable evidence (see, for example, Super, 1970, 1974) that vocational maturity is not a unitary construct, that instead it has both affective and cognitive aspects, and that the satisfactory measurement of these different aspects requires different measurement instruments. Similar remarks could be made about other psychological constructs, such as general achievement, aptitude, and anxiety. Each has several clearly discernible aspects that are sufficiently distinct from one another to suggest their separate measurement. In some research, of course, we may wish to study these several aspects *individually*. In other cases, the *general* construct might be of prime interest and we would then find it necessary to obtain an appropriate measure of it, perhaps by combining in some way measures of its separate aspects.

Now suppose we are concerned with a general construct, X, but have, say, three measures, X_1, X_2, and X_3, of different aspects of the general construct. We might, of course, compute the average of X_1, X_2, and X_3 to obtain a single measure which we could analyze by means of ANOVA. The question which might arise, however, is how to weight the three variables in obtaining the average. In some situations there may exist a defensible rationale for a particular set of weights. For example, we might wish to define a general measure, $X = X_1 + 2X_2 + X_3$, which implies equal weighting of X_1 and X_3, with X_2 receiving twice as much weight as either of the other variables. Such a measure would be appropriate if theory and past research suggested strongly that X_2 was the most "important" aspect of the general construct and that it, therefore, should receive the greatest emphasis in the combined measure, X.

Cases in which there is sufficient justification for assigning weights on theoretical or empirical grounds seem to be rare in educational and psychological research. The MANOVA method employs a different rationale, based on the separation between group centroids. Weights are assigned that maximize the variability *between* groups, relative to the variability *within* groups. Note that this is exactly what was done in discriminant analysis, discussed in Chapter 6. In that case, maximizing the discriminant criterion, λ, was equivalent to maximizing the ratio of the between-groups variability and the within-groups variability. As we shall see, one-way MANOVA is equivalent to testing the significance of the discriminant criterion, λ.

We deal first with the one-way case and show its equivalence to discriminant analysis. We limit our discussion of two-way MANOVA to the case of equal (or proportional) cell frequencies primarily because our presentation follows the "classical" approach rather than the regression approach, discussed briefly in Chapter 4. For a more general treatment along the lines of the *general linear hypothesis*, the reader is referred to Bock (1975), Searle (1971), or Timm (1975).

7.1 ONE-WAY MANOVA

As we suggested above, the one-way MANOVA is an extension of the one-way univariate analysis of variance to the case in which each individual is measured on at least two variables (or at least twice on the same variable). As an example, suppose that we conduct an experiment to evaluate the relative effectiveness of four different methods of teaching reading to young children. Ten children, matched by sex, age, and initial level of ability, are assigned to each of four treatment groups, 1, 2, 3, and 4. Each group is then taught by a different method. After the instructional period, a test measuring two aspects of general reading achievement, namely, Reading Speed (X_1) and Reading Comprehension (X_2), is administered to all forty subjects. The raw scores on the two tests are shown in Table 7.1.1. We wish to test the null hypothesis, H_0: $\boldsymbol{\mu}_1 = \boldsymbol{\mu}_2 = \boldsymbol{\mu}_3 = \boldsymbol{\mu}_4$, that the mean vectors in the populations representing the four methods are all

TABLE 7.1.1 DATA AND SUMMARY STATISTICS FOR ONE-WAY MANOVA EXAMPLE

Method:	*I*		*II*		*III*		*IV*		
Variable:	X_1	X_2	X_1	X_2	X_1	X_2	X_1	X_2	
	7	9	43	21	24	14	27	17	
	37	16	39	17	30	15	20	13	
	44	16	21	9	17	8	25	15	
	28	14	32	16	34	15	13	11	
	42	16	21	9	32	16	32	17	
	23	11	14	6	35	17	4	10	
	45	25	24	11	19	11	3	7	
	32	15	24	12	6	6	36	19	
	37	16	26	12	30	14	28	16	
Statistic	45	19	26	12	19	10	21	15	*Totals*
$\sum_{i=1}^{n} X_{1i}$	340		270		246		209		1065
$\sum_{i=1}^{n} X_{2i}$		157		125		126		140	548
$\sum_{i=1}^{n} X_{1i}^2$	12874		7976		6828		5493		33171
$\sum_{i=1}^{n} X_{2i}^2$		2633		1737		1708		2084	8162
$\sum_{i=1}^{n} X_{1i}X_{2i}$	5728		3716		3394		3289		16127
Mean	34.0	15.7	27.0	12.5	24.6	12.6	20.9	14.0	
Standard deviation	12.1	4.3	8.7	4.4	9.3	3.7	11.2	3.7	

equal. In this hypothesis, $\boldsymbol{\mu}_1'=[\mu_{11} \quad \mu_{21}]$, $\boldsymbol{\mu}_2'=[\mu_{12} \quad \mu_{22}]$, $\boldsymbol{\mu}_3'=[\mu_{13} \quad \mu_{23}]$, and $\boldsymbol{\mu}_4'=[\mu_{14} \quad \mu_{24}]$, in which μ_{jg} is the mean on the jth variable in the gth population. In this example $j=1,2$ and $g=1,2,3,4$. In the general case, however, $j=1,2,\ldots,p$, and $g=1,2,\ldots,G$.

Several different statistics have been suggested for testing the above hypothesis. The one most widely used is based on Wilks' (1932) likelihood ratio criterion, described in Chapters 5 and 6. To see how this criterion may be used in MANOVA, it will be helpful to consider briefly the one-way univariate ANOVA. In ANOVA the hypothesis H_0: $\mu_1=\ldots=\mu_G$ may be tested using

$$F=MS_A/MS_W=\frac{SS_A}{SS_W}\cdot\frac{N-G}{G-1}, \tag{7.1.1}$$

in which $N=\sum_{g=1}^{G}n_g$ (that is, the total number in all samples combined), $SS_A=\sum_{g=1}^{G}n_g(\bar{X}_g-\bar{X})^2$, and $SS_W=\sum_{g=1}^{G}\sum_{i=1}^{n_g}(X_{gi}-\bar{X}_g)^2$. In these expressions for SS_A and SS_W, X_{gi} is the value of the ith observation in sample g, $\bar{X}_g$ is the mean of sample g, and $\bar{X}$ is the overall mean (grand mean) of all of the N observations. Recall now that the total sum of squares, $SS_T=\sum_{g=1}^{G}\sum_{i=1}^{n_g}(X_{gi}-\bar{X})^2=SS_A+SS_W$. Then $SS_A=SS_T-SS_W$, and Equation 7.1.1 can be written

$$F=\frac{SS_T-SS_W}{SS_W}\left(\frac{N-G}{G-1}\right),$$

so that

$$F\left(\frac{G-1}{N-G}\right)+1=\frac{SS_T}{SS_W}. \qquad (7.1.2)$$

Wilks' criterion may be defined in the MANOVA context as

$$\Lambda=\frac{|\mathbf{W}|}{|\mathbf{T}|}, \qquad (7.1.3)$$

in which **W** is the *within*-groups sums of squares and cross-products matrix defined by Equation 6.2.4, and **T** is the *total* sums of squares and cross-products matrix, a general element of which is

$$t_{jk}=\sum_{g=1}^{G}\sum_{i=1}^{n_g}(X_{jgi}-\bar{X}_j)(X_{kgi}-\bar{X}_k);\,(j,k=1,\ldots,p).$$

$$=\sum_{g=1}^{G}\sum_{i=1}^{n_g}X_{jgi}X_{kgi}-N\bar{X}_j\bar{X}_k$$

In this expression, $\bar{X}_j$ is the grand mean (the mean for all groups combined) of variable X_j. Now in the univariate case, that is, the case where $p=1$, $|\mathbf{W}|=SS_W$ and $|\mathbf{T}|=SS_T$. Then $\Lambda=SS_W/SS_T$ so that $1/\Lambda=SS_T/SS_W$. Substitution for SS_T/SS_W in Equation 7.1.2 yields

$$F\left(\frac{G-1}{N-G}\right)+1=\frac{1}{\Lambda} \qquad (7.1.4)$$

which shows, for the univariate case, the relationship between F and Λ. Note that F and Λ are inversely related, so that large values of F, or small values of Λ, would lead to rejection of the null hypothesis. In other words, the greater the differences between group means, the larger the value of F and the smaller the value of Λ.

Solving Equation 7.1.4 for F, we see that in the univariate case,

$$F=\frac{1-\Lambda}{\Lambda}\left(\frac{N-G}{G-1}\right), \qquad (7.1.5)$$

which, under H_0, has the F distribution with $n_1=G-1$ and $n_2=N-G$ degrees of freedom. In other special cases, certain functions of Λ are known to be distributed exactly as F. These are given in Table 7.1.2.

TABLE 7.1.2 DISTRIBUTIONS OF FUNCTIONS OF Λ FOR CERTAIN VALUES OF G AND p IN ONE-WAY MANOVA

Value of G	*Value of* p	*Function distributed as F*	*Degrees of freedom of* $F(n_1,n_2)$
Any value	2	$\frac{1-\sqrt{\Lambda}}{\sqrt{\Lambda}}\left(\frac{N-G-1}{G-1}\right)$	$2(G-1), 2(N-G-1)$
2	Any value	$\frac{1-\Lambda}{\Lambda}\left(\frac{N-p-1}{p}\right)$	$p, N-p-1$
3	Any value	$\frac{1-\sqrt{\Lambda}}{\sqrt{\Lambda}}\left(\frac{N-p-2}{p}\right)$	$2p, 2(N-p-2)$

Thus, in many practical research situations, the test of equality of means in one-way MANOVA is an exact test.

When $p \geqslant 3$ and $G \geqslant 4$, the exact sampling distribution of Λ does not conform to that of any well-known model such as F or χ^2. Schatzoff (1964, 1966) showed that exact significance probabilities could be obtained feasibly only when p is an even number or G is an odd number. He prepared tables for use in such cases, based on a logarithmic function of Λ, namely,

$$V = [N - 1 - (p + G)/2] \ln \frac{1}{\Lambda} \cdot \qquad (7.1.6)$$

Because Schatzoff's tables are quite extensive, they are not included here. Instead, we suggest the use of either of two approximations to the exact significance probability of Λ. These are:

1. Bartlett's V. The statistic defined in Equation 7.1.6 was shown by Bartlett (1947) to be distributed approximately as chi-square with $p(G-1)$ degrees of freedom for large n. The approximation will be good if our recommendation in Chapters 4, 5, and 6 concerning sample size in multivariate research is followed*.

2. Rao's R. A slightly better approximation to the exact significance probability of Λ is based on a statistic due to Rao (1952):

$$R = \frac{1 - \Lambda^{1/s}}{\Lambda^{1/s}} \left(\frac{ms - p(G-1)/2 + 1}{p(G-1)} \right), \qquad (7.1.7)$$

in which $m = N - 1 - (p + G)/2$ and

$$s = \sqrt{\frac{p^2(G-1)^2 - 4}{p^2 + (G-1)^2 - 5}} \ .$$

R is distributed approximately as F with $n_1 = p(G-1)$ and $n_2 = ms - p(G-1)/2 + 1$ degrees of freedom, when the null hypothesis is true. If n_2 is not an integer, one may either interpolate or use the closest integral value. Note that when $p = 1$ or 2, *or* when $G = 2$ or 3, R reduces to a function that has an exact F distribution (see Table 7.1.2).

It will be convenient at this point to show that the test of the null hypothesis H_0: $\boldsymbol{\mu}_1 = \ldots = \boldsymbol{\mu}_G$ in MANOVA is equivalent to the test of significance of the discriminant criteria in discriminant analysis. If we take the reciprocals of both sides of Equation 7.1.3, we have

$$\frac{1}{\Lambda} = \frac{|\mathbf{T}|}{|\mathbf{W}|} = |\mathbf{W}^{-1}|\,|\mathbf{T}| = |\mathbf{W}^{-1}\mathbf{T}|.$$

Now, as indicated in Chapter 6, the total sums of squares and cross-products matrix, $\mathbf{T}$, is the sum of $\mathbf{W}$ and $\mathbf{A}$, the within-groups and among-groups sums of squares and cross-products matrices, respectively. Therefore,

$$\frac{1}{\Lambda} = |\mathbf{W}^{-1}(\mathbf{W} + \mathbf{A})| = |\mathbf{I} + \mathbf{W}^{-1}\mathbf{A}|. \qquad (7.1.8)$$

*We recommend that the total sample size be at least 100 *or* $20p$, whichever is larger.

Then according to Properties 3 and 7(d) of the eigenvalues of matrices given in Appendix A, we can state

$$\frac{1}{\Lambda}=(1+\lambda_1)(1+\lambda_2)\ldots(1+\lambda_r), \tag{7.1.9}$$

in which the λs are the eigenvalues of $\mathbf{W}^{-1}\mathbf{A}$. Note that this equation is identical to Equation 6.2.13, in which the λs are also the eigenvalues of $\mathbf{W}^{-1}\mathbf{A}$. Therefore, the tests of significance of Λ in MANOVA and in discriminant analysis are equivalent.

Returning now to our example involving the four methods of teaching reading, we shall illustrate the computation of the one-way MANOVA. To calculate Λ, we must first obtain $\mathbf{W}$ and $\mathbf{T}$. From Equation 6.2.4, we see that, in our example, $\mathbf{W}=\mathbf{S}_{xx}^{(1)}+\mathbf{S}_{xx}^{(2)}+\mathbf{S}_{xx}^{(3)}+\mathbf{S}_{xx}^{(4)}$. Using the summary data given in Table 7.1, we find:

$$\mathbf{S}_{xx}^{(1)}=\begin{bmatrix}1314.0 & 390.0\\ 390.0 & 168.1\end{bmatrix},\quad \mathbf{S}_{xx}^{(2)}=\begin{bmatrix}686.0 & 341.0\\ 341.0 & 174.5\end{bmatrix},$$

$$\mathbf{S}_{xx}^{(3)}=\begin{bmatrix}776.4 & 294.4\\ 294.4 & 120.4\end{bmatrix},\quad \text{and}\quad \mathbf{S}_{xx}^{(4)}=\begin{bmatrix}1124.9 & 363.0\\ 363.0 & 124.0\end{bmatrix}.$$

Then $\mathbf{W}=\begin{bmatrix}3901.3 & 1388.4\\ 1388.4 & 587.0\end{bmatrix}$ and we find that

$$\mathbf{T}=\begin{bmatrix}4815.375 & 1536.500\\ 1536.500 & 654.400\end{bmatrix}.$$

To use Equation 7.1.3, we need to find the determinants of $\mathbf{W}$ and $\mathbf{T}$. These are: $|\mathbf{W}|=362408.54$ and $|\mathbf{T}|=790349.15$. Then $\Lambda=|\mathbf{W}|/|\mathbf{T}|=.458542329$.

Because $p=2$ in this example, we can use the exact test given in Table 7.2 for testing the significance of Λ. The function

$$\frac{1-\sqrt{\Lambda}}{\sqrt{\Lambda}}\left(\frac{N-G-1}{G-1}\right)=5.5622$$

has the F distribution with $2(G-1)=6$ and $2(N-G-1)=70$ degrees of freedom when the null hypothesis is true. This result is significant beyond the .001 level.

We will also illustrate the use of Bartlett's V and Rao's R for this example. The tests are:

1. Using Bartlett's V we have

$$V=[40-1-(2+4)/2]\ln\frac{1}{.458542329}$$
$$=36(.7797)=28.07,$$

which, under H_0, has approximately a chi-square distribution with $p(G-1)=6$ degrees of freedom. Therefore, the observed value of V is significant beyond the .001 level.

2. To use Rao's R, we must first compute

$$m = N-1-(p+G)/2 = 40-1-3 = 36, \quad \text{and}$$

$$s = \sqrt{\frac{p^2(G-1)^2-4}{p^2+(G-1)^2-5}} = \sqrt{\frac{32}{8}} = 2.$$

Then

$$R = \frac{1-\sqrt{.458542329}}{\sqrt{.458542329}}\left(\frac{(36)(2)-2(3)/2+1}{2(3)}\right) = 5.5622,$$

which is equal to the value obtained from computing the appropriate function from Table 7.1.2 (because R reduces to a function having an exact F distribution when $p=2$). Based on this test, as well as the test using Bartlett's approximation, we conclude that the population mean vectors for the four teaching methods are not all equal.

When a one-way MANOVA results in rejection of the null hypothesis, the only conclusion that can be drawn *immediately* is that the populations differ from one another in terms of particular linear combinations of the Xs involved in the analysis. Remember that in MANOVA the weights in these linear combinations are those that produce maximum group separation. These linear combinations of the Xs define what we have earlier called "dimensions" (traits or characteristics) in terms of which the groups differ. It is usually of interest, following a MANOVA, to determine the dimension or dimensions along which the groups differ.

One set of such dimensions is defined by the individual Xs. For example, in our illustration concerning the four methods of teaching reading, the two criterion variables used were Reading Speed (X_1) and Reading Comprehension (X_2). It would probably be of interest to determine whether the groups differ significantly in terms of *each* of these two characteristics, that is, to answer the question, Do the four teaching methods differentially affect reading speed or comprehension, considered separately? To answer this question it would be appropriate to carry out two separate univariate ANOVAs, one on X_1 and the other on X_2. For our example, the results of these analyses are given in Table 7.1.3. We conclude that the four groups do not differ significantly in terms of Reading Speed (X_1) or Reading Comprehension (X_2), when these dimensions are considered separately.

TABLE 7.1.3 SUMMARY OF ANOVA TESTS OF READING SPEED AND COMPREHENSION FOR THE DATA OF TABLE 7.1.1

Source	*SS*	*df*	*MS*	*F*
	Reading speed (X_1)			
Between groups	914.075	3	304.692	2.81 ($p>.05$)
Within groups	3901.300	36	108.369	
Total	4815.375	39		
	Reading comprehension (X_2)			
Between groups	67.4	3	22.467	1.38 ($p>.05$)
Within groups	587.0	36	16.306	
Total	654.4	39		

Another approach to the determination of the dimension(s) along which the groups differ is provided by discriminant analysis, discussed in Chapter 6. Concerning our example, we know, at this point, that (1) the groups differ significantly in terms of dimensions defined by certain linear combinations of X_1 and X_2, and (2) the groups do *not* differ significantly in terms of the dimensions defined by X_1 and X_2 considered separately. Discriminant analysis should enable us to identify the dimensions defined by the linear combination(s) in terms of which the groups do in fact differ. Recall that the number of discriminant functions in a discriminant analysis is equal to the smaller of p, the number of variables (Xs) and $G-1$, where G is the number of groups. For our example, $p=2$ and $G-1=3$. Therefore, the number of discriminant functions is 2. To determine these, we first need to obtain the eigenvalues of $\mathbf{W}^{-1}\mathbf{A}$. We know that

$$\mathbf{W}=\begin{bmatrix} 3901.3 & 1388.4 \\ 1388.4 & 587.0 \end{bmatrix},$$

so that

$$\mathbf{W}^{-1}=\begin{bmatrix} .001619719 & -.003831036 \\ -.003831036 & .010764923 \end{bmatrix}.$$

Now, since $\mathbf{T}=\mathbf{W}+\mathbf{A}$, we find that

$$\mathbf{A}=\mathbf{T}-\mathbf{W}=\begin{bmatrix} 914.075 & 148.100 \\ 148.100 & 67.400 \end{bmatrix}.$$

Therefore, the determinantal equation to be solved for the roots, λ_1 and λ_2, of $\mathbf{W}^{-1}\mathbf{A}$ is

$$|\mathbf{W}^{-1}\mathbf{A}-\lambda\mathbf{I}|=\begin{vmatrix} .913168285-\lambda & -.018331411 \\ -1.907568735 & .158179440-\lambda \end{vmatrix}=0.$$

Then the characteristic equation may be written

$$\lambda^2-1.071347725\lambda+.109476021=0,$$

the roots of which are $\lambda_1=.956946300$ and $\lambda_2=.114401426$.

A partial check on computations can be made now using Equation 7.1.9. Substituting for λ_1 and λ_2, we find

$$\frac{1}{\Lambda}=(1+\lambda_1)(1+\lambda_2)=(1.95694630)(1.114401426)$$
$$=2.180823747.$$

Therefore, $\Lambda=.458542329$, which agrees with the value obtained using Equation 7.1.3.

Because we know already that Λ is significant, we may conclude that the two discriminant functions based on λ_1 and λ_2, *taken together*, are significant. Thus, we need only determine whether the function based on λ_2 is significant. For this purpose we shall use the procedure involving Bartlett's V statistic, used earlier in Chapters 5 and 6. Substituting in Equation 6.2.15 with $k=2$, we first obtain

$$\ln\left(\frac{1}{\Lambda_2}\right)=\ln(1+\lambda_2)=.108317423.$$

Then

$$V_2 = [N-1-(p+G)/2]\ln\left(\frac{1}{\Lambda_2}\right)$$
$$= 36(.108317423) = 3.90,$$

which has approximately a chi-square distribution with $(p-k+1)(G-k) = (2-2+1)(4-2) = 2$ degrees of freedom when the null hypothesis is true. The significance probability of this result is greater than .10, and so we conclude that the second discriminant function is not significant.

To determine the discriminant function associated with λ_1, we again follow procedures used in Chapters 5 and 6. We first substitute for λ_1 in the expression $\mathbf{W}^{-1}\mathbf{A} - \lambda_1 \boldsymbol{I}$, to obtain

$$\mathbf{W}^{-1}\mathbf{A} - \lambda_1\mathbf{I} = \begin{bmatrix} -.043778015 & -.018331411 \\ -1.907568735 & -.798766860 \end{bmatrix}.$$

Then we obtain the cofactors of the elements of either row of this expression. Using the first row, we find

$$\mathbf{f} = \begin{bmatrix} -.798766860 \\ +1.907568735 \end{bmatrix},$$

which, when normalized, becomes

$$\mathbf{u} = \begin{bmatrix} -.386240831 \\ .922397973 \end{bmatrix}.$$

Thus, the discriminant function based on λ_1 is

$$Z_{ig1} = -.38624083X_{i1g} + .922397973X_{i2g},$$

in which Z_{ig1} denotes the value of the discriminant function based on λ_1 for the ith individual in the gth group. The group means on this function may be obtained from $\bar{Z}_{1g} = -.38624083\bar{X}_{1g} + .922397973\bar{X}_{2g}$. We find that $\bar{Z}_{11} = 1.35$, $\bar{Z}_{21} = 1.10$, $\bar{Z}_{31} = 2.12$, and $\bar{Z}_{41} = 4.84$. Thus, we see that Group IV had a noticeably larger mean on the discriminant function than any of the other three groups. This appears to account for the significance of the first discriminant function.

In order to use the weights on X_1 and X_2 to determine the dimension measured by this discriminant function, we first need to transform the weights to standard form. For this purpose, we use Equation 6.2.18. The transformation matrix is (see Equation 6.2.18)

$$\mathbf{H} = \begin{bmatrix} 10.41 & 0 \\ 0 & 4.04 \end{bmatrix}.$$

Then $\mathbf{u}_s$, the vector of standardized weights, is

$$\mathbf{u}_s = \mathbf{H}\mathbf{u} = \begin{bmatrix} -4.02 \\ 3.72 \end{bmatrix}.$$

This result shows that X_1 and X_2 receive approximately equal weights in the linear combination that defines the first discriminant function. However, X_1 is negatively weighted, whereas X_2 has a positive weight. Thus, the larger values on Z_{ig1} tend to be associated with small scores on Speed (X_1) and moderate or large scores on Comprehension (X_2), whereas smaller values on Z_{ig1} arise when Speed is moderate or large and Comprehension is moderate or small. The result is that Group I has a relatively small mean ($\bar{Z}_{11}=1.35$) even though it had the largest means of all groups on *both* Speed ($\bar{X}_{11}=34.0$) and Comprehension ($\bar{X}_{21}=15.7$). On the other hand, Group IV had a much larger mean ($\bar{Z}_{41}=4.84$), primarily because its mean Speed score ($\bar{X}_{41}=20.9$) was much lower than any other group and its mean Comprehension score ($\bar{X}_{42}=14.0$) was moderate. The appropriateness of MANOVA for this example depends on whether the dimension measured by the first discriminant function provides a meaningful criterion measure for judging the relative effectiveness of the four methods. The MANOVA tells us simply that this measure (roughly, Comprehension minus Speed) is the one that produces maximum group separation. If we wish to use such a measure as a criterion, then MANOVA is a reasonable method of analysis. If we do not, then an alternative method for combining the X_1 and X_2 variables might be sought. For example, an equal positive weighting, that is, an average of the standard score forms such as

$$z=\frac{z_{X_1}+z_{X_2}}{2},$$

might be computed and analyzed by means of ANOVA. Alternatively, of course, the two variables could be analyzed separately, as in Table 7.1.3.

Following a significant MANOVA result, one may wish to determine the pairs (or other combinations) of groups between which the differences are significant. Methods for making such comparisons between group mean vectors are available in MANOVA, as well as in ANOVA. For example, both the Scheffé and Dunn (Bonferroni) methods have been generalized to the multivariate case. Because of limitations of space, we shall not discuss these here. For a treatment of these methods, we refer the reader to Harris (1975), Chapter 4, or Timm (1975), Chapter 5.

7.2. TWO-WAY MANOVA

As we mentioned earlier, we shall restrict our discussion of two-way MANOVA to the case of equal (or proportional) cell frequencies. This restriction is necessary in part because of space limitations. However, it is also consistent with our intent to emphasize concepts somewhat more than computational procedures. Furthermore, it results in a presentation consistent with other chapters in terms of level of mathematical sophistication required.

The two-way MANOVA may be viewed as an extension of two-way ANOVA to the case in which each individual has observations on the variables $X_1,\ldots,X_p$. Thus, an individual's value on variable $X_j (j=1,\ldots,p)$

might be denoted by X_{jrci}, that is, the value of the ith individual $(i=1,\ldots,n)$ in the rth row $(r=1,\ldots,R)$ and the cth column $(c=1,\ldots,C)$, on variable X_j. We shall first assume that cell frequencies are equal. Then the number in each cell may be denoted by n, and we may define, for variable X_j:

$$T_{jrc.}=\sum_{i=1}^{n} X_{jrci}, \text{ the total in cell } rc,$$

$$T_{jr..}=\sum_{c=1}^{C} T_{jrc.}, \text{ the total in row } r,$$

$$T_{j.c.}=\sum_{r=1}^{R} T_{jrc.}, \text{ the total in column } c,$$

and

$$T_{j...}=\sum_{r=1}^{R} T_{jr..}=\sum_{c=1}^{C} T_{j.c.}, \text{ the total over all cells.}$$

The corresponding means are:

$$\bar{X}_{jrc.}=\frac{T_{jrc.}}{n}, \text{ the mean of cell } rc,$$

$$\bar{X}_{jr..}=\frac{T_{jr..}}{nC}, \text{ the mean of row } r,$$

$$\bar{X}_{j.c.}=\frac{T_{j.c.}}{nR}, \text{ the mean of column } c,$$

and

$$\bar{X}_{j...}=\frac{T_{j...}}{nRC}, \text{ the mean of all observations for variable } X_j, \text{ that is, the grand mean.}$$

Recall that in the two-way ANOVA, the total sum of squares, SS_T, can be expressed as the sum of four components, namely,

$$SS_T=SS_R+SS_C+SS_{RC}+SS_W \tag{7.2.1}$$

where SS_R, SS_C, SS_{RC}, and SS_W are the sums of squares for rows, columns, interaction, and error (within groups), respectively. Now, for the multivariate context, the analogous components are:

1. Total sums of squares and cross-products matrix. We shall continue to denote this matrix by **T**, as we did in the one-way MANOVA. The jkth element of **T** is

$$t_{jk}=\sum_{r=1}^{R}\sum_{c=1}^{C}\sum_{i=1}^{n}\left(X_{jrci}-\bar{X}_{j...}\right)\left(X_{krci}-\bar{X}_{k...}\right);\ (j,k=1,\ldots,p)$$

$$=\sum_{r=1}^{R}\sum_{c=1}^{C}\sum_{i=1}^{n} X_{jrci}X_{krci}-T_{j...}T_{k...}/nRC \tag{7.2.2}$$

2. Row sums of squares and cross-products matrix. We shall denote this matrix by $\mathbf{S}_r$. Its jkth element is

$$S_{r(jk)} = nC \sum_{r=1}^{R} (\bar{X}_{jr..} - \bar{X}_{j...})(\bar{X}_{kr..} - \bar{X}_{k...}); \; (j,k=1,\ldots,p)$$

$$= \sum_{r=1}^{R} T_{jr..}T_{kr..}/nC - T_{j...}T_{k...}/nRC \qquad (7.2.3)$$

3. Column sums of squares and cross-products matrix. We shall denote this matrix by $\mathbf{S}_c$. Its jkth element is

$$S_{c(jk)} = nR \sum_{c=1}^{C} (\bar{X}_{j.c.} - \bar{X}_{j...})(\bar{X}_{k.c.} - \bar{X}_{k...}); \; (j,k=1,\ldots,p)$$

$$= \sum_{c=1}^{C} T_{j.c.}T_{k.c.}/nR - T_{j...}T_{k...}/nRC \qquad (7.2.4)$$

4. Interaction sums of squares and cross-products matrix. We shall denote this matrix by $\mathbf{S}_{rc}$. Its jkth element is

$$S_{rc(jk)} = n \sum_{r=1}^{R} \sum_{c=1}^{C} \big[(\bar{X}_{jrc.} - \bar{X}_{jr..} - \bar{X}_{j.c.} + \bar{X}_{j...})$$

$$\cdot (\bar{X}_{krc.} - \bar{X}_{kr..} - \bar{X}_{k.c.} + \bar{X}_{k...}) \big]; \; (j,k=1,\ldots,p)$$

$$= \sum_{r=1}^{R} \sum_{c=1}^{C} T_{jrc.}T_{krc.}/n - \sum_{r=1}^{R} T_{jr..}T_{kr..}/nC$$

$$- \sum_{c=1}^{C} T_{j.c.}T_{k.c.}/nR + T_{j...}T_{k...}/nRC \qquad (7.2.5)$$

5. Within-cells sums of squares and cross-products matrix. We shall denote this matrix by $\mathbf{S}_w$. Its jkth element is

$$S_{w(jk)} = \sum_{r=1}^{R} \sum_{c=1}^{C} \left[\sum_{i=1}^{n} (X_{jrci} - \bar{X}_{jrc.})(X_{krci} - \bar{X}_{krc.}) \right]; \; (j,k=1,\ldots,p)$$

$$= \sum_{r=1}^{R} \sum_{c=1}^{C} \left[\sum_{i=1}^{n} X_{jrci}X_{krci} - T_{jrc.}T_{krc.}/n \right] \qquad (7.2.6)$$

Then we can state the multivariate analogue of Equation 7.2.1 as

$$\mathbf{T} = \mathbf{S}_r + \mathbf{S}_c + \mathbf{S}_{rc} + \mathbf{S}_w. \qquad (7.2.7)$$

Note that $\mathbf{S}_w$ here is analogous to $\mathbf{W}$ in the one-way MANOVA.

The matrices of Equation 7.2.7, together with the degrees of freedom associated with each, are shown in Table 7.2.1.

In MANOVA, as in the two-way ANOVA, we ordinarily wish to test hypotheses concerning row effects, column effects, and the interaction effects. To simplify the discussion of these tests, we shall define a general form of Wilks' Λ criterion that may be used in all of the MANOVA tests discussed in this chapter. Recall that Λ was defined in Equation 7.1.3 as

TABLE 7.2.1 SUMMARY TABLE FOR TWO-WAY MANOVA

Source	*Matrix*	*Degrees of freedom*
Rows	$\mathbf{S}_r$	$R-1$
Columns	$\mathbf{S}_c$	$C-1$
Interaction	$\mathbf{S}_{rc}$	$(R-1)(C-1)$
Within cells	$\mathbf{S}_w$	$RC(n-1)$
Total	$\mathbf{T}$	$nRC-1$

$\Lambda = |\mathbf{W}|/|\mathbf{T}|$, which, because $\mathbf{T} = \mathbf{A} + \mathbf{W}$, may be written

$$\Lambda = \frac{|\mathbf{W}|}{|\mathbf{A}+\mathbf{W}|}. \tag{7.2.8}$$

In one-way MANOVA, $\mathbf{W}$ is sometimes referred to as the *error* sums of squares and cross-products matrix, whereas $\mathbf{A}$ is the sums of squares and cross-products matrix associated with the null hypothesis being tested. Therefore, Equation 7.2.8 may be written as

$$\Lambda = \frac{|\mathbf{S}_e|}{|\mathbf{S}_h + \mathbf{S}_e|}, \tag{7.2.9}$$

in which $\mathbf{S}_e$ takes the place of $\mathbf{W}$ and $\mathbf{S}_h$ of $\mathbf{A}$. This expression is not only equivalent to Λ as defined by Equation 7.1.3, but also provides a definition of Λ that may be applied generally, with appropriate definitions of $\mathbf{S}_e$ and $\mathbf{S}_h$. As in one-way MANOVA, the significance of Λ may be tested either through computation of a function of it (analogous to those given in Table 7.1.2) that has an exact F distribution, or by means of Bartlett's or Rao's approximations. We shall illustrate an appropriate test procedure for a specific case and then indicate how the tests may be carried out in other types of two-way designs.

In the following example we illustrate the MANOVA method applied to a fixed-effects design with equal cell frequencies. Twenty children were randomly selected from each of three grades: Kindergarten, Grade 1 and Grade 2. Within each grade, 10 children were assigned to each of two methods of teaching reading, Phonetic (I) and Sight (II). Achievement was measured in terms of Reading Speed (X_1) and Reading Comprehension (X_2). The data are given in Table 7.2.2 along with results of a number of intermediate calculations. Using those results and Equations 7.2.2 through 7.2.6, we obtain:

$$\mathbf{T} = \begin{bmatrix} 6451.73333 & 7349.80000 \\ 7349.80000 & 16227.60000 \end{bmatrix},$$

$$\mathbf{S}_r = \begin{bmatrix} 4.26667 & 26.66667 \\ 26.66667 & 166.66667 \end{bmatrix},$$

$$\mathbf{S}_c = \begin{bmatrix} 4994.13333 & 6953.60000 \\ 6953.60000 & 9811.20000 \end{bmatrix},$$

$$\mathbf{S}_{rc} = \begin{bmatrix} 810.13333 & 546.53333 \\ 546.53333 & 1473.73333 \end{bmatrix}, \quad \text{and}$$

$$\mathbf{S}_w = \begin{bmatrix} 643.20000 & -177.00000 \\ -177.00000 & 4776.00000 \end{bmatrix}.$$

TABLE 7.2.2 DATA AND SUMMARY STATISTICS FOR TWO-WAY MANOVA EXAMPLE

	Grades									
	K		1		2		*Totals*		*Row means*	
Method	X_1	X_2	X_1	X_2	X_1	X_2	X_1	X_2	X_1	X_2
	14	31	25	50	49	80				
	20	25	29	52	43	79				
	16	24	27	63	46	66				
	21	23	31	51	46	57				
I	20	27	27	62	44	55				
(Phonics)	14	11	34	56	43	60				
	21	25	32	46	50	65				
	23	22	34	66	43	70				
	14	22	35	75	48	64				
	15	37	28	43	52	67				
ΣX_1	178		302		464		944		31.5	
ΣX_2		247		564		663		1474		49.1
ΣX_1^2	3280		9230		21624		34134			
ΣX_2^2		6503		32700		44581		83784		
$\Sigma X_1 X_2$	4389		17163		30791		52343			
Mean	17.8	24.7	30.2	56.4	46.4	66.3				
Standard deviation	3.5	6.7	3.5	9.9	3.2	8.3				
	17	40	35	58	34	66				
	22	41	36	68	33	78				
	19	45	40	56	34	60				
II	20	54	37	55	39	42				
(Sight)	26	43	37	38	38	43				
	18	43	41	49	33	65				
	26	23	42	51	35	62				
	18	47	33	54	42	43				
	25	45	34	60	42	66				
	23	36	43	67	38	76				
ΣX_1	214		378		368		960		32.0	
ΣX_2		417		556		601		1574		52.5
ΣX_1^2	4688		14398		13652		32738			
ΣX_2^2		17979		31600		37703		87282		
$\Sigma X_1 X_2$	8806		21013		21911		51730			
Mean	21.4	41.7	37.8	55.6	36.8	60.1				
Standard deviation	3.5	8.1	3.5	8.7	3.5	13.3				
Column means	19.6	33.2	34.0	56.0	41.6	63.2			31.7	50.8
Column totals										
ΣX_1	392		680		832		1904			
ΣX_2		664		1120		1264		3048		
ΣX_1^2	7968		23628		35276		66872			
ΣX_2^2		24482		64300		82284		171066		
$\Sigma X_1 X_2$	13195		38176		52702		104073			

TABLE 7.2.3 MANOVA SUMMARY TABLE FOR THE TWO-WAY FIXED-EFFECTS EXAMPLE OF TABLE 7.2.2

Source	*Matrix*	*df*	$\lvert \mathbf{S}_h + \mathbf{S}_e \rvert$*	Λ_h
Rows (Method)	$\mathbf{S}_r$	$R-1=1$	$\begin{vmatrix} 647.46667 & -150.33333 \\ -150.33333 & 4942.66667 \end{vmatrix}$	$\Lambda_r = \dfrac{3040594.2}{3177611.8} = .956880321$
Columns (Grade)	$\mathbf{S}_c$	$C-1=2$	$\begin{vmatrix} 5637.33333 & 6776.60000 \\ 6776.60000 & 14587.20000 \end{vmatrix}$	$\Lambda_c = \dfrac{3040594.2}{36310601.2} = .08373847$
Interaction (Method by grade)	$\mathbf{S}_{rc}$	$(R-1)(C-1) = 2$	$\begin{vmatrix} 1453.33333 & 369.53333 \\ 369.53333 & 6249.73333 \end{vmatrix}$	$\Lambda_{rc} = \dfrac{3040594.2}{8946390.9} = .33986825$
Within cells (Error)	$\mathbf{S}_e$	$RC(n-1) = 54$		
Total	$\mathbf{T}$	$nRC-1=59$		

$^*\lvert \mathbf{S}_e \rvert = \lvert \mathbf{S}_w \rvert = 3040594.2$

TABLE 7.2.4 DISTRIBUTIONS OF FUNCTIONS OF Λ_h FOR CERTAIN VALUES OF ν^*_h AND p IN TWO-WAY MANOVA

Value of ν_h	*Value of* p	*Function distributed as F*	*Degrees of freedom* of* $F(n_1, n_2)$
Any value	2	$\dfrac{1-\sqrt{\Lambda_h}}{\sqrt{\Lambda_h}}\left(\dfrac{\nu_e - 1}{\nu_h}\right)$	$2\nu_h, 2(\nu_e - 1)$
1	Any value	$\dfrac{1-\Lambda_h}{\Lambda_h}\left(\dfrac{\nu_e + \nu_h - p}{p}\right)$	$p, \nu_e + \nu_h - p$
2	Any value	$\dfrac{1-\sqrt{\Lambda_h}}{\sqrt{\Lambda_h}}\left(\dfrac{\nu_e + \nu_h - p - 1}{p}\right)$	$2p, 2(\nu_e + \nu_h - p - 1)$

*ν_h is the number of degrees of freedom associated with S_h and ν_e is the number of degrees of freedom associated with S_e.

Although **T** is not required for carrying out the tests of row, column, and interaction effects, its separate computation provides a partial check on the computations of the other matrices. The reader may confirm that $\mathbf{T} = \mathbf{S}_r + \mathbf{S}_c + \mathbf{S}_{rc} + \mathbf{S}_w$.

The computation of $\Lambda_h = \lvert \mathbf{S}_e \rvert / \lvert \mathbf{S}_h + \mathbf{S}_e \rvert$, in which the subscript h may stand for r, c, or rc, depending on the hypothesis being tested, is summarized in Table 7.2.3.

Note that, for all three tests in the two-way fixed effects design, $\mathbf{S}_e = \mathbf{S}_w$. To test the significance of Λ_h in this example, we can compute a function that has an exact F distribution, because the number of variables is $p = 2$. The necessary functions are given in Table 7.2.4.

For our example, the values of F are:

Rows (Methods)

$$F_{2,106} = \frac{1-\sqrt{\Lambda_r}}{\sqrt{\Lambda_r}}\left(\frac{\nu_e - 1}{\nu_r}\right) = 1.18\ (p > .05);$$

Columns (Grades)

$$F_{4,106} = \frac{1-\sqrt{\Lambda_c}}{\sqrt{\Lambda_c}}\left(\frac{\nu_e - 1}{\nu_c}\right) = 65.08\ (p < .001);$$

Interaction (Method by Grade)

$$F_{4,106} = \frac{1-\sqrt{\Lambda_{rc}}}{\sqrt{\Lambda_{rc}}}\left(\frac{\nu_e - 1}{\nu_{rc}}\right) = 18.96\ (p < .001).$$

From these results we may conclude that there are significant Grade and Method-by-Grade (Interaction) effects.

When an exact test cannot be carried out, either Bartlett's V or Rao's R may be used to yield an approximate test. More general forms of each of these statistics may be obtained by replacing N by $\nu_h + \nu_e + 1$ and G by $\nu_h + 1$ in Equations 7.1.6 and 7.1.7. The results are:

$$V = [\nu_h + \nu_e - (p + \nu_h + 1)/2]\ln\left(\frac{1}{\Lambda_h}\right), \tag{7.2.10}$$

which has an approximate chi-square distribution under the null hypothesis, with $p\nu_h$ degrees of freedom, and

$$R = \frac{1-\Lambda_h^{1/s}}{\Lambda_h^{1/s}}\left(\frac{ms - p\nu_h/2 + 1}{p\nu_h}\right), \tag{7.2.11}$$

in which $m = \nu_h + \nu_e - (p + \nu_h + 1)/2$ and

$$s = \sqrt{\frac{(p\nu_h)^2 - 4}{p^2 + \nu_h^2 - 5}}\ .$$

R has approximately an F distribution with $n_1 = p\nu_h$ and $n_2 = ms - p\nu_h/2 + 1$ degrees of freedom when H_0 is true. As before, if n_2 is not an integer, one may either interpolate or use the nearest integral value.

Again, we may wish to determine the dimension or dimensions in terms of which the groups differ. Because the Methods effect was not significant, we shall not investigate it further. However, we can use discriminant analysis, as we did in the one-way case, to investigate further the Grade effect and the Method-by-Grade interaction effect. We first find that

$$\mathbf{S}_e^{-1} = \mathbf{S}_w^{-1} = \begin{bmatrix} .001570746 & .000058212 \\ .000058212 & .000211538 \end{bmatrix}.$$

Then, for the column effect (Grades), the determinantal equation to be solved for λ is

$$|\mathbf{S}_w^{-1}\mathbf{S}_c - \lambda\mathbf{I}| = \begin{vmatrix} 9.24929811-\lambda & 11.49346927 \\ 1.761667874 & 3.480222793-\lambda \end{vmatrix} = 0.$$

The resulting characteristic equation may be written

$$\lambda^2 - 12.72952090\lambda + 11.94194255 = 0,$$

the roots of which are $\lambda_1 = 11.70968636$ and $\lambda_2 = 1.019834535$.

To test the significance of λ_2, we again use Bartlett's V statistic, which now becomes

$$\begin{aligned} V_2 &= [\nu_h + \nu_e - (p + \nu_h + 1)/2]\ln(1+\lambda_2) \\ &= [2+54-(2+2+1)/2](.703015595) \\ &= 37.61. \end{aligned}$$

This statistic has approximately a chi-square distribution with $(p-1)(\nu_h + 1 - 2) = 1$ degree of freedom. This result is significant well beyond the .001 level. Thus, both λ_1 and λ_2 are significantly different from zero.

To determine the discriminant function associated with λ_1, we first obtain

$$\mathbf{S}_w^{-1}\mathbf{S}_c - \lambda_1\mathbf{I} = \begin{bmatrix} -2.46038825 & 11.49346927 \\ 1.761667874 & -8.229463567 \end{bmatrix}.$$

The cofactors of the first row of this matrix are

$$\mathbf{f} = \begin{bmatrix} -8.229463567 \\ -1.761667874 \end{bmatrix},$$

which, when it is normalized and the signs are reversed, becomes

$$\mathbf{u} = \begin{bmatrix} .977845935 \\ .209325888 \end{bmatrix}.$$

Thus, one of the discriminant functions on which the columns (Grades) differ significantly is

$$Z_{ic1} = .977845935X_1 + .209325888X_2,$$

in which the subscript, c, stands for "column."

The means of this function for the three grades are:

Grade	*Mean of discriminant function,* $\bar{Z}_{c1}$
K	26.12
1	44.97
2	53.91

Therefore, it is clear that reading achievement means, as measured by this discriminant function, increase from Kindergarten to Grade 2, as one might expect.

To clarify the dimension measured by this discriminant function, we need to transform to standard form, again using Equation 6.2.18. The transformation matrix is

$$\mathbf{H}=\begin{bmatrix} 3.45 & 0 \\ 0 & 9.40 \end{bmatrix}.$$

Then

$$\mathbf{u}_s=\mathbf{Hu}=\begin{bmatrix} 3.37 \\ 1.97 \end{bmatrix}.$$

Thus, we see that the two variables, Reading Speed (X_1) and Reading Comprehension (X_2), which define the dimension measured by the discriminant function, Z_{ic1}, are weighted in the ratio of about 1.7 to 1, with Reading Speed receiving the greater emphasis. Of course, as in the one-way MANOVA, it might be of interest to follow up this analysis with ANOVAs to determine whether the Grades differ significantly on X_1 and/or X_2, considered separately.

We may also wish to determine the discriminant function associated with λ_2, which was also highly significant. Using the same procedure as we used above for λ_1, we first compute

$$\mathbf{S}_w^{-1}\mathbf{S}_c-\lambda\mathbf{I}=\begin{bmatrix} 8.229463575 & 11.49346927 \\ 1.761667874 & 2.460388258 \end{bmatrix}.$$

The cofactors of the first row of this matrix are

$$\mathbf{f}=\begin{bmatrix} 2.460388258 \\ -1.761667874 \end{bmatrix},$$

which, when normalized, yields the desired weights

$$\mathbf{u}=\begin{bmatrix} .813069025 \\ -.582167296 \end{bmatrix}.$$

Then the second discriminant function on which the grades differ significantly is

$$Z_{ic2}=.813069025X_1-.582167296X_2.$$

The means of this function for the three grades are:

Grade	*Mean of Discriminant Function,* $\bar{Z}_{c2}$
K	−3.39
1	−4.96
2	−2.97

To aid interpretation, we may wish to reverse the signs of the elements of **u** to obtain

$$\mathbf{u}=\begin{bmatrix} -.813069025 \\ .582167296 \end{bmatrix}.$$

Then $Z_{ic2}=-.813069025X_1+.582167296X_2$ becomes the second discriminant function and the means are: Kindergarten, 3.39; Grade 1, 4.96; and Grade 2, 2.97. It is apparent that the significance of λ_2 is due to the difference on this function between Grade 1 on the one hand and Kindergarten and Grade 2 on the other. That is, Grade 1 had a noticeably larger mean than either Kindergarten or Grade 2.

To understand better the dimension that this second function measures, we need to transform **u** to standard form. The transformation matrix (Equation 6.2.18) is again

$$\mathbf{H}=\begin{bmatrix} 3.45 & 0 \\ 0 & 9.40 \end{bmatrix},$$

so that

$$\mathbf{u}_s=\mathbf{H}\mathbf{u}=\begin{bmatrix} -2.81 \\ 5.47 \end{bmatrix}.$$

Therefore, Reading Comprehension (X_2) is weighted about twice as heavily as Reading Speed (X_1) with the latter receiving a negative weight. A relatively large mean on this dimension, such as in Grade 1, would result from a relatively high mean on X_2 and only a moderate value on X_1. Note that the dimension measured by the second discriminant function, Z_{ic2}, is quite different from that measured by the first, Z_{ic1}

A discriminant analysis completely analogous to that just illustrated may be carried out for the Method-by-Grade interaction to determine the dimension or dimensions which produced this significant effect. In such an analysis $\mathbf{S}_{rc}$ replaces $\mathbf{S}_c$, but the steps in the analysis remain the same as for the column effect. We first compute:

$$|\mathbf{S}_w^{-1}\mathbf{S}_{rc}-\lambda\mathbf{I}|=\begin{vmatrix} 2.304328486-\lambda & .944254007 \\ .162772049 & 1.343565399-\lambda \end{vmatrix}=0,$$

to obtain $\lambda^2-3.647893885\lambda+2.942317863=0$. Then the two roots are $\lambda_1=2.443999026$ and $\lambda_2=1.203894859$. Because the test of the interaction effect, described above, resulted in significance, we may conclude that λ_1 is significantly different from zero. The test of λ_2 may be carried out by computing

$$V_2=[\nu_h+\nu_e-(p+\nu_h+1)/2]\ln(1+\lambda_2)$$
$$=[2+54-(2+2+1)/2](.790226185)=42.28.$$

This statistic has approximately a chi-square distribution with $(p-1)(\nu_h+1-2)=(2-1)(2+1-2)=1$ degree of freedom, when the null hypothesis is true. Therefore, in this case we conclude that both λ_1 and λ_2 are significantly different from zero.

Again, we may wish to carry out a discriminant analysis to determine the dimensions on which significant interactions occur and to see how the treatment groups differ on these dimensions. Considering first the function based on λ_1, we find that

$$\mathbf{S}_w^{-1}\mathbf{S}_{rc}-\lambda_1\mathbf{I}=\begin{bmatrix} -.139670540 & .944254007 \\ .162772049 & -1.100433627 \end{bmatrix}.$$

The cofactors of the first row of this matrix are

$$\mathbf{f}=\begin{bmatrix} -1.100433627 \\ -\ .162772049 \end{bmatrix},$$

which, when normalized and with the signs reversed, becomes

$$\mathbf{u}=\begin{bmatrix} .989236689 \\ .146324212 \end{bmatrix}.$$

Then the discriminant function based on λ_1 is

$$Z_{irc1}=.989236689X_{i1}+.146324212X_{i2}.$$

Substituting the group means on X_1 and X_2, we find that the means, $\bar{Z}_{1rc}$, for the six groups are

	Grade		
Method	*K*	1	2
I	21.22	38.13	55.60
II	27.27	45.53	45.20

We see that the Method II means were higher in Kindergarten and Grade 1, compared with those for Method I, while the Method I mean was higher in Grade 2. These differences account for the significant interaction between Grade and Method on this function.

To define the dimension measured by this discriminant function, we first transform **u** to standard form to obtain

$$\mathbf{u}_s=\mathbf{Hu}=\begin{bmatrix} 3.45 & 0 \\ 0 & 9.40 \end{bmatrix}\cdot\begin{bmatrix} .989236689 \\ .146324212 \end{bmatrix}=\begin{bmatrix} 3.42 \\ 1.38 \end{bmatrix}.$$

Thus, Z_{irc1} measures a dimension in which Reading Speed (X_1) is weighted about two and one-half times more heavily than Reading Comprehension (X_2). We may conclude that, on this dimension, Method I resulted in larger means in Grade 2, whereas Method II resulted in larger means in Kindergarten and Grade 1.

The function based on λ_2 measures a quite different dimension from that based on λ_1. For λ_2, we find that

$$\mathbf{S}_w^{-1}\mathbf{S}_{rc}-\lambda_2\mathbf{I}=\begin{bmatrix} 1.100433627 & .944254007 \\ .162772049 & .139670540 \end{bmatrix},$$

so that

$$\mathbf{f}=\begin{bmatrix} .139670540 \\ -.162772049 \end{bmatrix},$$

which, when it is normalized and the signs are reversed, is

$$\mathbf{u}=\begin{bmatrix} -.651198850 \\ .758907147 \end{bmatrix}.$$

In standard form, we have

$$\mathbf{u}_s=\mathbf{Hu}=\begin{bmatrix} 3.45 & 0 \\ 0 & 9.40 \end{bmatrix}\times\begin{bmatrix} -.651198850 \\ .758907147 \end{bmatrix}=\begin{bmatrix} -2.25 \\ 7.13 \end{bmatrix},$$

which shows that Comprehension (X_2) is weighted about three times more heavily than Speed (X_1), with the latter receiving a negative weight. The discriminant function is $Z_{irc2} = -.651198850X_1 + .758907147X_2$, which yields the following group means:

	Grade		
Method	*K*	1	2
I	7.15	23.14	20.10
II	17.71	17.58	21.65

We see that in terms of Z_{irc2}, which measures primarily Comprehension and gives negative weight to Speed (larger Speed scores would *decrease* Z_{2rc}), the Method I mean is larger in Grade 1, whereas the Method II mean is larger in Kindergarten. The means for the two methods are about equal in Grade 2.

Although the procedures discussed in connection with the above example apply to a two-way *fixed* effects design, they can also be used with other models having equal cell frequencies if $\mathbf{S}_e$ is properly defined in accordance with $\mathbf{S}_h$ (that is, with $\mathbf{S}_r$, $\mathbf{S}_c$, or $\mathbf{S}_{rc}$). The appropriate forms of Equation 7.2.9 for testing row, column, and interaction effects in two-way random and mixed models are given in Table 7.2.5.

TABLE 7.2.5 EXPRESSIONS FOR Λ_h IN RANDOM AND MIXED MANOVA MODELS

Type of model	*Effect*	Λ_h
Random	Row	$\Lambda_r = \dfrac{\lvert\mathbf{S}_{rc}\rvert}{\lvert\mathbf{S}_r + \mathbf{S}_{rc}\rvert}$
	Column	$\Lambda_c = \dfrac{\lvert\mathbf{S}_{rc}\rvert}{\lvert\mathbf{S}_c + \mathbf{S}_{rc}\rvert}$
	Interaction	$\Lambda_{rc} = \dfrac{\lvert\mathbf{S}_{w}\rvert}{\lvert\mathbf{S}_{rc} + \mathbf{S}_{w}\rvert}$
Mixed (rows fixed, columns random)	Row	$\Lambda_r = \dfrac{\lvert\mathbf{S}_{rc}\rvert}{\lvert\mathbf{S}_r + \mathbf{S}_{rc}\rvert}$
	Column	$\Lambda_c = \dfrac{\lvert\mathbf{S}_{w}\rvert}{\lvert\mathbf{S}_c + \mathbf{S}_{w}\rvert}$
	Interaction	$\Lambda_{rc} = \dfrac{\lvert\mathbf{S}_{w}\rvert}{\lvert\mathbf{S}_{rc} + \mathbf{S}_{w}\rvert}$
Mixed (rows random, columns fixed)	Row	$\Lambda_r = \dfrac{\lvert\mathbf{S}_{w}\rvert}{\lvert\mathbf{S}_r + \mathbf{S}_{w}\rvert}$
	Column	$\Lambda_c = \dfrac{\lvert\mathbf{S}_{rc}\rvert}{\lvert\mathbf{S}_c + \mathbf{S}_{rc}\rvert}$
	Interaction	$\Lambda_{rc} = \dfrac{\lvert\mathbf{S}_{w}\rvert}{\lvert\mathbf{S}_{rc} + \mathbf{S}_{w}\rvert}$

The procedures given above for equal cell frequencies can easily be adapted for use when cell frequencies are proportional to marginal, that is, total row and column, frequencies. Cell frequencies in the two-way design are proportional if, for every cell, $n_{rc} = n_{r.}n_{.c}/N$, in which $n_{r.}$ and $n_{.c}$ are the frequencies in row r and column c, respectively, and $N = \Sigma_{r=1}^{R}\Sigma_{c=1}^{C} n_{rc}$, that is, the sum of the frequencies in all RC cells. The formulas given earlier for the case of equal cell frequencies may be adapted as shown in Table 7.2.6. The tests for row, column, and interaction effects, as well as the discriminant analysis procedures, may be carried out using the same methods we have given for the case of equal cell frequencies. We shall not give a numerical example of the proportional case here. Instead, we refer the reader to Exercise 7.2.

TABLE 7.2.6 COMPUTATIONAL FORMULAS FOR TWO-WAY MANOVA IN THE CASE OF PROPORTIONAL CELL FREQUENCIES

For Variable X_j:

Totals:	$T_{jrc.} = \sum_{i=1}^{n_{rc}} X_{jrci}$	(Total in cell rc)
	$T_{jr..} = \sum_{c=1}^{C} T_{jrc.}$	(Total in row r)
	$T_{j.c.} = \sum_{r=1}^{R} T_{jrc.}$	(Total in column c)
	$T_{j...} = \sum_{r=1}^{R} T_{jr..} = \sum_{c=1}^{C} T_{j.c.}$	(Total over all cells)
Means:	$\bar{X}_{jrc.} = \frac{T_{jrc.}}{n_{rc}}$	(Mean of cell rc)
	$\bar{X}_{jr..} = \frac{T_{jr..}}{n_{r.}}$	(Mean of row r)
	$\bar{X}_{j.c.} = \frac{T_{j.c.}}{n_{.c}}$	(Mean of column c)
	$\bar{X}_{j...} = \frac{T_{j...}}{N}$	(Mean of all observations on X_j)

General (jkth) Elements of $\mathbf{T}$, $\mathbf{S}_r$, $\mathbf{S}_c$, $\mathbf{S}_{rc}$, and $\mathbf{S}_w$:

$$\mathbf{T}\text{: } t_{jk} = \sum_{r=1}^{R}\sum_{c=1}^{C}\sum_{i=1}^{n_{rc}} X_{jrci}X_{krci} - T_{j...}T_{k...}/N$$

$$\mathbf{S}_r\text{: } S_{r(jk)} = \sum_{r=1}^{R} (T_{jr..}T_{kr..}/n_{r.}) - T_{j...}T_{k...}/N$$

$$\mathbf{S}_c\text{: } S_{c(jk)} = \sum_{c=1}^{C} (T_{j.c.}T_{k.c.}/n_{.c}) - T_{j...}T_{k...}/N$$

$$\mathbf{S}_{rc}\text{: } S_{rc(jk)} = \sum_{r=1}^{R}\sum_{c=1}^{C} (T_{jrc.}T_{krc.}/n_{rc}) - \sum_{r=1}^{R} (T_{jr..}T_{kr..}/n_{r.}) - \sum_{c=1}^{C} (T_{j.c.}T_{k.c.}/n_{.c}) + T_{j...}T_{k...}/N$$

$$\mathbf{S}_w\text{: } S_{w(jk)} = \sum_{r=1}^{R}\sum_{c=1}^{C}\Big(\sum_{i=1}^{n_{rc}} X_{jrci}X_{krci} - T_{jrc.}T_{krc.}/n_{rc}\Big)$$

The extension of the MANOVA methods to higher-order designs in the case of equal (or proportional) cell frequencies is quite straightforward. The student familiar with ANOVA should have no difficulty in applying the procedures discussed here to such problems. Equations 7.2.9, 7.2.10, and 7.2.11, for example, are general in form and may be used for a variety of designs with proper choice of $\mathbf{S}_h$ and $\mathbf{S}_e$, and proper determination of ν_h and ν_e. The procedures for discriminant analysis are also generally applicable and we strongly recommend their use in defining the dimensions on which groups are found to differ significantly. A particular discriminant function does not always provide an appropriate basis for differentiating among groups, even though the groups may differ significantly in terms of it.

As we pointed out earlier, both the Scheffé and Dunn (Bonferroni) methods for testing the significance of contrasts in ANOVA have been adapted for use in MANOVA. These methods can be used with two-way and higher order MANOVA designs as well as with one-way designs. Because of space limitations we shall not discuss them here. The interested reader might wish to consult the references given at the end of Section 7.1 for a discussion of these procedures.

Finally, as we stated at the beginning of this chapter, our discussion has been limited to the case of equal (or proportional) cell frequencies. When cell frequencies are disproportionate, other computational procedures must be used. For a treatment of these methods, the reader is again referred to Bock (1971), Searle (1971), or Timm (1975).

7.3 EXERCISES

7.1. A study was carried out to determine whether college sophomores in three different programs (A, B, and C) differed in terms of freshman year grade-point average (X_1) and SAT scores (X_2), obtained when they were high school seniors. Data on X_1 and X_2 obtained from samples of $n_A=30$, $n_B=26$, and $n_C=21$ were:

Program A		B		C	
X_1	X_2	X_1	X_2	X_1	X_2
3.38	490	2.82	700	2.92	370
2.77	410	2.02	580	2.70	440
3.35	770	2.28	540	2.83	640
3.31	660	1.81	680	2.12	610
2.38	540	2.51	660	2.00	600
3.77	690	1.86	720	2.90	510
2.67	590	3.03	510	3.38	790
2.57	740	2.90	380	2.57	630
3.30	550	1.81	570	3.33	590
2.75	610	2.50	690	3.29	370
3.50	720	2.87	730	2.36	410
3.54	760	2.54	660	2.31	390
2.18	625	3.00	540	3.10	570

Program					
A		*B*		*C*	
X_1	X_2	X_1	X_2	X_1	X_2
3.20	590	3.00	650	1.70	640
2.40	360	2.50	580	2.50	650
3.00	770	3.42	450	2.80	450
3.20	570	2.42	510	3.10	570
3.46	740	2.68	480	2.70	460
3.40	510	2.68	580	2.55	490
2.87	650	3.50	690	2.90	650
2.69	470	2.56	700	3.13	580
2.42	430	3.87	740		
2.62	730	2.60	490		
3.45	670	3.31	650		
3.10	680	2.60	600		
2.90	630	2.60	490		
3.13	450				
3.10	650				
2.26	580				
2.27	600				

(a) Carry out a multivariate analysis of variance and two separate univariate analyses of variance on these data to test the null hypotheses H_0: $\boldsymbol{\mu}_A = \boldsymbol{\mu}_B = \boldsymbol{\mu}_C$, H_0: $\mu_{1A} = \mu_{1B} = \mu_{1C}$, and H_0: $\mu_{2A} = \mu_{2B} = \mu_{2C}$. Use $\alpha = .05$ in each test.
(b) Use discriminant analysis to determine the dimension(s) on which the groups differ and show by computing the means of the discriminant function(s) how they differ.
(c) Summarize and interpret the results of (a) and (b).

7.2. A study was conducted to determine whether two different methods of speech therapy were equally effective in treating children who had three different types of speech disorders. Children regularly attending a speech clinic were classified as having either Articulation (A), Fluency (B), or Voice (C) disorders. The numbers in each group were $n_A = 10$, $n_B = 16$, and $n_C = 12$. Within each of these three groups, subjects were randomly divided into two subgroups; one subgroup was treated by Method I and the other by Method II. At the conclusion of the treatment period, scores on a Speech Assessment Scale consisting of ratings of Technical Ability (X_1) and Feelings about one's own speech (X_2) were obtained from all subjects. The results were:

Method of therapy	Type of disorder					
	Articulation		Fluency		Voice	
	X_1	X_2	X_1	X_2	X_1	X_2
	87	52	86	62	77	68
	91	67	82	69	74	80
	80	63	88	67	85	66
I	100	58	77	54	88	70
	96	60	92	53	83	73
			81	51	72	76
			75	65		
			90	58		
	99	40	88	58	85	54
	90	59	73	65	83	68
	108	46	78	60	97	55
II	104	55	83	67	94	58
	97	50	93	64	81	65
			90	57	92	63
			85	56		
			81	68		

(a) Analyze these data by means of multivariate analysis of variance, testing for row, column, and interaction effects. Also conduct separate univariate analyses of variance of each of the two variables, X_1 and X_2. Use $\alpha = .05$.
(b) Use discriminant analysis, where appropriate, to define the dimensions on which methods or types differ. Compute group means and describe how they differ.
(c) Discuss and interpret the results in (a) and (b).

8/ Factor Analysis

Factor analysis is a generic term for a number of different but related mathematical and statistical techniques designed to investigate the nature of the relationships between variables in a specified set, in a somewhat different way than the techniques discussed in earlier chapters. The basic problem is to determine whether the *n* variables in a set exhibit patterns of relationships with one another, such that the set could be broken down into, say, *m* subsets, each consisting of a group of variables tending to be more highly related to others within the subset than to those in other subsets. The major purposes of factor analysis are to determine whether a set of variables can be described in terms of a number of "dimensions" or "factors" smaller than the number of variables, and to determine what these dimensions (factors) are, that is, to indicate what trait or characteristic each of them represents.

The earliest work in factor analysis was done by the psychologist Charles Spearman (1904b), generally considered to be the "father" of the method, and by Karl Pearson (1901b), who proposed the "method of principal axes," later more fully developed by Harold Hotelling (1933). Spearman's purpose was to provide a mathematical model to represent his theory of general intelligence. He believed that "all branches of intellectual activity have in common one fundamental function (or group of functions) whereas the remaining or specific elements of the activity seem in every case to be wholly different from that in all others" (Spearman, 1904, p. 202). Translated into factor analytic language, this statement means that each test in a battery that deals with intellectual activity measures both a general factor, measured in common by all the tests, and a specific factor, measured only by that test and by no others in the battery. Spearman's theory of general intelligence and his interest in mathematical models led to the development of his two-factor (general and specific) method, which we shall discuss later in this chapter.

Much of the early work in factor analysis, that is, during the period from 1900 to 1930, was devoted to the application of Spearman's model to a great variety of applied problems and to investigations of conditions under which the model was appropriate. Among the important contributors to the factor analysis literature during this period were Cyril Burt, Karl Holzinger, Truman Kelly, Karl Pearson, and Godfrey Thomson, as well as Spearman himself. By 1930 it was becoming clear that the single general factor model was not always adequate to describe the relationships between variables in a set. Several writers suggested that perhaps not all tests in a battery necessarily measured general intelligence and that even the "unique" portions of tests might be measuring something in common with other tests in a set. Thus, Holzinger's (1937) *bi-factor* theory made allowance for a general factor, measured by all variables in the set; for several group factors, each measured by a subset of variables; and for a number of specific factors equal to the number of variables in the entire set. We shall briefly discuss Holzinger's model later in this chapter.

The multiple-factor theory of L. L. Thurstone (1931, 1938, 1947) represented a significant departure from that of Spearman and Holzinger, as well as that of Thomson and Kelly, in that Thurstone did not believe that there was a general ability factor. Neither did he believe that mental abilities were organized hierarchically, from general to specific, as suggested by several British psychologists (for example, Cyril Burt, 1941, 1948) as well as by Holzinger's bi-factor theory. Instead he believed that there were several major group factors, which he called the *primary mental abilities*. The Thurstone model postulates several specific factors (as E. L. Thorndike did earlier). However, Thurstone was unable to demonstrate empirically that the primary mental abilities represented anything but a collection of group and specific factors.

Thurstone was perhaps the most prominent of the early modern factor analysts, and he has had considerable influence on the development of the method from the 1930s to the present. He was responsible for the development of the centroid method (1931), used extensively prior to the advent of high-speed computers. He was also responsible for the concept of *simple structure*, which has been regarded by most factor analysts as representing an ideal factor-analytic solution. In addition, as Harman (1976) states, "The truly remarkable contribution of Thurstone was the generalization of Spearman's tetrad-difference criterion to the *rank* of the correlation matrix as the basis for determining the number of common factors. He saw that a zero tetrad-difference corresponded to the vanishing of a second-order determinant, and extended this notion to the vanishing of higher order determinants as the condition for more than a single factor." We shall discuss the importance of this contribution in a later section.

The early work in factor analysis carried out by the various psychologists we have mentioned tended to be theory oriented, even though no methods were available for statistical tests of specific hypotheses concerning the factorial structures of particular sets of variables. However, when large high-speed computers became quite generally available in universities, in the mid to late 1950s, there was a move away from the theoretical orientation and toward what has come to be called *exploratory* factor analysis. This move was clearly encouraged by Thurstone's common-factor theory and facilitated by Hotelling's (1933) general formulation of the mathematics of principal components, which had not been generally applied before then because of the extreme computational labor required. Thus, in the late 1950s and 1960s it seemed that nearly everyone was factor analyzing nearly everything, in the hope that apparently complex relationships between variables in a set could be simplified and thus interpreted more easily. During this period also, the number of factor-analytic methods increased noticeably with the addition of *image analysis* (Guttman, 1953), *canonical factor analysis* (Rao, 1955, and Harris, 1962), *alpha factor analysis* (Kaiser and Caffrey, 1965) and the *minres* method (Harman and Jones, 1966). Although there were a great many cases in which meaningful structures were found through the use of these methods, there were others (see the examples in Cooley and Lohnes, 1962; Mulaik, 1964; and Armstrong and Soelberg, 1968) in which the factor-analytic results were highly questionable, primarily because of the indeterminacy of the factors obtained. Furthermore, as one might have expected, the methods of exploratory analysis failed to be very helpful in testing and refining psychological theory.

Although Hotelling's (1933) paper on *principal component analysis* represented the first notable contribution of a *statistician* to factor analysis, it was not until Lawley's (1940) paper on the method of maximum likelihood that factor analysis could be regarded as a respectable statistical technique. There was some reaction to Lawley's paper (for example, Young, 1941, and Kendall and Lawley, 1956), and some further attention to the factor analysis problem from statisticians such as Anderson and Rubin (1956), Danford (1953), Kendall (1954), Rao (1955) and Wold (1953). However, there were few important contributions to the method from statisticians until the middle and late 1960s when Bock and Bargmann (1966) and Jöreskog (1966, 1967, 1969, 1970) considered testing specific hypotheses about parameters of the factor analysis model. Jöreskog's work was based essentially on the method of maximum likelihood (Lawley, 1940), but he solved many of the very difficult problems of computation and interpretation with which Lawley did not deal. The Bock and Bargmann and the Jöreskog procedures, because of the emphasis on hypothesis testing, are classified as *confirmatory* factor analysis methods. They have clear advantages over exploratory factor analysis in terms of developing and testing theory, even though it is often difficult to specify the particular hypotheses to be tested. Exploratory factor analysis can often be used successfully to generate such hypotheses, which can then be tested with confirmatory analyses.

It should be obvious at this point that we cannot deal with all, or even with very many, of the myriad methods of factor analysis currently available. As in other chapters, we have had to make some difficult choices about what to include. In this chapter we have decided to discuss methods which are most widely used and most widely available in computer program packages such as SPSS and BMD, and which illustrate best, in our view, the basic concepts of factor analysis. Accordingly, we shall first discuss factor analysis models and define some of the basic concepts important in using any factor analysis method. We shall then, for heuristic as well as historical reasons, discuss two of the early models, Spearman's and Holzinger's bi-factor, followed by the principal-component method of exploratory factor analysis. We conclude with a treatment of orthogonal and oblique rotation and a brief mention of factor scores and their uses. Our omission of confirmatory factor analysis methods does not reflect a bias on our part, but merely a choice intended to meet the criteria of wide use and availability and to maintain consistency of level and approach with other chapters.

8.1 MODELS IN EXPLORATORY FACTOR ANALYSIS

Before we define the basic factor analysis model, a word about notation will be helpful. Because of the wide variety of exploratory factor analysis methods available, notation is far from standard. In order to avoid, as much as possible, adding to its diversity, we have tried to use the notation of Harman (1976), not necessarily because it is "best," but because Harman's book *Modern Factor Analysis* has, through three editions, been one of the standard texts in the field. Because of this decision, the notation used in this chapter is not entirely consistent with that in other chapters. However, because the material in this chapter overlaps minimally with that in previous chapters, we do not believe that the few such inconsistencies in notation will interfere with understanding of the concepts and procedures presented.

The term factor analysis is used here as a generic term for two distinctly different kinds of analysis, namely, *component analysis* and *common-factor analysis*. They differ essentially in terms of the amount and kind of variance of each X_j (the jth variable) accounted for by the factors in the model. In component analysis the factors account for *all* of the variance in each variable, including that "shared" with other variables in the set (covariance), and that specific to the jth variable. In theory, then, the number of factors (or components) in a component analysis must be equal to the number of variables, because all the variance of each X_j must be accounted for by the factors. On the other hand, in common-factor analysis the factors account only for variance that is "shared" (or common) to other (not necessarily all) variables in the set. Thus, the number of *common* factors may be less than the number of variables, which is desirable in terms of a *parsimonious* description of the original set of n variables in terms of $m(m<n)$ common factors.

The above distinction leads to different models for component analysis and for common-factor analysis. For component analysis, we have

$$z_{ji} = a_{11}F_{1i} + a_{22}F_{2i} + \cdots + a_{nn}F_{ni}, \qquad (8.1.1)$$

in which the a_{jp}s $(j=1,\ldots,n;p=1,\ldots,n)$ are weights on the F_{pi}s. The F_{pi}s are assumed to be uncorrelated, that is, $r_{F_pF_q}=0(p,q=1,\ldots,n;p\neq q)$. The model states that the value $z_{ji}=(X_{ji}-\bar{X}_j)/s_j$, of the ith $(i=1,\ldots,N)$ individual on the jth variable (in standard form) is a linear combination of n uncorrelated components, the factors, with the F_{ji}s weighted according to the a_{jp}s. Here we define s_j with divisor N, that is, $s_j^2=\Sigma_i(X_{ji}-\bar{X}_j)^2/N$. Note that the number of factors, m, is equal to the number of variables, n, in this model.

For common-factor analysis, the model may be written

$$z_{ji}=a_{j1}F_{1i}+a_{j2}F_{2i}+\cdots+a_{jm}F_{mi}+u_jY_{ji}. \tag{8.1.2}$$

Here the F_{pi}s $(p=1,\ldots,m)$ are the m (usually less than n) common factors, and the additional unique factor, Y_{ji}, accounts for the unique contribution of variable X_j to z_{ji}. We also assume in this model that $r_{F_pY_j}=0$ $(p=1,\ldots,m;$ $j=1,\ldots,n)$. Note that the a_{jp}s and F_{pi}s in Equation 8.1.2 are not equivalent to the a_{jp}s and F_{pi}s of Equation 8.1.1. It should also be understood that, as we indicated previously, the component-analysis model implies that the factors account for the *variance* of the z_js. However, the factors in the common-factor analysis model account for the *relationships* (correlations) between variables.

8.1.1 Partition of Variance in Factor Analysis Models

In order to understand better the concept of a "factor" in factor analysis, it will be helpful to show how the variance of a variable, z_j, can be expressed in terms of the elements of the model. At this point we will deal with the common-factor model, Equation 8.1.2. The analogous result based on Equation 8.1.1 will be obvious. First we note that, by definition,

$$\mathrm{Var}(z_j)=\frac{\sum_{i=1}^{N}z_{ji}^2}{N}=1,$$

that is, the variance of the standardized variable, $z_j=(X_j-\bar{X}_j)/s_j$, equals one. Then, if we write Equation 8.1.2 as

$$z_{ji}=\sum_{p=1}^{m}a_{jp}F_{pi}+u_jY_{ji},$$

it follows that

$$z_{ji}^2=\sum_{p=1}^{m}a_{jp}^2F_{pi}^2+u_j^2Y_{ji}^2+\sum_{p,q=1}^{m}a_{jp}a_{jq}F_{pi}F_{qi}+2u_jY_{ji}\sum_{p=1}^{m}a_{jp}F_{pi}$$

$$(p=1,\ldots,m;q=1,\ldots,m;p\neq q),$$

and that

$$1=\frac{\sum_{i=1}^{N} z_{ji}^2}{N}=\sum_{p=1}^{m} a_{jp}^2\left[\frac{\sum_{i=1}^{N} F_{pi}^2}{N}\right]+u_j^2\left[\frac{\sum_{i=1}^{N} Y_{ji}^2}{N}\right]$$

$$+\sum_{p,q=1}^{m} a_{jp}a_{jq}\left[\frac{\sum_{i=1}^{N} F_{pi}F_{qi}}{N}\right]+2u_j\sum_{p=1}^{m} a_{jp}\left[\frac{\sum_{i=1}^{N} Y_{ji}F_{pi}}{N}\right]$$

$$(p=1,\ldots,m;q=1,\ldots,m;p\neq q). \tag{8.1.3}$$

Because all F_{pi}s and Y_{ji}s are standardized, that is, in standard score form,

$$\frac{\sum_{i=1}^{N} F_{pi}^2}{N}=\frac{\sum_{i=1}^{N} Y_{ji}^2}{N}=1.$$

Furthermore,

$$\frac{\sum_{i=1}^{N} F_{pi}F_{qi}}{N}=r_{F_pF_q}=0$$

and

$$\frac{\sum_{i=1}^{N} Y_{ji}F_{pi}}{N}=r_{Y_jF_p}=0,$$

because all F_ps and Y_js are assumed to be uncorrelated. Then Equation 8.1.3 reduces to

$$1=\sum_{p=1}^{m} a_{jp}^2+u_j^2, \tag{8.1.4}$$

which indicates that the total variance of variable z_j may be expressed as the sum of m components, the a_{jp}^2s, and u_j^2. Thus, the a_{jp}^2s as well as u_j^2 may be regarded as variances.

Similarly, for component analysis, we may write

$$1=\sum_{j=1}^{n} a_j^2, \tag{8.1.5}$$

because $u_j=0$ in the component analysis model.

Now, to represent all n variables in a common-factor analysis, there would be a series of equations such as 8.1.4, that is,

$$\left.\begin{array}{l} a_{11}^2+a_{12}^2+\cdots+a_{1m}^2+u_1^2=1 \\ a_{21}^2+a_{22}^2+\cdots+a_{2m}^2+u_2^2=1 \\ \vdots \\ a_{n1}^2+a_{n2}^2+\cdots+a_{nm}^2+u_n^2=1 \end{array}\right\} \tag{8.1.6}$$

Because all of the terms in these equations are variances, the jth equation indicates what portion of the total variance of the jth variable is accounted for by each factor. The sum of the a_{jp}^2s for each variable is called the *communality*, h_j^2, of the jth variable, that is,

$$h_j^2 = \sum_{p=1}^{m} a_{jp}^2 = 1 - u_j^2. \tag{8.1.7}$$

The communality of the jth variable is thus a measure of the extent to which its variance can be accounted for on the basis of the common factors. In that sense, the communality indicates the extent to which the jth variable is measuring something in common with the other variables in the set. On the other hand, u_j^2 indicates the *uniqueness* of the jth variable. It is sometimes broken down into two parts, namely, *specificity*, b_j^2, and *error*, e_j^2, so that $u_j^2 = b_j^2 + e_j^2$. Because, in classical measurement theory (see, for example, Lord and Novick, 1968) the reliability of the jth variable may be defined as $r_{jJ} = 1 - e_j^2$, we may write $r_{jJ} = h_j^2 + b_j^2$. Therefore, $h_j^2 = r_{jJ} - b_j^2$, which shows that $h_j^2 \leqslant r_{jJ}$. In other words, the reliability is the upper limit for the communality.

If we examine the collections of terms in Equations 8.1.6 that refer to each of the factors, we can determine the extent to which each factor is accounting for the total variance of the n variables. If we ignore the "+" signs for the moment and think in terms of columns of a_{jp}^2s, we see that the sum of the a_{j1}^2s in column one,

$$V_1 = \sum_{j=1}^{n} a_{j1}^2,$$

is the variance accounted for by factor F_1. In general,

$$V_p = \sum_{j=1}^{n} a_{jp}^2 \tag{8.1.8}$$

is the variance accounted for by factor F_p. Because the n variables are in standard form, their total variance is n. Therefore, V_p/n is the proportion of variance accounted for by factor F_p and

$$\frac{V}{n} = \frac{\sum_{p=1}^{m} V_p}{n} \tag{8.1.9}$$

is the proportion of the total variance accounted for by all m factors. In a common-factor analysis, this number should be less than one, assuming that there is something uniquely measured by many, if not all, of the n variables. In a component analysis, however, V/n should equal one, because the factors account for all the variance, there being no uniqueness associated with any of the variables in the model.

In addition to providing information about the proportion of variance accounted for by factor F_p, the values of a_{jp}^2 also aid in determining what the factors measure. A factor is a variable in the same sense that the Xs are variables, and, as such, it is a measure of some trait or characteristic, just as are the Xs. However, the factor measurement is not obtained directly, but rather indirectly, in terms of the Xs. One of the problems in factor analysis is to specify what a given factor measures, and the a_{jp}^2s are helpful in doing this. Their magnitudes indicate which variables contribute most and which least to a given factor. The factor is then usually regarded as measuring the characteristic or characteristics measured by the Xs having the largest associated values of the a_{jp}^2s. We shall discuss this point further later in this chapter.

The major problem in common-factor analysis is to determine the values of the a_{jp}s, the factor weights or *loadings*. Before discussing methods for solving this problem, we will find it helpful to consider how the model can be expressed in matrix terms.

8.1.2 Factor Analysis Models in Matrix Terms

It will be convenient for work in later sections of this chapter to express the models defined in Equations 8.1.1 and 8.1.2 in matrix notation. We first define the following matrices:

$$\mathbf{Z}=\begin{bmatrix} z_{11} & z_{12} & \cdots & z_{1N} \\ z_{21} & z_{22} & \cdots & z_{2N} \\ \vdots & \vdots & & \vdots \\ z_{n1} & z_{n2} & \cdots & z_{nN} \end{bmatrix}, \qquad (8.1.10)$$

$$\mathbf{F}=\begin{bmatrix} F_{11} & F_{12} & \cdots & F_{1N} \\ F_{21} & F_{22} & \cdots & F_{2N} \\ \vdots & \vdots & & \vdots \\ F_{m1} & F_{m2} & \cdots & F_{mN} \end{bmatrix}, \qquad (8.1.11)$$

$$\mathbf{Y}=\begin{bmatrix} Y_{11} & Y_{12} & \cdots & Y_{1N} \\ Y_{21} & Y_{22} & \cdots & Y_{2N} \\ \vdots & \vdots & & \vdots \\ Y_{n1} & Y_{n2} & \cdots & Y_{nN} \end{bmatrix}, \qquad (8.1.12)$$

$$\mathbf{A}=\begin{bmatrix} a_{11} & a_{12} & \cdots & a_{1m} \\ a_{21} & a_{22} & \cdots & a_{2m} \\ \vdots & \vdots & & \vdots \\ a_{n1} & a_{n2} & \cdots & a_{nm} \end{bmatrix}, \qquad (8.1.13)$$

and

$$\mathbf{U}=\begin{bmatrix} u_{1} & 0 & \cdots & 0 \\ 0 & u_{2} & \cdots & 0 \\ \vdots & \vdots & & \vdots \\ 0 & 0 & \cdots & u_{n} \end{bmatrix}. \qquad (8.1.14)$$

Then the component-analysis model of Equation 8.1.1 can be expressed, in terms of all of the n variables, as

$$\mathbf{Z}=\mathbf{AF}, \tag{8.1.15}$$

where $\mathbf{A}$ is an $n\times n$ matrix and $\mathbf{F}$ is $n\times N$, because the number of factors is equal to the number of variables. The common-factor analysis model of Equation 8.1.2 can be written in matrix form as

$$\mathbf{Z}=\mathbf{AF}+\mathbf{UY}, \tag{8.1.16}$$

in which all matrices are as defined in Equations 8.1.10 through 8.1.14. The Equations 8.1.15 and 8.1.16 are often referred to as the *fundamental equations* of component analysis and common-factor analysis, respectively. The matrix $\mathbf{A}$ in these equations is called the *pattern* matrix. It is always an essential part of the results of factor analysis computations because its elements are the *loadings* of each of the n variables on each of the m factors. The matrix $\mathbf{F}$ consists of *factor scores* of each of N individuals on each of the m factors. It may also be an important part of the results of a factor analysis. We shall discuss both $\mathbf{A}$ and $\mathbf{F}$ in more detail later in this chapter.

Until this point we have considered the case in which the m common factors are uncorrelated. More generally, however, we must deal with the possibility of correlated factors. The distinction is between methods that produce *orthogonal* solutions (uncorrelated factors) and those that produce *oblique* solutions (correlated factors). We first define a matrix, $\mathbf{S}$, called a *structure* matrix,

$$\mathbf{S}=\begin{bmatrix} r_{11} & r_{12} & \cdots & r_{1m} \\ r_{21} & r_{22} & \cdots & r_{2m} \\ \vdots & \vdots & & \vdots \\ r_{n1} & r_{n2} & \cdots & r_{nm} \end{bmatrix}, \tag{8.1.17}$$

the elements of which, the r_{jp}s, are the correlations between the n variables and the m factors. By definition, because the n variables and m factors are in standard form,

$$\mathbf{S}=\frac{1}{N}(\mathbf{ZF}'). \tag{8.1.18}$$

Considering the model of Equation 8.1.15, we may write

$$\frac{1}{N}(\mathbf{ZF}')=\mathbf{A}\left(\frac{1}{N}\right)\mathbf{FF}'. \tag{8.1.19}$$

If we now define

$$\mathbf{C}=\left(\frac{1}{N}\right)\mathbf{FF}', \tag{8.1.20}$$

as the matrix of correlations between factors, we may write Equation 8.1.18 as

$$\mathbf{S}=\mathbf{AC}, \tag{8.1.21}$$

which indicates that the structure matrix equals the pattern matrix postmultiplied by the matrix of correlations between factors. Note that if $\mathbf{C}=\mathbf{I}$, then

S = **A**, that is, if the factors are orthogonal, then the structure and pattern matrices are equivalent. Thus, the elements of **A** in the orthogonal solution are the correlations of the variables with the factors. We shall illustrate these concepts in later sections of this chapter.

8.2 THE SPEARMAN AND HOLZINGER MODELS

Although the Spearman two-factor (general and specific) model and the Holzinger bi-factor (general, group, and specific) model are not now in general use as a basis for the development of factor analysis methods, their conceptual simplicity, as well as their historical significance, makes it appropriate and useful to discuss each of them briefly. In particular, the Spearman model is important because its applicability depends on meeting the conditions necessary to postulate a single general factor. We shall discuss these conditions later in this Section. Furthermore, a research situation occasionally arises in which a single general factor is postulated, so the Spearman model still has some limited practical application in educational and psychological research.

8.2.1 The Spearman Two-factor Model

The Spearman two-factor model may be written

$$z_{ji} = a_{jo}F_{oi} + u_j Y_{ji} \quad (i=1,\ldots,N; j=1,\ldots,n), \tag{8.2.1}$$

in which F_{oi} denotes the value of the ith individual on the single general factor and Y_{ji} is the value of the ith individual on the uniqueness of variable X_j. What we wish to show first is how the general factor coefficients, the a_{jo}s, can be expressed in terms of the bivariate correlation coefficients, the r_{jk}s $(j \neq k)$. Recall that

$$r_{jk} = \frac{\sum_{i=1}^{N} z_{ji} z_{ki}}{N}.$$

Substituting for z_{ji} and z_{ki} on the basis of Equation 8.2.1 yields

$$r_{jk} = \frac{\sum_{i=1}^{N} (a_{jo}F_{oi} + u_j Y_{ji})(a_{ko}F_{oi} + u_k Y_{ki})}{N}$$

$$= \frac{\sum_{i=1}^{N} a_{jo}a_{ko}F_{oi}^2}{N} = a_{jo}a_{ko}\left[\frac{\sum_{i=1}^{N} F_{oi}^2}{N}\right] = a_{jo}a_{ko}, \tag{8.2.2}$$

because expressions like $\Sigma_i F_{oi} Y_{ji}/N$, that is, the correlations between the general and unique factors, are assumed to be zero and the correlation of F_o with itself is equal to $+1$. Now suppose we wish to find a_{eo}, the general factor coefficient for variable z_e. Multiplying $r_{jk}=a_{jo}a_{ko}$ by a_{eo}^2 gives

$$a_{eo}^2 r_{jk}=a_{eo}^2 a_{jo}a_{ko}=(a_{eo}a_{jo})(a_{eo}a_{ko})=r_{ej}r_{ek}.$$

Summing over all r_{jk}s except those in which $j=e$ or $k=e$, we have

$$a_{eo}^2 \sum_{j<k=1}^{n} r_{jk}=\sum_{j<k=1}^{n} r_{ej}r_{ek} \quad (j,k\neq e).$$

Then

$$a_{eo}^2=\frac{\sum_{j<k=1}^{n} r_{ej}r_{ek}}{\sum_{j<k=1}^{n} r_{jk}} \quad (j,k\neq e). \tag{8.2.3}$$

It can be shown that an equivalent computational formula is

$$a_{eo}^2=\frac{\left(\sum_{e\neq j=1}^{n} r_{ej}\right)^2-\sum_{e\neq j=1}^{n} r_{ej}^2}{2\left(\sum_{j<k=1}^{n} r_{jk}-\sum_{e\neq j=1}^{n} r_{ej}\right)}. \tag{8.2.4}$$

Now, because Equations 8.2.3 and 8.2.4 were obtained by assuming that the Spearman two-factor model is appropriate, it is necessary to demonstrate the appropriateness of the model, Equation 8.2.1, before proceeding to determine the values of the a_{jo}s. According to Spearman's (1927) theory, the necessary and sufficient conditions for describing the relationships between n variables in terms of one general factor are that the *tetrads* vanish, that is,

$$r_{jk}r_{lm}-r_{lk}r_{jm}=0 \quad (j,k,l,m=1,\ldots,n;\, j\neq k\neq l\neq m\neq j). \tag{8.2.5}$$

As Harman (1976, Section 5.3) points out, not all of these conditions are independent; actually, there are $n(n-3)/2$ such conditions that must be satisfied before a single general factor can be assumed. Note that we can always assume a single general factor if $n\leqslant 3$ because then the number of such conditions to be satisfied is less than or equal to zero.

To illustrate the applications of Spearman's two-factor model, we use the matrix of correlations in Table 8.2.1. In this case, we have $4(4-3)/2=2$ conditions (Equation 8.2.5) to check in order to determine whether the model may be applied. Thus, we have

$$r_{21}r_{43}-r_{41}r_{32}=(.611)(.738)-(.660)(.576)=.07$$

TABLE 8.2.1 CORRELATIONS AMONG FOUR MEASURES OF VERBAL ABILITY

Variable	1	2	3	4
1. Verbal completion	—			
2. Understanding paragraphs	.611	—		
3. Reading vocabulary	.642	.576	—	
4. General information	.660	.545	.738	—

and

$$r_{42}r_{31} - r_{32}r_{41} = (.545)(.642) - (.576)(.660) = -.03.$$

Because these differences are both close to zero and their mean, .02, is also near zero, we conclude that the Spearman model is appropriate. Spearman and Holzinger (1925) developed formulas for the sampling errors of tetrad differences that could be used in testing the hypothesis that they were zero in the population. We shall not describe these tests here.

Having decided that the model is appropriate, we may proceed to compute the a_{jo}s by means of Equation 8.2.4. The computations are shown in Table 8.2.2, together with the resulting pattern matrix.

The large values of the a_{jo}s, which in this model are the correlations between the variables and the general factor (see Section 8.1.2), confirm that a single general factor accounts for the relationships among the four variables quite well, a result we might have expected, because all four variables measure cognitive, primarily verbal, characteristics. The communalities in this case are $h_j^2 = a_{jo}^2$, because we have only one factor; the values are $h_1^2 = .656$, $h_2^2 = .490$, $h_3^2 = .699$, and $h_4^2 = .684$. With the exception of the value for Variable 2, these are all quite large. This suggests that the general factor accounts for a large percentage of the non-error variance in the four variables. In fact, if we compute V_o (see Equation 8.1.8) for this factor, we find

$$V_o = \sum_{j=1}^{n} a_{jo}^2 = 3.173.$$

Because the total variance is 4, we find that the percentage of variance accounted for by the single general factor, V_o/n, is about 79%.

TABLE 8.2.2 SPEARMAN TWO-FACTOR SOLUTION BASED ON THE CORRELATIONS IN TABLE 8.1.1

	—	.611	.642	.660
	.611	—	.576	.545
	.642	.576	—	.738
	.660	.545	.738	—
Σr_{ej}	1.913	1.732	1.956	1.943
Σr_{ej}^2	1.221	1.002	1.289	1.277
$2\left(\sum_{j<k=1}^{n} r_{jk} - \sum_{e\neq j=1}^{n} r_{ej}\right)$	3.718	4.080	3.632	3.658
a_{eo}^2	.656	.490	.699	.683
a_{eo}	.810	.700	.836	.827

Pattern:

Variable	a_{jo}	u_j^*
1	.810	.587
2	.700	.714
3	.836	.549
4	.827	.563

*Because there is a single common factor, $u_j = \sqrt{1 - a_{jo}^2}$.

8.2.2 Holzinger's Bi-Factor Model

As discussed earlier, Holzinger's bi-factor model was developed when it became clear that there were problems for which the Spearman two-factor model was inappropriate. The bi-factor model incorporates several group factors in addition to the general and specific factors of the Spearman model. Before stating the model and indicating how the method proceeds, we shall consider differences in the general appearance of the correlation matrix when different models are appropriate.

Figure 8.2.1 shows three correlation matrices for a set of 10 variables that differ in the sizes of their bivariate correlations. In Figures 8.2.1(b) and 8.2.1(c), the variables have been arranged in order so that the first three, the next four, and the last three can be considered as belonging to three distinct

FIGURE 8.2.1 Three Configurations of Correlation Matrices Suggesting Three Different Factor M Models*

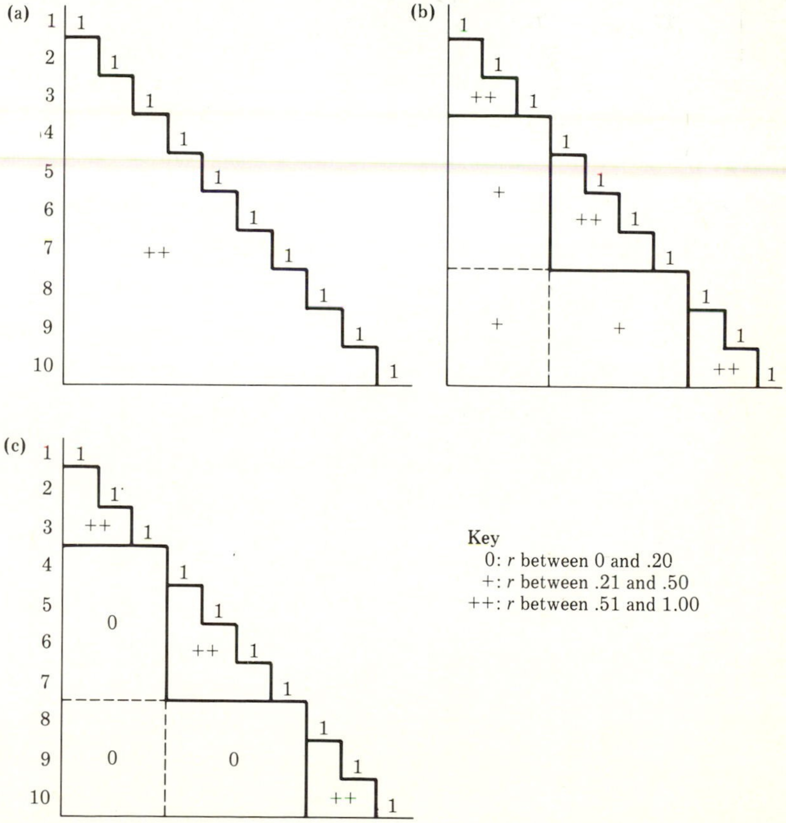

*Adapted from Harman (1976), pp. 95-96

groups. The configuration that would suggest a Spearman model is given in Figure 8.2.1(a). In this figure all the correlations are quite large, suggesting that a single general factor might well account for the relationships between the 10 variables. Figure 8.2.1(b), on the other hand suggests a case in which the bi-factor model might be appropriate. There are moderately sized correlations between all pairs of variables, which indicates the presence of a general factor, but there seem also to be several group factors, as suggested by the large correlations among variables within the three groups. Figure 8.2.1(c), however, indicates several group factors, all distinct, but no general factor, suggesting a multiple-factor model akin to that of Thurstone. Of course, it would be unusual in practice to encounter configurations as clear cut as those in Figure 8.2.1, but it seems useful to try to relate such configurations, even though they are idealized, to the various models which have been proposed.

The bi-factor model may be written

$$z_{ji} = a_{jo}F_{oi} + a_{j1}B_{1i} + \cdots + a_{jm}B_{mi} + u_j Y_{ji} \qquad (8.2.6)$$

in which F_{oi} denotes the general factor and the B_{pi}s ($p = 1,\ldots,m$) the various group factors. For any z_{ji} there is only one nonzero value of a_{jp} ($p = 1,\ldots,m$), because each variable is a member of only one group. If we denote the group to which variable X_j belongs by p, then the model may be written

$$z_{ji} = a_{jo}F_{oi} + a_{jp}B_{pi} + u_j Y_{ji}. \qquad (8.2.7)$$

In order to compute the factor coefficients, a_{jo} and a_{jp} for variable X_j, it is first necessary to specify the group to which variable X_j belongs. This can sometimes be done on logical or theoretical grounds. For example, one might believe that verbal measures should be in one group and quantitative measures in another. Sometimes past research with a particular set of variables suggests such a basis for grouping. When no such basis is available, the correlation coefficients themselves may provide information for grouping. One might attempt, for example, to arrange variables in a configuration similar to one of those shown in Figure 8.2.1; however, such a procedure would be difficult if the number of variables were large. Harman (1976) proposes the use of a "coefficient of belongingness" for accomplishing the grouping process. This coefficient provides a mathematical basis for grouping but is not entirely free of subjective judgment.

When the variables have been grouped, the solution proceeds as follows:

1. Compute the general factor coefficient for variable X_e using

$$a_{eo}^2 = \frac{\sum r_{ej} r_{ek}}{\sum r_{jk}} \qquad \begin{array}{l}(e \text{ is in } G_r, j \text{ is in } G_p, k \text{ is in } G_q;\\ p,q = 1,\ldots,m; p < q; p,q \neq r),\end{array}$$

in which the sums are taken over correlations in which j and k assume all values within their respective variable groups. For example, suppose variables 1 through 4 are in G_1, 5 through 8 in G_2, and 9 through 12 in G_3. Then

$$a_{10}^2 = \frac{\sum_{j=5}^{8} \sum_{k=9}^{12} r_{1j} r_{1k}}{\sum_{j=5}^{8} \sum_{k=9}^{12} r_{jk}} = \frac{r_{1,5} r_{1,9} + r_{1,5} r_{1,10} + \cdots + r_{1,8} r_{1,12}}{r_{5,9} + r_{5,10} + \cdots + r_{8,12}}.$$

2. Compute the residual correlations, with the relationship due to the general factor removed, for each pair, jk, in G_p $(p = 1, \ldots, m)$. These are:

$$\dot{r}_{jk} = r_{jk} - a_{jo} a_{ko} \quad (j, k = 1, \ldots, n).$$

The result of this should be a residual correlation matrix that looks like Figure 8.2.1(c), that is, the correlations between variables in different groups should all be close to zero, while those between pairs of variables within groups should be at least moderately large.

3. Within each group use the method of Section 8.2.1 (Spearman), based on the residual correlations. This method should be applicable because within each group there is a single group factor that is analogous to a single general factor for the group.

TABLE 8.2.3. MEANS, STANDARD DEVIATIONS, AND RELIABILITY COEFFICIENTS FOR 24 PSYCHOLOGICAL TESTS ADMINISTERED TO 145 7TH AND 8TH GRADE PUPILS

Test X_j	*Mean* $\bar{X}_j$	*Standard deviation* s_j	*Reliability coefficient* r_{jJ}
1. Visual perception	29.60	6.90	.756
2. Cubes	24.84	4.50	.568
3. Paper form board	15.65	3.07	.544
4. Flags	36.31	8.38	.922
5. General information	44.92	11.75	.808
6. Paragraph comprehension	9.95	3.36	.651
7. Sentence completion	18.79	4.63	.754
8. Word classification	28.18	5.34	.680
9. Word meaning	17.24	7.89	.870
10. Addition	90.16	23.60	.952
11. Code	68.41	16.84	.712
12. Counting dots	109.83	21.04	.937
13. Straight-curved capitals	191.81	37.03	.889
14. Word recognition	176.14	10.72	.648
15. Number recognition	89.45	7.57	.507
16. Figure recognition	103.43	6.74	.600
17. Object-number	7.15	4.57	.725
18. Number-figure	9.44	4.49	.610
19. Figure-word	15.24	3.58	.569
20. Deduction	30.38	19.76	.649
21. Numerical puzzles	14.46	4.82	.784
22. Problem reasoning	27.73	9.77	.787
23. Series completion	18.82	9.35	.931
24. Arithmetic problems	25.83	4.70	.836

TABLE 8.2.4. PRODUCT-MOMENT CORRELATIONS AMONG 24 PSYCHOLOGICAL TESTS

Test: *j*	1	2	3	4	5	6	7	8	9	10	11	12	13	14	15	16	17	18	19	20	21	22	23	24
1	—	—	—	—	—	—	—	—	—	—	—	—	—	—	—	—	—	—	—	—	—	—	—	—
2	.318	—	—	—	—	—	—	—	—	—	—	—	—	—	—	—	—	—	—	—	—	—	—	—
3	.403	.317	—	—	—	—	—	—	—	—	—	—	—	—	—	—	—	—	—	—	—	—	—	—
4	.468	.230	.305	—	—	—	—	—	—	—	—	—	—	—	—	—	—	—	—	—	—	—	—	—
5	.321	.285	.247	.227	—	—	—	—	—	—	—	—	—	—	—	—	—	—	—	—	—	—	—	—
6	.335	.234	.268	.327	.622	—	—	—	—	—	—	—	—	—	—	—	—	—	—	—	—	—	—	—
7	.304	.157	.223	.335	.656	.722	—	—	—	—	—	—	—	—	—	—	—	—	—	—	—	—	—	—
8	.332	.157	.382	.391	.578	.527	.619	—	—	—	—	—	—	—	—	—	—	—	—	—	—	—	—	—
9	.326	.195	.184	.325	.723	.714	.685	.532	—	—	—	—	—	—	—	—	—	—	—	—	—	—	—	—
10	.116	.057	−.075	.099	.311	.203	.246	.285	.170	—	—	—	—	—	—	—	—	—	—	—	—	—	—	—
11	.308	.150	.091	.110	.344	.353	.232	.300	.280	.484	—	—	—	—	—	—	—	—	—	—	—	—	—	—
12	.314	.145	.140	.160	.215	.095	.181	.271	.113	.585	.428	—	—	—	—	—	—	—	—	—	—	—	—	—
13	.489	.239	.321	.327	.344	.309	.345	.395	.280	.408	.535	.512	—	—	—	—	—	—	—	—	—	—	—	—
14	.125	.103	.177	.066	.280	.292	.236	.252	.260	.172	.350	.131	.195	—	—	—	—	—	—	—	—	—	—	—
15	.238	.131	.065	.127	.229	.251	.172	.175	.248	.154	.240	.173	.139	.370	—	—	—	—	—	—	—	—	—	—
16	.414	.272	.263	.322	.187	.291	.180	.296	.242	.124	.314	.119	.281	.412	.325	—	—	—	—	—	—	—	—	—
17	.176	.005	.177	.187	.208	.273	.228	.255	.274	.289	.362	.278	.194	.341	.345	.324	—	—	—	—	—	—	—	—
18	.368	.255	.211	.251	.263	.167	.159	.250	.208	.317	.350	.349	.323	.201	.334	.344	.448	—	—	—	—	—	—	—
19	.270	.112	.312	.137	.190	.251	.226	.274	.274	.190	.290	.110	.263	.206	.192	.258	.324	.358	—	—	—	—	—	—
20	.365	.292	.297	.339	.398	.435	.451	.427	.446	.173	.202	.246	.241	.302	.272	.388	.262	.301	.167	—	—	—	—	—
21	.369	.306	.165	.349	.318	.263	.314	.362	.266	.405	.399	.355	.425	.183	.232	.348	.173	.357	.331	.413	—	—	—	—
22	.413	.232	.250	.380	.441	.386	.396	.357	.483	.160	.304	.193	.279	.243	.246	.283	.273	.317	.342	.463	.374	—	—	—
23	.474	.348	.383	.335	.435	.431	.405	.501	.504	.262	.251	.350	.382	.242	.256	.360	.287	.272	.303	.509	.451	.503	—	—
24	.282	.211	.203	.248	.420	.433	.437	.388	.424	.531	.412	.414	.358	.304	.165	.262	.326	.405	.374	.366	.448	.375	.434	—

4. Compute the final residuals by

$$\bar{r}_{jk} = \dot{r}_{jk} - a_{jp} a_{kp}.$$

These serve as a check on the "fit" of the model and the accuracy of grouping. All of these values should be close to zero. If some are not, adjustments might have to be made in the number or composition of groups.

We shall not give an example here which shows the application of each of these steps in the solution of a problem. Instead, we present the result of the application of the bi-factor model to the data of an example in Harman (1976). The example involves a set of 24 psychological tests administered to 145 children in the 1930s. Because of its wide use by several authors for illustrating different factor analysis methods, the example has become a classic in the factor analysis literature. The basic statistics are given in Table 8.2.3, the correlations in Table 8.2.4, and the bi-factor solution in Table 8.2.5.

Note from Table 8.2.5 that there appears to be an identifiable general factor (General Deduction) as well as 5 reasonably clear group factors (Spatial Relations, Verbal, Perceptual Speed, Recognition, and Associative Memory). Several variables appear not to be associated with any group, and the

TABLE 8.2.5 BI-FACTOR PATTERN FOR 24 PSYCHOLOGICAL TESTS

Test j	*General deduction* F	*Spatial relations* B_1	*Verbal* B_2	*Perceptual speed* B_3	*Recog-nition* B_4	*Associative memory* B_5	*Doublet* D_1	*Unique* Y_j
1	.589	.484	—	—	—	—	—	.647
2	.357	.285	—	—	—	—	—	.889
3	.401	.479	—	—	—	—	—	.781
4	.463	.317	—	—	—	—	—	.828
5	.582	—	.574	—	—	—	—	.576
6	.575	—	.559	—	—	—	—	.597
7	.534	—	.708	—	—	—	—	.463
8	.624	—	.375	—	—	—	—	.686
9	.560	—	.628	—	—	—	—	.540
10	.388	—	—	.594	—	—	.377	.595
11	.521	—	—	.478	—	—	—	.707
12	.404	—	—	.642	—	—	—	.652
13	.576	—	—	.438	—	—	—	.690
14	.388	—	—	—	.545	—	—	.743
15	.351	—	—	—	.476	—	—	.806
16	.496	—	—	—	.353	—	—	.793
17	.422	—	—	—	.361	.493	—	.670
18	.515	—	—	—	—	.468	—	.718
19	.442	—	—	—	—	.278	—	.853
20	.644	—	—	—	—	—	—	.765
21	.645	—	—	—	—	—	—	.764
22	.644	—	—	—	—	—	—	.765
23	.734	—	—	—	—	—	—	.679
24	.712	—	—	—	—	—	.377	.592
Contribution of factor	6.874	0.645	1.678	1.185	0.779	0.539	0.284	—

group called "Doublet" does not seem to be definable (it accounts for only 1.2% of the total variance). To determine what proportion of the total variance is accounted for by all seven common factors, we simply add their contributions and divide by $n=24$. Thus,

$$\frac{V}{n}=\frac{6.874+.645+\cdots+.284}{24}=.50,$$

which, as we shall see, is relatively small. Furthermore, three of the group factors, B_1, B_4 and B_5, each accounts for less than 5% of the variance, also relatively small. On the whole, although the bi-factor model provides some insight into the relationships among the variables in this set, it does not provide an entirely satisfactory solution. In fact, the results shown in Table 8.2.5 were obtained by making certain "adjustments" following the application of the bi-factor solution described above. For a more complete discussion and interpretation of the results in Table 8.2.5, see Harman (1976, Section 7.6).

8.3 PRINCIPAL-COMPONENT ANALYSIS

As we indicated earlier, the distinction between component analysis and common-factor analysis may be described in terms of the amount and kinds of variability of the variables being analyzed. In the component analysis model of Equation 8.1.1, no distinction is made between common and unique factors, as contrasted with the common-factor analysis model of Equation 8.1.2. The methods of obtaining the a_{jp}s are different for these models. In component analysis we analyze the "complete" correlation matrix, **R**, with 1s in the diagonal. In common-factor analysis, we analyze the "reduced" correlation matrix, $\mathbf{R}_h$, with estimated communalities in the diagonal. Analysis of **R** implies an analysis of the total variance, because $\mathrm{Var}(z_j)=1$, whereas analysis of $\mathbf{R}_h$ implies an analysis only of the variance of each variable that is common or shared by other variables, that is, h_j^2.

We shall be concerned in this section with *principal-component* analysis, a term derived from the *principal-axes* method of extracting factors. We have chosen to discuss component analysis instead of common-factor analysis for several reasons: (1) It usually produces results that do not differ markedly from those of common-factor analysis. (2) It avoids the necessity of estimating communalities and selecting one of a number of methods of varying theoretical and practical validity for estimating them. (3) It allows factor scores to be computed directly, rather than requiring that they be estimated, as is the case if $\mathbf{R}_h$ is analyzed. (4) An effective theoretically and empirically based method is available for determining the number of common factors. Actually, much of the development here can be applied to common-factor analysis as well as to component analysis. However, component analysis is somewhat neater mathematically, and produces results that are more objectively interpretable than those of common-factor analysis.

The mathematical formulation of the principal component method is quite straightforward, being similar to, but simpler than, the formulation of the canonical correlation coefficients discussed in Chapter 5. The basic principle is that of (1) finding a linear combination of the variables, the z_js, with maximum variance, (2) finding a second linear combination of the variables, orthogonal to (independent of) the first, with maximal remaining variance, and so on. Note that this objective is similar to that of Chapter 5 in which we were interested in obtaining independent pairs of linear combinations of variables in *two* sets which had, successively, the largest, the next largest, etc., correlations with each other. The problem here is simpler because we have only one set of variables with which to deal.

We first denote the matrix of correlations between the n variables by

$$\mathbf{R}=\begin{bmatrix} 1 & r_{12} & \cdots & r_{1n} \\ r_{21} & 1 & \cdots & r_{2n} \\ \vdots & \vdots & & \vdots \\ r_{n1} & r_{n2} & \cdots & 1 \end{bmatrix} \tag{8.3.1}$$

and the vector of coefficients (weights) on the n variables for the pth component by

$$\mathbf{a}_p=\begin{bmatrix} a_{1p} \\ a_{2p} \\ \vdots \\ a_{np} \end{bmatrix}. \tag{8.3.2}$$

The problem of determining the vectors $\mathbf{a}_p$ which maximize (1) the variance accounted for by the first component, (2) the variance accounted for by the second component, orthogonal to the first, etc., is analogous to the problem of solving Equation 5.3.1. In this case, the equation to be solved for $\mathbf{a}_p$ is

$$(\mathbf{R}-\lambda_p\mathbf{I})\mathbf{a}_p=\mathbf{0} \tag{8.3.3}$$

in which the λ_ps are the *characteristic roots* or *eigenvalues* of $\mathbf{R}$ and the $\mathbf{a}_p$s are the associated *eigenvectors*. Just as in Chapter 5, we may solve this equation for the λ_ps by the method discussed in Appendix A. We first obtain the characteristic equation of $\mathbf{R}$, that is,

$$|\mathbf{R}-\lambda_p\mathbf{I}|=0 \tag{8.3.4}$$

which leads to a polynomial equation in λ_p. The number of different values of λ_p obtained is equal to the number of variables, n.

Each value of λ_p (equivalent to Harman's V_p in Equation 8.1.8) may be substituted in Equation 8.3.3 and the associated eigenvector, $\mathbf{a}_p$, may be obtained by the method described in Appendix A. The result will be a set of weights that have the correct proportionality to one another. However, they will not necessarily be scaled to satisfy Equation 8.1.8, which states that V_p equals the sum of squares of the elements of $\mathbf{a}_p$. If we denote an initial solution of $\mathbf{a}_p$ as $\mathbf{a}_p^*$, then the desired vector can be obtained by substituting for $\mathbf{a}_p^*$ in

$$\mathbf{a}_p = \mathbf{a}_p^* \sqrt{\lambda_p}\ . \tag{8.3.5}$$

Then the resulting elements of $\mathbf{a}_p$, that is, $a_{1p}, a_{2p}, \ldots, a_{np}$, will satisfy Equation 8.1.8.

The solution of Equation 8.3.4 by direct methods when n is large is a difficult task even with the aid of electronic computers. One of Thurstone's important contributions in factor analysis was the development of the *centroid* method, which provided a good approximation to the solution of 8.3.4 and was a widely employed factor-analytic computational method during the 1940s and 1950s. Hotelling's (1933) contribution was a method of computation by an iterative procedure that considerably reduced the labor. Several other methods, of more recent application, have also been used, including one based on a paper of Jacobi (1846) and one proposed by Francis (1961, 1962), known as the *Q-R* method. Because it is unlikely that applied researchers in the behavioral sciences will be inclined to use any of these methods without the aid of a computer, we shall not consider them in detail here. Computer programs are available (for example, in SPSS, BMD, etc.) that use one of these methods for determining the λ_ps and $\mathbf{a}_p$s. For a discussion of Hotelling's iterative procedure, see Tatsuoka (1971), and for the Jacobian and *Q-R* methods, see Mulaik (1972).

To illustrate the results of a principal-component analysis we return to the example involving the 24 psychological tests. Table 8.3.1 shows the eigenvalues (λ_ps in Equation 8.3.4, but here denoted by V_p in accordance with Equation 8.1.8) and eigenvectors (the values in the columns headed by F_1, F_2, etc.) for the first ten principal components, together with the percentage of the total variance accounted for $[100(V_p/n)]$ by each.

We may note several features of Table 8.3.1. One is that the components are ordered, from largest to smallest, in terms of the sizes of their eigenvalues, or, equivalently, in terms of the percentage of the total variance for which they account. Using the percentages, we can easily determine the total variance accounted for by the first two, or three, or four, etc., components. We may also note that all of the elements of the first eigenvector are positive, which reflects the fact that almost all correlations in **R** were positive. Subsequent components have about equal numbers of positive and negative elements in their eigenvectors. It can easily be verified that the sum of the products of the elements of any pair of columns, for example, (.616) (−.005) + (.400) (−.079) + · · · + (.673) (.196), is equal to zero (within rounding error), a consequence of the fact that all of the components are orthogonal (independent).

Although occasionally the results of a principal-component analysis will constitute a final solution, and can be interpreted directly, it is ordinarily necessary to employ one of the transformation procedures discussed in the next section, in order to make interpretation easier. Thus, the principal-component analysis results, such as in Table 8.3.1, may be considered an

TABLE 8.3.1 FIRST TEN PRINCIPAL COMPONENTS FOR 24 PSYCHOLOGICAL TESTS

Test	F_1	F_2	F_3	F_4	F_5	F_6	F_7	F_8	F_9	F_{10}
1	.616	−.005	.428	−.204	−.009	.070	.199	.220	−.153	.156
2	.400	−.079	.400	−.202	.348	.089	−.506	−.024	−.263	−.256
3	.445	−.191	.476	−.106	−.375	.329	−.083	−.356	−.060	.021
4	.510	−.178	.335	−.216	−.010	−.192	.461	.142	.135	−.266
5	.695	−.321	−.335	−.053	.079	.078	−.123	.028	−.232	−.010
6	.690	−.418	−.265	.081	−.008	.124	.001	.129	−.054	−.129
7	.677	−.425	−.355	−.072	−.040	.011	.081	.009	.001	−.091
8	.694	−.243	−.144	−.116	−.141	.119	.158	−.172	.115	−.065
9	.694	−.451	−.291	.080	−.005	−.071	−.009	.117	−.124	.029
10	.576	.542	−.446	−.202	.079	−.085	−.013	−.080	.079	−.099
11	.434	.474	−.210	.034	.002	.301	−.043	.320	−.004	.062
12	.482	.549	−.127	−.340	.099	.039	.158	−.301	−.132	.159
13	.618	.279	.035	−.366	−.075	.364	.130	.175	−.040	.104
14	.448	.093	−.055	.555	.156	.383	−.084	−.126	.262	.056
15	.416	.142	.078	.526	.306	−.057	.126	.072	−.304	.208
16	.534	.091	.392	.327	.171	.172	.081	.128	.297	−.212
17	.488	.276	−.052	.469	−.255	−.107	.248	−.214	−.152	−.137
18	.544	.386	.198	.152	−.104	−.252	−.019	−.003	−.344	−.287
19	.475	.138	.122	.193	−.604	−.139	−.341	.192	.102	.104
20	.643	−.186	.132	.070	.285	−.191	.026	−.294	.176	.057
21	.622	.232	.100	−.202	.174	−.226	−.161	.176	.323	−.000
22	.640	−.146	.110	.056	−.023	−.331	−.045	.131	.022	.368
23	.712	−.105	.150	−.103	.064	−.111	−.081	−.248	.067	.300
24	.673	.196	−.233	−.062	−.097	−.170	−.228	−.119	.154	−.185
V_p	8.137	2.097	1.692	1.501	1.025	.943	.900	.817	.790	.707
$100V_p/24$	33.9	8.7	7.0	6.2	4.3	3.9	3.8	3.4	3.3	2.9

initial solution, which in most cases must be subjected to further treatment. However, there are two important kinds of information to be gleaned from this initial solution. The first is the number of common factors. It may seem strange, in view of the earlier statement that the number of components must be equal to the number of variables, n, to say that the number of *common* factors can be determined from a solution such as that in Table 8.3.1. However, Kaiser (1960) suggested that the number of common factors should be equal to the number of principal components having eigenvalues greater than one. There is logical, theoretical, and empirical support for this criterion. As Harman (1976, p. 86) states, this criterion "is quite plausible since the sum of all n roots [eigenvalues] is precisely n, so that a value of one is merely par and surely if another dimension [factor] is to be added, it would be desirable to have it account for at least an average contribution." Guttman (1954, 1956) proved that the weakest lower bound for the rank (the number of common factors) of a reduced correlation matrix (with communalities instead of ones in the principal diagonal) is the number of principal components of **R** having eigenvalues greater than one. Kaiser (1965) showed that the number of principal components having eigenvalues greater than one corresponds to the number of common factors that have a positive generalizability in the sense of Cronbach's alpha (see, for example, Mehrens and Lehman, 1973, for a discussion of Cronbach's alpha). Kaiser

(1961) has also provided empirical evidence in support of this criterion, namely that, in a large number of different factor-analytic studies, the factors retained by applying the criterion are usually interpretable, whereas those not retained are normally difficult to interpret. Using this criterion, we should conclude that in our example involving the 24 psychological tests, the number of common factors is 5.

A number of other proposals have been made for determining the number of common factors. Cattell (1966) suggested a graphical procedure called the *scree* test, which is based on the size of the differences between eigenvalues of adjacent components. Significance tests have been proposed by a number of statisticians, including Lawley (1940), Bartlett (1950, 1951), Burt (1952), Jöreskog (1962), and McNemar (1942). In general these are large-sample tests, based on rather stringent assumptions, and they have not been widely used in expoloratory factor analysis. Summaries of several of them may be found in a paper of Maxwell (1959) and also in a paper of Velicer (1976).

In addition to information about the number of factors, the initial principal-components solution can be used to estimate the reliability of the principal components obtained. These estimates require that good estimates of the reliabilities of the individual variables be available. Then the reliability, r_P, of the pth principal component may be estimated from

$$r_p = 1 - \frac{\sum_{j=1}^{n} a_{jp}^2 (1 - r_{jJ})}{\lambda_p},$$

in which a_{jp} is the jth element of the pth eigenvector, r_{jJ} is the estimated reliability of test X_j, and λ_p is the pth eigenvalue. The values of r_p can be interpreted as any reliability coefficient. The interpretation depends, of course, on how the test reliabilities were estimated.

For the data of our example, the reliabilities of the first five principal components are (see Tables 8.2.1 and 8.3.1 for the necessary reliabilities and eigenvectors): $r_1 = .75$, $r_2 = .79$, $r_3 = .73$, $r_4 = .70$, and $r_5 = .60$. These values suggest a moderately high reliability for all five common factors, that is, for those principal components with eigenvalues greater than one.

8.4 TRANSFORMED (ROTATED) SOLUTIONS

One of the major contributions of L. L. Thurstone (1947) to factor analysis was the definition of the concept of *simple structure*. Thurstone's conditions for simple structure, for orthogonal factors (in which case the pattern matrix, **A**, is equivalent to the structure matrix, **S**), are:

1. Each row of the pattern matrix **A** should have at least one zero. In other words, for each variable, there should be at least one factor that makes no contribution at all to the variable.
2. If there are m common factors, each column of **A** should have at least m zeros.
3. For every pair of columns of **A**, there should be several variables with elements equal to zero in one column, but not in the other.
4. For every pair of columns of **A**, a large proportion of the variables should have elements equal to zero in both columns, if there are 4 or more common factors.
5. For every pair of columns of **A** there should be only a small number of variables with elements not equal to zero in both columns.

This definition of simple structure may seem complicated, but it really isn't. The ideal that Thurstone was suggesting can be expressed easily if we look first at the factor coefficients, the a_{jp}s, for a single variable, and then look at the a_{jp}s for a single factor. He thought that, in an ideal solution, only one factor should contribute substantially to variable X_j, and the other factors should not contribute at all, which would suggest that the jth row of **A** should look something like this:

	Factor						
Variable	1	2	3	...	p	...	m
j	0	0	0		.8		0

Of course, the value .8 is arbitrary (it could have been .5 or .6 or .7) and is meant to imply some moderately large value. In terms of a column of **A**, the configuration should be something like this:

Variable	*Factor* F_p
1	0
2	0
⋮	⋮
j	.8
$j+1$	.8
$j+2$	.8
$j+3$	.8
⋮	⋮
n	0

Again, the choice of .8 is meant to suggest only relatively large loadings on several (not more than $n-m$) variables, and loadings of 0 (or close to 0) on the others. Thus, a pattern matrix, based on 10 variables and three factors, and exhibiting simple structure, might look something like this:

Variable	Factor 1	Factor 2	Factor 3
1	*xx*	0	0
2	*xx*	0	0
3	*xx*	0	0
4	0	*xx*	0
5	0	*xx*	0
6	0	*xx*	0
7	0	0	*xx*
8	0	0	*xx*
9	0	0	*xx*
10	0	0	*xx*

in which the symbol *xx* indicates a moderately large value of a_{jp}, and 0 indicates a value close to zero. Not only does such a pattern suggest a distinct grouping of variables, but it also facilitates *definition* (or *naming*) of factors, which should be done on the basis of the variables with which the factors are highly correlated. We shall consider the basis on which factors are named in a later section.

Now, obviously, the solution of Table 8.3.1 does not meet the criteria for simple structure suggested by Thurstone and widely accepted by factor analysts. Apparently, some type of transformation of this solution is required. To illustrate the logic of such transformations, we shall use the initial principal-component solution given in Table 8.4.1, based on six variables.

Using the pair of values a_{j1} and a_{j2} as coordinates with respect to F_1 and F_2, each variable can be represented as a point on a graph, as shown in Figure 8.4.1.

For example, Variable 1, with coordinates $a_{11}=.86$ and $a_{12}=-.02$, is represented by the point labelled "1" in Figure 8.4.1. An inspection of Figure 8.4.1 shows that there are two distinct groups of variables, but that

TABLE 8.4.1 INITIAL AND ROTATED SOLUTIONS FOR A FACTOR ANALYSIS PROBLEM INVOLVING SIX VARIABLES AND TWO FACTORS

	Initial solution		Rotated solutions $\theta=-15°$		$\theta=-30°$		$\theta=-55°$	
Variable	F_1	F_2	F_1'	F_2'	F_1'	F_2'	F_1'	F_2'
1	.86	−.02	.84	.20	.76	.41	.51	.69
2	.83	−.16	.85	.06	.80	.28	.60	.59
3	.86	−.38	.93	−.14	.93	.10	.80	.49
4	.58	.40	.46	.54	.30	.64	.00	.70
5	.50	.34	.40	.46	.26	.54	.01	.60
6	.33	.23	.26	.31	.17	.36	.00	.40

FIGURE 8.4.1 GRAPHIC REPRESENTATION OF INITIAL FACTOR ANALYSIS SOLUTION SHOWN IN TABLE 8.4.1 FOR SIX VARIABLES AND TWO FACTORS

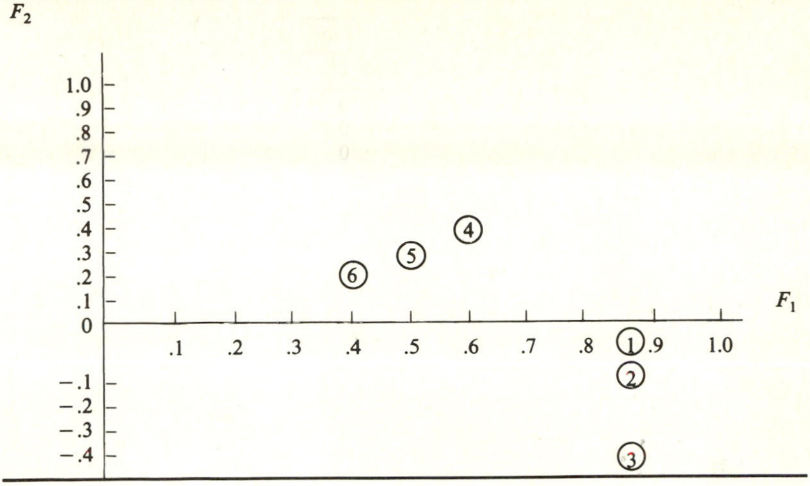

the original axes do not represent these groups of variables optimally in terms of the simple structure criteria. To achieve simple structure, it would be desirable for the axes to pass as close as possible to the two groups of variables. Then the coordinates of each variable would be large with respect to one axis but small with respect to the other. An approach to this goal can be made by rotating the axes to a new position, maintaining their orthogonality (that is, keeping the angle between them equal to 90 degrees). It seems clear that such a rotation should be in a clockwise direction, but it is not so clear what the angle of rotation should be. Consequently, we show in Table 8.4.1 the pattern matrices that result when the original axes, F_1 and F_2, are rotated through $-15°$, $-30°$, and $-55°$, respectively. Before discussing these results, we consider the procedure for obtaining them.

Suppose we denote by b_{j1} and b_{j2} the coordinates of variable X_j with respect to a set of *new* (rotated) axes, F_1' and F_2'. Using elementary trigonometry, it can be shown that b_{j1} and b_{j2} can be computed from

$$b_{j1} = a_{j1}\cos\theta + a_{j2}\sin\theta$$

and

$$b_{j2} = a_{j2}\cos\theta - a_{j1}\sin\theta \qquad (8.4.1)$$

in which θ is the angle of rotation*. We shall illustrate the application of Equations 8.4.1 by computing the new coordinates, b_{11} and b_{12}, of variable X_1 when $\theta = -15°$. Then

$$b_{j1} = (.86)(.97) + (-.02)(-.26) = .84,$$

*Positive angles indicate rotation in the counter-clockwise direction.

FIGURE 8.4.2 Graphic Representations of Three Orthogonally Rotated Solutions Given in Table 8.4.1 for Six Variables and Two Factors

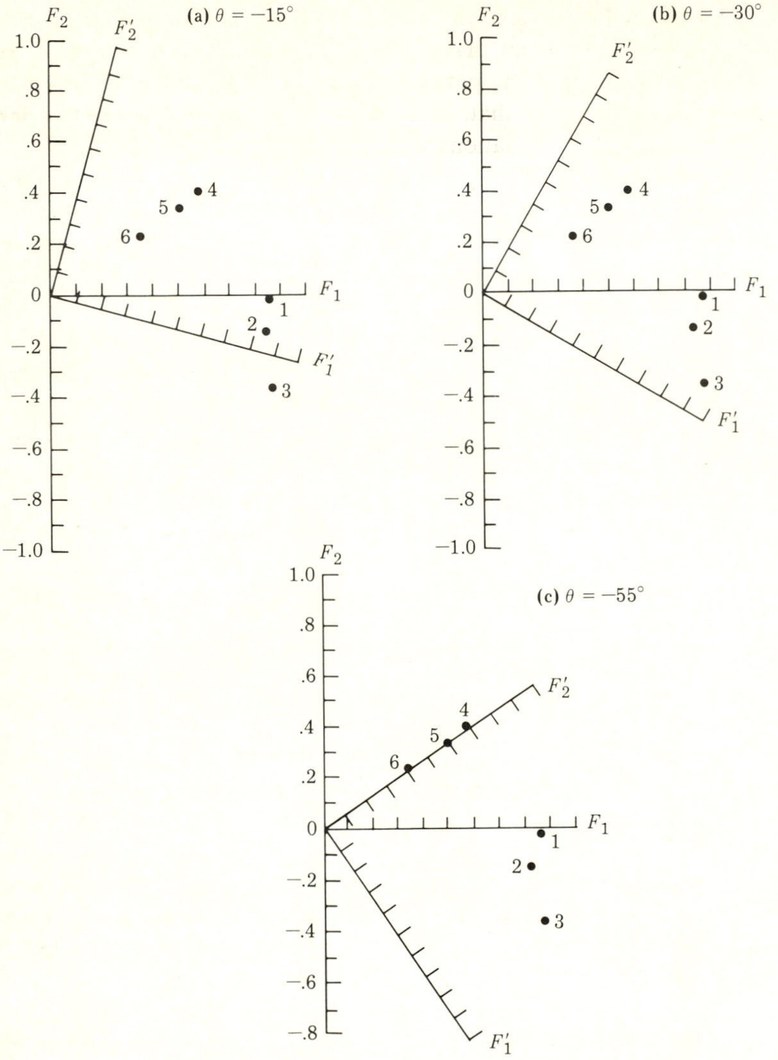

and

$$b_{j2} = (-.02)(.97) - (.86)(-.26) = .20.$$

These two values are recorded for variable X_1 in the column of Table 8.4.1 headed $\theta = -15°$. The other values in Table 8.4.1 were obtained similarly. The resulting graphical configurations of the six variables with respect to each set of rotated axes are shown in Figure 8.4.2.

In terms of simple structure, it appears that the rotation through $\theta = -30°$ is the best of the three. Two quite distinct factors appear, the first having relatively large correlations with the first three variables and much smaller ones with the last three. The pattern for the second factor is essentially reversed, except that variable X_6 does not have a large correlation with either factor. The patterns when $\theta = -15°$ and $-55°$ are not as satisfactory, in the sense that, in both of these solutions, several variables have relatively large correlations with both factors.

In the early days of factor analysis, rotated solutions were obtained in essentially the manner we have just described. Figures such as Figure 8.4.1 were constructed and inspected to determine the direction and degree of rotation required. Of course, two different factor analysts working with the same data sometimes came to quite different conclusions concerning which solution was "best." Furthermore, when the number of factors was more than two or three, the task became very difficult because the type of rotation deemed appropriate for Factors 1 and 2, say, might worsen the picture for Factors 2 and 3. Thus, not only were such methods highly subjective, but quite tedious as well.

When computers became generally available, there was a great deal of interest in the development of more objective, analytical (mathematical) procedures for carrying out rotations of axes. In the next section we shall discuss one of these methods that has been widely applied in factor analysis.

8.4.1 The Varimax Method of Orthogonal Rotation

Before discussing the varimax method we shall describe briefly the rationale behind a method of rotation, now called the *quartimax* criterion, that was suggested independently and nearly concurrently by several writers, including Carroll (1953), Saunders (1953), Ferguson (1954), and Neuhaus and Wrigley (1954). The logic of the procedure is quite straightforward:

1. The simplest description of a variable, X_j, in terms of a pair of factors (components), is achieved when one of the axes passes directly through the point that represents the variable.

2. For the rotated factors, F_p' and F_q' $(p \neq q = 1, \ldots m)$, let us denote the factor coefficients of variable X_j by b_{jp} and b_{jq}. If either F_p' or F_q' passes directly through the point representing variable X_j, then the product $b_{jp}^2 b_{jq}^2$ will be minimized.

3. Therefore, we want to minimize the sum

$$\sum_{j=1}^{n} \sum_{p,q=1}^{m} b_{jp}^2 b_{jq}^2 \qquad (p<q).$$

It can be shown that the communality, h_j^2, of a variable remains constant under an orthogonal rotation, that is, $\sum_{p=1}^{m} b_{jp}^2 = \sum_{p=1}^{m} a_{jp}^2 = h_j^2$. In the case of two factors, for example, we have from Equation 8.4.1,

$$\begin{aligned} h_j^2 = b_{j1}^2 + b_{j2}^2 &= (a_{j1}\cos\theta + a_{j2}\sin\theta)^2 + (a_{j2}\cos\theta - a_{j1}\sin\theta)^2 \\ &= a_{j1}^2\cos^2\theta + a_{j2}^2\sin^2\theta + 2a_{j1}a_{j2}\sin\theta\cos\theta \\ &\quad + a_{j2}^2\cos^2\theta + a_{j1}^2\sin^2\theta - 2a_{j1}a_{j2}\sin\theta\cos\theta \\ &= a_{j1}^2(\sin^2\theta + \cos^2\theta) + a_{j2}^2(\sin^2\theta + \cos^2\theta) \\ &= a_{j1}^2 + a_{j2}^2, \end{aligned}$$

because $\sin^2\theta + \cos^2\theta = 1$. Then $(h_j^2)^2$ is also constant under such a transformation. Thus,

$$\begin{aligned} (h_j^2)^2 &= \left(\sum_{p=1}^{m} b_{jp}^2\right)^2 = (b_{j1}^2 + \ldots + b_{jm}^2)^2 \\ &= \sum_{p=1}^{m} b_{jp}^4 + 2\sum_{p,q=1}^{m} b_{jp}^2 b_{jq}^2 = \text{constant} \qquad (p<q). \end{aligned}$$

Summing over the n variables, we have

$$\sum_{j=1}^{n}\sum_{p=1}^{m} b_{jp}^4 + 2\sum_{j=1}^{n}\sum_{p,q=1}^{m} b_{jp}^2 b_{jq}^2 = \text{constant} \qquad (p<q). \tag{8.4.2}$$

Now the second term on the left is the one we wish to minimize, but because the sum of these two terms is constant under an orthogonal rotation, minimizing the second is equivalent to maximizing the first, namely, $\Sigma_{j=1}^{n}\Sigma_{p=1}^{m} b_{jp}^4$. The term quartimax thus refers to *max*imizing the sum of the *fourth* powers of the rotated factor coefficients. It can easily be shown that this is equivalent to maximizing the sum of the variances of the squared factor loadings in the *rows* of the pattern matrix, **A**.

The *varimax* method of orthogonal rotation originally proposed by Kaiser (1956), but later modified and improved (Kaiser, 1958), is regarded as an improvement over quartimax in the sense of achieving more nearly Thurstone's simple structure criteria. It differs essentially from quartimax in that it operates on the column loadings instead of the row loadings. It maximizes the sum, over the m factors, of the column variances of the squared coefficients of the pattern matrix. It is easy to see that if all coefficients in each column were either 0 or 1, then one important feature of Thurstone's simple structure criteria would be realized. Such a configuration, with all coefficients either 0 or 1, would clearly correspond to maximum variance. The actual quantity to be maximized is

$$V = \sum_{p=1}^{m}\left[\frac{n\sum_{j=1}^{n}(b_{jp}/h_j)^4 - \left(\sum_{j=1}^{n} b_{jp}^2/h_j^2\right)^2}{n^2}\right]. \tag{8.4.3}$$

Kaiser (1958) calls this the "normal" varimax criterion to distinguish it from the earlier (1956) "raw" varimax criterion, which did not incorporate the communalities. When the term varimax is used today, it is understood to mean "normal" varimax. Note that the expression in brackets is the variance of $(b_{jp}/h_j)^2$ for factor F_p', so that V is the sum of these variances over the m common factors.

We shall once again omit a detailed description of computational procedures, and simply show what the results of a varimax rotation look like. We return to the example concerning the 24 psychological tests, and show both a quartimax and a varimax solution based on the first four common factors. These results (see Table 8.4.2) are taken from Harman (1976), who chose in this example to rotate only the first four, rather than all five, of the common factors.

In both the quartimax and varimax solutions, the coefficients greater than or equal to .4 are shown in bold-face for emphasis. The choice of .4, rather than .3 or .5, may seem arbitrary, but is supported by quite a bit of empirical evidence that variables with coefficients of at least .4 (in absolute value) make meaningful contributions to defining factors, whereas those with lesser values usually do not. In a sense, this rule of thumb implies that we are regarding coefficients less than .4 (in absolute value) as 0, at least for the purpose of factor definition.

TABLE 8.4.2 QUARTIMAX AND VARIMAX ORTHOGONAL SOLUTIONS FOR 24 PSYCHOLOGICAL TESTS

Test	*Quartimax*				*Varimax*			
j	F_1'	F_2'	F_3'	F_4'	F_1'	F_2'	F_3'	F_4'
1	.37	.19	**.60**	.07	.14	.19	**.67**	.17
2	.24	.07	.38	.04	.10	.07	**.43**	.10
3	.31	.01	**.48**	.01	.15	.02	**.54**	.08
4	.36	.07	**.46**	−.01	.20	.09	**.54**	.07
5	**.81**	.14	−.02	−.04	**.75**	.21	.22	.13
6	**.81**	.03	.00	.06	**.75**	.10	.23	.21
7	**.85**	.07	−.04	−.10	**.82**	.16	.21	.08
8	**.66**	.20	.20	−.04	**.54**	.26	.38	.12
9	**.86**	−.06	−.02	.10	**.80**	.01	.22	.25
10	.23	**.70**	−.12	.11	.15	**.70**	−.06	.24
11	.31	**.62**	.01	.23	.17	**.60**	.08	.36
12	.16	**.69**	.19	−.01	.02	**.69**	.23	.11
13	.35	**.57**	.32	−.08	.18	**.59**	**.41**	.06
14	.32	.19	−.03	**.42**	.22	.16	.04	**.50**
15	.25	.11	.10	**.45**	.12	.07	.14	**.50**
16	.29	.13	.37	.37	.08	.10	**.41**	**.43**
17	.28	.24	.02	**.57**	.14	.18	.06	**.64**
18	.22	.32	.30	**.47**	.00	.26	.32	**.54**
19	.28	.18	.19	.32	.13	.15	.24	.39
20	**.52**	.09	.35	.14	.35	.11	**.47**	.25
21	.35	.38	.35	.14	.15	.38	**.42**	.26
22	**.53**	.04	.30	.26	.36	.04	**.41**	.36
23	**.55**	.19	**.44**	.09	.35	.21	**.57**	.22
24	**.49**	**.43**	.10	.20	.34	**.44**	.22	.34
V_p	5.59	2.42	1.96	1.42	3.50	2.44	3.08	2.36

Before looking further at Table 8.4.2, we should say a word about naming (or defining) factors. As pointed out earlier, a factor is a variable in the same sense that a test, say X, is a variable. In other words, a score on a factor is a measurement of some trait, characteristic, or construct. The difference between a test score and a factor score is the difference between a direct measurement and an indirect measurement. The factor measurement must be based on the values of the variables to which the factor makes the greatest contribution. Hence, in order to determine what the factor score is a measurement of, we must look at what is measured by such variables. These variables allow us to define (or name) the factor. That is, what *they* measure, in combination with one another, is what the factor measures.

Looking now at Table 8.4.2, we see first that, in terms of meeting the simple structure criteria, both solutions represent marked improvements over the initial solution of Table 8.3.1, with the varimax solution being slightly better. Both show clearly that there are several distinct groups of variables, each defining one of the four factors. Of course, the picture might have looked somewhat different had all five factors been rotated. The two methods seem to produce very similar results, except in terms of Variables 20, 22, 23, and 24, which contribute to different factors in the two solutions. We would tend to have a slight preference for the varimax solution in terms of these variables. In terms of what they measure, these variables seem to be more like Variables 1 through 4 than Variables 5 through 9. The names suggested by Harman for the four factors are: F_1', Verbal; F_2', Speed; F_3', Deduction; and F_4', Memory.

One of the important properties of a successful factorial method is that of factorial invariance, a principle defined by Thurstone (1947, p. 361). In fact, Thurstone stated, "It is a fundamental criterion of a valid method of isolating primary abilities that the weights of the primary abilities for a test must remain invariant when it is moved from one test battery to another test battery. The same principle can be stated as a fundamental requirement of a successful factorial method, that the factorial description of a test must remain invariant when the test is moved from one battery to another which involves the same common factors." This does not mean that the factor structure should be the same for all populations. As Thurstone said, "Limiting ourselves here to analyses that are made on the same population, we should expect to find that if several samples are drawn from the same population and if independent factor analyses are made with the same battery of tests for the several samples, then the factor loadings should remain invariant for the different samples within sampling errors if the simple structure is complete and overdetermined" (Thurstone, 1947, p. 361). The varimax method tends to have this invariance property; Kaiser (1958) proved that it does for the special case of only two factors. The importance of this property is, as Harman (1976) states, that "it permits the drawing of inferences about the factors in an (indefinite) domain of psychological content from a varimax solution based on a sample of n tests. This does not ascribe greater 'psychological meaning' to the varimax factors than to factors obtained on some other basis, but it does mean that varimax factors obtained in a sample will have a greater likelihood of portraying the universe varimax factors" (Harman, 1976, p. 298).

A number of other procedures and criteria have been suggested for obtaining orthogonal rotated solutions, among them the *transvarimax* methods (including equamax and ratiomax, Saunders, 1962), *parsimax* (Crawford, 1967), and *simultaneous orthogonal varimax and parsimax* (Horst, 1965, and independently, Sherin, 1966). For a discussion of these methods, see Mulaik (1972), Chapter 10.

Until this point we have considered only rotations that maintain the orthogonality of the factors. It may have occurred to the reader that perhaps there are cases in which simple structure cannot be achieved under such a restraint. In the next section we consider what are called *oblique* rotation methods. In these, F_1 and F_2, for example, can be rotated through different angles, so that the rotated factors are correlated.

8.4.2 Methods of Oblique Rotation

We will again find it helpful to consider the problem of rotation in a geometric setting before we discuss specific analytical methods. We return, therefore, to the initial solution in Table 8.4.1 and to Figure 8.4.1. It appears from inspecting the configuration of points in Figure 8.4.1 that the "best" solution in terms of the simple structure criterion would be obtained by rotating F_2 through $\theta_2 = -55°$ (approximately) and F_1 through $\theta_1 = -15°$ (approximately). Then one axis would pass directly through the points representing Variables 4, 5, and 6, and the other should come quite close to Variables 1, 2, and 3. Such a solution, in which the factors are correlated, is called an oblique solution. It often provides the best representation of the data in the simple structure sense.

We may express the Equations 8.4.1 in matrix form as

$$\mathbf{b}_j = \mathbf{a}_j'\mathbf{T}, \tag{8.4.4}$$

in which

$$\mathbf{b}_j = \begin{bmatrix} b_{j1} \\ b_{j2} \end{bmatrix}, \quad \mathbf{a}_j = \begin{bmatrix} a_{j1} \\ a_{j2} \end{bmatrix}, \quad \text{and } \mathbf{T} = \begin{bmatrix} \cos\theta & -\sin\theta \\ \sin\theta & \cos\theta \end{bmatrix}.$$

In the oblique case we wish to rotate F_1 and F_2 so that the rotated axes, which we now denote by P_1 and P_2 to distinguish them from F_1' and F_2', have angles with F_1 of θ_1 and θ_2, respectively. In other words, we select one of the initial coordinate axes and express the angles of rotation in terms of the selected axis. Then the transformation matrix, which we now denote by

$\mathbf{T}_{ob}$, is

$$\mathbf{T}_{ob} = \begin{bmatrix} \cos\theta_1 & \cos\theta_2 \\ \sin\theta_1 & \sin\theta_2 \end{bmatrix}.$$

Note that the sums of squares of the columns of this matrix are equal to one. It can be shown that the sum of the products of the row elements, that is, $(\cos\theta_1)(\cos\theta_2)+(\sin\theta_1)(\sin\theta_2)$, is equal to the cosine of the angle between the new coordinate axes, P_1 and P_2.

We will now find it helpful to distinguish between the orthogonal pattern matrix, **A**, and the oblique pattern matrix, which we shall denote by **P**. Then Equation 8.1.21 will become, in the oblique case,

$$\mathbf{S}=\mathbf{PC}, \qquad (8.4.5)$$

in which **C** is the matrix of factor correlations defined earlier. It can be shown that

$$\mathbf{C}=\mathbf{T}'_{ob}\mathbf{T}_{ob}. \qquad (8.4.6)$$

It can also be shown that the structure matrix, **S**, can be calculated from

$$\mathbf{S}=\mathbf{AT}_{ob}. \qquad (8.4.7)$$

Therefore, we can write

$$\mathbf{PC}=\mathbf{AT}_{ob},$$

so that

$$\mathbf{P}=\mathbf{AT}_{ob}\mathbf{C}^{-1}=\mathbf{SC}^{-1}. \qquad (8.4.8)$$

Because we will ordinarily be interested in **S**, the matrix of correlations between the variables and the factors, we shall use Equation 8.4.8 to obtain the oblique pattern, **P**, first obtaining **S** and **C**.

For the six variables we have used as our example, we have*

$$\mathbf{T}_{ob} = \begin{bmatrix} \cos -15^\circ & \cos 35^\circ \\ \sin -15^\circ & \sin 35^\circ \end{bmatrix} = \begin{bmatrix} .96593 & .81915 \\ -.25882 & .57358 \end{bmatrix}.$$

Then

$$\mathbf{S}=\mathbf{AT}_{ob} = \begin{bmatrix} .86 & -.02 \\ .83 & -.16 \\ .86 & -.38 \\ .58 & .40 \\ .50 & .34 \\ .33 & .23 \end{bmatrix} \begin{bmatrix} .96593 & .81915 \\ -.25882 & .57358 \end{bmatrix} = \begin{bmatrix} .84 & .69 \\ .84 & .59 \\ .93 & .49 \\ .46 & .70 \\ .39 & .60 \\ .26 & .40 \end{bmatrix}.$$

The matrix **C** is calculated as

$$\mathbf{C}=\mathbf{T}'_{ob}\mathbf{T}_{ob} = \begin{bmatrix} 1.00000 & .64279 \\ .64279 & 1.00000 \end{bmatrix},$$

which has inverse

$$\mathbf{C}^{-1} = \begin{bmatrix} 1.70410 & -1.09538 \\ -1.09538 & 1.70410 \end{bmatrix}.$$

*Rotation of F_2 through $-55°$ is equivalent to rotation of F_1 through $+35°$. Therefore, $\theta_2=35°$.

Then

$$\mathbf{P}=\mathbf{S}\mathbf{C}^{-1}=\begin{bmatrix} .68 & .26 \\ .79 & .09 \\ 1.05 & -.18 \\ .02 & .69 \\ .01 & .60 \\ .00 & .40 \end{bmatrix}.$$

Note that this solution meets the simple structure criteria considerably better than any of the orthogonal solutions of Table 8.4.1. Note also that the elements of **P**, the factor loadings, can exceed 1.00, because they are not correlation coefficients.

The Oblimin Method. Just as in the case of orthogonal transformations, a number of objective mathematical procedures have been developed for oblique rotation. We shall discuss only one of them here, the oblimin method due to Carroll (1960). This method has been among those most widely used by factor analysts in psychology and education, perhaps because of its inclusion in the SPSS computer program package. Our selection of it for discussion here was based primarily on availability and does not imply that it is the "best" method available. In fact, Hakstian and Abell (1974), after a thorough examination of a number of different oblique rotation procedures, state, "One conclusion that appears inescapable is that no *single* computing procedure—general paradigm or specialization—at least of those examined in the present study, can be expected to yield uniformly optimal oblique solutions *for all kinds of data.*"

In order to discuss the oblimin method, we need to consider briefly the notion of *reference axes*, upon which Carroll's original development of the oblimin method was based. The system of reference axes was introduced by Thurstone (1947) in an effort to satisfy more closely the principles of simple structure in oblique solutions. In Thurstone's system the pth reference axis, say Λ_p $(p=1,\ldots,m)$ corresponds to the pth oblique factor, say P_p $(p=1,\ldots,m)$, and is orthogonal to the hyperplane formed by the remaining factors. For example, when $m=3$, Λ_1 is orthogonal to the plane formed by P_2 and P_3, Λ_2 is orthogonal to the plane formed by P_1 and P_3, and Λ_3 is orthogonal to the plane formed by P_1 and P_2. When $m=2$, Λ_1 is perpendicular to P_2 and Λ_2 is perpendicular to P_1. An oblique solution based on the reference axes yields a pattern (called a reference pattern) and a structure (called a reference structure) that are different from, but related to, **P** and **S** defined earlier. If we denote the reference pattern by **W** and the reference structure by **V**, it can be shown that $\mathbf{W}=\mathbf{S}\mathbf{D}^{-1}$ and $\mathbf{V}=\mathbf{P}\mathbf{D}$, in which **D** is a diagonal matrix of correlations between Λ_p and P_p $(p=1,\ldots,m)$. To distinguish between the P_ps and the Λ_ps, Thurstone referred to the P_ps as *primary* factors and to the Λ_ps merely as reference axes. Harman (1976) calls **S** the structure, but refers to **P** as the *primary* pattern.

Carroll's original analytical procedure for oblique rotation, introduced in 1953, was an adaptation of the orthogonal quartimax method to permit correlated factors. The criterion he suggested was the minimization of

$$N = \sum_{j=1}^{n} \sum_{p,q=1}^{m} v_{jp}^2 v_{jq}^2 \qquad (p<q), \tag{8.4.9}$$

in which v_{jp} is an element of the reference structure matrix, **V**, defined in the preceding paragraph, and is the correlation between the jth variable and the pth reference axis. Note that this criterion bears a striking resemblance to the second term on the left in Equation 8.4.2. Carroll suggested calling this the *quartimin* criterion.

In 1957, Carroll, following Kaiser's (1956) oblique version of the varimax criterion, suggested a somewhat different, but related, criterion which he called *covarimin*. The quantity to be minimized was

$$C^* = \sum_{p,q=1}^{m} \left[\sum_{j=1}^{n} v_{jp}^2 v_{jq}^2 - \frac{1}{n} \sum_{j=1}^{n} v_{jp}^2 \sum_{j=1}^{n} v_{jq}^2 \right] \qquad (p<q). \tag{8.4.10}$$

Experience with both quartimin and covarimin suggested that the former generally gave solutions that were too oblique (factors too highly correlated) while solutions from the latter were usually too orthogonal (factors not sufficiently correlated). The result was the oblimin criterion, proposed by Carroll (1960), of which quartimin and covarimin are special cases. The oblimin criterion is the minimization of

$$B^* = \sum_{p,q=1}^{m} \left[\sum_{j=1}^{n} v_{jp}^2 v_{jq}^2 - \frac{\gamma}{n} \sum_{j=1}^{n} v_{jp}^2 \sum_{j=1}^{n} v_{jq}^2 \right] \qquad (p<q). \tag{8.4.11}$$

Note that when $\gamma = 1$, $B^* = C^*$ and when $\gamma = 0$, $N = C^*$. Carroll also suggested using $\gamma = .5$, which he called the *biquartimin* criterion. This has been found to yield the most satisfactory solutions in terms of simple structure, although the optimum value of γ for a particular data set may be some other value between 0 and 1. Unfortunately, there is no known a priori analytical method for determining the optimum value of γ for a given set of data.

When the v_{jp}^2s are "normalized" through division by communalities, the result is

$$B = \sum_{p,q=1}^{m} \left[\sum_{j=1}^{n} \left(\frac{v_{jp}^2}{h_j^2} \right) \left(\frac{v_{jq}^2}{h_j^2} \right) - \frac{\gamma}{n} \sum_{j=1}^{n} \frac{v_{jp}^2}{h_j^2} \sum_{j=1}^{n} \frac{v_{jq}^2}{h_j^2} \right] \qquad (p<q). \tag{8.4.12}$$

Note that in the orthogonal case and when $\gamma = 1$, this is equivalent to Equation 8.4.3, because then the pattern and structure matrices are identical, so that $v_{jp}^2 = b_{jp}^2$.

Direct Oblimin. The oblimin criteria based on Equations 8.4.11 and 8.4.12 are expressed in terms of the elements of the reference structure, **V**. In terms of interpretation, interest is usually centered on the pattern matrix, **P**, rather than on **V**, but because $\mathbf{P} = \mathbf{V}\mathbf{D}^{-1}$, simple structure expressed in terms of **V** results in simple structure in terms of **P**. Jennrich and Sampson (1966), however, proposed a criterion expressed directly in terms of the elements of

P, namely, the minimization of

$$F(P)=\sum_{p,q=1}^{m}\left[\sum_{j=1}^{n} b_{jp}^2 b_{jq}^2-\frac{\delta}{n}\sum_{j=1}^{n} b_{jp}^2\sum_{j=1}^{n} b_{jq}^2\right] \quad (p<q) \qquad (8.4.13)$$

in which b_{jp} is an element of **P** and δ is the analogue of γ in Equation 8.4.11. Because this criterion is closely related to 8.4.11, but is expressed directly in terms of the elements of **P**, it is called the *direct oblimin* criterion*. There is no simple relationship between the value of γ in Equation 8.4.11 and that of δ in Equation 8.4.13. For $\delta=0$, the factors are highly correlated and they become less highly correlated as δ becomes negative. We suggest that for most problems the optimum value of δ lies between 0 and -5, although some trial and error may have to be employed to determine where the optimum value might be located.

We can make no general statement concerning the relative effectiveness of indirect and direct oblimin methods. When feasible, both should be tried

*A "normalized" version of Equation 8.4.13 may be obtained by dividing b_{jp}^2 and b_{jq}^2 by h_j^2.

TABLE 8.4.3 BIQUARTIMIN SOLUTION FOR 24 PSYCHOLOGICAL TESTS

Test	*Reference structure*: **V**				*Primary Pattern*: **P**			
					Verbal	*Speed*	*Deduction*	*Memory*
j	Λ_1	Λ_2	Λ_3	Λ_4	P_1	P_2	P_3	P_4
1	.014	.094	.598	.051	.015	.103	**.666**	.058
2	.028	.007	.392	.029	.031	.008	**.436**	.033
3	.067	−.052	.498	.000	.074	−.057	**.555**	.000
4	.106	.011	.490	−.022	.117	.012	**.545**	−.025
5	**.675**	.110	.075	−.008	**.748**	.120	.083	−.009
6	**.670**	−.016	.089	.084	**.742**	−.018	.099	.095
7	**.755**	.063	.068	−.061	**.836**	.069	.076	−.069
8	**.449**	.159	.256	−.019	**.497**	.174	.285	−.021
9	**.722**	−.111	.079	.130	**.800**	−.122	.087	.146
10	.080	**.650**	−.186	.132	.089	**.711**	−.208	.149
11	.074	**.525**	−.054	.245	.082	**.574**	−.060	.276
12	−.072	**.641**	.132	−.011	−.080	**.701**	.146	−.012
13	.074	**.524**	.307	−.082	.082	**.573**	.341	−.092
14	.135	.070	−.069	**.437**	.149	.077	−.076	**.492**
15	.029	−.019	.052	**.452**	.032	−.021	.058	**.518**
16	−.032	−.007	.330	.359	−.035	−.008	.368	**.404**
17	.034	.079	−.053	**.579**	.038	.086	−.059	**.652**
18	−.130	.161	.217	**.460**	−.144	.176	.242	**.518**
19	.035	.066	.151	.324	.039	.072	.168	.365
20	.240	−.002	.378	.143	.265	−.002	**.421**	.161
21	.038	.289	.318	.141	.042	**.316**	**.353**	.159
22	.250	−.071	.311	.258	.277	−.078	**.346**	.291
23	.223	.095	**.464**	.085	.247	.104	**.516**	.095
24	.234	.341	.082	.211	.260	**.373**	.091	.237

Factor	*Correlations among factors*			
1	1.000	.262	.341	.337
2		1.000	.295	.338
3			1.000	.329
4				1.000

to see which yields the most easily interpretable solution. It seems clear that neither is superior for all data sets.

Examples. Again we shall refrain from describing computational procedures for these methods on the assumption that computations will be carried out with the aid of a computer. It will be of interest, however, to look at the oblique solution based on the previous example of the 24 psychological tests. The biquartimin solution is given in Table 8.4.3 and the direct oblimin with $\delta = 0$ in Table 8.4.4, each again based on the first four common factors.

Again, only loadings (or correlations) greater than or equal to .4 are shown in bold face. Although the two oblique solutions do not differ markedly from each other, they both seem somewhat better than the varimax solution of Table 8.4.2 by the criteria of simple structure.

Other Methods of Oblique Rotation. Although a number of other methods of oblique rotation have been proposed, space does not permit a full discussion of them here. Of particular interest because of its elegance and

TABLE 8.4.4 DIRECT OBLIMIN SOLUTION FOR 24 PSYCHOLOGICAL TESTS ($\delta = 0$)

Test	*Structure* : **S**				*Primary pattern* : **P**			
j	r_{jP_1}	r_{jP_2}	r_{jP_3}	r_{jP_4}	P_1	P_2	P_3	P_4
1	.353	.331	**.731**	.344	.008	.113	**.680**	.035
2	.234	.170	**.478**	.198	.029	.026	**.458**	−.001
3	.283	.104	**.574**	.244	.053	−.095	**.564**	.039
4	.364	.202	**.577**	.232	.149	.014	**.523**	−.037
5	**.793**	.345	.365	.347	**.760**	.107	.009	−.011
6	**.816**	.225	.383	**.412**	**.785**	−.069	.024	.103
7	**.849**	.283	.361	.284	**.872**	.041	.009	−.096
8	**.674**	.361	**.481**	.345	**.547**	.131	.211	−.012
9	**.857**	.208	.376	**.404**	**.850**	−.094	.004	.085
10	.289	**.829**	.048	.308	.118	**.856**	−.273	.045
11	.336	**.617**	.251	**.503**	.074	**.493**	−.045	**.305**
12	.190	**.734**	.294	.259	−.084	**.734**	.124	−.029
13	.359	**.621**	.515	.292	.069	**.519**	**.359**	−.070
14	.318	.199	.177	**.591**	.126	−.032	−.095	**.588**
15	.244	.186	.220	**.554**	.022	−.029	.005	**.554**
16	.265	.205	**.501**	**.596**	−.083	−.069	.357	**.518**
17	.284	.333	.198	**.631**	.031	.122	−.086	**.606**
18	.220	**.449**	**.406**	**.554**	−.136	.267	.221	**.425**
19	.282	.279	.335	**.438**	.049	.096	.162	.319
20	.523	.252	**.529**	**.449**	.304	−.016	.323	.204
21	.361	**.528**	**.503**	.386	.061	.376	.329	.092
22	**.511**	.274	**.516**	**.445**	.290	.017	.308	.199
23	**.545**	.377	**.627**	**.425**	.279	.125	**.433**	.096
24	**.511**	**.588**	.346	**.465**	.294	**.421**	.026	.175

Factor	*Correlations among factors*			
P_1	1.000	.316	.432	.414
P_2		1.000	.296	.374
P_3			1.000	.387
P_4				1.000

flexibility is the *orthoblique* method of Harris and Kaiser (1964), so named because an analytical oblique solution is obtained through a series of orthogonal transformations. They treat several cases, including the orthogonal Case I, the *independent-clusters* and the *proportional* (Case II) solutions, and the completely general Case III. The independent-clusters solution has been found very effective when, as the name suggests, the variables can be expected (or observed) to fall into several quite distinct subsets. The proportional solution, on the other hand, is useful when factors are based on overlapping clusters of variables, that is, when the structure of the set is more complex.

Other methods which have been found useful are Kaiser's (1970) *second generation Little Jiffy*, the *oblimax* method of Saunders (1961), *promax* (Hendrickson and White, 1964), and *maxplane* (Cattell and Muerle, 1960; Eber, 1966). Again, we emphasize that no particular method is best for all data sets and refer the reader to the Hakstian and Abell (1974) article for a thorough critique and comparison of these methods, as well as of those discussed in more detail in this section.

8.5 FACTOR SCORES

Before concluding this chapter we should indicate the procedures for obtaining what are usually termed *factor scores*. As we explained earlier, a factor may be considered a variable that is indirectly rather than directly measured. Because it represents some definable property of an object or person, that is, a trait, characteristic, or construct, it should be possible to associate with an individual a value, or score, on the factor. Such a value must be based on the individual's values or scores on the directly measured variables, the X_js.

For the component analysis model of Equation 8.1.1, the factor scores for the ith individual were denoted by F_{ij} $(i=1,\ldots,N;\ j=1,\ldots,n)$. The matrix formulation of that model was given by Equation 8.1.15 for all N individuals, that is,

$$\mathbf{Z}=\mathbf{AF},$$

or

$$\mathbf{AF}=\mathbf{Z}.$$

Since $\mathbf{A}$ in this case is an $n\times n$ non-singular matrix, it has an inverse*. Premultiplying by $\mathbf{A}^{-1}$ yields

$$\mathbf{A}^{-1}\mathbf{AF}=\mathbf{A}^{-1}\mathbf{Z},$$

so that

$$\mathbf{F}=\mathbf{A}^{-1}\mathbf{Z}. \tag{8.5.1}$$

*This would not be true if some eigenvalues of **A** were equal to zero. Zero eigenvalues occur when one X is a multiple of another, or when one X is the sum of other Xs. If such variables are removed from the analysis, the remaining variables will yield a non-singular **A**.

Thus, when a complete component-analysis solution is obtained, that is, when all n components are calculated, factor scores can be computed directly by means of Equation 8.5.1. The values obtained, in standard score form, are the elements of Equation 8.1.11 with $m=n$.

In most realistic problems, however, the number of common factors to be retained is considerably less than the number of variables. In that case, $\mathbf{A}$ will be of order $n \times m (m<n)$. If we premultiply both sides of Equation 8.1.15 by $\mathbf{A}'$ (of order $m \times n$), the result is

$$\mathbf{A}'\mathbf{Z} = \mathbf{A}'\mathbf{A}\mathbf{F},$$

which can be written

$$\mathbf{A}'\mathbf{A}\mathbf{F} = \mathbf{A}'\mathbf{Z}. \tag{8.5.2}$$

Now the product $\mathbf{A}'\mathbf{A}$ is a square matrix of order $m \times m$ and Equation 8.5.2 can be solved for $\mathbf{F}$ by premultiplying by $(\mathbf{A}'\mathbf{A})^{-1}$. Thus, we have

$$(\mathbf{A}'\mathbf{A})^{-1}\mathbf{A}'\mathbf{A}\mathbf{F} = (\mathbf{A}'\mathbf{A})^{-1}\mathbf{A}'\mathbf{Z},$$

so that

$$\mathbf{F} = (\mathbf{A}'\mathbf{A})^{-1}\mathbf{A}'\mathbf{Z}. \tag{8.5.3}$$

Again, the values obtained are the elements, in standard form, of Equation 8.1.11. The inverse of $\mathbf{A}'\mathbf{A}$ is quite easy to obtain because the m components retained are orthogonal. As pointed out in Section 8.3, this means that the sum of the products of elements of any pair of columns of $\mathbf{A}$ will be equal to zero, while the sum of squares of elements in any column of $\mathbf{A}$ is equal to the eigenvalue for that component. Therefore, $\mathbf{A}'\mathbf{A}$ is a diagonal matrix of eigenvalues of the components retained and its inverse is a diagonal matrix of the reciprocals of those eigenvalues.

When an initial solution, represented by $\mathbf{A}$, has been rotated, for example, by the varimax method, factor scores can also be computed for the rotated factors. Let $\mathbf{B}$ denote the $n \times m$ $(m<n)$ matrix of loadings for the rotated solution (for example, those shown in Table 8.4.2). Also let $\mathbf{G}$ denote the $m \times N$ matrix of factor scores for the N individuals on the m factors, based on the rotated solution. Then the analogue of Equation 8.1.15 is

$$\mathbf{Z} = \mathbf{B}\mathbf{G}. \tag{8.5.4}$$

Premultiplying both sides by $\mathbf{B}'$ and transposing yields

$$\mathbf{B}'\mathbf{B}\mathbf{G} = \mathbf{B}'\mathbf{Z}.$$

Then, solving for $\mathbf{G}$, we have

$$\mathbf{G} = (\mathbf{B}'\mathbf{B})^{-1}\mathbf{B}'\mathbf{Z}, \tag{8.5.5}$$

which is analogous to Equation 8.5.3.

The factor scores obtained indirectly by means of Equations 8.5.3 or 8.5.5 may be analyzed in the same ways as direct measurements, using, for example analysis of variance, correlations, or t tests. In fact, a procedure frequently employed is to reduce a large number of related variables to a factor measurement for purposes of analysis. Such measures are often superior to direct measures because of increased reliability. We should point out, however, that the use of Equations 8.5.3 and 8.5.5 is appropriate only when the initial solution is based on a principal-component analysis of the correlation matrix, **R**. When $\mathbf{R}_h$, with estimated communalities in the diagonal, is analyzed, factor scores must be estimated by multiple regression or other appropriate methods. A discussion of procedures that may be used in this case can be found in Harman (1976, pp. 368–76).

8.6 EXERCISES

8.1. Refer to the data matrices for the Rhode Island Pupil Identification Scale (RIPIS), Parts I and II, given in Exercise 5.3.

(a) Perform principal-component analyses (ones in the diagonal of **R**) on X_1 through X_5 (Part I), Y_1 through Y_5 (Part I), X_1 through X_4 (Part II), and Y_1 through Y_4 (Part II). How many common factors are obtained in each of these analyses?

(b) The combination of a principal-components analysis followed by a varimax rotation of factors having eigenvalues greater than or equal to one is commonly referred to as "Little Jiffy." The term seems to be due to Kaiser, who used it to describe what he most often advised students to do when they needed to carry out an exploratory factor analysis. Using the results of (a), complete the "Little Jiffy" analysis for each of the four sets of variables. What names would you give to the factors obtained? (Note: The names of the variables are given in Exercise 5.4.)

(c) Interpret the results in (a) and (b), keeping in mind that the data were obtained in a study of the reliability of the RIPIS.

8.2. Below is a 9×9 correlation matrix based on 205 independent observations.

(a) Perform a principal-components analysis (ones in the diagonal of **R**) on these data. How many common factors are obtained?

(b) Perform a varimax rotation of the common factors obtained in (a).

(c) Also perform an oblique rotation using the biquartimin method (Section 8.4.2). Does inspection of the correlations between factors indicate that an oblique rotation is warranted?

(d) Compare the pattern matrices obtained in (b) and (c). Also compare the structure and pattern matrices obtained in (c). Do these comparisons support your answer to the question in (c)?

	X_1	X_2	X_3	X_4	X_5	X_6	X_7	X_8	X_9
X_1	1.000								
X_2	.836	1.000							
X_3	.793	.852	1.000						
X_4	.586	.634	.649	1.000					
X_5	.610	.643	.667	.907	1.000				
X_6	.603	.639	.684	.894	.908	1.000			
X_7	.376	.393	.411	.356	.344	.359	1.000		
X_8	.424	.421	.451	.382	.413	.413	.690	1.000	
X_9	.469	.503	.525	.458	.483	.519	.705	.806	1.000

8.3. In a study by Merenda, Novack, and Bonaventura (1976), the California Test of Mental Ability, Primary Form, was administered to a sample of 716 children in grades *K*, 1, and 2. The correlations obtained were:

Test	*LR*	*NR*	*VC*	*M*
Logical Reasoning (LR)	1.00			
Numerical Reasoning (NR)	.47	1.00		
Verbal Concepts (VC)	.45	.47	1.00	
Memory (M)	.36	.33	.48	1.00

(a) Check the conditions (Equation 8.2.5) that must be met for describing these relationships in terms of a single general factor. Does it appear that the Spearman model (Section 8.2.1) would be appropriate for factor analyzing this matrix?
(b) Regardless of your answer to (a), apply the Spearman model to obtain a solution. For what percent of the total variance does the single factor account?
(c) Carry out a component analysis of the same data. How does the result compare with that in (b)?
(d) What is your conclusion about the factorial composition of these variables, based on (a), (b), and (c)?

8.4. In a study by Merenda, Novack, and Bonaventura (1972) the California Test of Mental Maturity (CTMM), Form S, Level 1, and the Stanford Achievement Test (SAT), Form W, Primary 1, were administered to a sample of 279 first-grade children. The authors used canonical correlation analysis to determine the nature and degree of overlap between these two batteries. The matrix of correlations was:

	CTMM				*SAT*				
Test	1	2	3	4	5	6	7	8	9
(CTMM)									
1. Logical Reasoning									
2. Numerical Reasoning	.55								
3. Verbal Concepts	.51	.41							
4. Memory	.39	.32	.51						
(SAT)									
5. Word Meaning	.38	.46	.31	.19					
6. Paragraph Meaning	.37	.45	.29	.20	.80				
7. Vocabulary	.42	.41	.49	.42	.46	.44			
8. Spelling	.33	.42	.29	.14	.74	.68	.46		
9. Word Study Skills	.40	.45	.37	.22	.71	.65	.55	.73	
10. Arithmetic	.42	.60	.38	.31	.59	.55	.53	.55	.65

(a) Factor analyze this matrix using Little Jiffy (see Exercise 8.1). Describe the factorial composition using this method.

(b) Repeat the factor analysis using principal components followed by an oblique rotation such as biquartimin (see Section 8.4.2). Compare this solution with that in (a).
(c) Based on (a) and (b), what conclusion would you reach concerning the relationship between these two test batteries? Compare your conclusion based on factor analysis methods to that of Merenda, Novack, and Bonaventura (1972) using canonical correlation methods.

9/ Multivariate Analysis of Categorical Data

The previous chapters of this book have been devoted for the most part to methods appropriate for measuring and interpreting relationships among variables whose values are determined by taking measurements of some kind. The purpose of this chapter is to consider questions concerning relationships among variables that are *qualitative* rather than quantitative. That is, the result of taking an observation on such a variable is a *category* rather than a number. Of course, the categories may be determined by a measurement; for example, the outcome "mastery" may mean a score of 80% or more on an achievement test, while "nonmastery" is assigned to scores below 80%.

In many investigations in the behavioral sciences, the data of interest concern variables that are qualitative, or categorical, by their nature. Various types of questions of interest lead to the collection and analysis of such data. For example, in the late 1960s the alarming increase in both crime rates and drug abuse prompted the Bureau of Narcotics and Dangerous Drugs (BNDD) of the U.S. Department of Justice to undertake a study in order to establish whether the tendency of the general public to associate the increase in drug usage with that of crime could be substantiated by empirical data. Crimes of violence cause particular consternation on the part of the public, whose stereotype of a "dope addict" includes a willingness to go to any extremes in order to obtain the wherewithal for a "fix." The BNDD-sponsored research, carried out by the Research Triangle Institute, included a study of new arrests, excluding individuals arrested for minor violations such as disorderly conduct, drunken driving, etc. Arrestees charged with *only* drug-related violations were also excluded. Data were collected in six cities: Chicago, New Orleans, San Antonio, St. Louis, Los Angeles, and New York. Details of the background for the study, the methods used, and the findings can be found in the Report of the BNDD (Eckerman et al., 1971).

In Table 9.01, adapted from Tables VIII-3 and VIII-4 of the BNDD report, 1889 new arrestees in all six cities are cross-classified by heroin use and crime categories.

TABLE 9.0.1 CROSS-CLASSIFICATION OF 1889 ARRESTEES IN SIX CITIES BY LEVEL OF HEROIN USE AND TYPE OF CRIME

Level of heroin use	*Crime classification*											
	Serious crimes against persons		*Robbery*		*Less serious crimes against persons*		*Property crimes*		*All other crimes*		*Total in heroin use category*	
	N	%	*N*	%	*N*	%	*N*	%	*N*	%	*N*	%
Current user	30	6.5	94	20.4	14	3.0	237	51.4	86	18.7	461	100.0
Past user	14	9.9	20	14.2	5	3.5	75	53.2	27	19.1	141	99.9
Other drug user	93	15.2	94	15.4	46	7.5	253	41.5	124	20.3	610	99.9
Non-drug user	163	24.1	79	11.7	77	11.4	265	39.1	93	13.7	677	100.0
Total in crime category	300	15.9	287	15.2	142	7.5	830	43.9	330	17.5	1889	100.0

Detailed definitions of the categories may be found in the report. These data do not in themselves provide information concerning the incidence of serious crimes against persons, such as homicide or forcible rape, among heroin users *in general* as compared with that among nonusers in general, although the report *does* deal with this question. The data do, however, provide some information about the relationship between drug habits and type of crime *among arrestees*.

The familiar χ^2 statistic was computed for the data and found to be 121.90, which for 12 degrees of freedom has a significance probability of less than .001, indicating that among arrestees there is a relationship between crime patterns and heroin use. The table shows that, although "property crimes" is the modal crime category for each of the four categories of heroin use (39.1% for nonusers to 53.2% for past users), the greatest differences are to be found in the two categories of crimes against persons. The arrested non-drug users committed 35.5% of their crimes in these categories, in contrast to 9.5% for the current heroin users.

The χ^2 statistic used here has the usual structure:

$$\chi^2 = \sum_{\substack{\text{over} \\ \text{all} \\ \text{cells}}} \frac{(O-E)^2}{E},$$

in which O represents the observed cell frequency and E the "expected" cell frequency under statistical independence of the classificatory variables. The estimated cell probabilities are products of estimated "row" (heroin) probabilities by estimated "column" (crime) probabilities. The estimates are the maximum likelihood estimates of these probabilities, that is, the corresponding observed row and column relative frequencies. Thus, the estimated probability of the first cell cross-classification (current users $\times$ serious crime against persons) is $(461/1889)(300/1889) = .03876$, and the expected cell frequency for this combination of drug $\times$ crime categories is $1889\,(.03876) = 73.2$. The contribution of this particular cell to the total χ^2 statistic is

$(30-73.2)^2/73.2=25.50$. Contributions of the other cells to the total χ^2 are determined in a similar manner.

Simply constructed statistics like the χ^2 computed above help to solve many problems that arise in the analysis of categorical data, but there are other interesting problems that require somewhat more complicated techniques. For example, in the heroin-crime problem we might be interested in questions relating to the conditional independence of type of crime and heroin use, given the city of the arrest, or to the conditional independence of type of crime and heroin use, given the age of the arrestee, or perhaps given both the city of the arrest and the age of the arrestee. Such questions are analogous to questions concerning partial correlation in the case of continuous variables. Thus, problems of interest may involve more than two classificatory variables and their solution may require more than the simple kind of calculations we have illustrated.

Although interest in the multivariate analysis of categorical data goes back at least to the work of Bartlett (1935) it was in the 1960s that the subject became a very popular one for theoretical research, as well as an important tool for the practical researcher, in the large-sample case. It is the purpose of this chapter to consider some of the techniques that can be applied to problems that may not have easily accessible simple solutions. It is by no means intended to cover the entire field. Many papers have been written and several books have recently appeared that are devoted to various aspects of the analysis of multivariate categorical data in the large-sample case. Recent books by Fienberg (1977), Reynolds (1977), Bishop, Fienberg, and Holland (1975), and Haberman (1974) consider a vast number of problems and provide an extensive bibliography on the subject.

We shall discuss some of the techniques that have been proposed for solving a number of problems, providing in each case data to which the techniques have been applied, along with numerical solutions. Before discussing the problems for which, in some cases, more than one method of solution is proposed, we review briefly the definition of multinomial distributions and some of their properties, because the underlying models consist in general of a single multinomial or a sequence of multinomial distributions.

We hope that the reader will be patient with the calculations performed along the way. They will be used later in the chapter, although they are inserted where we hope they will most easily be followed.

9.1 THE MULTINOMIAL DISTRIBUTIONS: DEFINITIONS AND NOTATION

Consider a random trial capable of c different outcomes: $O_1, O_2, \ldots, O_c$. Assume that the probability that outcome O_j will occur on each of n

independent trials is p_j $(0<p_j<1,\ \Sigma_{j=1}^{c} p_j=1)$. Note that an "outcome" is not necessarily restricted to a single variable, but may consist of results on more than one variable, such as "current heroin user who committed robbery" (representing the joint outcome on each of the two variables, heroin use and crime classification).

Let $X_j^{(n)}$ be the number of times outcome O_j is observed in n independent repeated trials, $(j=1,2,\ldots,c)$; $\Sigma_{j=1}^{c} X_j^{(n)}=n$. The random vector

$$\mathbf{X}^{(n)}=\begin{bmatrix} X_1^{(n)} \\ X_2^{(n)} \\ \vdots \\ X_c^{(n)} \end{bmatrix} \tag{9.1.1}$$

is called a *multinomial vector* with c classes and n trials. The $X_j^{(n)}$s are called *multinomial variables* and the p_js $(j=1,2,\ldots,c)$ are called the *class probabilities*. The distribution of $\mathbf{X}^{(n)}$, that is, the joint distribution of the components of $\mathbf{X}^{(n)}$, is called the *multinomial distribution* with parameters $n,p_1,\ldots,p_c$, where $\Sigma_{j=1}^{c} p_j=1$, and n is a positive integer. The distribution is given by

$$P(X_1^{(n)}=x_1,\ X_2^{(n)}=x_2,\ldots,X_c^{(n)}=x_c)$$

$$=\begin{cases} \dfrac{n!}{x_1!x_2!\ldots x_c!}\,p_1^{x_1}p_2^{x_2}\cdots p_c^{x_c},\ 0<p_j<1,\ \displaystyle\sum_{j=1}^{c} p_j=1, \\ \quad \text{for } x_j\text{s nonnegative integers} \\ \quad \text{such that } \displaystyle\sum_{j=1}^{c} x_j=n, \\ 0, \text{ otherwise.} \end{cases} \tag{9.1.2}$$

The reader may recognize the probabilities in Equation 9.1.2 as the terms in the expansion of $(p_1+p_2+\ldots+p_c)^n$. For the case $n=2$, we have the binomial distribution, for $n=3$, the trinomial, and so on.

It is easily verified that

$$\left.\begin{aligned} &E(X_j^{(n)})=np_j,\ (j=1,2,\ldots,c), \\ &\text{Var}(X_j^{(n)})=np_j(1-p_j),\ (j=1,2,\ldots,c), \\ &\text{Cov}(X_j^{(n)},\ X_k^{(n)})\equiv\sigma(X_j^{(n)},\ X_k^{(n)})=-np_jp_k,\ (k\neq j), \\ &\text{and} \\ &\rho(X_j^{(n)},\ X_k^{(n)})=\frac{\sigma(X_j^{(n)},\ X_k^{(n)})}{\sigma_{X_j^{(n)}}\sigma_{X_k^{(n)}}}=-\sqrt{\frac{p_jp_k}{(1-p_j)(1-p_k)}}\ . \end{aligned}\right\} \tag{9.1.3}$$

If we let $\hat{p}_j^{(n)} = X_j^{(n)}/n$ be the observed proportion in the jth category, $j = 1,2,\ldots,c$, then

$$\left.\begin{aligned}
&E(\hat{p}_j^{(n)}) = p_j, \ (j=1,2,\ldots,c),\\
&\mathrm{Var}(\hat{p}_j^{(n)}) = \frac{p_j(1-p_j)}{n}, \ (j=1,2,\ldots,c),\\
&\mathrm{Cov}(\hat{p}_j^{(n)}, \hat{p}_k^{(n)}) = -\frac{p_j p_k}{n}, \ (k \neq j),\\
&\text{and}\\
&\rho(\hat{p}_j^{(n)}, \hat{p}_k^{(n)}) = -\sqrt{\frac{p_j p_k}{(1-p_j)(1-p_k)}}\ .
\end{aligned}\right\} \tag{9.1.4}$$

Thus, the underlying model for the 1889 cross-classified arrestees considered in the example of Table 9.0.1 is a multinomial distribution with 20 classes ($c = 20$), each class consisting of a heroin use $\times$ crime pair, and with $n = 1889$. It may be appropriate to point out here that we might be confronted with the same data set if the sampling scheme had been different. Suppose that the investigators had considered dividing all the arrestees into heroin use categories and then taken a random sample of 461 current users, a different random sample of 141 past users, and so on. (We are ignoring the fact that *all* arrestees except for the previously mentioned cases have been included.) We would then have *four* multinomial distributions, one corresponding to each heroin-use category, each multinomial distribution consisting of five classes corresponding to the crime classifications. The question of interest would then be whether the multinomials were homogeneous, that is, whether the probability of each class (serious crimes against persons, robbery, and so on) was the same in each of the five multinomial distributions, one for each of the five heroin-use patterns. This question of homogeneity could be answered by the very same χ^2 test as that outlined in the above example. This is fortunate, for in many cases we are confronted with data that come to us in the form of an $I \times J$ contingency table without sufficient information to enable us to decide whether the appropriate model is a single multinomial with IJ classes or a sequence of I independent multinomials, each with J classes. In the following, we will not make the distinction unless it is important for the analysis or the interpretation of the results.

9.2 LARGE-SAMPLE APPROXIMATIONS FOR A MULTINOMIAL DISTRIBUTION

A very important property of a multinomial distribution is that when the number of trials, n, is very large, we may use a well-chosen multivariate normal distribution to approximate the probabilities of interest in the multinomial case.

Recall that a random vector

$$\mathbf{X} = \begin{bmatrix} X_1 \\ X_2 \\ \vdots \\ X_c \end{bmatrix} \tag{9.2.1}$$

is said to have a c-dimensional normal distribution if the value of the joint density function at the point

$$\mathbf{x} = \begin{bmatrix} x_1 \\ x_2 \\ \vdots \\ x_c \end{bmatrix} \tag{9.2.2}$$

has the value

$$f(\mathbf{x}) = (2\pi)^{-c/2} |\boldsymbol{\Sigma}_X|^{-1/2} \exp\left[-\frac{1}{2} (\mathbf{x} - \boldsymbol{\mu})' \boldsymbol{\Sigma}_X^{-1} (\mathbf{x} - \boldsymbol{\mu}) \right], \tag{9.2.3}$$

in which $\mathbf{x}$ is any c-dimensional real vector,

$$\boldsymbol{\mu} = \begin{bmatrix} \mu_1 \\ \mu_2 \\ \vdots \\ \mu_c \end{bmatrix} = \begin{bmatrix} E(X_1) \\ E(X_2) \\ \vdots \\ E(X_c) \end{bmatrix} \tag{9.2.4}$$

is a vector of constants, and $\boldsymbol{\Sigma}_X$ is the variance-covariance matrix of the components of $\mathbf{X}$,

$$\boldsymbol{\Sigma}_X = \begin{bmatrix} \sigma_{11} & \sigma_{12} & \cdots & \sigma_{1c} \\ \sigma_{21} & \sigma_{22} & \cdots & \sigma_{2c} \\ \vdots & \vdots & & \vdots \\ \sigma_{c1} & \sigma_{c2} & \cdots & \sigma_{cc} \end{bmatrix}. \tag{9.2.5}$$

$\boldsymbol{\Sigma}_X$ is of rank c with a general element being

$$\sigma_{jk} = E(X_j - \mu_j)(X_k - \mu_k);\ j,k = 1,2,\ldots,c.$$

In particular, $\sigma_{jj} = \mathrm{Var}(X_j)$.

Let

$$Q = (\mathbf{X} - \boldsymbol{\mu})' \boldsymbol{\Sigma}_X^{-1} (\mathbf{X} - \boldsymbol{\mu}). \tag{9.2.6}$$

Then Q has the χ^2 distribution with c degrees of freedom (see, for example, Anderson, 1958).

Now, let $\mathbf{X}^{(n)}$ be a multinomial vector, as in Equation 9.1.1, and let

$$Z_j^{(n)} = \frac{X_j^{(n)} - np_j}{\sqrt{np_j(1-p_j)}},\ j = 1,2,\ldots,c.$$

Then $E(Z_j^{(n)})=0$, $\text{Var}(Z_j^{(n)})=1$, and, for $j \neq k$,

$$\text{Cov}(Z_j^{(n)}, Z_k^{(n)}) = \frac{\text{Cov}(X_j^{(n)}, X_k^{(n)})}{n\sqrt{p_j(1-p_j)p_k(1-p_k)}} = -\sqrt{\frac{p_j p_k}{(1-p_j)(1-p_k)}} = \rho_{jk}.$$

Thus, if $\boldsymbol{\Sigma}_{Z^{(n)}}$ represents the variance-covariance matrix of the vector

$$\mathbf{Z}^{(n)} = \begin{bmatrix} Z_1^{(n)} \\ Z_2^{(n)} \\ \vdots \\ Z_{c-1}^{(n)} \end{bmatrix}, \tag{9.2.7}$$

then

$$\boldsymbol{\Sigma}_{Z^{(n)}} = \begin{bmatrix} 1 & \rho_{12} & \cdots & \rho_{1,c-1} \\ \rho_{12} & 1 & \cdots & \rho_{2,c-1} \\ \vdots & \vdots & & \vdots \\ \rho_{1,c-1} & \rho_{2,c-1} & \cdots & 1 \end{bmatrix}. \tag{9.2.8}$$

Note that we have deleted the cth component corresponding to $X_c^{(n)}$ because $\sum_{j=1}^{c} X_j^{(n)} = n$. This insures that the $Z_j^{(n)}$s $(j=1,2,\ldots,c-1)$ are not linearly dependent, and that the rank of $\boldsymbol{\Sigma}_{Z^{(n)}}$ is $c-1$.

Now, if we let

$$\mathbf{Z} = \begin{bmatrix} Z_1 \\ Z_2 \\ \vdots \\ Z_{c-1} \end{bmatrix} \tag{9.2.9}$$

be a random vector, with a $(c-1)$-dimensional normal distribution having mean vector $\boldsymbol{\mu}=\mathbf{0}$ and variance covariance matrix $\boldsymbol{\Sigma}_Z = \boldsymbol{\Sigma}_{Z^{(n)}}$, then, as n increases indefinitely, the distribution of the vector $\mathbf{Z}^{(n)}$ tends to that of the vector $\mathbf{Z}$ (see, for example, Gnedenko, 1968).

In terms of the $X_j^{(n)}$s, we may interpret this result as follows. For large n, the distribution of the vector

$$\mathbf{X}^{(n)*} = \begin{bmatrix} X_1^{(n)} \\ X_2^{(n)} \\ \vdots \\ X_{c-1}^{(n)} \end{bmatrix} \tag{9.2.10}$$

can be approximated by the $(c-1)$-dimensional normal density, with mean vector

$$\boldsymbol{\mu}^* = \begin{bmatrix} np_1 \\ np_2 \\ \vdots \\ np_{c-1} \end{bmatrix} \tag{9.2.11}$$

and variance-covariance matrix

$$\boldsymbol{\Sigma}^*_{X^{(n)}} = \begin{bmatrix} np_1(1-p_1) & -np_1p_2 & \cdots & -np_1p_{c-1} \\ -np_1p_2 & np_2(1-p_2) & \cdots & -np_2p_{c-1} \\ \vdots & \vdots & & \vdots \\ -np_1p_{c-1} & -np_2p_{c-1} & \cdots & np_{c-1}(1-p_{c-1}) \end{bmatrix}. \tag{9.2.12}$$

Note that the rank of $\boldsymbol{\Sigma}^*_{X^{(n)}}$ is $c-1$.

Now let

$$\mathbf{x}^* = \begin{bmatrix} x_1 \\ x_2 \\ \vdots \\ x_{c-1} \end{bmatrix} \tag{9.2.13}$$

be a vector of nonnegative integers such that $\sum_{j=1}^{c-1} x_j \leq n$. Then, for large n, the value of the joint density function of the components of $\mathbf{X}^{(n)*}$ at the point $\mathbf{x}^*$ may be approximated by

$$f(\mathbf{x}^*) = (2\pi)^{-\frac{c-1}{2}} |\boldsymbol{\Sigma}^*_{X^{(n)}}|^{-\frac{f-1}{2}} \exp\left(-\frac{1}{2}Q^*\right), \tag{9.2.14}$$

where

$$Q^* = (\mathbf{x}^* - \boldsymbol{\mu}^*)' \boldsymbol{\Sigma}^{*-1}_{X^{(n)}} (\mathbf{x}^* - \boldsymbol{\mu}^*). \tag{9.2.15}$$

It follows that, for large n, the approximate distribution of Q^* is the χ^2 distribution with $c-1$ degrees of freedom (see Chernoff, 1956).

It is easily verified that

$$\boldsymbol{\Sigma}^{*-1}_{X^{(n)}} = \frac{1}{np_c} \begin{bmatrix} \frac{p_1+p_c}{p_1} & 1 & \cdots & 1 \\ 1 & \frac{p_2+p_c}{p_2} & \cdots & 1 \\ \vdots & \vdots & & \vdots \\ 1 & 1 & \cdots & \frac{p_{c-1}+p_c}{p_{c-1}} \end{bmatrix}. \tag{9.2.16}$$

Thus,

$$Q^* = \frac{1}{np_c}[x_1 - np_1, x_2 - np_2, \ldots, x_{c-1} - np_{c-1}]$$

$$\cdot \begin{bmatrix} \frac{p_1+p_c}{p_1} & 1 & \cdots & 1 \\ 1 & \frac{p_2+p_c}{p_2} & \cdots & 1 \\ \vdots & \vdots & & \vdots \\ 1 & 1 & \cdots & \frac{p_{c-1}+p_c}{p_{c-1}} \end{bmatrix} \cdot \begin{bmatrix} x_1 - np_1 \\ x_2 - np_2 \\ \vdots \\ x_{c-1} - np_{c-1} \end{bmatrix}$$

$$= \frac{1}{np_c}\left[\Sigma_{j=1}^{c-1}(x_j - np_j) + \frac{p_c}{p_1}(x_1 - np_1) \ldots \Sigma_{j=1}^{c-1}(x_j - np_j) + \frac{p_c}{p_{c-1}}(x_{c-1} - np_{c-1})\right] \cdot \begin{bmatrix} x_1 - np_1 \\ x_2 - np_2 \\ \vdots \\ x_{c-1} - np_{c-1} \end{bmatrix}$$

$$= \frac{1}{np_c}\left\{\left[\Sigma_{j=1}^{c-1}(x_j - np_j)\right]^2 + p_c \Sigma_{j=1}^{c-1} \frac{(x_j - np_j)^2}{p_j}\right\}$$

$$= \frac{(\Sigma_{j=1}^{c-1} x_j - n\Sigma_{j=1}^{c-1} p_j)^2}{np_c} + \sum_{j=1}^{c-1} \frac{(x_j - np_j)^2}{np_j}$$

$$= \frac{[n - x_c - n(1-p_c)]^2}{np_c} + \sum_{j=1}^{c-1} \frac{(x_j - np_j)^2}{np_j}$$

$$= \sum_{j=1}^{c} \frac{(x_j - np_j)^2}{np_j}. \tag{9.2.17}$$

Thus, an algebraic reduction of the quadratic form Q^*, in the exponent of the asymptotic distribution of the vector, $\mathbf{X}^{(n)*}$, yields the well-known χ^2 statistic having the familiar $\Sigma \frac{(O-E)^2}{E}$ form. This particular version is the "goodness-of-fit" χ^2 that would be appropriate for testing the hypothesis that the multinomial probabilities are given constants. This goodness-of-fit χ^2 was first introduced by Karl Pearson (1900).

9.3 ESTIMATION OF MULTINOMIAL CLASS PROBABILITIES AND TESTS OF HYPOTHESES

The problems of interest to us here do not generally involve questions about the *values* of the class probabilities. In most cases of interest, one must estimate the class probabilities under various conditions, for example, conditions imposed by a statistical hypothesis. A statistical hypothesis may not specify the values of the class probabilities, but rather a condition or set of conditions that the class probabilities satisfy. If the underlying model is simply a multinomial distribution and the hypothesis is true, then the statistic formed by replacing the p_js by certain types of estimators still has, for large samples, approximately the χ^2 distribution with d degrees of freedom. Here, d is equal to the number of restrictions imposed by the hypothesis in addition to those imposed by the model, or, in other words, to the reduction in the number of unknown parameters if the hypothesis is true.

There are various ways of estimating the unknown class probabilities when they are subject to restrictions. Some of these involve finding the estimators that minimize the χ^2 expression; others minimize a modified version of the χ^2 expression. Perhaps the most familiar principle of estimation is the maximum likelihood principle proposed by R. A. Fisher (1912). In a fundamental paper on the theory of χ^2 tests, Neyman (1949) showed that a large class of estimators, called best asymptotically normal (BAN) could be used, and that replacing the ps in Equation 9.2.17 by these estimators would give rise to a statistic having a limiting χ^2 distribution. In particular, one type of estimator considered by Neyman minimizes the expression

$$\chi^2_{(N)} = \sum_{j=1}^{c} \frac{(X_j - np_j)^2}{X_j}, \tag{9.3.1}$$

subject to the restrictions imposed by the hypothesis, and he showed that when such estimators of the p_js were substituted in Equation 9.3.1, the resulting statistic had the same limiting distribution, namely, χ^2 with d degrees of freedom. In fact, he showed that as n increases the values of variously computed χ^2 statistics tend to be equal. In the recent literature on the multivariate analysis of categorical data, this particular χ^2 statistic is often referred to as the Neyman χ^2 statistic (see, for example, Bhapkar, 1966) and will be denoted here by $\chi^2_{(N)}$. For those readers interested in the early development of the theory of χ^2 tests, Bhapkar (1966) and Cramér (1946) may be consulted. Killion and Zahn (1974) have provided an extensive bibliography of contingency-table literature from 1900 to 1973.

9.3.1 Estimation by Fisher's Maximum Likelihood and by Minimizing Neyman's χ^2

When there are no restrictions on the class probabilities except that their sum be equal to one, it is easily shown that the estimators that maximize the likelihood function, L, for the outcome $\mathbf{X}$, as defined by Equation 9.1.1 (the superscript n has been omitted for convenience), are the observed class relative frequencies, $\hat{p}_j = X_j/n$. The likelihood function,

$$L = \frac{n!}{X_1!X_2!\ldots X_c!} p_1^{X_1} p_2^{X_2} \ldots p_{c-1}^{X_{c-1}} (1 - p_1 - p_2 - \ldots - p_{c-1})^{X_c} \quad (9.3.2)$$

is maximized for the same values of

$$\mathbf{p} = \begin{bmatrix} p_1 \\ p_2 \\ \vdots \\ p_c \end{bmatrix} \quad (9.3.3)$$

that maximize

$$\ln L = \ln K + X_1 \ln p_1 + X_2 \ln p_2 + \ldots + X_{c-1} \ln p_{c-1} + X_c \ln(1 - p_1 - \ldots - p_{c-1}),$$

where $K = n!/X_1!X_2!\ldots X_c!$.

To maximize $\ln L$, we take successive partial derivatives of $\ln L$ with respect to the p_js $(j = 1, 2, \ldots, c-1)$ and set these equal to zero, giving

$$\frac{\partial \ln L}{\partial p_j} = \frac{X_j}{p_j} - \frac{X_c}{p_c} = 0 \; (i = 1, 2, \ldots, c-1) \quad (9.3.4)$$

or

$$X_j p_c = X_c p_j \; (j = 1, 2, \ldots, c-1).$$

It follows that

$$\Sigma_{j=1}^{c-1} X_j p_c = \Sigma_{j=1}^{c-1} X_c p_j$$

and

$$p_c(n - X_c) = X_c(1 - p_c),$$

which, when solved for p_c, yields

$$\hat{p}_c = \frac{X_c}{n}, \quad (9.3.5)$$

the observed proportion in class c. Because the labelling of the outcome is arbitrary, that is, *any* of the outcomes could be called the cth outcome,

it follows that

$$\hat{p}_j = \frac{X_j}{n}. \quad (9.3.6)$$

To find the estimators of the class probabilities when they are subject to no restrictions other than that their sum is one, we minimize Neyman's chi-square statistic, given by Equation 9.3.1. First, we note that

$$\chi^2_{(N)} = \sum_{j=1}^{c} \frac{(X_j - np_j)^2}{X_j} = \sum_{j=1}^{c} \left(X_j - 2np_j + \frac{n^2 p_j^2}{X_j} \right)$$

$$= n - 2n + n^2 \sum_{j=1}^{c} \frac{p_j^2}{X_j}$$

$$= -n + n^2 \left[\sum_{j=1}^{c-1} \frac{p_j^2}{X_j} + \frac{(1 - p_1 - p_2 - \dots - p_{c-1})^2}{X_c} \right].$$

Then, to find the estimators that minimize $\chi^2_{(N)}$,

we set the partial derivatives of $\chi^2_{(N)}$ with respect to the ps equal to zero, which gives

$$\frac{\partial \chi^2_{(N)}}{\partial p_j} = n^2 \left[\frac{2p_j}{X_j} - \frac{2p_c}{X_c} \right] = 0$$

or

$$p_j X_c = p_c X_j \quad (j = 1, 2, \dots, c-1).$$

Because these are the same equations we obtained for finding the maximum likelihood estimators, the estimators obtained by minimizing Neyman's $\chi^2_{(N)}$, under no restrictions except that $\sum_{j=1}^{c} p_j = 1$, are, in fact, identical to the maximum likelihood estimators.

9.3.2 Testing a Multinomial Hypothesis by Two Methods: Fisher's and Neyman's $\chi^2_{(N)}$

We consider now a well-known problem that arises in the analysis of a quadrinomial whose four categories are determined by the cross-classification of individuals according to two dichotomous characteristics, each of which, for example, may be present or not. The quadrinomial probabilities, as well as the outcomes of n independent trials, may be conveniently arranged in a four-fold, or 2×2, contingency table, as in Table 9.3.1.

TABLE 9.3.1 PROBABILITIES OF PRESENCE AND ABSENCE OF TWO CHARACTERISTICS, *A* AND *B*

	Characteristic A		
Characteristic B	Present	Absent	Total
Present	p_1	p_2	p_1+p_2
Absent	p_3	p_4	p_3+p_4
Total	p_1+p_3	p_2+p_4	1

The question of interest is whether the probability that A is present is the same as the probability that B is present. In terms of the class probabilities, we are interested in whether $p_1+p_2=p_1+p_3$, or, equivalently, whether $p_2=p_3$.

We shall present several ways of testing H_0: $p_3=p_2$, using the same data to illustrate each of the methods. The data consist of the responses of 46 individuals to each of three drugs and the question of interest was whether the probability of a favorable response was the same for all three drugs. Cochran (1950) presented the data and derived a statistical test for the hypothesis that the probability of a favorable response is the same for all three drugs. The same data have been analyzed by others, including Grizzle, Starmer, and Koch (1969). We shall present these data in full in a later section of this chapter, along with a method of answering the question of interest to Cochran. For the present, we consider only the responses to two of the drugs, A and C. The data are given in Table 9.3.2.

We have said in Section 9.1 that an appropriate test for H_0: $p_2=p_3$ is

$$\chi^2=\sum_{j=1}^{4}\frac{(X_j-n\hat{p}_j)^2}{n\hat{p}_j},$$

in which X_j is the number observed in class j; the classes are defined by the subscripts of the ps in Table 9.3.1, and $\hat{p}_j$ is the maximum likelihood estimate of p_j if H_0 is true. This statistic has approximately the χ^2 distribution with one degree of freedom, because the hypothesis imposes one additional restriction, $p_2=p_3$, to those imposed by the assumptions $\Sigma_{j=1}^{4}p_j=1$.

To find the maximum likelihood estimators of the p_js under H_0, we have to maximize

$$L=\frac{n!}{X_1!X_2!X_3!X_4!}p_1^{X_1}p_2^{X_2}p_2^{X_3}(1-p_1-2p_2)^{X_4} \tag{9.3.7}$$

because $p_2=p_3$ or, equivalently, to maximize

$$\ln L=K'+X_1\ln p_1+(X_2+X_3)\ln p_2+X_4\ln(1-p_1-2p_2) \tag{9.3.8}$$

where K' does not depend on the p_js.

TABLE 9.3.2 RESPONSES OF 46 INDIVIDUALS TO TWO DRUGS *A* AND *C*

	Response to Drug A		
Response to Drug C	Favorable	Unfavorable	Total
Favorable	8	8	16
Unfavorable	20	10	30
Total	28	18	46

Taking partial derivatives of $\ln L$ with respect to p_1 and p_2 and setting them equal to zero gives

$$\frac{\partial \ln L}{\partial p_1}=\frac{X_1}{p_1}-\frac{X_4}{1-p_1-2p_2}=0 \tag{9.3.9}$$

and

$$\frac{\partial \ln L}{\partial p_2}=\frac{X_2+X_3}{p_2}-\frac{2X_4}{1-p_1-2p_2}=0. \tag{9.3.10}$$

Because we have already taken the derivatives subject to $\Sigma_{j=1}^{4} p_j=1$, we may rewrite these equations as $X_1/p_1-X_4/p_4=0$ and $(X_2+X_3)/p_2-2X_4/p_4=0$, or $X_1p_4=X_4p_1$ and $(X_2+X_3)p_4=2X_4p_2$, which, when added, give $p_4(X_1+X_2+X_3)=X_4(p_1+2p_2)$. Then, using the relations $X_1+X_2+X_3+X_4=n$ and $p_1+2p_2+p_4=1$, we have the solution

$$\hat{p}_4=\frac{X_4}{n}. \tag{9.3.11}$$

Substituting this for p_4 in Equations 9.3.9 and 9.3.10 gives

$$\left.\begin{aligned}\hat{p}_1&=\frac{X_1}{n}\\ &\text{and}\\ \hat{p}_2&=\frac{X_2+X_3}{2n}\end{aligned}\right\} \tag{9.3.12}$$

To test H_0: $p_2=p_3$, we substitute these estimators in the formula for the χ^2 statistic, giving

$$\begin{aligned}\chi^2&=\sum_{j=1}^{4}\frac{(X_j-n\hat{p}_j)^2}{n\hat{p}_j}\\
&=\frac{(X_1-X_1)^2}{X_1}+\frac{\left(X_2-\dfrac{X_2+X_3}{2}\right)^2}{\dfrac{X_2+X_3}{2}}+\frac{\left(X_3-\dfrac{X_2+X_3}{2}\right)^2}{\dfrac{X_2+X_3}{2}}+\frac{(X_4-X_4)^2}{X_4}\\
&=\frac{2}{X_2+X_3}\left[\left(\frac{X_2-X_3}{2}\right)^2+\left(\frac{X_3-X_2}{2}\right)^2\right]\\
&=\frac{(X_2-X_3)^2}{X_2+X_3},\end{aligned} \tag{9.3.13}$$

the well-known McNemar statistic (see McNemar, 1947).

For the Cochran data, the value of the statistic is

$$\chi^2=\frac{(8-20)^2}{28}=\frac{144}{28}=5.14,$$

which, for one degree of freedom, has a significance probability of about .02. Thus, the conclusion at the .02 level or greater would be that H_0 is rejected, that is, that the probability of a favorable response is not the same under Drug A as under Drug C. Apparently, from the data, Drug A is more effective.

It is of interest to compute Neyman's $\chi^2_{(N)}$ statistic for testing H_0: $p_2 = p_3$. First, we must find the values of p_1 and p_2 ($p_4 = 1 - p_1 - 2p_2$) under H_0 that minimize

$$\chi^2_{(N)} = \frac{(X_1 - np_1)^2}{X_1} + \frac{(X_2 - np_2)^2}{X_2} + \frac{(X_3 - np_2)^2}{X_3} + \frac{[X_4 - n(1 - p_1 - 2p_2)]^2}{X_4}$$

$$= -n + n^2\left[\frac{p_1^2}{X_1} + p_2^2\left(\frac{1}{X_2} + \frac{1}{X_3}\right) + \frac{(1 - p_1 - 2p_2)^2}{X_4}\right]. \tag{9.3.14}$$

Taking partial derivatives with respect to p_1 and p_2 and setting them equal to zero, gives

$$n^2\left[\frac{2p_1}{X_1} - \frac{2(1 - p_1 - 2p_2)}{X_4}\right] = 0$$

or

$$\frac{p_1}{X_1} - \frac{p_4}{X_4} = 0, \tag{9.3.15}$$

and

$$n^2\left[2p_2\left(\frac{1}{X_2} + \frac{1}{X_3}\right) - \frac{4(1 - p_1 - 2p_2)}{X_4}\right] = 0$$

or

$$p_2\left(\frac{1}{X_2} + \frac{1}{X_3}\right) - \frac{2p_4}{X_4} = 0. \tag{9.3.16}$$

Thus, the solution for the p_js must satisfy

$$p_1 = p_4\frac{X_1}{X_4} \tag{9.3.17}$$

and

$$p_2 = 2p_4\frac{X_2X_3}{X_4(X_2 + X_3)} \tag{9.3.18}$$

and, because $p_1 + 2p_2 + p_4 = 1$,

$$p_4\left(\frac{X_1}{X_4} + \frac{4X_2X_3}{X_4(X_2 + X_3)} + \frac{X_4}{X_4}\right) = 1. \tag{9.3.19}$$

Solving Equation 9.3.19 gives

$$\tilde{p}_4 = \frac{X_4(X_2 + X_3)}{n(X_2 + X_3) - (X_2 - X_3)^2}, \tag{9.3.20}$$

where the notation $\tilde{p}$ is used to distinguish the minimum $\chi^2_{(N)}$ estimator from the maximum likelihood estimator denoted by $\hat{p}$. Substituting $\tilde{p}_4$ for p_4 in Equations 9.3.17 and 9.3.18 gives the estimators for p_1 and p_2:

$$\tilde{p}_1 = \frac{X_1(X_2+X_3)}{n(X_2+X_3)-(X_2-X_3)^2},$$

and

$$\tilde{p}_2 = \frac{2X_2X_3}{n(X_2+X_3)-(X_2-X_3)^2} \tag{9.3.21}$$

Substituting these values for the p_js in Equation 9.3.14 and simplifying, gives

$$\chi^2_{(N)} = \frac{(X_2-X_3)^2}{(X_2+X_3)-\dfrac{(X_2-X_3)^2}{n}}. \tag{9.3.22}$$

This χ^2 statistic is not identical to the McNemar statistic for testing H_0: $p_2 = p_3$ based on maximum likelihood estimation of the parameters. The term $(X_2-X_3)^2/n$ in the denominator of Equation 9.3.22 is not present in the McNemar statistic. Nevertheless, for large n, the probability is high that the difference between these statistics will be small, if H_0 is true.

For the Cochran example,

$$\chi^2_{(N)} = \frac{144}{28-(12)^2/46} = \frac{144}{28-3.13} = 5.79.$$

compared to the χ^2 value of 5.14 obtained previously, but leading to the same conclusion with a somewhat smaller significance probability.

9.4 ASYMPTOTIC DISTRIBUTION OF ESTIMATORS OF LINEAR COMBINATIONS OF MULTINOMIAL CLASS PROBABILITIES

In the preceding section, we considered the problem of testing the hypothesis that two class probabilities in a quadrinomial are equal. This is an example of a class of problems of general interest, namely, testing the hypothesis that a linear combination, or a set of linear combinations, of the class probabilities of a multinomial has the value zero. In the example of Section 9.3.2, the linear combination of interest was $L = p_2 - p_3$. We have illustrated two tests of the hypothesis H_0: $L = 0$. We wish now to deal more generally with the problem of testing hypotheses about linear combinations of multinomial probabilities. Our approach is based on the asymptotic normality of multinomial distributions discussed in Section 9.2, and on what is known about linear combinations of jointly distributed normal variables.

9.4.1 Estimators of Linear Combinations of Multinomial Class Probabilities

First consider the multivariate-normal random vector **X**, defined by Equation 9.2.1, with density function $f(\mathbf{x})$ defined by Equation 9.2.3. Let $\boldsymbol{\mu}$ and $\boldsymbol{\Sigma}_X$ be the vector of means and the variance-covariance matrix of the components of **X** defined by Equations 9.2.4 and 9.2.5, respectively. Let

$$\mathbf{Y}=\mathbf{B}\mathbf{X}, \tag{9.4.1}$$

where **B** is a $t\times c$ matrix of constants of rank $t \leqq c$. Then it is well known (see Anderson, 1958) that **Y** has the t-dimensional normal distribution, with mean vector

$$E(\mathbf{Y})=\mathbf{B}\boldsymbol{\mu} \tag{9.4.2}$$

and variance-covariance matrix

$$\boldsymbol{\Sigma}_Y=\mathbf{B}\boldsymbol{\Sigma}_X\mathbf{B}'. \tag{9.4.3}$$

It follows (see Equation 9.2.6) that

$$Q(\mathbf{Y})=(\mathbf{Y}-E(\mathbf{Y}))'\boldsymbol{\Sigma}_Y^{-1}(\mathbf{Y}-E(\mathbf{Y}))$$

has the χ^2 distribution with t degrees of freedom.

Now, consider a set of linear combinations of multinomial class probabilities, $\mathbf{L}=\mathbf{A}\mathbf{p}$, where **A** is a matrix of constants of rank $t<c$, and **p** is the vector of multinomial class probabilities defined by Equation 9.3.3. Let $\hat{\mathbf{L}}=\mathbf{A}\hat{\mathbf{p}}$, where

$$\hat{\mathbf{p}}=\begin{bmatrix}\hat{p}_1\\ \hat{p}_2\\ \cdot\\ \cdot\\ \cdot\\ \hat{p}_c\end{bmatrix}. \tag{9.4.4}$$

In the vector $\hat{\mathbf{p}}$, $\hat{p}_j$ is the observed relative frequency in the jth class of the multinomial. It follows (see Chernoff, 1956) that the limiting distribution of $\hat{\mathbf{L}}$ is the t-dimensional normal density with mean vector, $E(\hat{\mathbf{L}})=\mathbf{L}=\mathbf{A}\mathbf{p}$, and variance-covariance matrix, $\boldsymbol{\Sigma}_L=\mathbf{A}\boldsymbol{\Sigma}_{\hat{p}}\mathbf{A}'$, where the elements of $\boldsymbol{\Sigma}_{\hat{p}}$ are defined by Equation 9.1.4. It also follows (see Chernoff, 1956) that

$$Q(\hat{\mathbf{L}})=(\hat{\mathbf{L}}-\mathbf{L})'\boldsymbol{\Sigma}_L^{-1}(\hat{\mathbf{L}}-\mathbf{L}) \tag{9.4.5}$$

has as its limiting distribution the χ^2 distribution with t degrees of freedom. (Recall that in Section 9.2 we deleted the cth component of **p**, we need not do so here because the rank of **A** is $c-1$ or less, and premultiplication of $\boldsymbol{\Sigma}_{\hat{p}}$ by **A** will yield a nonsingular matrix of rank $t \leqq c-1$. The rank of $\boldsymbol{\Sigma}_L=\mathbf{A}\boldsymbol{\Sigma}_{\hat{p}}\mathbf{A}'$ will also be $t\leqq c-1$; its order will be $t\times t$; and it will, therefore, have an inverse.)

Statistics formed in the manner of $Q(\hat{\mathbf{L}})$ have been called *Wald statistics* (see, for example, Bhapkar, 1966), because the results are based on the work of Wald (1943). Grizzle, Starmer, and Koch (1969) have outlined a general approach derived from results of Wald and Neyman that can be used to solve a variety of problems that arise in the analysis of categorical data. They have suggested a general solution for problems involving tests of hypotheses that a set of linear combinations of the multinomial class probabilities are zero. It is based on the $Q(\hat{\mathbf{L}})$ statistic and will be discussed further in a later section. The method suggested by Grizzle, Starmer, and Koch has sometimes been referred to as the *GSK method* in subsequent literature.

We return now to the problem of testing H_0: $p_2 = p_3$ in a quadrinomial representing a cross-classification of two dichotomous characteristics, as displayed in Table 9.3.1, and we seek a solution with the help of the Wald statistic, $Q(\hat{\mathbf{L}})$.

Here we are interested in testing whether a *single* linear combination of the class probabilities is 0, that is, the case where $\mathbf{L} = p_2 - p_3$, $\mathbf{A} = [0\,1\,-1\,0]$,

$$\mathbf{p} = \begin{bmatrix} p_1 \\ p_2 \\ p_3 \\ p_4 \end{bmatrix}$$

and

$$\hat{\mathbf{p}} = \begin{bmatrix} \hat{p}_1 \\ \hat{p}_2 \\ \hat{p}_3 \\ \hat{p}_4 \end{bmatrix}$$

is the vector of observed class relative frequencies in classes 1 through 4. Here,

$$\mathbf{\Sigma}_{\hat{p}} = \frac{1}{n}\begin{bmatrix} p_1(1-p_1) & -p_1p_2 & -p_1p_3 & -p_1p_4 \\ -p_1p_2 & p_2(1-p_2) & -p_2p_3 & -p_2p_4 \\ -p_1p_3 & -p_2p_3 & p_3(1-p_3) & -p_3p_4 \\ -p_1p_4 & -p_2p_4 & -p_3p_4 & p_4(1-p_4) \end{bmatrix},$$

$$\mathbf{A\Sigma}_{\hat{p}} = \frac{1}{n}[p_1(p_3-p_2) \quad p_2(1-p_2+p_3) \quad -p_3(1+p_2-p_3) \quad p_4(p_3-p_2)],$$

and

$$\mathbf{\Sigma}_{\hat{L}} = \mathbf{A\Sigma}_{\hat{p}}\mathbf{A}' = \frac{p_2(1-p_2+p_3)+p_3(1+p_2-p_3)}{n} = \frac{p_2+p_3-(p_2-p_3)^2}{n}.$$

Because the class probabilities are unknown, $\mathbf{\Sigma}_{\hat{L}}$ is unknown, but Wald (1943) has shown that the class probabilities may be replaced by their maximum likelihood estimators under the assumptions without changing the limiting distribution of $Q(\hat{\mathbf{L}})$. Thus, we may estimate $\mathbf{\Sigma}_{\hat{L}}$ by

$$\hat{\mathbf{\Sigma}}_{\hat{L}} = \frac{\hat{p}_2+\hat{p}_3-(\hat{p}_2-\hat{p}_3)^2}{n}$$

and $\mathbf{\Sigma}_{\hat{L}}^{-1}$ by

$$\hat{\mathbf{\Sigma}}_{\hat{L}}^{-1} = \frac{n}{\hat{p}_2+\hat{p}_3-(\hat{p}_2-\hat{p}_3)^2}.$$

Substituting $\hat{\Sigma}_{\hat{L}}^{-1}$ for $\Sigma_{\hat{L}}^{-1}$ in $Q(\hat{\mathbf{L}})$, and denoting the result of this substitution by $\hat{Q}(\hat{\mathbf{L}})$, we have

$$\begin{aligned}\hat{Q}(\hat{\mathbf{L}}) &= (\hat{\mathbf{L}}-0)'\hat{\Sigma}_{\hat{L}}^{-1}(\hat{\mathbf{L}}-0)\\ &= (\hat{p}_2-\hat{p}_3)\frac{n}{\hat{p}_2+\hat{p}_3-(\hat{p}_2-\hat{p}_3)^2}(\hat{p}_2-\hat{p}_3)\\ &= \frac{n(\hat{p}_2-\hat{p}_3)^2}{\hat{p}_2+\hat{p}_3-(\hat{p}_2-\hat{p}_3)^2} = \frac{n\left(\dfrac{X_2}{n}-\dfrac{X_3}{n}\right)^2}{\left(\dfrac{X_2+X_3}{n}\right)-\left(\dfrac{X_2}{n}-\dfrac{X_3}{n}\right)^2}\\ &= \frac{(X_2-X_3)^2}{(X_2+X_3)-\dfrac{(X_2-X_3)^2}{n}},\end{aligned}$$

where X_j is the observed frequency in the jth class of the quadrinomial and n is the number of trials.

Note that this statistic is identical with Neyman's $\chi^2_{(N)}$ statistic as defined by Equation 9.3.22. In fact, Bhapkar (1966) has shown that the Wald statistic denoted by $\hat{Q}(\hat{\mathbf{L}})$ here and Neyman's $\chi^2_{(N)}$ statistic are, in general, equivalent.

The example we discussed in detail earlier concerned the question of whether a single linear combination of the class probabilities ($\mathbf{L}=p_2-p_3$) has the value zero. An interesting extension of this particular example arises in the case of a multinomial with eight classes, each of which is characterized by the presence or absence of each of *three* characteristics. If we let

$$X_i=\begin{cases}1 \text{ if characteristic } i \text{ is present}\\ 0 \text{ if characteristic } i \text{ is absent, } (i=1,2,3)\end{cases}$$

then the eight multinomial classes may be represented as in Table 9.4.1.

TABLE 9.4.1 EIGHT MULTINOMIAL CLASSES DEFINED BY THE PRESENCE OR ABSENCE OF EACH OF THREE CHARACTERISTICS

	Value of X_i			
Class	x_1	x_2	x_3	*Class probability*
1	1	1	1	p_1
2	1	1	0	p_2
3	1	0	1	p_3
4	1	0	0	p_4
5	0	1	1	p_5
6	0	1	0	p_6
7	0	0	1	p_7
8	0	0	0	p_8

Let

$$\pi_i = P(X_i = 1), \quad (i = 1, 2, 3). \tag{9.4.6}$$

Then the hypothesis that each of the three characteristics occurs with the same probability is

$$H_0\colon \pi_1 = \pi_2 = \pi_3. \tag{9.4.7}$$

This is an extension to the $2 \times 2 \times 2$ case of the McNemar problem. In terms of the class probabilities, H_0 states that $\pi_1 - \pi_2 = p_1 + p_2 + p_3 + p_4 - p_1 - p_2 - p_5 - p_6 = 0$, and thus,

$$L_1 = p_3 + p_4 - p_5 - p_6, \tag{9.4.8}$$

and that $\pi_1 - \pi_3 = p_2 + p_4 - p_5 - p_7 = 0$, and thus,

$$L_2 = p_2 + p_4 - p_5 - p_7. \tag{9.4.9}$$

The hypothesis of interest is H_0: $\mathbf{L} = \mathbf{0}$, or

$$\begin{bmatrix} L_1 \\ L_2 \end{bmatrix} = \begin{bmatrix} 0 \\ 0 \end{bmatrix},$$

that is, $L_1 = L_2 = 0$. Cochran (1950) considered this problem and derived a statistic, usually referred to as *Cochran's Q*, that is asymptotically distributed as χ^2 with two degrees of freedom. However, the derivation assumed that the covariances between the Xs are equal. If this condition is not met, the asymptotic distribution of Cochran's statistic is not χ^2 with two degrees of freedom (see, for example, Berger and Gold, 1973).

Grizzle, Starmer and Koch also considered the problem of testing H_0: $\pi_1 = \pi_2 = \pi_3$, but with the help of an appropriate Wald statistic. In this example, we take

$$\mathbf{A} = \begin{bmatrix} 0 & 0 & 1 & 1 & -1 & -1 & 0 & 0 \\ 0 & 1 & 0 & 1 & -1 & 0 & -1 & 0 \end{bmatrix}. \tag{9.4.10}$$

The data they used, and that we will also use to illustrate the method for the case of more than one linear combination of the class probabilities are Cochran's (1950). For these data, $X_i = 1$ represents a favorable response to the ith drug, and 0 an unfavorable response, $(i = 1, 2, 3)$. The observed frequencies in the eight classes as defined in Table 9.4.1 were those in Table 9.4.2.

TABLE 9.4.2 OBSERVED FREQUENCIES IN EIGHT MULTINOMIAL CLASSES, COCHRAN'S (1950) DATA

Class	Observed class frequency
1	6
2	16
3	2
4	4
5	2
6	4
7	6
8	6
	Total: $n = 46$

(Note that the data of Table 9.3.2 represent the joint responses to X_1 and X_3 only.)

Here we have

$$\hat{\Sigma}_{\hat{p}} = \frac{1}{(46)^3} \cdot \begin{bmatrix} 240 & -96 & -12 & -24 & -12 & -24 & -36 & -36 \\ -96 & 480 & -32 & -64 & -32 & -64 & -96 & -96 \\ -12 & -32 & 88 & -8 & -4 & -8 & -12 & -12 \\ -24 & -64 & -8 & 168 & -8 & -16 & -24 & -24 \\ -12 & -32 & -4 & -8 & 88 & -8 & -12 & -12 \\ -24 & -64 & -8 & -16 & -8 & 168 & -24 & -24 \\ -36 & -96 & -12 & -24 & -12 & -24 & 240 & -36 \\ -36 & -96 & -12 & -24 & -12 & -24 & -36 & 240 \end{bmatrix}, \tag{9.4.11}$$

where $\hat{\Sigma}_{\hat{p}}$ is $\Sigma_{\hat{p}}$, the variance-covariance matrix of the $\hat{p}_j$s, with the p_js replaced by the corresponding $\hat{p}_j$s, (see Equation 9.1.4). Then

$$\mathbf{A}\hat{\Sigma}_{\hat{p}} = \frac{1}{(46)^3} \cdot \begin{bmatrix} 0 & 0 & 92 & 184 & -92 & -184 & 0 & 0 \\ -72 & 544 & -24 & 136 & -116 & -48 & -348 & -72 \end{bmatrix}, \tag{9.4.12}$$

$$\hat{\Sigma}_{\hat{L}} = \mathbf{A}\hat{\Sigma}_{\hat{p}}\mathbf{A}' = \frac{1}{(46)^3} \begin{bmatrix} 552 & 276 \\ 276 & 1144 \end{bmatrix}, \tag{9.4.13}$$

$$|\hat{\Sigma}_{\hat{L}}| = \frac{555312}{(46)^3}, \tag{9.4.14}$$

and

$$\hat{\Sigma}_{\hat{L}}^{-1} = \frac{(46)^3}{555312} \begin{bmatrix} 1144 & -276 \\ -276 & 552 \end{bmatrix}.$$

Because $\hat{L}_1 = \hat{\pi}_1 - \hat{\pi}_2 = 0$ and $\hat{L}_2 = \hat{\pi}_1 - \hat{\pi}_3 = \frac{12}{46}$, where the $\hat{\pi}_i$s are equal to the corresponding π_is with the p_js in the π_i replaced by the corresponding $\hat{p}_j$s, the value of the test statistic is:

$$\begin{aligned} \hat{Q}(\hat{\mathbf{L}}) &= \hat{\mathbf{L}}'\hat{\Sigma}_{\hat{L}}^{-1}\hat{\mathbf{L}} \\ &= \begin{bmatrix} 0 & \frac{12}{46} \end{bmatrix} \cdot \left(\frac{(46)^3}{555312} \right) \begin{bmatrix} 1144 & -276 \\ -276 & 552 \end{bmatrix} \cdot \begin{bmatrix} 0 \\ \frac{12}{46} \end{bmatrix} \\ &= \frac{(46)^3}{555312} \left(\frac{12}{46} \right)^2 (552) = 6.584, \end{aligned}$$

as obtained by Grizzle, Starmer, and Koch (1969). The significance probability of this value of χ^2 for two degrees of freedom is less than .05, so that $H_0: \mathbf{L} = \mathbf{O}$ would be rejected at a level of significance of .05 or greater. We conclude that the probability of a favorable response is not the same for the three drugs.

9.4.2 Linear Combinations of Class Probabilities for More Than One Multinomial Distribution

There are many other statistical problems of interest that can be solved with the help of a Wald statistic, as well as by the more familiar χ^2 techniques. The method is appropriate when the questions of interest concern linear combinations of the class probabilities of more than one multinomial

TABLE 9.4.3 FORMAT FOR DISPLAYING DATA FOR TEST OF $H_0: P_1 = P_2$

	Number of successes	*Number of failures*	*Total*
Treatment A	X	$n_1 - X$	n_1
Treatment B	Y	$n_2 - Y$	n_2
Total	$S = X + Y$	$F = n - S$	$n = n_1 + n_2$

distribution. Perhaps the simplest problem of this kind concerns the question of whether two binomial distributions have the same probability of success. The data from an experiment to answer such a question may be displayed in the familiar four-fold table format of Table 9.4.3.

If we assume that the n_1 trials under Treatment A are independent, each with probability p_1 of success, and that the n_2 trials under Treatment B, each with probability p_2 of success, are independent of each other and of the n_1 trials under Treatment A, the statistical hypothesis of interest is $H_0: p_1 - p_2 = 0$. The usual one-degree-of-freedom χ^2 statistic for testing H_0 is

$$\chi^2 = \frac{(\hat{p}_1 - \hat{p}_2)^2}{\left(\frac{S}{n}\right)\left(\frac{F}{n}\right)\left(\frac{1}{n_1} + \frac{1}{n_2}\right)}, \tag{9.4.15}$$

where $\hat{p}_1 = X/n_1$ and $\hat{p}_2 = Y/n_2$.

To solve the same problem with the help of a Wald χ^2 statistic, as suggested in the Grizzle, Starmer, and Koch (1969) paper, we have

$$\left.\begin{aligned} \mathbf{p} &= \begin{bmatrix} p_1 \\ q_1 \\ p_2 \\ q_2 \end{bmatrix} \\ &\text{and} \\ \hat{\mathbf{p}} &= \begin{bmatrix} \hat{p}_1 \\ \hat{q}_1 \\ \hat{p}_2 \\ \hat{q}_2 \end{bmatrix}, \end{aligned}\right\} \tag{9.4.16}$$

where $q_j = 1 - p_j$, and $\hat{q}_j = 1 - \hat{p}_j$. Because $\hat{p}_1$ and $\hat{q}_1$ are independent of $\hat{p}_2$ and $\hat{q}_2$, the variance-covariance matrix of the components of $\hat{\mathbf{p}}$ is

$$\boldsymbol{\Sigma}_{\hat{p}} = \begin{bmatrix} \frac{p_1 q_1}{n_1} & -\frac{p_1 q_1}{n_1} & 0 & 0 \\ -\frac{p_1 q_1}{n_1} & \frac{p_1 q_1}{n_1} & 0 & 0 \\ 0 & 0 & \frac{p_2 q_2}{n_2} & -\frac{p_2 q_2}{n_2} \\ 0 & 0 & -\frac{p_2 q_2}{n_2} & \frac{p_2 q_2}{n_2} \end{bmatrix}. \tag{9.4.17}$$

We may take

$$L=p_1-p_2,$$
$$\hat{L}=\hat{p}_1-\hat{p}_2,$$
$$\mathbf{A}=[1 \quad 0 \quad -1 \quad 0],$$

and

$$\boldsymbol{\Sigma}_{\hat{L}}=\mathbf{A}\boldsymbol{\Sigma}_{\hat{p}}\mathbf{A}'=\frac{p_1q_1}{n_1}+\frac{p_2q_2}{n_2}.$$

Then

$$\boldsymbol{\Sigma}_{\hat{L}}^{-1}=\frac{1}{\dfrac{p_1q_1}{n_1}+\dfrac{p_2q_2}{n_2}}.$$

Again, because p_1 and p_2 are unknown, we may estimate them by their respective sample analogues and let

$$\hat{\boldsymbol{\Sigma}}_{\hat{L}}^{-1}=\frac{1}{\dfrac{\hat{p}_1\hat{q}_1}{n_1}+\dfrac{\hat{p}_2\hat{q}_2}{n_2}}.$$

The Wald χ^2 statistic is then

$$\hat{Q}(\hat{\mathbf{L}})=(\hat{\mathbf{L}}-\mathbf{0})'\hat{\boldsymbol{\Sigma}}_{\hat{L}}^{-1}(\hat{\mathbf{L}}-\mathbf{0})=\frac{(\hat{p}_1-\hat{p}_2)^2}{\dfrac{\hat{p}_1\hat{q}_1}{n_1}+\dfrac{\hat{p}_2\hat{q}_2}{n_2}}. \tag{9.4.18}$$

Note that this is not the same as the usual χ^2 statistic, (see Equation 9.4.15), however, if we estimate p_1 and p_2 in $\boldsymbol{\Sigma}_{\hat{L}}^{-1}$ under $H_0{:}p_1=p_2$, the obvious estimator for the common probability would be S/n, and substituting this for $\hat{p}_1$ and $\hat{p}_2$ in the denominator of Equation 9.4.18 will result in the familiar χ^2 statistic of Equation 9.4.15.

Note also the structure of $\boldsymbol{\Sigma}_{\hat{p}}$ when we are considering more than one multinomial distribution, say r c-nomials. If all the trials are mutually independent, then $\boldsymbol{\Sigma}_{\hat{p}}$ will be a square matrix or order rc consisting of r $c\times c$ submatrices along the diagonal, the ith submatrix being the variance-covariance matrix of the ith multinomial, and all other elements being equal to zero. (The rank of the square matrix will be $r(c-1)$. However, because this matrix will be pre- and postmultiplied by $\mathbf{A}$ and $\mathbf{A}'$, respectively, to obtain $\boldsymbol{\Sigma}_{\hat{L}}$, we need concern ourselves only with the rank of the former matrix.) For example, if we are concerned with four trinomials and let p_{ij} be the probability of the jth class and n_i the number of trials in the ith multinomial, $\boldsymbol{\Sigma}_{\hat{p}}$ will be a 12×12 matrix, as indicated in Table 9.4.4.

As an illustration, consider the following example involving three binomials. The Health Professions Educational Assistance Act (PL 94–484), enacted in 1976, had as an aim reversing the trend toward postgraduate training in specialties not identified with primary care of patients. A study in the *New England Journal of Medicine* by Wechsler, Dorsey, and Bovey

TABLE 9.4.4 DESCRIPTION OF THE 12 × 12 VARIANCE-COVARIANCE MATRIX, $\Sigma_{\hat{p}}$, IN THE CASE OF FOUR TRINOMIALS

Rows 1–3											
$\frac{p_{11}(1-p_{11})}{n_1}$	$\frac{-p_{11}p_{12}}{n_1}$	$\frac{-p_{11}p_{13}}{n_1}$	0	0	0	0	0	0	0	0	0
$\frac{-p_{11}p_{12}}{n_1}$	$\frac{p_{12}(1-p_{12})}{n_1}$	$\frac{-p_{12}p_{13}}{n_1}$	0	0	0	0	0	0	0	0	0
$\frac{-p_{11}p_{13}}{n_1}$	$\frac{-p_{12}p_{13}}{n_1}$	$\frac{p_{13}(1-p_{13})}{n_1}$	0	0	0	0	0	0	0	0	0
Rows 4–6											
0	0	0	$\frac{p_{21}(1-p_{21})}{n_2}$	$\frac{-p_{21}p_{22}}{n_2}$	$\frac{-p_{21}p_{23}}{n_2}$	0	0	0	0	0	0
0	0	0	$\frac{-p_{21}p_{22}}{n_2}$	$\frac{p_{22}(1-p_{22})}{n_2}$	$\frac{-p_{22}p_{23}}{n_2}$	0	0	0	0	0	0
0	0	0	$\frac{-p_{21}p_{23}}{n_2}$	$\frac{-p_{22}p_{23}}{n_2}$	$\frac{p_{23}(1-p_{23})}{n_2}$	0	0	0	0	0	0
Rows 7–9											
0	0	0	0	0	0	$\frac{p_{31}(1-p_{31})}{n_3}$	$\frac{-p_{31}p_{32}}{n_3}$	$\frac{-p_{31}p_{33}}{n_3}$	0	0	0
0	0	0	0	0	0	$\frac{-p_{31}p_{32}}{n_3}$	$\frac{p_{32}(1-p_{32})}{n_3}$	$\frac{-p_{32}p_{33}}{n_3}$	0	0	0
0	0	0	0	0	0	$\frac{-p_{31}p_{33}}{n_3}$	$\frac{-p_{32}p_{33}}{n_3}$	$\frac{p_{33}(1-p_{33})}{n_3}$	0	0	0
Rows 10–12											
0	0	0	0	0	0	0	0	0	$\frac{p_{41}(1-p_{41})}{n_4}$	$\frac{-p_{41}p_{42}}{n_4}$	$\frac{-p_{41}p_{43}}{n_4}$
0	0	0	0	0	0	0	0	0	$\frac{-p_{41}p_{42}}{n_4}$	$\frac{p_{42}(1-p_{42})}{n_4}$	$\frac{-p_{42}p_{43}}{n_4}$
0	0	0	0	0	0	0	0	0	$\frac{-p_{41}p_{43}}{n_4}$	$\frac{-p_{42}p_{43}}{n_4}$	$\frac{p_{43}(1-p_{43})}{n_4}$

(1978) reported the results of a questionnaire survey of physicians who had been residents in Massachusetts hospitals during the period 1967–72 in three specialties identified with primary care, namely, internal medicine, pediatrics, and obstetrics-gynecology. One of the purposes of the survey was to study the time spent on "primary care" by the physicians who had received their training in the three specialties, but who were, at the time of the study, in practice as "primary care physicians." Such a physician is defined in the paper as "a physician who establishes a relation with an individual or family for which he or she provides *continual surveillance* of their health needs and comprehensive care for the acute and chronic disorders that he or she is qualified to care for and access to the health-delivery system for those disorders requiring the services of other specialists."

The number and percentage of responding physicians in the three specialties who provided primary care in their primary-care field or subspecialty are shown in Table 9.4.5.

The study then analyzed the proportion of time spent in primary-care activities among those physicians who reported engaging in at least some such activity. The results are given in Table 9.4.6.

Let p_i be the probability that over 50% of the practitioner's time is spent in primary-care service in group $i (i=1,2,3)$. The hypothesis of interest is H_0: $p_1 = p_2 = p_3$. The usual χ^2 statistic with two degrees of freedom was

TABLE 9.4.5 NUMBER AND PERCENTAGE OF PHYSICIANS IN THREE SPECIALTIES BY AMOUNT OF PRIMARY CARE PROVIDED

Specialty	*Some primary care*		*No primary care*		*Total*
	No.	*%*	*No.*	*%*	
Internal medicine	219	62	136	38	355
Pediatrics	83	66	43	34	126
Obstetrics-Gynecology	32	82	7	18	39
Total	334	64	186	36	520

TABLE 9.4.6 PRACTITIONERS PROVIDING SOME PRIMARY CARE, BY PERCENTAGE OF TIME SPENT, AND BY SPECIALTY

	Percent time in primary care activities				
Specialty	*Over 50%*		*50% or Less*		*Total**
	No.	*%*	*No.*	*%*	
1. Internal medicine	91	42.13	125	57.87	216
2. Pediatrics	56	67.47	27	32.53	83
3. Obstetrics-Gynecology	15	53.57	13	46.43	28
Total	162	49.54	165	50.46	327

*Seven physicians reporting time devoted to primary-care activities did not specify the amount of time.

calculated, to test for the homogeneity of the three groups, and it was found to be 15.601, which has significance probability less than .001.

We now indicate how to compute the Wald statistic for testing H_0, as proposed by the GSK method. For the data of Table 9.4.6,

$$\hat{\mathbf{p}} = \begin{bmatrix} \hat{p}_1 \\ \hat{q}_1 \\ \hat{p}_2 \\ \hat{q}_2 \\ \hat{p}_3 \\ \hat{q}_3 \end{bmatrix} = \begin{bmatrix} 91/216 \\ 125/216 \\ 56/83 \\ 27/83 \\ 15/28 \\ 13/28 \end{bmatrix},$$

$$\hat{\boldsymbol{\Sigma}}_{\hat{p}} = \begin{bmatrix} \frac{91(125)}{216^3} & \frac{-91(125)}{216^3} & 0 & 0 & 0 & 0 \\ \frac{-91(125)}{216^3} & \frac{91(125)}{216^3} & 0 & 0 & 0 & 0 \\ 0 & 0 & \frac{56(27)}{83^3} & \frac{-56(27)}{83^3} & 0 & 0 \\ 0 & 0 & \frac{-56(27)}{83^3} & \frac{56(27)}{83^3} & 0 & 0 \\ 0 & 0 & 0 & 0 & \frac{15(13)}{28^3} & \frac{-15(13)}{28^3} \\ 0 & 0 & 0 & 0 & \frac{-15(13)}{28^3} & \frac{15(13)}{28^3} \end{bmatrix},$$

and

$$\hat{\mathbf{L}} = \begin{bmatrix} \hat{p}_1 - \hat{p}_2 \\ \hat{p}_1 - \hat{p}_3 \end{bmatrix} = \begin{bmatrix} -.2534 \\ -.1144 \end{bmatrix}.$$

Here, the hypothesis H_0: $p_1 = p_2 = p_3$ may be stated as $\mathbf{L} = \mathbf{O}$, where

$$\mathbf{L} = \begin{bmatrix} p_1 - p_2 \\ p_1 - p_3 \end{bmatrix}.$$

Then

$$\mathbf{A} = \begin{bmatrix} 1 & 0 & -1 & 0 & 0 & 0 \\ 1 & 0 & 0 & 0 & -1 & 0 \end{bmatrix}$$

and

$$\hat{\boldsymbol{\Sigma}}_{\hat{L}} = \mathbf{A}\hat{\boldsymbol{\Sigma}}_{\hat{p}}\mathbf{A}' = \begin{bmatrix} \frac{11375}{216^3} + \frac{1512}{83^3} & \frac{11375}{216^3} \\ \frac{11375}{216^3} & \frac{11375}{216^3} + \frac{195}{28^3} \end{bmatrix}$$

$$= \begin{bmatrix} .00377307 & .00112873 \\ .00112873 & .01001176 \end{bmatrix}.$$

Then

$$\hat{\mathbf{\Sigma}}_{\hat{L}}^{-1} = \begin{bmatrix} 274.28698 & -30.923228 \\ -30.923228 & 103.36883 \end{bmatrix}$$

and the Wald χ^2 statistic has the value

$$\hat{Q}(\hat{\mathbf{L}}) = \hat{\mathbf{L}}' \hat{\mathbf{\Sigma}}_{\hat{L}}^{-1} \hat{\mathbf{L}} = [-.2534 \quad -.1144] \hat{\mathbf{\Sigma}}_{\hat{L}}^{-1} \begin{bmatrix} -.2534 \\ -.1144 \end{bmatrix}$$

$$= 17.172,$$

compared to the value of the "usual" homogeneity χ^2, 15.601.

9.5 ASYMPTOTIC DISTRIBUTION OF ESTIMATORS OF NONLINEAR COMBINATIONS OF MULTINOMIAL CLASS PROBABILITIES

Not all statistical problems of interest concern linear combinations of the multinomial class probabilities. In fact, the question of independence in the 2×2 contingency table is one that frequently arises in practice. This problem has already been considered in Chapter 3, but it serves our purpose to take another look at it in the light of the current discussion. As you will recall, the problem is as follows. A population is cross-classified with respect to two binary variables. We arbitrarily assign the values 1 and 0 to each of the two possible outcomes for each variable, say X and Y. Thus, the joint probability function can be represented as in Table 9.5.1.

TABLE 9.5.1 JOINT PROBABILITY DISTRIBUTION OF TWO BINARY VARIABLES, *X* AND *Y*.

	Value of X		
1	p_1	p_2	p_1+p_2
Value of Y	1	0	$P(Y=y)$
0	p_3	p_4	p_3+p_4
$P(X=x)$	p_1+p_3	p_2+p_4	1

It is easy to show that X and Y are independent if and only if $p_1p_4 = p_2p_3$:

If X and Y are independent, then $P(X=1, Y=1) = P(X=1)P(Y=1)$, that is

$$p_1 = (p_1+p_3)(p_1+p_2) = p_1^2 + p_1p_2 + p_1p_3 + p_2p_3 = p_1(p_1+p_2+p_3) + p_2p_3$$
$$= p_1(1-p_4) + p_2p_3,$$

and

$$p_1 - p_1(1-p_4) = p_2p_3,$$

giving

$$p_1p_4 = p_2p_3.$$

On the other hand, if $p_1p_4=p_2p_3$, then

$$p_1(1-p_1-p_2-p_3)=p_2p_3,$$

and therefore

$$\begin{aligned} p_1 &= p_1^2+p_1p_2+p_1p_3+p_2p_3 \\ &= p_1(p_1+p_2)+p_3(p_1+p_2)=(p_1+p_3)(p_1+p_2) \\ &= P(X=1)P(Y=1). \end{aligned}$$

Similar arguments hold for p_2, p_3 and p_4.

Thus, if we are interested in testing the statistical hypothesis of independence,

$$H_0: p_1p_4-p_2p_3=0, \tag{9.5.1}$$

we are no longer dealing with a linear combination of the four class probabilities, p_1, p_2, p_3, and p_4. If we perform n independent trials, with outcomes as indicated in Table 9.5.2, the familiar one-degree-of-freedom χ^2 statistic to test H_0 is

$$\chi^2=\frac{(ad-bc)^2n}{(a+b)(a+c)(b+c)(c+d)}. \tag{9.5.2}$$

Recall that in Section 9.3.2 we were concerned with the question of whether the probabilities of a favorable response to two drugs were equal, that is, the hypothesis dealt with there was $H_0: p_2=p_3$. In the same example, it is reasonable to ask whether the responses to the two drugs, A and C, are independent. Using the data of Table 9.3, the value of χ^2 statistic appropriate for testing the hypothesis of independence (Equation 9.5.1) is

$$\chi^2=\frac{[8(10)-8(20)]^2(46)}{(16)(30)(28)(18)}=1.217,$$

which suggests that the responses to the two drugs are independent.

9.5.1 A Nonlinear Combination of Multinomial Class Probabilities

In order to test the above hypothesis with the help of a Wald statistic, we take advantage of the "delta method." A detailed discussion of the applicability of this method, along with other large-sample approximations, may be found in Bishop, Fienberg, and Holland (1975, Chapter 14). We now

TABLE 9.5.2 OBSERVED FREQUENCIES OF TWO BINARY VARIABLES, *X* AND *Y*, ON *n* INDEPENDENT TRIALS

	Value of X		
Value of Y	$X=1$	$X=0$	*Total*
$Y=1$	a	b	$a+b$
$Y=0$	c	d	$c+d$
Total	$a+c$	$b+d$	n

describe the application of the method to nonlinear combinations of the c class probabilities of a single multinomial with n independent trials.

Let $\mathbf{p}$ be the vector of the c multinomial class probabilities and $\hat{\mathbf{p}}$ the corresponding vector of observed class relative frequencies. Let $F(\hat{\mathbf{p}})$ be a function of $\hat{\mathbf{p}}$ having continuous partial derivatives up to the second order with respect to each of the components of $\hat{\mathbf{p}}$ in a neighborhood of $\mathbf{p}$. Then we can expand $\sqrt{n}\ F(\hat{\mathbf{p}})$ in a Taylor series about the point $\hat{\mathbf{p}}=p$:

$$\sqrt{n}\ F(\hat{\mathbf{p}})=\sqrt{n}\ F(\mathbf{p})+\sqrt{n}\sum_{i=1}^{c} F_i'(\mathbf{p})(\hat{p}_i-p_i)+R, \tag{9.5.3}$$

where p_i is the probability of the ith multinomial class, $\hat{p}_i$ is the observed relative frequency in the ith class,

$$F_i'(\mathbf{p})=\left.\frac{\partial F(\hat{\mathbf{p}})}{\partial \hat{P}_i}\right|_{\hat{\mathbf{p}}=\mathbf{p}},$$

and R is the remainder term, which in this case can be shown to tend to zero with high probability as n increases. We then have the approximation,

$$\sqrt{n}\,[F(\hat{\mathbf{p}})-F(\mathbf{p})]\cong\sqrt{n}\sum_{i=1}^{c} F_i'(\mathbf{p})(\hat{p}_i-p_i). \tag{9.5.4}$$

The left hand side of Equation 9.5.4 then has the same limiting distribution as that of the right hand side. In stating the result it is necessary to include $\sqrt{n}$ as a factor because the limiting distribution must be *fixed,* that is, independent of n. In fact, the limiting distribution of the right hand side of Equation 9.5.4 is normal, because it is a linear combination of asymptotically normal variables, with mean zero. To calculate the variance of the limiting distribution, first consider the structure of the right hand member of 9.5.4,

$$\sqrt{n}\sum_{i=1}^{c} F_i'(\mathbf{p})(\hat{p}_i-p_i)=g_1(\hat{p}_1-p_1)+g_2(\hat{p}_2-p_2)+\ldots+g_c(\hat{p}_c-p_c). \tag{9.5.5}$$

The variance of the right hand member of Equation 9.5.5 is

$$\begin{aligned}\mathrm{Var}[g_1(\hat{p}_1-p_1)+\ldots+g_c(\hat{p}_c-p_c)]&=\sum_{i=1}^{c} g_i^2\,\mathrm{Var}(\hat{p}_i)+\sum_{i\neq j} g_i g_j\,\mathrm{Cov}(\hat{p}_i,\hat{p}_j)\\&=\sum_{i=1}^{c} g_i^2\frac{p_i(1-p_i)}{n}-\sum_{i\neq j} g_i g_j\frac{p_i p_j}{n},\end{aligned} \tag{9.5.6}$$

where $g_i=\sqrt{n}\ F_i'(\mathbf{p})$. This may be written as the quadratic form

$$\mathbf{G}\mathbf{V}(\hat{\mathbf{p}})\mathbf{G}', \tag{9.5.7}$$

where $\mathbf{G}=[g_1\ g_2 \cdots g_c]$ and

$$\mathbf{V}(\hat{\mathbf{p}})=\begin{bmatrix} \frac{p_1(1-p_1)}{n} & \frac{-p_1p_2}{n} & \cdots & \frac{-p_1p_c}{n} \\ \frac{-p_1p_2}{n} & \frac{p_2(1-p_2)}{n} & \cdots & \frac{-p_2p_c}{n} \\ \vdots & \vdots & & \vdots \\ \frac{-p_1p_c}{n} & \frac{-p_2p_c}{n} & \cdots & \frac{p_c(1-p_c)}{n} \end{bmatrix}. \qquad (9.5.8)$$

Note that $\mathbf{GV}(\hat{\mathbf{p}})\mathbf{G}'$ is independent of n.

To apply this result we consider $F(\hat{\mathbf{p}})$. For large samples the asymptotic result obtained above for $\sqrt{n}\,[F(\hat{\mathbf{p}})-F(\mathbf{p})]$ permits us to consider $F(\hat{\mathbf{p}})$ as approximately normal, with mean $F(\mathbf{p})$ and variance (in the GSK notation)

$$V[F(\hat{\mathbf{p}})]=\mathbf{HV}(\hat{\mathbf{p}})\mathbf{H}', \qquad (9.5.9)$$

where $\mathbf{H}=[h_1\ h_2 \ldots h_c]$ and $h_i=F_i'(\mathbf{p})$. Because both $\mathbf{H}$ and $\mathbf{V}(\hat{\mathbf{p}})$ involve the unknown class probabilities, we will, as usual, substitute the appropriate observed relative frequencies and indicate this as follows:

$$\hat{V}[F(\hat{\mathbf{p}})]=\hat{\mathbf{H}}\hat{\mathbf{V}}(\hat{\mathbf{p}})\hat{\mathbf{H}}'. \qquad (9.5.10)$$

It follows that, for large samples, the ratio

$$\frac{F(\hat{\mathbf{p}})-F(\mathbf{p})}{\sqrt{\hat{V}[F(\hat{\mathbf{p}})]}} \qquad (9.5.11)$$

is distributed approximately as a standard normal variable. Alternatively,

$$[F(\hat{\mathbf{p}})-F(\mathbf{p})]'\hat{V}[F(\hat{\mathbf{p}})]^{-1}[F(\hat{\mathbf{p}})-F(p)] \qquad (9.5.12)$$

has, for large samples, approximately the χ^2 distribution with one degree of freedom.

Now let us apply this result to the case of independence in the 2×2 contingency table. Recall that in this case $F(\mathbf{p})=p_1p_4-p_2p_3$, the null hypothesis is $H_0: F(\mathbf{p})=0$, and, if the two variables are independent, then

$$\chi^2=F(\hat{\mathbf{p}})'\hat{V}[F(\hat{\mathbf{p}})]^{-1}F(\hat{\mathbf{p}})$$

has the χ^2 distribution with one degree of freedom. For the data of Table 9.3.2 we have

$$F(\hat{\mathbf{p}})=\hat{p}_1\hat{p}_4-\hat{p}_2\hat{p}_3=\left(\frac{8}{46}\right)\left(\frac{10}{46}\right)-\left(\frac{8}{46}\right)\left(\frac{20}{46}\right),$$

$$\hat{\mathbf{V}}(\hat{\mathbf{p}})=\frac{1}{46^3}\begin{bmatrix} (8)(38) & (-8)(8) & (-8)(20) & (-8)(10) \\ (-8)(8) & (8)(38) & (-8)(20) & (-8)(10) \\ (-8)(20) & (-8)(20) & (20)(26) & (-20)(10) \\ (-8)(10) & (-8)(10) & (-20)(10) & (10)(36) \end{bmatrix}$$

$$=\frac{1}{46^3}\begin{bmatrix} 304 & -64 & -160 & -80 \\ -64 & 304 & -160 & -80 \\ -160 & -160 & 520 & -200 \\ -80 & -80 & -200 & 360 \end{bmatrix}.$$

To calculate **H**, we have

$$h_1 = \left.\frac{\partial F(\hat{\mathbf{p}})}{\partial \hat{p}_1}\right|_{\hat{\mathbf{p}}=\mathbf{p}} = p_4, \quad \text{and similarly,} \quad h_2 = -p_3, h_3 = -p_2, \quad \text{and} \quad h_4 = p_1.$$

Therefore, in this example, we estimate **H** by $\hat{\mathbf{H}}$ as follows:

$$\hat{\mathbf{H}} = [\hat{p}_4 \quad -\hat{p}_3 \quad -\hat{p}_2 \quad \hat{p}_1] = \left[\frac{10}{46} \quad -\frac{20}{46} \quad -\frac{8}{46} \quad \frac{8}{46}\right],$$

$$\hat{V}[F(\hat{\mathbf{p}})] = \hat{\mathbf{H}}\hat{\mathbf{V}}(\hat{\mathbf{p}})\hat{\mathbf{H}}' = \frac{246720}{46^5},$$

and

$$\hat{V}[F(\hat{\mathbf{p}})]^{-1} = \frac{46^5}{246720}.$$

Then

$$\chi^2 = F(\hat{\mathbf{p}})' V[F(\hat{\mathbf{p}})]^{-1} F(\hat{\mathbf{p}}) = \frac{46^5}{246720}\left[\left(\frac{8}{46}\right)\left(\frac{10}{46}\right) - \left(\frac{8}{46}\right)\left(\frac{20}{46}\right)\right]^2$$

$$= \frac{294400}{246720} = 1.193.$$

compared to the value of 1.217 for the "usual" χ^2 statistic previously calculated.

It is worthwhile to consider yet another statistic that is appropriate for testing the hypothesis of independence, that is $H_0: p_1p_4 = p_2p_3$. This is equivalent to the question of whether the ratio, $p_1p_4/p_2p_3 = 1$. The ratio p_1p_4/p_2p_3 is the well-known *odds ratio*, that is, the odds in favor of a positive response to Drug A, given a positive response to Drug C, divided by the odds in favor of a positive response to Drug A, given a negative response to Drug C. In the notation of Table 9.5.1, this ratio is

$$\frac{P(X=1|Y=1)}{P(X=0|Y=1)} \bigg/ \frac{P(X=1|Y=0)}{P(X=0|Y=0)}$$

$$= \frac{P(X=1, Y=1)/P(Y=1)}{P(X=0, Y=1)/P(Y=1)} \bigg/ \frac{P(X=1, Y=0)/P(Y=0)}{P(X=0, Y=0)/P(Y=0)}$$

$$= \frac{P(X=1, Y=1)}{P(X=0, Y=1)} \bigg/ \frac{P(X=1, Y=0)}{P(X=0, Y=0)} = \frac{p_1/p_2}{p_3/p_4} = \frac{p_1/p_3}{p_2/p_4},$$

that is, the ratio is symmetrical. We should note that the odds ratio has a long history as an important tool in the analysis of the 2×2 contingency table (see, for example, Mosteller, 1968). Note also that $p_1p_4/p_2p_3 = 1$ if and only if $\ln(p_1p_4/p_2p_3) = 0$.

We may use the machinery provided above to answer the question of independence in the 2×2 table, with the help of the logarithm of the odds ratio.

Then

$$F(\mathbf{p}) = \ln p_1 - \ln p_2 - \ln p_3 + \ln p_4$$

and the null hypothesis is $H_0: F(\mathbf{p}) = 0$. In this case,

$$\mathbf{H} = \begin{bmatrix} \frac{1}{p_1} & -\frac{1}{p_2} & -\frac{1}{p_3} & \frac{1}{p_4} \end{bmatrix},$$

$$V(\hat{\mathbf{p}}) = \boldsymbol{\Sigma}_{\hat{p}} = \frac{1}{n}\begin{bmatrix} p_1(1-p_1) & -p_1p_2 & -p_1p_3 & -p_1p_4 \\ -p_1p_2 & p_2(1-p_2) & -p_2p_3 & -p_2p_4 \\ -p_1p_3 & -p_2p_3 & p_3(1-p_3) & -p_3p_4 \\ -p_1p_4 & -p_2p_4 & -p_3p_4 & p_4(1-p_4) \end{bmatrix},$$

$$\mathbf{HV}(\hat{\mathbf{p}}) = \frac{1}{n}[1 \quad -1 \quad -1 \quad 1],$$

and

$$V[F(\hat{\mathbf{p}})] = \mathbf{HV}(\hat{p})\mathbf{H}' = \frac{1}{n}\left(\frac{1}{p_1} + \frac{1}{p_2} + \frac{1}{p_3} + \frac{1}{p_4}\right).$$

For the data of Table 9.3.2,

$$\hat{V}[F(\hat{\mathbf{p}})] = \hat{\mathbf{H}}\hat{\mathbf{V}}(\hat{\mathbf{p}})\hat{\mathbf{H}}' = \frac{1}{46}\left(\frac{46}{8} + \frac{46}{8} + \frac{46}{20} + \frac{46}{10}\right) = .4,$$

and

$$\hat{V}[F(\hat{\mathbf{p}})]^{-1} = \frac{1}{.4} = 2.5.$$

Then $F(\hat{\mathbf{p}}) = \ln(\hat{p}_1\hat{p}_4/\hat{p}_2\hat{p}_3) = \ln(1/2) = -\ln 2 = -.69315$ and $\chi^2 = F(\hat{p})'\hat{V}(\hat{\mathbf{p}})^{-1}F(\hat{p}) = (.69315)^2(2.5) = 1.201$. This result is in close agreement with the values obtained for the alternative χ^2 statistics computed to answer the question of whether the responses to the two treatments are independent.

9.5.2 More Than One Nonlinear Combination of Multinomial Class Probabilities

The data in Table 9.5.3 are taken from a paper by Cochran (1954). It compares the previous infant-loss histories of mothers of Baltimore school children who had been identified by their teachers as behavior problems with the infant-loss histories of a comparable group whose children had not been so designated. The birth order of the "index" child is an important variable because the incidence of birth accidents such as stillbirths increases with birth order.

The model that one might assume here is that one is dealing with six independent sequences of binomial trials, that is, three pairs, with one pair associated with each birth order category, and probabilities as indicated in Table 9.5.4.

There are various questions that might have prompted the collection of the data of Table 9.5.3. For example, for each birth order, is the probability

TABLE 9.5.3 NUMBER OF MOTHERS WITH PREVIOUS INFANT LOSSES BY TYPE AND BIRTH ORDER OF INDEX CHILD

		Number of mothers		
Birth order	*Type of index child*	*Losses*	*No losses*	*Total*
2	Problem	20	82	102
	Control	10	54	64
	Total	30	136	166
3–4	Problem	26	41	67
	Control	16	30	46
	Total	42	71	113
5+	Problem	27	22	49
	Control	14	23	37
	Total	41	45	86

TABLE 9.5.4 PROBABILITIES THAT MOTHERS HAD PREVIOUS INFANT LOSSES, ACCORDING TO BIRTH ORDER AND TYPE OF INDEX CHILD

		Loss category		
Birth Order	*Index Child*	*Losses*	*No Losses**	*Total*
2	Problem	p_{11}	q_{11}	1
	Control	p_{21}	q_{21}	1
3–4	Problem	p_{12}	q_{12}	1
	Control	p_{22}	q_{22}	1
5+	Problem	p_{13}	q_{13}	1
	Control	p_{23}	q_{23}	1

*$q_{ij}=1-p_{ij}$

of losses the same for mothers of problem children as for mothers of control children? That is, are $p_{11}=p_{21}$, $p_{12}=p_{22}$, and $p_{13}=p_{23}$? One possible way to answer this question is to consider each birth order separately and compute the usual χ^2 statistic for testing homogeneity. For example, for birth order 2, we have

$$\chi^2=\frac{[20(54)-82(10)]^2 166}{(30)(136)(102)(64)}=.4213,$$

for birth order 3–4 we have

$$\chi^2=\frac{[26(30)-41(16)]^2 113}{(42)(71)(67)(46)}=.1891,$$

and, for birth order 5+,

$$\chi^2=\frac{[27(23)-22(14)]^2 86}{(41)(45)(49)(37)}=2.5188.$$

Each of these χ^2s may be examined to answer the question of homogeneity separately for each birth-order group. Alternatively, because the three birth-order groups are independent, one might choose to add the three values and compare the result with percentiles of the χ^2 distribution with three degrees of freedom.

Berkson (1968) discussed another problem of interest concerning these data. This is the question of whether the relationship between type of child and loss history as measured by the odds ratio is the same for the three birth-order groups. For each pair within a birth order, one may consider the odds ratio

$$\frac{p_{1i}/q_{1i}}{p_{2i}/q_{2i}}, \ (i=1,2,3),$$

an odds ratio of one being equivalent to $p_{1i}=p_{2i}$. Thus, a statistical hypothesis of interest may be

$$H_0: \frac{p_{11}/q_{11}}{p_{21}/q_{21}} = \frac{p_{12}/q_{12}}{p_{22}/q_{22}} = \frac{p_{13}/q_{13}}{p_{23}/q_{23}}. \tag{9.5.13}$$

Bartlett (1935) first considered problems of this type and termed the null hypothesis of Equation 9.5.13 that of "no interaction." Berkson (1968) solved the problem for the data of Table 9.5.3 with his method of *logit analysis* (logit = log odds), obtaining a χ^2 value of .849107. Grizzle (1961) had previously considered the same problem. He calculated the usual χ^2 statistic (substituting the maximum likelihood estimates of the p_{ij}s to obtain the appropriate "expected" frequencies), and obtained a χ^2 value of .851. However, an iterative procedure that involves laborious calculations was used to obtain the maximum likelihood estimates.

We now solve the problem of testing the null hypothesis of Equation 9.5.13 by a simple extension of the method used in the previous section. In the current problem we have more than one multinomial as well as more than one nonlinear function of the class probabilities to consider. Because of the structure of the odds ratios, it is convenient to convert these functions to logarithms. We may then write H_0 as follows: $H_0: F_1(\mathbf{p}) = F_2(\mathbf{p}) = 0$, where

$$\begin{aligned} F_1(\mathbf{p}) &= \ln\left[\frac{p_{11}/q_{11}}{p_{21}/q_{21}}\right] - \ln\left[\frac{p_{12}/q_{12}}{p_{22}/q_{22}}\right] \\ &= \ln p_{11} - \ln q_{11} - \ln p_{21} + \ln q_{21} - \ln p_{12} + \ln q_{12} + \ln p_{22} - \ln q_{22} \end{aligned} \tag{9.5.14}$$

and

$$\begin{aligned} F_2(\mathbf{p}) &= \ln\left[\frac{p_{11}/q_{11}}{p_{21}/q_{21}}\right] - \ln\left[\frac{p_{13}/q_{13}}{p_{23}/q_{23}}\right] \\ &= \ln p_{11} - \ln q_{11} - \ln p_{21} + \ln q_{21} - \ln p_{13} + \ln q_{13} + \ln p_{23} - \ln q_{23}. \end{aligned} \tag{9.5.15}$$

Here

$$\mathbf{p}=\begin{bmatrix} p_{11} \\ q_{11} \\ p_{21} \\ q_{21} \\ p_{12} \\ q_{12} \\ p_{22} \\ q_{22} \\ p_{13} \\ q_{13} \\ p_{13} \\ q_{23} \end{bmatrix}, \tag{9.5.16}$$

and $\hat{\mathbf{p}}$, as usual, has the structure of $\mathbf{p}$, with the class probabilities replaced by the corresponding observed relative frequencies. In this case,

$$\mathbf{V}(\hat{\mathbf{p}})=\begin{bmatrix} \mathbf{V}_{11} & & & & & \\ & \mathbf{V}_{21} & & 0 & & \\ & & \mathbf{V}_{12} & & & \\ & & & \mathbf{V}_{22} & & \\ & 0 & & & \mathbf{V}_{13} & \\ & & & & & \mathbf{V}_{23} \end{bmatrix}, \tag{9.5.17}$$

where

$$\mathbf{V}_{ij}=\begin{bmatrix} \dfrac{p_{ij}q_{ij}}{n_{ij}} & \dfrac{-p_{ij}q_{ij}}{n_{ij}} \\ \dfrac{-p_{ij}q_{ij}}{n_{ij}} & \dfrac{p_{ij}q_{ij}}{n_{ij}} \end{bmatrix},$$

and n_{ij} is the number of trials in the binomial whose probabilities have the pair ij of subscripts.

$\mathbf{H}$ is the 2×12 matrix whose first row consists of the derivatives of $F_1(\hat{\mathbf{p}})$ with respect to the components of $\hat{\mathbf{p}}$ (in order) evaluated at $\hat{\mathbf{p}}=\mathbf{p}$, and whose second row consists similarly of the derivatives of $F_2(\hat{\mathbf{p}})$.

Thus,

$$\mathbf{H}=\begin{bmatrix} \dfrac{1}{p_{11}} & \dfrac{-1}{q_{11}} & \dfrac{-1}{p_{21}} & \dfrac{1}{q_{21}} & \dfrac{-1}{p_{12}} & \dfrac{1}{q_{12}} & \dfrac{1}{p_{22}} & \dfrac{-1}{q_{22}} & 0 & 0 & 0 & 0 \\ \dfrac{1}{p_{11}} & \dfrac{-1}{q_{11}} & \dfrac{-1}{p_{21}} & \dfrac{1}{q_{21}} & 0 & 0 & 0 & 0 & \dfrac{-1}{p_{13}} & \dfrac{1}{q_{13}} & \dfrac{1}{p_{23}} & \dfrac{-1}{q_{23}} \end{bmatrix} \tag{9.5.18}$$

and the variance-covariance matrix of the components of

$$\mathbf{F}(\hat{\mathbf{p}})=\begin{bmatrix} F_1(\hat{\mathbf{p}}) \\ F_2(\hat{\mathbf{p}}) \end{bmatrix}$$

is

$$\boldsymbol{\Sigma}_{\hat{F}} = \mathbf{HV}(\hat{\mathbf{p}})\mathbf{H}' = \begin{bmatrix} a+b & a \\ a & a+c \end{bmatrix}, \tag{9.5.19}$$

where

$$a = \frac{1}{n_{11}p_{11}q_{11}} + \frac{1}{n_{21}p_{21}q_{21}},$$

$$b = \frac{1}{n_{12}p_{12}q_{12}} + \frac{1}{n_{22}p_{22}q_{22}},$$

and

$$c = \frac{1}{n_{13}p_{13}q_{13}} + \frac{1}{n_{23}p_{23}q_{23}}.$$

It is easily seen that

$$(\mathbf{HV}(\hat{\mathbf{p}})\mathbf{H}')^{-1} = \boldsymbol{\Sigma}_{\hat{F}}^{-1} = \frac{\begin{bmatrix} a+c & -a \\ -a & a+b \end{bmatrix}}{ab+bc+ac}.$$

We have, for the data of Table 9.5.3,

$$F_1(\hat{\mathbf{p}}) = \ln\left[\frac{\hat{p}_{11}\hat{q}_{21}\hat{q}_{12}\hat{p}_{22}}{\hat{q}_{11}\hat{p}_{21}\hat{p}_{12}\hat{q}_{22}}\right] = \ln\left(\frac{20}{82}\right)\left(\frac{54}{10}\right)\left(\frac{41}{26}\right)\left(\frac{16}{30}\right)$$

$$= \ln\frac{72}{65} = .102279,$$

and

$$F_2(\hat{\mathbf{p}}) = \ln\left[\frac{\hat{p}_{11}\hat{q}_{21}\hat{q}_{13}\hat{p}_{23}}{\hat{q}_{11}\hat{p}_{21}\hat{p}_{13}\hat{q}_{23}}\right] = \ln\left[\frac{616}{943}\right] = -.425819.$$

Putting $\hat{a} = 1/n_{11}\hat{p}_{11}\hat{q}_{11} + 1/n_{21}\hat{p}_{21}\hat{q}_{21}$, we have

$$\hat{a} = \frac{1}{(102)\left(\frac{20}{102}\right)\left(\frac{80}{102}\right)} + \frac{1}{(64)\left(\frac{10}{64}\right)\left(\frac{54}{64}\right)} = .182268,$$

and similarly, $\hat{b} = .158685$ and $\hat{c} = .197398$. Then,

$$\hat{\boldsymbol{\Sigma}}_{\hat{F}}^{-1} = (\hat{\mathbf{H}}\hat{\mathbf{V}}(\hat{\mathbf{p}})\hat{\mathbf{H}}')^{-1} = \frac{1}{.096227}\begin{bmatrix} .399666 & -.182268 \\ -.182268 & .340954 \end{bmatrix},$$

and the value of the χ^2 statistic for testing H_0 is $\mathbf{F}(\hat{\mathbf{p}})'\hat{\boldsymbol{\Sigma}}_{\hat{F}}^{-1}\mathbf{F}(\hat{\mathbf{p}}) = .84873$, close to the values obtained by Berkson and Grizzle. This χ^2 has two degrees of freedom and has a significance probability that is greater than .30.

Although the method of testing a statistical hypothesis concerning more than one nonlinear combination of more than one multinomial distribution was illustrated by only one specific problem, it is easy to see how such a problem can be solved with the help of a Wald χ^2 statistic in general.

9.5.3 Hypotheses About Multinomial Class Probabilities Having the Traditional ANOVA Structure

It has long been of interest to treat qualitative variables by the familiar regression and ANOVA techniques that have been found so useful in the case of quantitative normal variables. As a matter of fact, Cochran (1954) did suggest ANOVA methods for solving certain categorical data problems. Cox (1970) discusses the difficulties that arise in the case of binary data. He points out that, in the case of an underlying categorical variable, the variances of the observed counts in the various classes depend on the probabilities of the outcomes, so that the assumption of homoscedasticity that is usually made in the quantitative case is necessarily violated. In addition, least-squares or normal-theory maximum likelihood estimation will not necessarily yield estimates of probabilities that fall between zero and one.

The lack of homoscedasticity can be overcome with the help of weighted least-squares procedures, as we will explain later. Consider again the question of interaction that we have discussed previously for the Cochran data. We may propose the following reparametrization of the six unknown binomial probabilities for the model of Table 9.5.4, that is, we may replace the original unknown parameters by new ones, as follows.

Let p_{ij} represent the probability of previous loss history for a mother with type of child i, ($i=1$ for a problem child, $i=2$ for a control child), and birth order j($j=1$ for birth order 2, $j=2$ for birth order 3 or 4, $j=3$ for birth order 5 or more). The model proposed is $p_{ij}=\mu+\alpha_i+\beta_j+(\alpha\beta)_{ij}$, $i=1,2$, $j=1,2,3$, where μ is a constant, α_i represents the main effect of type of child i, β_j represents the main effect of birth order j, and $(\alpha\beta)_{ij}$, the interaction of type of child i with birth order j, and where the constants are subject to the usual ANOVA restraints: $\Sigma_i\alpha=\Sigma_j\beta=\Sigma_i(\alpha\beta)_{ij}=\Sigma_j(\alpha\beta)_{ij}=0$. Under this model, we may ask whether there is no interaction by testing the hypothesis $H_0:(\alpha\beta)_{ij}=0$, $i=1,2;\ j=1,2,3$. Thus, if H_0 is true, the appropriate model is

$$p_{ij}=\mu+\alpha_i+\beta_j,\ (i=1,2;\ j=1,2,3), \tag{9.5.21}$$

a model involving only four unknown parameters because of the restrictions on the α_is and β_js. The test for the hypothesis of no interaction that is proposed in the GSK paper is referred to there as a test of the goodness of fit of the model in Equation 9.5.21, and is based upon standard weighted least-squares theory (see for example Draper and Smith 1968) and a Wald statistic as defined by Equation 9.4.5. It is also possible, assuming that Equation 9.5.21 holds, to test the hypothesis that one or both of the main effects can be zero. The GSK methods will be carried out using the type-of-child × birth-order × previous-loss-history data.

First, because some readers may not be familiar with the theory used in the GSK paper, we review the main results needed, first for the normal case (with and without the assumption of equal variances), and then for the asymptotically normal case. The latter is of importance because we are dealing with multinomial distributions, and the variance of the observed proportion in a multinomial class depends on the class probability, as well as on the number of trials.

Least-Squares Theory for Independent Normal Variables with Equal Variances. In Chapter 4 we showed that if

$$\mathbf{Y}=\begin{bmatrix} Y_1 \\ Y_2 \\ \vdots \\ Y_n \end{bmatrix}$$

is a vector random variable with independent normally distributed components having the same variance, σ^2, and

$$E(\mathbf{Y})=\mathbf{X}\boldsymbol{\beta},$$

where $\mathbf{X}$ is a matrix of known constants with elements X_{ij}, $(i=1,2,\ldots,n;$ $j=1,2,\ldots,p)$, of rank $p<n$, and $\boldsymbol{\beta}$ a vector of unknown constants, then the least-squares estimators of the components of $\boldsymbol{\beta}$ are obtained by solving a set of normal equations whose matrix form is

$$\mathbf{X}'\mathbf{Y}=\mathbf{X}'\mathbf{X}\boldsymbol{\beta}. \tag{9.5.22}$$

The solution is

$$\hat{\boldsymbol{\beta}}=(\mathbf{X}'\mathbf{X})^{-1}\mathbf{X}'\mathbf{Y}. \tag{9.5.23}$$

By the Gauss-Markov Theorem (see for example Rao, 1973),

$$\frac{\sum_i \left(Y_i-X_{i1}\hat{\beta}_1-X_{i2}\hat{\beta}_2-\ldots-X_{ip}\hat{\beta}_p\right)^2}{\sigma^2} \tag{9.5.24}$$

is distributed as χ^2 with $n-p$ degrees of freedom. Now, the numerator of 9.5.24 is equal to

$$\sum_i Y_i\left(Y_i-X_{i1}\hat{\beta}_1-X_{i2}\hat{\beta}_2-\ldots-X_{ip}\hat{\beta}_p\right)=\mathbf{Y}'\mathbf{Y}-\hat{\boldsymbol{\beta}}'\mathbf{X}'\mathbf{Y}, \tag{9.5.25}$$

because

$$\sum_i X_{ij}\left(Y_i-X_{i1}\hat{\beta}_1-X_{i2}\hat{\beta}_2-\ldots-X_{ip}\hat{\beta}_p\right)=0,\ (j=1,2,\ldots,n). \tag{9.5.26}$$

This follows from the fact that $\hat{\boldsymbol{\beta}}$ is the solution to Equation 9.5.22, so that

$$\mathbf{X}'\mathbf{Y}-\mathbf{X}'\mathbf{X}\hat{\boldsymbol{\beta}}=0, \tag{9.5.27}$$

and the left hand side of Equation 9.5.26 is the jth row of the left hand side of Equation 9.5.27. Thus, by the Gauss-Markov Theorem, the distribution of Equation 9.5.24, which may be written as

$$(\mathbf{Y}'\mathbf{Y}-\hat{\boldsymbol{\beta}}'\mathbf{X}'\mathbf{Y})/\sigma^2, \tag{9.5.28}$$

is the χ^2 distribution with $n-p$ degrees of freedom. A goodness-of-fit test for $H_0: E(\mathbf{Y})=\mathbf{X}\boldsymbol{\beta}$ could therefore be performed by calculating the value of 9.5.28 and comparing it with the appropriate percentile of the χ^2 distribution with $n-p$ degrees of freedom. Of course, it rarely happens that the value of σ^2 is known, so that the test usually performed for H_0 involves an F ratio.

Now suppose that one is ready to assume that H_0 is true and wishes to test the hypothesis that a set of known linear combinations of the β_js is equal to zero, say for example $\mathbf{C}\boldsymbol{\beta}=\mathbf{0}$ where $\mathbf{C}$ is the following matrix of rank r:

$$\mathbf{C}=\begin{bmatrix} 1 & 0 & 0 & \cdots & 0 & 0 & \cdots & 0 \\ 0 & 1 & 0 & \cdots & 0 & 0 & \cdots & 0 \\ \vdots & \vdots & \vdots & \vdots & \vdots & \vdots & \vdots & \vdots \\ 0 & 0 & 0 & \cdots & 1 & 0 & \cdots & 0 \end{bmatrix},$$

a matrix of r rows and p columns ($r<p$), whose first r columns form the identity matrix of rank r. In this case the hypothesis $H_0:\mathbf{C}\boldsymbol{\beta}=\mathbf{0}$ says that the first r β_js are equal to zero. This hypothesis can be tested with the help of a Wald statistic, as follows.

Recall that

$$\hat{\boldsymbol{\beta}}=\mathbf{AY}, \text{ where } \mathbf{A}=(\mathbf{X}'\mathbf{X})^{-1}\mathbf{X}', \tag{9.5.29}$$

that is, $\hat{\boldsymbol{\beta}}$ is a set of known linear combinations of the components of $\mathbf{Y}$, and therefore the components of $\hat{\boldsymbol{\beta}}$ have a multivariate normal distribution with

$$E(\hat{\boldsymbol{\beta}})=(\mathbf{X}'\mathbf{X})^{-1}\mathbf{X}'E(\mathbf{Y})=(\mathbf{X}'\mathbf{X})^{-1}\mathbf{X}'\mathbf{X}\boldsymbol{\beta}=\boldsymbol{\beta} \tag{9.5.30}$$

and variance-covariance matrix

$$\boldsymbol{\Sigma}_{\hat{\beta}}=\mathbf{A}\boldsymbol{\Sigma}_{\mathbf{Y}}\mathbf{A}'=(\mathbf{X}'\mathbf{X})^{-1}\mathbf{X}'\mathbf{I}\sigma^2\mathbf{X}(\mathbf{X}'\mathbf{X})^{-1}=(\mathbf{X}'\mathbf{X})^{-1}\sigma^2. \tag{9.5.31}$$

Now, because $\mathbf{C}\hat{\boldsymbol{\beta}}$ is a set of known linear combinations of the components of $\hat{\boldsymbol{\beta}}$, the components of $\mathbf{C}\hat{\boldsymbol{\beta}}$ in turn have a multivariate normal distribution with

$$E(\mathbf{C}\hat{\boldsymbol{\beta}})=\mathbf{C}\boldsymbol{\beta}=\mathbf{0}, \tag{9.5.32}$$

if $H_0:\mathbf{C}\boldsymbol{\beta}=\mathbf{0}$ is true, and with variance-covariance matrix

$$\boldsymbol{\Sigma}_{\mathbf{C}\hat{\beta}}=\mathbf{C}\boldsymbol{\Sigma}_{\hat{\beta}}\mathbf{C}'=\mathbf{C}(\mathbf{X}'\mathbf{X})^{-1}\mathbf{C}'\sigma^2. \tag{9.5.33}$$

The Wald statistic derived from the exponent of this multivariate normal density is therefore

$$Q(\mathbf{C}\hat{\boldsymbol{\beta}})=\frac{(\mathbf{C}\hat{\boldsymbol{\beta}})'\left[\mathbf{C}(\mathbf{X}'\mathbf{X})^{-1}\mathbf{C}'\right]^{-1}(\mathbf{C}\hat{\boldsymbol{\beta}})}{\sigma^2}, \tag{9.5.34}$$

which has the χ^2 distribution, if $\mathbf{C}\boldsymbol{\beta}=\mathbf{0}$, with r degrees of freedom. Thus $H_0:\mathbf{C}\boldsymbol{\beta}=\mathbf{0}$ may be tested by calculating the value of $Q(\mathbf{C}\hat{\boldsymbol{\beta}})$ and comparing it with the desired percentile of the appropriate χ^2 distribution.

Weighted Least-Squares Theory for Variables with Unequal Variances. Suppose now that $\mathbf{Y}$ is a vector satisfying all the conditions as stated in the previous section except that the variance of the ith component of $\mathbf{Y}$, $\sigma^2(Y_i)=\sigma_i^2$ is not necessarily the same for all i.

Let $w_i = 1/\sigma_i$, and let

$$\mathbf{W} = \begin{bmatrix} w_1 & 0 & 0 & \dots & 0 \\ 0 & w_2 & 0 & \dots & 0 \\ \vdots & \vdots & \vdots & & \vdots \\ 0 & 0 & 0 & \dots & w_n \end{bmatrix}, \qquad (9.5.35)$$

and $\mathbf{Z} = \mathbf{WY}$. If $E(\mathbf{Y}) = \mathbf{X}\boldsymbol{\beta}$, then $E(\mathbf{Z}) = \mathbf{WX}\boldsymbol{\beta}$, and we can follow the steps of the previous section with $\mathbf{Y}$ replaced by $\mathbf{Z}$ and $\mathbf{X}$ replaced by $\mathbf{WX}$. Here, however, in place of σ^2, we have $\sigma^2(Z_i) = 1$. Equation 9.5.22 is replaced by

$$(\mathbf{WX})'\mathbf{Z} = (\mathbf{WX})'(\mathbf{WX})\boldsymbol{\beta} = \mathbf{X}'\mathbf{W}'\mathbf{WX}\boldsymbol{\beta},$$

or

$$\mathbf{X}'\mathbf{W}'\mathbf{Z} = \mathbf{X}'\boldsymbol{\Sigma}_{\mathbf{Y}}^{-1}\mathbf{X}\boldsymbol{\beta}. \qquad (9.5.36)$$

The solution to Equation 9.5.36 is obtained by premultiplying both sides by $(\mathbf{X}'\boldsymbol{\Sigma}_{\mathbf{Y}}^{-1}\mathbf{X})^{-1}$, giving

$$(\mathbf{X}'\boldsymbol{\Sigma}_{\mathbf{Y}}^{-1}\mathbf{X})^{-1}\mathbf{X}'\mathbf{W}'\mathbf{Z} = \hat{\boldsymbol{\beta}}, \qquad (9.5.37)$$

and because $\mathbf{WY} = \mathbf{Z}$,

$$\hat{\boldsymbol{\beta}} = (\mathbf{X}'\boldsymbol{\Sigma}_{\mathbf{Y}}^{-1}\mathbf{X})^{-1}\mathbf{X}'\mathbf{W}'\mathbf{WY} = (\mathbf{X}'\boldsymbol{\Sigma}_{\mathbf{Y}}^{-1}\mathbf{X})^{-1}\mathbf{X}'\boldsymbol{\Sigma}_{\mathbf{Y}}^{-1}\mathbf{Y} = \mathbf{AY}, \qquad (9.5.38)$$

with $\mathbf{A} = (\mathbf{X}'\boldsymbol{\Sigma}_{\mathbf{Y}}^{-1}\mathbf{X})^{-1}\mathbf{X}'\boldsymbol{\Sigma}_{\mathbf{Y}}^{-1}$.

The goodness-of-fit χ^2 statistic, given by Equation 9.5.28 becomes

$$\frac{\mathbf{Z}'\mathbf{Z} - \hat{\boldsymbol{\beta}}'(\mathbf{WX})'(\mathbf{WX})\hat{\boldsymbol{\beta}}}{1} = (\mathbf{WY})'(\mathbf{WY}) - \hat{\boldsymbol{\beta}}'\mathbf{X}'\mathbf{W}'\mathbf{WX}\hat{\boldsymbol{\beta}}$$
$$= \mathbf{Y}'\boldsymbol{\Sigma}_{\mathbf{Y}}^{-1}\mathbf{Y} - \hat{\boldsymbol{\beta}}'\mathbf{X}'\boldsymbol{\Sigma}_{\mathbf{Y}}^{-1}\mathbf{X}\hat{\boldsymbol{\beta}}. \qquad (9.5.39)$$

In this case, the variance-covariance matrix of the components of $\hat{\boldsymbol{\beta}}$ is

$$\boldsymbol{\Sigma}_{\hat{\boldsymbol{\beta}}} = \mathbf{A}\boldsymbol{\Sigma}_{\mathbf{Y}}\mathbf{A}' = (\mathbf{X}'\boldsymbol{\Sigma}_{\mathbf{Y}}^{-1}\mathbf{X})^{-1}\mathbf{X}'\boldsymbol{\Sigma}_{\mathbf{Y}}^{-1}\boldsymbol{\Sigma}_{\mathbf{Y}}\left[(\mathbf{X}'\boldsymbol{\Sigma}_{\mathbf{Y}}^{-1}\mathbf{X})^{-1}\mathbf{X}'\boldsymbol{\Sigma}_{\mathbf{Y}}^{-1}\right]' = (\mathbf{X}'\boldsymbol{\Sigma}_{\mathbf{Y}}^{-1}\mathbf{X})^{-1}. \qquad (9.5.40)$$

Thus, in this case, to test $H_0: \mathbf{C}\boldsymbol{\beta} = \mathbf{0}$, with $\mathbf{C}$ a known matrix of rank $r < p$, the Wald statistic is obtained by using

$$\boldsymbol{\Sigma}_{\mathbf{C}\hat{\boldsymbol{\beta}}} = \mathbf{C}\boldsymbol{\Sigma}_{\hat{\boldsymbol{\beta}}}\mathbf{C}' = \mathbf{C}(\mathbf{X}'\boldsymbol{\Sigma}_{\mathbf{Y}}^{-1}\mathbf{X})^{-1}\mathbf{C}', \qquad (9.5.41)$$

and is equal to

$$Q(\mathbf{C}\hat{\boldsymbol{\beta}}) = (\mathbf{C}\hat{\boldsymbol{\beta}})'\left[\mathbf{C}(\mathbf{X}'\boldsymbol{\Sigma}_{\mathbf{Y}}^{-1}\mathbf{X})^{-1}\mathbf{C}'\right]^{-1}\mathbf{C}\hat{\boldsymbol{\beta}}, \qquad (9.5.42)$$

which has the χ^2 distribution with r degrees of freedom if in fact $\mathbf{C}\boldsymbol{\beta} = \mathbf{0}$.

Weighted Least-Squares Theory Applied to Hypotheses About Multinomials. If $\mathbf{Y}$ is not normal, but only asymptotically so, the results stated in the preceding section hold asymptotically for analogous hypotheses (see Chernoff, 1956). We now return to the data of Table 9.5.3 to test some hypotheses about the underlying model for these data. The role of **Y** is taken by

$$\mathbf{Y}=\hat{\mathbf{p}}=\begin{bmatrix} \hat{p}_{11}=20/102=.19607843 \\ \hat{p}_{21}=10/64=.15625000 \\ \hat{p}_{12}=26/67=.38805970 \\ \hat{p}_{22}=16/46=.34782608 \\ \hat{p}_{13}=27/49=.55102040 \\ \hat{p}_{23}=14/37=.37837837 \end{bmatrix}, \tag{9.5.43}$$

$$\boldsymbol{\beta}=\begin{bmatrix} \mu \\ \alpha \\ \beta_1 \\ \beta_2 \end{bmatrix},$$

and

$$\mathbf{X}=\begin{bmatrix} 1 & 1 & 1 & 0 \\ 1 & -1 & 1 & 0 \\ 1 & 1 & 0 & 1 \\ 1 & -1 & 0 & 1 \\ 1 & 1 & -1 & -1 \\ 1 & -1 & -1 & -1 \end{bmatrix}.$$

Because $\boldsymbol{\Sigma}_\mathbf{Y}$ is unknown, we may replace it by a consistent estimator, $\hat{\boldsymbol{\Sigma}}_\mathbf{Y}$, without changing the limiting distributions involved. Replacing the unknown class probabilities, the p_{ij}s, by the corresponding observed class relative frequencies, the $\hat{p}_{ij}$s, the matrix $\hat{\boldsymbol{\Sigma}}_\mathbf{Y}^{-1}=\hat{\boldsymbol{\Sigma}}_{\hat{p}}^{-1}$ is the diagonal matrix

$$\hat{\boldsymbol{\Sigma}}_\mathbf{Y}^{-1}=\begin{bmatrix} \dfrac{n_{11}}{\hat{p}_{11}(1-\hat{p}_{11})} & & 0 \\ & \dfrac{n_{21}}{\hat{p}_{21}(1-\hat{p}_{21})} & \\ 0 & & \dfrac{n_{23}}{\hat{p}_{23}(1-\hat{p}_{23})} \end{bmatrix}.$$

The diagonal elements have the following values:

$$n_{11}/\hat{p}_{11}(1-\hat{p}_{11})=(102)^3/20(82)=647.07804$$
$$n_{21}/\hat{p}_{21}(1-\hat{p}_{21})=(64)^3/10(54)=485.45185$$
$$n_{12}/\hat{p}_{12}(1-\hat{p}_{12})=(67)^3/26(41)=282.14165$$
$$n_{22}/\hat{p}_{22}(1-\hat{p}_{22})=(46)^3/30(16)=202.78333$$
$$n_{13}/\hat{p}_{13}(1-\hat{p}_{13})=(47)^3/27(22)=198.06228$$
$$n_{23}/\hat{p}_{23}(1-\hat{p}_{23})=(37)^3/14(23)=157.30745.$$

The weighted least squares estimates of the elements of $\boldsymbol{\beta}$ under the hypothesis of no interaction, that is, under the conditions of Equation

9.5.21, are as follows:

$$\hat{\boldsymbol{\beta}} = (\mathbf{X}'\boldsymbol{\Sigma}_{\mathbf{Y}}^{-1}\mathbf{X})^{-1}\mathbf{X}'\boldsymbol{\Sigma}_{\mathbf{Y}}^{-1}\mathbf{Y} = \begin{bmatrix} 0.337 \\ 0.032 \\ -0.163 \\ 0.029 \end{bmatrix}.$$

The value of the χ^2 statistic calculated by using Equation 9.5.39 is 1.264, which for two degrees of freedom has an associated significance probability greater than .5. We would therefore conclude that there is no interaction between birth order and type of child.

If we are interested in testing further that only birth order may matter, that is that $\alpha = 0$, we would have,

$$\mathbf{C} = [0 \quad 1 \quad 0 \quad 0],$$

and the value of the χ^2 statistic for testing $H_0: \mathbf{C}\boldsymbol{\beta} = 0$, substituting the appropriate values in Equation 9.5.42, is 1.981. For one degree of freedom, this value of χ^2 has a significance probability greater than .10, supporting H_0. We therefore conclude that the probability of having a previous loss history does not depend on the type of index child.

The extension of the methods outlined in this section to higher-order contingency tables, as well as to more complicated functions of the class probabilities, is straightforward.

9.6 LOG-LINEAR MODELS

In the preceding section, we explored methods of testing hypotheses about linear combinations of the logarithms of class probabilities in one or more multinomials. There are many problems of interest in the analysis of categorical data that can be solved by assuming initially a *structure* of the logarithms of the class probabilities called a *log-linear model* that resembles the familiar ANOVA model in form, as the method outlined in Section 9.5.3 did. In fact, some of the ANOVA language has been adopted to describe the components of the model. There is a wealth of literature in this area, and it is constantly growing. We refer the reader to the texts by Fienberg (1977), Reynolds (1977), Bishop, Fienberg and Holland (1975) and Haberman (1974), and to the paper of Goodman (1970) both for more extensive discussion of the subject and for further references. We use the notation of Goodman in this section. We consider only the so-called hierarchical models, which are explained later. Consider, for example, an $I \times J \times K$ contingency table corresponding to the cross-classification of individuals according to variables A, B, and C, respectively.

Let p_{ijk} be the probability of simultaneously falling in the ith class of variable A, the jth class of variable B, and the kth class of variable G, ($i = 1,2,\ldots,I$, $j = 1,2,\ldots,J$, and $k = 1,2,\ldots,K$). Let $v_{ijk} = \ln p_{ijk}$. The model then is

$$v_{ijk} = \mu + \lambda_i^A + \lambda_j^B + \lambda_k^C + \lambda_{ij}^{AB} + \lambda_{ik}^{AC} + \lambda_{jk}^{BC} + \lambda_{ijk}^{ABC},$$

where μ is a constant that insures that $\Sigma_{ijk} p_{ijk} = 1$ in which Σ_{ijk} is equivalent to $\Sigma_i \Sigma_j \Sigma_k$. The symbols λ_i^A, λ_j^B, and λ_k^C denote the main effects, respectively, of the ith category of variable A, the jth category of variable B, and the kth category of variable C. The symbols λ_{ij}^{AB}, λ_{ik}^{AC}, and λ_{jk}^{BC} denote, respectively, the two-factor interactions of A_i and B_j, A_i and C_k, and B_j and C_k, where A_i represents the ith class of variable A, B_j the jth class of variable B, and C_k the kth class of variable C. Finally, λ_{ijk}^{ABC} denotes the three-factor interaction of A_i, B_j, and C_k. The notation, $\lambda^{AB} = 0$, means that $\lambda_{ij}^{AB} = 0$ for all pairs, ij, $i = 1, \ldots, I$ and $j = 1, \ldots, J$. Similarly, $\lambda^{AC} = 0$ means $\lambda_{ik}^{AC} = 0$ for all ik pairs and $\lambda^{BC} = 0$ means $\lambda_{jk}^{BC} = 0$ for all jk pairs.

The hierarchical model states that if any λ with no subscripts is zero, then every λ of higher order (containing more superscripts) that contains the *same* superscript(s) must also be zero. Thus, if $\lambda^{AB} = 0$, then also $\lambda^{ABC} = 0$, but λ^{AC} and λ^{BC} are not necessarily zero. Similarly, $\lambda^A = 0$ implies that $\lambda^{AB} = \lambda^{AC} = \lambda^{ABC} = 0$.

It is best to explain the interpretation of the main effects and the interactions in the context of the order of the contingency table. For example, $\lambda^{AB} = 0$ in the $I \times J$ contingency table does not mean the same thing as $\lambda^{AB} = 0$ in the $I \times J \times K$ case. Thus, we will discuss the model, the interpretation of the λs, and the hypotheses in terms of the order of the table. We consider first the two-way contingency table.

9.6.1 The Two-way ($I \times J$) Contingency Table

Consider again the data of Table 9.0.1 concerning the cross-classification of arrestees by heroin use and type of crime. We have here an example of an $I \times J$ contingency table, with $I = 4$ (the heroin-use categories) and $J = 5$ (the crime categories). Let A represent the heroin-use variable and B the crime-classification variable, so that p_{ij} is the probability that an arrestee falls in the ith category of A and the jth category of B. For simplicity, let us denote this probability by

$$p_{ij} = P(A = i, B = j). \tag{9.6.1}$$

The question of interest raised previously concerns the independence of A and B. The statistical hypothesis tested was

$$H_0 : p_{ij} = P(A = i)P(B = j); (i = 1, \ldots, I, j = 1, \ldots, J). \tag{9.6.2}$$

Let

$$v_{ij} = \ln p_{ij}. \tag{9.6.3}$$

The full log-linear model, corresponding to the full model in a two-way ANOVA is

$$v_{ij} = \mu + \lambda_i^A + \lambda_j^B + \lambda_{ij}^{AB}, \tag{9.6.4}$$

or

$$p_{ij}=\exp(\mu+\lambda_i^A+\lambda_j^B+\lambda_{ij}^{AB})=\exp(\mu)\exp(\lambda_i^A+\lambda_j^B+\lambda_{ij}^{AB}). \quad (9.6.5)$$

Because $\sum_i \sum_j p_{ij}=1$, μ must satisfy

$$\exp(\mu)\sum_i \sum_j \exp(\lambda_i^A+\lambda_j^B+\lambda_{ij}^{AB})=1. \quad (9.6.6)$$

The definitions of the λs involve averages of the v_{ij}s. The notation will reflect an average of v_{ij}s by placing a dot in the position of the variable that has been "averaged out." Thus,

$$\left.\begin{aligned} v_{i\cdot}&=\frac{1}{J}\sum_j v_{ij}\\ v_{\cdot j}&=\frac{1}{I}\sum_i v_{ij}\\ v_{\cdot\cdot}&=\frac{1}{IJ}\sum_i\sum_j v_{ij}\end{aligned}\right\} \quad (9.6.7)$$

and (the last line above)

We now define the λs,

$$\left.\begin{aligned}\lambda_i^A&=v_{i\cdot}-v_{\cdot\cdot}\\ \lambda_j^B&=v_{\cdot j}-v_{\cdot\cdot}\end{aligned}\right\} \quad (9.6.8)$$

where λ_i^A and λ_j^B are called the main effects of A_i and B_j, respectively, and

$$\begin{aligned}\lambda_{ij}^{AB}&=v_{ij}-v_{i\cdot}-v_{\cdot j}+v_{\cdot\cdot}\\ &=(v_{ij}-v_{\cdot\cdot})-\left[(v_{i\cdot}-v_{\cdot\cdot})+(v_{\cdot j}-v_{\cdot\cdot})\right]\end{aligned} \quad (9.6.9)$$

is called the two-factor interaction of A_i and B_j. Note that if we call $v_{ij}-v_{\cdot\cdot}$ the joint effect of A_i and B_j, then λ_{ij}^{AB} is the excess of the joint effect of A_i and B_j over the sum of the main effects of A_i and B_j, just as in the traditional ANOVA case. Note also that it follows from the above definitions that

$$\sum_i \lambda_i^A=\sum_j \lambda_j^B=\sum_i \lambda_{ij}^{AB}=\sum_j \lambda_{ij}^{AB}=\sum_i\sum_j \lambda_{ij}^{AB}=0. \quad (9.6.10)$$

In order to examine the roles of components of the model (Equation 9.6.4) in terms of the class probabilities, consider first a question of great interest in the analogous traditional ANOVA case, that is, $\lambda^{AB}=0$. If $\lambda_{ij}^{AB}=0$ for all values of i and j, what does this tell us about the class probabilities? If $\lambda^{AB}=0$, then from Equation 9.6.4,

$$v_{ij}=\ln p_{ij}=\mu+\lambda_i^A+\lambda_j^B, \quad (9.6.11)$$

that is, we have the "additive" case. Because $\Sigma_i\lambda_i^A=\Sigma_j\lambda_j^B=0$, in this case the two-way table of v_{ij}s would be as given in Table 9.6.1

TABLE 9.6.1 TWO-WAY TABLE OF v_{ij}s WHEN $v_{ij}=\mu+\lambda_i^A+\lambda_j^B$: THE ADDITIVE CASE

	Category B(j)						
Category A(i)	1	2	...	j	...	J	*Total*
1	$\mu+\lambda_1^A+\lambda_1^B$	$\mu+\lambda_1^A+\lambda_2^B$	...	$\mu+\lambda_1^A+\lambda_j^B$	...	$\mu+\lambda_1^A+\lambda_J^B$	$J(\mu+\lambda_1^A)$
2	$\mu+\lambda_2^A+\lambda_1^B$	$\mu+\lambda_2^A+\lambda_2^B$	...	$\mu+\lambda_2^A+\lambda_j^B$	...	$\mu+\lambda_2^A+\lambda_J^B$	$J(\mu+\lambda_2^A)$
⋮	⋮	⋮		⋮		⋮	⋮
i	$\mu+\lambda_i^A+\lambda_1^B$	$\mu+\lambda_i^A+\lambda_2^B$	...	$\mu+\lambda_i^A+\lambda_j^B$	...	$\mu+\lambda_i^A+\lambda_J^B$	$J(\mu+\lambda_i^A)$
⋮	⋮	⋮		⋮		⋮	⋮
I	$\mu+\lambda_I^A+\lambda_1^B$	$\mu+\lambda_I^A+\lambda_2^B$	...	$\mu+\lambda_I^A+\lambda_j^B$	...	$\mu+\lambda_I^A+\lambda_J^B$	$J(\mu+\lambda_I^A)$
Total	$I(\mu+\lambda_1^B)$	$I(\mu+\lambda_2^B)$	...	$I(\mu+\lambda_j^B)$	...	$I(\mu+\lambda_J^B)$	$IJ\mu$

Note that in Table 9.6.1, for two v_{ij}s in the same *row*, say i, we have $v_{ij}-v_{ij'}=\lambda_j^B-\lambda_{j'}^B$, where $j\neq j'$ represent two columns. Thus, the difference between two v_{ij}s in the same *row* depends only on the *columns*, and for a given pair of columns is the same for *every* row. But if $v_{ij}-v_{ij'}=\ln p_{ij}-\ln p_{ij'}$ is constant for all rows, it follows that

$$\frac{p_{1j}}{p_{1j'}}=\frac{p_{2j}}{p_{2j'}}=\ldots=\frac{p_{Ij}}{p_{Ij'}}=\frac{\sum_i p_{ij}}{\sum_i p_{ij'}}. \tag{9.6.12}$$

Similarly, the difference between two v_{ij}s in the same *column* will depend only on the *rows*, so that

$$\frac{p_{i1}}{p_{i'1}}=\frac{p_{i2}}{p_{i'2}}=\ldots=\frac{p_{iJ}}{p_{i'J}}=\frac{\sum_j p_{ij}}{\sum_j p_{i'j}}. \tag{9.6.13}$$

It follows from Equations 9.6.12 and 9.6.13 that if $\lambda^{AB}=0$, then the two-way table of probabilities, the table of $p_{ij}=P(A=i,B=j)$ must have the structure indicated in Table 9.6.2 where $a_1=b_1=1$, and $a_2,a_3,\ldots,a_I,b_2,b_3,\ldots,b_J$ are constants and $a=\Sigma_{i=2}^I a_i$, and $b=\Sigma_{j=2}^J b_j$. We let $p_{i+}=\Sigma_j p_{ij}$ and $p_{+j}=\Sigma_i p_{ij}$; thus,

$$p_{i+}=P(A=i)$$

and

$$p_{+j}=P(B=j) \tag{9.6.13}$$

Now from Table 9.6.2 we see that if $\lambda^{AB}=0$,

$$P(A=i,B=j)=p_{ij}=a_ib_jp_{11}=\frac{p_{11}(1+b)a_ip_{11}(1+a)b_j}{p_{11}(1+a)(1+b)}$$

$$=p_{i+}p_{+j}=P(A=i)P(B=j), \tag{9.6.14}$$

because $p_{11}(1+a)(1+b)=1$, and thus $\lambda^{AB}=0$, or "no interaction", implies that A and B are independent.

TABLE 9.6.2 TWO-WAY TABLE OF PROBABILITIES $p_{ij} = P(A = i, B = j)$, WHEN $\lambda^{AB} = 0$

	Category $B(j)$						
Category $A(i)$	1	2	...	j	...	J	Total*
1	p_{11}	b_2p_{11}	...	b_jp_{11}	...	b_Jp_{11}	$p_{11}(1+b)$
2	a_2p_{11}	$a_2b_2p_{11}$	...	$a_2b_jp_{11}$	...	$a_2b_Jp_{11}$	$p_{11}(1+b)a_2$
⋮	⋮	⋮		⋮		⋮	⋮
i	a_ip_{11}	$a_ib_2p_{11}$	...	$a_ib_jp_{11}$	...	$a_ib_Jp_{11}$	$p_{11}(1+b)a_i$
⋮	⋮	⋮		⋮		⋮	⋮
I	a_Ip_{11}	$a_Ib_2p_{11}$	...	$a_Ib_jp_{11}$	...	$a_Ib_Jp_{11}$	$p_{11}(1+b)a_I$
Total*	$p_{11}(1+a)$	$p_{11}(1+a)b_2$	...	$p_{11}(1+a)b_j$	...	$p_{11}(1+a)b_J$	$p_{11}(1+a)(1+b) = 1$

*$a_2, a_3, \ldots, a_I$, and $b_2, b_3, \ldots, b_J$ are constants, $a_1 = b_1 = 1$, and $a = \sum_{i=2}^{I} a_i$ and $b = \sum_{j=2}^{J} b_j$.

On the other hand, suppose that A and B are independent. Then $p_{ij} = p_{i+}p_{+j}$, ($i = 1, \ldots, I$ and $j = 1, \ldots, J$), and it follows that $v_{ij} = \ln p_{ij} = \ln p_{i+} + \ln p_{+j}$. Therefore,

$$v_{i\cdot} = \frac{1}{J}\sum_j v_{ij} = \ln p_{i+} + \sum_j \ln p_{+j}/J$$

and

$$v_{\cdot j} = \frac{1}{I}\sum_i v_{ij} = \sum_i \ln p_{i+}/I + \ln p_{+j} \qquad (9.6.15)$$

$$v_{\cdot\cdot} = \frac{1}{IJ}\sum_i\sum_j v_{ij} = \sum_i \ln p_{i+}/I + \sum_j \ln p_{+j}/J$$

In that case, $\lambda_{ij}^{AB} = v_{ij} - v_{i\cdot} - v_{\cdot j} + v_{\cdot\cdot} = 0$, and thus, in the two-way contingency table, "no AB interaction," that is, the additive model, is equivalent to independence of A and B. Thus, to test the hypothesis $H_0: \lambda^{AB} = 0$ in the log-linear model, against the alternative of the full model, we may use any of the previously discussed tests for independence in the $I \times J$ contingency table.

We now consider the interpretation of the main effects, the λ_i^As and λ_j^Bs, in the log-linear model for the $I \times J$ contingency table (Equation 9.6.4). Consider the ith category of variable A. The main effect of this category has been defined as

$$\lambda_i^A = v_{i\cdot} - v_{\cdot\cdot} = \sum_j \ln p_{ij}/J - \sum_i\sum_j \ln p_{ij}/IJ. \qquad (9.6.16)$$

It is difficult to associate a practical meaning with this function in terms of the underlying multinomial probabilities. In the traditional ANOVA, we are interested in whether main effects are equal to zero. Analogously, then, let us assume that $\lambda^A=0$. Because $\lambda_i^A=0$ for all i means that

$$\frac{1}{J}\sum_j \ln p_{ij} - \frac{1}{IJ}\sum_i\sum_j \ln p_{ij}=0, (i=1,\ldots,I), \qquad (9.6.17)$$

this means that for every i, $\Sigma_j \ln p_{ij}$ is the same. This does *not* imply anything about the equality of any class probabilities or sums of class probabilities. On the other hand, if we limit ourselves to hierarchical models, then

$$\lambda^A=0 \Rightarrow \lambda^{AB}=0. \qquad (9.6.18)$$

Let us examine the consequences of Equation 9.6.18. Under the hierarchical scheme, $\lambda^A=\lambda^{AB}=0$ means that

$$v_{ij}=\mu+\lambda_j^B, \qquad (9.6.19)$$

that is, that in the two-way table of log probabilities, the log probability for every ij pair depends only on the j. This, in turn, implies that in the two-way table of probabilities, Table 9.6.2, $p_{ij}=P(A=i,B=j)=p_{i'j}=P(A=i',B=j)$ for $i\neq i'$, and, therefore,

$$\sum_j p_{ij}=\sum_j p_{i'j}$$

or,

$$p_{i+}=p_{i'+} \qquad (9.6.20)$$

that is, $P(A=i)=P(A=i')$, which says that the marginal distribution of A is *uniform*. Because there are I categories of A and because each of these must have the same probability under the conditions $\lambda^A=\lambda^{AB}=0$ and their sum must be one, it follows that

$$\lambda^A=\lambda^{AB}=0 \Rightarrow p_{i+}=\frac{1}{I}=P(A=i), (i=1,\ldots,I). \qquad (9.6.21)$$

Thus, under the hierarchical model, the question of whether a main effect is zero would usually be of no interest in categorical data analysis of a two-way table, unlike the analogue in the two-way ANOVA. If it *is* of interest to test such a hierarchical hypothesis, maximum likelihood estimates of the class probabilities can easily be found from which appropriate "expected" cell frequencies can be calculated, and a χ^2 of the familiar $\Sigma(O-E)^2/E$ form can then be used as a test statistic.

We should note that the conditions $p_{i+}=1/I$, along with $p_{ij}=1/I(p_{+j})$ for all ij, implies $\lambda^A=\lambda^{AB}=0$. Thus, uniform marginals for variable A along with independence, and $\lambda^A=\lambda^{AB}=0$ in the log-linear model for the $I\times J$ contingency table are equivalent.

9.6.2 The Three-Way Table

Consider now a model under which there are three classificatory variables, A, B, and C, with A having I categories, B having J, and C having K categories. The probability that an observation falls in the ith category of A

will be denoted by $P(A_i)$ or $P(A=i)$, in the jth category of B by $P(B_j)$ or $P(B=j)$, and in the kth category of C by $P(C_k)$ or $P(C=k)$.

Let

$$p_{ijk}=P(A=i,B=j,C=k);\,(i=1,\ldots,I,j=1,\ldots,J,k=1,\ldots,K). \tag{9.6.22}$$

In order to focus our attention on a particular problem, let us consider again the data concerning type of child (Variable A), history of previous infant losses of mothers (Variable B), and birth order (Variable C), presented in Table 9.5.3. Let

$$i=\begin{cases}1 \text{ if the child is a problem child}\\ 2 \text{ if the child is a control child}\end{cases}$$

$$j=\begin{cases}1 \text{ if the mother suffered previous infant losses}\\ 2 \text{ if there were no previous infant losses}\end{cases}$$

$$k=\begin{cases}1 \text{ if the birth order is 2}\\ 2 \text{ if the birth order is 3 or 4}\\ 3 \text{ if the birth order is 5 or more.}\end{cases}$$

For the general case of the three-way table, we define functions of the p_{ijk}s that are analogous to those defined for the two-way case. In the following definitions, the index i varies from 1 to I, j varies from 1 to J, and k varies from 1 to K.

$$\left.\begin{aligned}
v_{ijk}&=\ln p_{ijk}\\
v_{\ldots}&=\sum_i\sum_j\sum_k v_{ijk}/IJK\\
v_{i..}&=\sum_j\sum_k v_{ijk}/JK\\
v_{.j.}&=\sum_i\sum_k v_{ijk}/IK\\
v_{..k}&=\sum_i\sum_j v_{ijk}/IJ\\
v_{ij.}&=\sum_k v_{ijk}/K\\
v_{i.k}&=\sum_j v_{ijk}/J\\
v_{.jk}&=\sum_i v_{ijk}/I
\end{aligned}\right\} \tag{9.6.23}$$

Again, a dot in the position of a subscript indicates averaging over the variable(s) associated with the position(s).

The full log-linear model for the $I\times J\times K$ table is again analogous to the full model in a three-way ANOVA,

$$v_{ijk}=\mu+\lambda_i^A+\lambda_j^B+\lambda_k^C+\lambda_{ij}^{AB}+\lambda_{ik}^{AC}+\lambda_{jk}^{BC}+\lambda_{ijk}^{ABC}, \tag{9.6.24}$$

where the "main effects" are

$$\left.\begin{aligned}\lambda_i^A &= v_{i..} - v_{...}, (i=1,\ldots,I)\\ \lambda_j^B &= v_{.j.} - v_{...}, (j=1,\ldots,J)\\ \lambda_k^C &= v_{..k} - v_{...}, (k=1,\ldots,K)\end{aligned}\right\} \quad (9.6.25)$$

the two-factor interactions are

$$\left.\begin{aligned}\lambda_{ij}^{AB} &= v_{ij.} - v_{i..} - v_{.j.} + v_{...}, (i=1,\ldots,I; j=1,\ldots,J)\\ \lambda_{ik}^{AC} &= v_{i.k} - v_{i..} - v_{..k} + v_{...}, (i=1,\ldots,I; k=1,\ldots,K)\\ \lambda_{jk}^{BC} &= v_{.jk} - v_{.j.} - v_{..k} + v_{...}, (j=1,\ldots,J; k=1,\ldots,K)\end{aligned}\right\} \quad (9.6.26)$$

and the three-factor interaction is

$$\lambda_{ijk}^{ABC} = v_{ijk} - v_{ij.} - v_{i.k} - v_{.jk} + v_{i..} + v_{.j.} + v_{..k} - v_{...};$$
$$(i=1,\ldots,I; j=1,\ldots,J; k=1,\ldots,K). \quad (9.6.27)$$

Once again it follows from the definitions that the sum(s) over any of the subscripts that index a λ is zero, so that, for example, $\Sigma_i \lambda_i^A = \Sigma_{ik} \lambda_{ik}^{AC} = \Sigma_{ijk} \lambda_{ijk}^{ABC} = 0$.

In order to interpret the terms of the model, preferably in terms of the class probabilities, let us consider first the three-factor interaction λ_{ijk}^{ABC}. Take a fixed value of variable C, say k, and consider the $I \times J$ table of v_{ijk}s for that value of k. (In our example, we might consider those cases for which $C=2$, that is, the birth order is 3 or 4, and consider the 2×2 table of v_{ijk}s, where $i=1,2$ and $j=1,2$.) The resulting v_{ijk}s are shown in Table 9.6.3.

Let us denote the *conditional* two-way interaction for Table 9.6.3 by $\lambda_{ij}^{AB|k}$. Then, using the definition of a two-factor interaction in a two-way table, as given by Equation 9.6.9, for Table 9.6.3, we have

$$\lambda_{ij}^{AB|k} = v_{ijk} - v_{i.k} - v_{.jk} + v_{..k}. \quad (9.6.28)$$

Then, denoting the average of all such conditional two-way interactions by $\lambda_{ij}^{AB|C}$, we have

$$\lambda_{ij}^{AB|C} = \frac{1}{K}\sum_k \lambda_{ij}^{AB|k} = \frac{1}{K}\sum_k (v_{ijk} - v_{i.k} - v_{.jk} + v_{..k})$$
$$= v_{ij.} - v_{i..} - v_{.j.} + v_{...}. \quad (9.6.29)$$

Now the deviation of the conditional interaction, $\lambda_{ij}^{AB|k}$ from the average of all such conditional interactions is

$$\lambda_{ij}^{AB|k} - \lambda_{ij}^{AB|C} = v_{ijk} - v_{ij.} - v_{i.k} - v_{.jk} + v_{i..} + v_{.j.} + v_{..k} - v_{...}$$
$$= \lambda_{ijk}^{ABC}. \quad (9.6.30)$$

The same result would be obtained if we conditioned on A or B. Thus, the three-factor interaction, λ_{ijk}^{ABC}, is the deviation of the conditional two-factor interaction, given the other variable, from the average of all such conditional two-factor interactions, given that variable.

TABLE 9.6.3 $I \times J$ TABLE OF v_{ijk}s WITH $C=k$

	Variable $B(j)$						
Variable $A(i)$	1	2	...	j	...	J	Total
1	v_{11k}	v_{12k}	...	v_{1jk}	...	v_{1Jk}	$Jv_{1.k}$
2	v_{21k}	v_{22k}	...	v_{2jk}	...	v_{2Jk}	$Jv_{2.k}$
⋮	⋮	⋮		⋮		⋮	⋮
i	v_{i1k}	v_{i2k}	...	v_{ijk}	...	v_{iJk}	$Jv_{i.k}$
⋮	⋮	⋮		⋮		⋮	⋮
I	v_{I1k}	v_{I2k}	...	v_{Ijk}	...	v_{IJk}	$Jv_{I.k}$
Total	$Iv_{.1k}$	$Iv_{.2k}$	...	$Iv_{.jk}$	...	$Iv_{.Jk}$	$IJv_{..k}$

It follows from the above that $\lambda^{ABC}=0$ means that the conditional two-factor interactions, given the other variable, are all equal, because they must then all be equal to their average. That is, for every category of any one of the three variables, the conditional second-order interactions of the other two variables are the same. These conditional second-order interactions are related to the conditional independence (or dependence) of two of the variables, given the third, so that one possible interpretation of $\lambda^{ABC}=0$ is that, for every value of one of the variables, the degree of conditional dependence of the other two is the same.

In the example concerning the cross-classification of child type, infant loss, and birth order, we may be interested in such a question. Is the degree of relationship (as measured by the second order interaction) between birth order and previous infant-loss history the same for problem children as for control children? Or, given the previous infant-loss history of the mother, is the conditional two-factor interaction of type of child (problem or control) and birth order constant? The former interpretation may be more reasonable, because a more appropriate model here than a single multinomial under which the three variables are cross-classified is that of two multinomials, one for the problem children and one for the controls, under each of which there is a cross-classification according to the other two variables, or perhaps three pairs of binomials, one pair for each birth order.

One way of testing the hypothesis $H_0:\lambda^{ABC}=0$, is to obtain maximum likelihood estimates (MLEs), or other best asymptotic normal (BAN) estimates, of the class probabilities in order to compute the "expected" cell frequencies, and form the appropriate χ^2 statistic of the $\Sigma((O-E)^2/E)$ form. The direct calculation of the MLEs in this case is not possible. However, there is an iterative procedure due to Birch (1963) that converges to the "expected" frequency, that is, to the product of n (the total number of observations) and the MLE of the class probability.

To carry out the iteration, we first examine the superscripts that *remain* in the model. In the case of $H_0:\lambda^{ABC}=0$, the superscripts remaining are AB, AC, BC, A, B, and C. The method requires that the remaining group of superscripts "fit" the observed marginals. So, for the data of Table 9.5.3, the six observed marginal frequencies corresponding to *AB*, *AC*, etc., must also be the marginal frequencies for the corresponding "expected" marginal frequencies. In our example, the observed *AB* marginal frequencies, obtained by summing over Variable *C* (birth order), are given in Table 9.6.4. Also given in Table 9.6.4 are the observed *AC* and *BC* marginal frequencies, obtained by summing over Variables *B* (infant-loss history) and *A* (type of child), respectively.

Note that if the "expected" frequencies "fit" the observed *AB* marginal totals, they necessarily fit the observed *A* marginal frequencies and the observed *B* marginal frequencies; the same is true for the observed *AC* and *BC* marginal totals. Thus, in general, fitting the marginal totals for combined variables necessarily results in fitting the observed totals of any subset of those combined variables; that is, if an r-way table of "expected" marginal frequencies corresponding to variables $A_1, A_2 \ldots, A_r$ fit the observed r-way marginal table, then any m-way table of "expected" frequencies ($m<r$) corresponding to a subset of $A_1, A_2, \ldots, A_r$ will automatically "fit" the corresponding m-way observed-marginal table. If the total number of marginals that the "expected" frequencies must fit is m, then each cycle of iteration consists of m steps, "fitting" in any order once to each of these m marginals.

TABLE 9.6.4 OBSERVED *AB, AC,* AND *BC* MARGINAL FREQUENCIES BASED ON THE DATA OF TABLE 9.5.3

Observed AB Marginal Frequencies (obtained by summing over Variable *C*, birth order)

Variable *A*	Variable *B* (Infant Loss History)		
(Type of Child)	Loss ($j=1$)	No Loss ($j=2$)	Total
Problem ($i=1$)	73	145	218
Control ($i=2$)	40	107	147
Total	113	252	365

Observed AC Marginal Frequencies (obtained by summing over Variable *B*, infant loss history)

Variable *A*	Variable *C* (Birth Order)			
(Type of Child)	$2(k=1)$	$3-4(k=2)$	$5+(k=3)$	Total
Problem ($i=1$)	102	67	49	218
Control ($i=2$)	64	46	37	147
Total	166	113	86	365

Observed BC Marginal Frequencies (obtained by summing over Variable *A*, type of child)

Variable *B*	Variable *C* (Birth Order)			
(Infant Loss)	$2(k=1)$	$3-4(k=2)$	$5+(k=3)$	Total
Loss ($j=1$)	30	42	41	113
No Loss ($j=2$)	136	71	45	252
Total	166	113	86	365

Using a modification of Goodman's (1970) notation, let

$$\left.\begin{aligned}
f_{ijk} &= \text{observed frequency in cell } A_iB_jC_k \\
f_{ij+} = \sum_k f_{ijk} &= \text{observed number for which } A=i \text{ and } B=j \\
f_{i+k} = \sum_j f_{ijk} &= \text{observed number for which } A=i \text{ and } C=k \\
f_{+jk} = \sum_i f_{ijk} &= \text{observed number for which } B=j \text{ and } C=k \\
f_{i++} = \sum_k \sum_j f_{ijk} &= \text{observed number for which } A=i \\
f_{+j+} = \sum_k \sum_i f_{ijk} &= \text{observed number for which } B=j \\
f_{++k} = \sum_j \sum_i f_{ijk} &= \text{observed number for which } C=k \\
f_{+++} = \sum_k \sum_j \sum_i f_{ijk} &= \text{total number of observations, } n.
\end{aligned}\right\} \quad (9.6.31)$$

Thus, for the data of our example (Table 9.5.3), f_{11+} = observed number of cases for which $A=1$ (problem child) and $B=1$ (previous infant loss) $= f_{111} + f_{112} + f_{113} = 20+26+27=73$; f_{2+2} = observed number of cases for which $A=2$ (control child) and $C=2$ (birth order 3 or 4) $= f_{212}+f_{222} = 16+30=46$, and so on.

Let $\hat{F}_{ijk}^{(s)}$ represent the estimated frequency at the sth stage. Then the notation of Goodman (1970), also used here, is

$$\left.\begin{aligned}
\hat{F}_{ij+}^{(s)} &= \sum_k \hat{F}_{ijk}^{(s)} \\
\hat{F}_{i+k}^{(s)} &= \sum_j \hat{F}_{ijk}^{(s)} \\
\hat{F}_{+jk}^{(s)} &= \sum_i \hat{F}_{ijk}^{(s)} \\
\hat{F}_{i++}^{(s)} &= \sum_j \sum_k \hat{F}_{ijk}^{(s)} \\
\hat{F}_{+j+}^{(s)} &= \sum_k \sum_i \hat{F}_{ijk}^{(s)} \\
\hat{F}_{++k}^{(s)} &= \sum_j \sum_i \hat{F}_{ijk}^{(s)}
\end{aligned}\right\} \quad (9.6.32)$$

As Goodman suggested, it is convenient to start with initial guesses of 1 for each cell, that is, take $\hat{F}_{ijk}^{(0)} = 1$. At the sth stage of the iteration, we obtain $\hat{F}_{ijk}^{(s)}$ by the following formula when fitting to the observed AB marginal table:

$$\hat{F}_{ijk}^{(s)} = \hat{F}_{ijk}^{(s-1)} f_{ij+} / \hat{F}_{ij+}^{(s-1)}. \quad (9.6.33)$$

If the next stage "fits" to the observed AC marginal table, then

$$\hat{F}_{ijk}^{(s+1)} = \hat{F}_{ijk}^{(s)} f_{i+k} / \hat{F}_{i+k}^{(s)},$$

and so on. The iteration continues until the "fitted" marginal totals are as close to the corresponding observed totals as we wish, or until additional cycles of iteration result in small enough differences in the $\hat{F}_{ijk}$s (as arbitrarily decided in advance). For the example we are using here, we decided to continue the process of iteration until $|\hat{F}_{ijk}^{(s)} - \hat{F}_{ijk}^{(s+1)}| \leqq 0.1$ (rounding to one decimal place). When the iteration stops, then the test of the hypothesis is carried out with the help of the following χ^2 statistic:

$$\chi^2 = \sum_k \sum_j \sum_i \frac{(f_{ijk} - \hat{F}_{ijk})^2}{\hat{F}_{ijk}} \tag{9.6.34}$$

where $\hat{F}_{ijk}$ is the "expected" frequency as of the last iteration. We now illustrate the calculations involved for the example under consideration.

Stage 1. As we indicated above, we shall start with initial values of 1 for each cell, that is, $\hat{F}_{ijk}^{(0)} = 1$. Thus, we have

	$k=1$		$k=2$		$k=3$	
	$j=1$	$j=2$	$j=1$	$j=2$	$j=1$	$j=2$
$i=1$	1	1	1	1	1	1
$i=2$	1	1	1	1	1	1

For each cycle we will "fit" first to the AB marginal table, then to the AC marginal table, and last to the BC marginal table (see Table 9.6.4). Therefore, for the first stage, we need $\hat{F}_{ij+}^{(0)} = \Sigma_{k=1}^{3} \hat{F}_{ijk}^{(0)}, (i,j=1,2)$. The table of $\hat{F}_{ij+}^{(0)}$ is:

	$j=1$	$j=2$
$i=1$	3	3
$i=2$	3	3

In order to calculate $\hat{F}_{ijk}^{(1)} = \hat{F}_{ijk}^{(0)} f_{ij+} / \hat{F}_{ij+}^{(0)}$, we proceed as follows. The table of $\hat{F}_{ijk}^{(1)}$s is

	$k=1$		$k=2$		$k=3$	
	$j=1$	$j=2$	$j=1$	$j=2$	$j=1$	$j=2$
$i=1$	$(1)\left(\frac{73}{3}\right) = 24.3333$	$(1)\left(\frac{145}{3}\right) = 48.3333$	$*_1$	$*_2$	$*_1$	$*_2$
$i=2$	$(1)\left(\frac{40}{3}\right) = 13.3333$	$(1)\left(\frac{107}{3}\right) = 35.6667$	$*_3$	$*_4$	$*_3$	$*_4$

$*_1 = 24.3333, *_2 = 48.3333, *_3 = 13.3333, *_4 = 35.6667$

Note that because $\hat{F}_{ijk}^{(0)} = 1$ for $k = 1,2,3$, and because neither f_{ij+} nor $\hat{F}_{ij+}, (i,j=1,2)$, depend upon k, $\hat{F}_{ijk}^{(1)}$ will also not depend upon k.

Stage 2. In order to proceed to the next iteration, that is, "fitting" to the *AC* marginal table, we need the table of $\hat{F}^{(1)}_{i+k} = \Sigma^2_{j=1} \hat{F}^{(1)}_{ijk}$:

	$k=1$	$k=2$	$k=3$
$i=1$	72.6666	72.6666	72.6666
$i=2$	49.0000	49.0000	49.0000

Then we can obtain $\hat{F}^{(2)}_{ijk} = \hat{F}^{(1)}_{ijk} f_{i+k} / \hat{F}^{(2)}_{i+k}$:

$k=1$	$j=1$	$j=2$
$i=1$	$24.3333\left(\frac{102}{72.6666}\right) = 34.1559$	$48.3333\left(\frac{102}{72.6666}\right) = 67.8441$
$i=2$	$13.3333\left(\frac{64}{49.0000}\right) = 17.4149$	$35.6667\left(\frac{64}{49.0000}\right) = 46.5851$
$k=2$		
$i=1$	$24.3333\left(\frac{67}{72.6666}\right) = 22.4358$	$48.3333\left(\frac{67}{72.6666}\right) = 44.5642$
$i=2$	$13.3333\left(\frac{46}{49.0000}\right) = 12.5170$	$35.6667\left(\frac{46}{49.0000}\right) = 33.4830$
$k=3$		
$i=1$	$24.3333\left(\frac{49}{72.6666}\right) = 16.4082$	$48.3333\left(\frac{49}{72.6666}\right) = 32.5918$
$i=2$	$13.3333\left(\frac{37}{49.0000}\right) = 10.0680$	$35.6667\left(\frac{37}{49.0000}\right) = 26.9320$

Stage 3. In order to complete the first cycle, we have to "fit" to the BC marginal table and need the table of $\hat{F}^{(2)}_{+jk} = \Sigma^2_{i=1} \hat{F}^{(2)}_{ijk}$:

	$k=1$	$k=2$	$k=3$
$j=1$	51.5708	34.9528	26.4762
$j=2$	114.4292	78.0472	59.5238

We now compute $\hat{F}^{(3)}_{ijk} = \hat{F}^{(2)}_{ijk} f_{+jk} / \hat{F}^{(2)}_{+jk}$:

$k=1$	$j=1$	$j=2$
$i=1$	$34.1559\left(\frac{30}{51.5708}\right) = 19.8693$	$67.8441\left(\frac{136}{114.4292}\right) = 80.6332$
$i=2$	$17.4149\left(\frac{30}{51.5708}\right) = 10.1307$	$46.5851\left(\frac{136}{114.4292}\right) = 55.3668$
$k=2$		
$i=1$	$22.4358\left(\frac{42}{34.9528}\right) = 26.9593$	$44.5642\left(\frac{71}{78.0472}\right) = 40.5403$
$i=2$	$12.5170\left(\frac{42}{34.9528}\right) = 15.0407$	$33.4830\left(\frac{71}{78.0472}\right) = 30.4597$

$k=3$		
$i=1$	$16.4082\left(\frac{41}{26.4762}\right)=25.4091$	$32.5918\left(\frac{45}{59.5238}\right)=24.6394$
$i=2$	$10.0680\left(\frac{41}{26.4762}\right)=15.5909$	$26.9320\left(\frac{45}{59.5238}\right)=20.3606$

This completes one cycle of iteration. To proceed to the second cycle, with the fourth stage of iteration, we begin by fitting to the AB marginal table first.

Stage 4. To fit to the AB marginal table, we need the table of $\hat{F}_{ij+}^{(3)} = \Sigma_{k=1}^{3}\hat{F}_{ijk}^{(3)}$:

	$j=1$	$j=2$
$i=1$	72.2377	145.8129
$i=2$	40.7623	106.1871

We can now obtain the table of $\hat{F}_{ijk}^{(4)} = \hat{F}_{ijk}^{(3)} f_{ij+} / \hat{F}_{ij+}^{(3)}$:

$k=1$	$j=1$	$j=2$
$i=1$	$19.8693\left(\frac{73}{72.2377}\right)=20.0790$	$80.6332\left(\frac{145}{145.8129}\right)=80.1837$
$i=2$	$10.1307\left(\frac{40}{40.7623}\right)=9.9412$	$55.3688\left(\frac{107}{106.1871}\right)=55.7907$
$k=2$		
$i=1$	$26.9593\left(\frac{73}{72.2377}\right)=27.2438$	$40.5403\left(\frac{145}{145.8129}\right)=40.3143$
$i=2$	$15.0407\left(\frac{40}{40.7623}\right)=14.7594$	$30.4597\left(\frac{107}{106.1871}\right)=30.6929$
$k=3$		
$i=1$	$25.4091\left(\frac{73}{72.2377}\right)=25.6772$	$24.6394\left(\frac{145}{145.8129}\right)=24.5020$
$i=2$	$15.5909\left(\frac{40}{40.7623}\right)=15.2993$	$20.3606\left(\frac{107}{106.1871}\right)=20.5165$

Note that the ratios $f_{ij+}/\hat{F}_{ij+}^{(3)}$ are substantially closer to one (73/72.2377, 40/40.7623, etc.) than were the previous multiplicative ratios, $f_{+jk}/\hat{F}_{+jk}^{(2)}$ (30/51.5708, 136/114.4292, etc.).

The process was continued until the tenth stage yielded the condition set, that is, $|\hat{F}_{ijk}^{(10)} - \hat{F}_{ijk}^{(9)}| \leqq 0.1$. Therefore, we took as our "expected" cell frequencies $\hat{F}_{ijk}^{(10)}$, which were "close" to the expected cell frequencies computed by using maximum likelihood estimation of the class probabilities. The table of expected cell frequencies, $\hat{F}_{ijk}$, is:

	$k=1$		$k=2$		$k=3$	
	$j=1$	$j=2$	$j=1$	$j=2$	$j=1$	$j=2$
$i=1$	20.6	81.3	27.2	39.9	25.2	23.8
$i=2$	9.4	54.6	14.8	31.1	15.8	21.2

The χ^2 statistic, Equation 9.6.34, is:

$$\chi^2 = \sum_{k=1}^{3} \sum_{j=1}^{2} \sum_{i=1}^{2} \frac{(f_{ijk} - \hat{F}_{ijk})^2}{\hat{F}_{ijk}} = .9104.$$

The number of degrees of freedom associated with this statistic is two, the number of restrictions imposed by H_0: $\lambda^{ABC} = 0$ (because $\Sigma_{i=1}^{2}\lambda_{ijk}^{ABC} = \Sigma_{j=1}^{2}\lambda_{ijk}^{ABC} = \Sigma_{k=1}^{3}\lambda_{ijk}^{ABC} = 0$, by definition). Note that we may also compute the number of degrees of freedom as the reduction in the number of unknown parameters if H_0 is true. To begin with there are $IJK-1$ unknown cell probabilities ($\Sigma_{ijk}p_{ijk} = 1$), which is also the number of unknown parameters in the full model,

$$v_{ijk} = \mu + \lambda_i^A + \lambda_j^B + \lambda_k^C + \lambda_{ij}^{AB} + \lambda_{ik}^{AC} + \lambda_{jk}^{BC} + \lambda_{ijk}^{ABC}.$$

The total number of unknown parameters may be broken down as follows:

$$\begin{aligned} &I-1 \text{ (for the } \lambda_i^A\text{s)} + J-1 \text{ (for the } \lambda_j^B\text{s)} \\ &+ K-1 \text{ (for the } \lambda_k^C\text{s)} + (I-1)(J-1) \text{ (for the } \lambda_{ij}^{AB}\text{s)} \\ &+ (I-1)(K-1) \text{ (for the } \lambda_{ik}^{AC}\text{s)} + (J-1)(K-1) \text{ (for the } \lambda_{jk}^{BC}\text{s)} \\ &+ (I-1)(J-1)(K-1) \text{ (for the } \lambda_{ijk}^{ABC}\text{s)} = IJK-1, \end{aligned}$$

because μ is determined by the condition $\Sigma_{ijk}p_{ijk} = 1$, and the definition of the λs guarantees that $\Sigma_i\lambda_i^A = 0 = \Sigma_i\lambda_{ij}^{AB} = \Sigma_j\lambda_{ij}^{AB} = \Sigma_k\lambda_{ijk}^{ABC} = \ldots$, etc. If H_0 is true, then the number of unknown parameters is the same as in the full model *less* the $(I-1)(J-1)(K-1)\lambda_{ijk}^{ABC}$s whose values under H_0 are known (that is, equal to zero). Therefore, the reduction in the number of unknown parameters if H_0 is true is $(I-1)(J-1)(K-1)$, and because, in our example, $I = J = 2$ and $K = 3$, $(I-1)(J-1)(K-1) = 2$.

In the example, $\chi^2 = .9104$, a value that is close to those of the "interaction" statistics previously computed for these data. For two degrees of freedom this value of χ^2 has a significance probability greater than .50, and so we conclude that there is no three-factor interaction. This means that the degree of relationship between type of child and loss history, as measured by the conditional AB interactions, is the same for all birth orders. Perhaps, in the light of the way the data were collected, a better interpretation would be that there is no evidence to contradict the equality of the odds ratios, that is,

$$\frac{p_{111}/p_{121}}{p_{211}/p_{221}}=\frac{p_{112}/p_{122}}{p_{212}/p_{222}}=\frac{p_{113}/p_{213}}{p_{213}/p_{223}}.$$

(Note that this notation differs from that used for odds ratios in Section 9.5, where we assumed a different model.)

For the purpose of finding the most "parsimonious" model, that is, the one involving the fewest unknown parameters, as well as for obtaining a reasonable explanation for the data, it is usually instructive to follow up the test of a hierarchical hypothesis with further tests. In the example under consideration, had we concluded that the three-factor interaction was not zero, we might then have separated the data according to one of the variables, perhaps birth order, and examined various hierarchical hypotheses for the three birth orders *separately*. Since, however, we have concluded that there is no three-factor interaction, let us examine some further models or hierarchical hypotheses. Our aim is to find the most parsimonious model for the data as well as to consider in an orderly fashion any relationships that might exist among the variables.

We consider first the hypothesis H_0: $\lambda^{AB}=\lambda^{ABC}=0$, examine its meaning for the three-way table, and test this hierarchical hypothesis for the type-of-child × loss-history × birth-order data.

Hierarchical Hypothesis H_0:$\lambda^{AB}=\lambda^{ABC}=0$. If H_0 is true, then

$$v_{ijk}=\mu+\lambda_i^A+\lambda_j^B+\lambda_k^C+\lambda_{ik}^{AC}+\lambda_{jk}^{BC},$$
$$i=1,\ldots,I;j=1,\ldots,J;k=1,\ldots,K. \quad (9.6.35)$$

Now, fix a category of one of the variables, C, say $C=k$, and consider the $I\times J$ table of v_{ijk}s for this value of C. Let the I rows be indexed by i (to denote the categories of A) and the J columns by j (to denote the categories of B). For two v_{ijk}s in the same row, say for columns j and j', $j\neq j'$, $v_{ijk}-v_{ij'k}=\lambda_j^B+\lambda_{jk}^{BC}-\lambda_{j'}^B-\lambda_{j'k}^{BC}$, independent of the row index, i;

therefore

$$p_{ijk}/p_{ij'k}=d_{j'jk}, \quad (9.6.36)$$

a constant depending on the two categories j and j', of B, and the fixed category, k, of C. But Equation 9.6.36 implies that

$$\frac{\sum_i p_{ijk}}{\sum_i p_{ij'k}}=d_{j'jk}=\frac{p_{+jk}}{p_{+j'k}}=\frac{P(B=j,C=k)}{P(B=j',C=k)}. \quad (9.6.37)$$

Similarly, for the same k,

$$\frac{p_{ijk}}{p_{i'jk}}=\delta_{i'ik}=\frac{p_{i+k}}{p_{i'+k}}=\frac{P(A=i,C=k)}{P(A=i',C=k)}, \quad (9.6.38)$$

where $\delta_{i'ik}$ is a constant depending on the two categories, i and i', of A and the fixed category, k, of C. It follows that

$$p_{12k}=d_{12k}p_{11k},\ p_{13k}=d_{13k}p_{11k},\ \text{etc.},$$

while

$$p_{21k}=\delta_{21k}p_{11k},\ p_{31k}=\delta_{31k}p_{11k},\ \text{etc.},$$

from which it follows that the $I\times J$ table of *probabilities*, p_{ijk}, for $C=k$, must have the structure shown in Table 9.6.5. (Note that in Table 9.6.5 the subscripts, 1 and k, on the d_{ijk}s and δ_{ijk}s have been deleted, because they appear in all such expressions; that is, d_j is written for d_{1jk} and δ_i is written for δ_{i1k}.)

Now, putting $d_1=\delta_1=1$,

$$p_{ijk}=\delta_i d_j p_{11k}=\frac{p_{i+k}p_{+jk}}{p_{++k}}, \tag{9.6.39}$$

from which it follows that the conditional probability that $A=i$ and $B=j$, given $C=k$, is

$$P(A=i,B=j|C=k)=\frac{p_{ijk}}{p_{++k}}=\frac{p_{i+k}p_{+jk}}{p_{++k}p_{++k}}=P(A=i|C=k)P(B=j|C=k), \tag{9.6.40}$$

that is, A and B are *conditionally* independent, given $C=k$. Note that this is *not* the same as unconditional independence, for which we must have $P(A=i,B=j)=P(A=i)P(B=j)$, for every pair i,j. On the other hand, if we assume that A and B are conditionally independent, given C,

that is, that

$$p_{ijk}=\frac{p_{i+k}p_{+jk}}{p_{++k}},(i=1,\dots,I;j=1,\dots,J;k=1,\dots,K),$$

then for $i=1,2,\dots,I$, $j=1,2,\dots,J$, and $k=1,2,\dots,K$,

$$v_{ijk}=\ln p_{ijk}=\ln p_{i+k}+\ln p_{+jk}-\ln p_{++k},$$

$$v_{ij.}=\frac{1}{K}\sum_{k=1}^{K}v_{ijk}=\frac{1}{K}\sum_k \ln p_{i+k}+\frac{1}{K}\sum_k \ln p_{+jk}-\frac{1}{K}\sum_k \ln p_{++k},$$

$$v_{i..}=\frac{1}{JK}\sum_{j=1}^{J}\sum_{k=1}^{K}v_{ijk}=\frac{1}{K}\sum_k \ln p_{i+k}+\frac{1}{JK}\sum_j\sum_k \ln p_{+jk}-\frac{1}{K}\sum_k \ln p_{++k},$$

$$v_{.j.}=\frac{1}{IK}\sum_{i=1}^{I}\sum_{k=1}^{K}v_{ijk}=\frac{1}{IK}\sum_i\sum_k \ln p_{i+k}+\frac{1}{K}\sum_k \ln p_{+jk}-\frac{1}{K}\sum_k \ln p_{++k},$$

$$v_{...}=\frac{1}{IJK}\sum_{i=1}^{I}\sum_{j=1}^{J}\sum_{k=1}^{K}v_{ijk}=\frac{1}{IK}\sum_i\sum_k \ln p_{i+k}+\frac{1}{JK}\sum_j\sum_k \ln p_{+jk}-\frac{1}{K}\sum_k \ln p_{++k},$$

and

$$\lambda_{ij}^{AB}=v_{ij.}-v_{i..}-v_{.j.}+v_{...}=0.$$

To show that $\lambda^{ABC}=0$ also, we need

$$v_{i.k}=\frac{1}{J}\sum_j v_{ijk}=\ln p_{i+k}+\frac{1}{J}\sum_j \ln p_{+jk}-\ln p_{++k},$$

$$v_{.jk}=\frac{1}{I}\sum_i v_{ijk}=\frac{1}{I}\sum_i \ln p_{i+k}+\ln p_{+jk}-\ln p_{++k},$$

TABLE 9.6.5. $I \times J$ TABLE OF p_{ijk}s FOR $\lambda^{AB} = 0 = \lambda^{ABC}$, k FIXED

Category of A (i)	*Category of B (j)* 1	2		j		J	*Total*
1	p_{11k}	$d_2 p_{11k}$	$\cdots$	$d_j p_{11k}$	$\cdots$	$d_J p_{11k}$	$p_{1+k} = p_{11k} \cdot (1+d)$
2	$\delta_2 p_{11k}$	$\delta_2 d_2 p_{11k}$	$\cdots$	$\delta_2 d_j p_{11k}$	$\cdots$	$\delta_2 d_J p_{11k}$	$p_{2+k} = p_{11k} \delta_2 \cdot (1+d)$
	$\vdots$	$\vdots$		$\vdots$		$\vdots$	$\vdots$
i	$\delta_i p_{11k}$	$\delta_i d_2 p_{11k}$	$\cdots$	$\delta_i d_j p_{11k}$	$\cdots$	$\delta_i d_J p_{11k}$	$p_{i+k} = p_{11k} \delta_i \cdot (1+d)$
	$\vdots$	$\vdots$		$\vdots$		$\vdots$	$\vdots$
I	$\delta_I p_{11k}$	$\delta_I d_2 p_{11k}$	$\cdots$	$\delta_I d_j p_{11k}$	$\cdots$	$\delta_I d_J p_{11k}$	$p_{I+k} = p_{11k} \delta_I \cdot (1+d)$
Total	$p_{+1k} =$ $p_{11k}(1+\delta)$	$p_{+2k} =$ $p_{11k} d_2(1+\delta)$	$\cdots$	$p_{+jk} =$ $p_{11k} d_j(1+\delta)$	$\cdots$	$p_{+Jk} =$ $p_{11k} d_J(1+\delta)$	$p_{++k} =$ $p_{11k}(1+\delta)(1+d)$

Note: $d_j = d_{1jk}, \delta_i = \delta_{i1k}, d = \sum_{j=2}^{J} d_j$, and $\delta = \sum_{i=2}^{I} \delta_i$.

and

$$v_{..k}=\frac{1}{IJ}\sum_{j}\sum_{k}v_{ijk}=\frac{1}{I}\sum_{i}\ln p_{i+k}+\frac{1}{J}\sum_{j}\ln p_{+jk}-\ln p_{++k}.$$

Then

$$\lambda_{ijk}^{ABC}=v_{ijk}-v_{ij.}-v_{i.k}-v_{.jk}+v_{i..}+v_{.j.}+v_{..k}-v_{...}=0;$$
$$(i=1,\ldots,I;j=1,\ldots,J;k=1,\ldots,K).$$

Thus, conditional independence of A and B, given C, and $\lambda^{AB}=\lambda^{ABC}=0$ are equivalent.

In the context of our example, $\lambda^{AB}=\lambda^{ABC}=0$ would mean that within each birth order, the variables type of child and previous-loss history, are independent, or more appropriately, *within each birth order*, the probability of previous-loss history is the same for mothers of problem children as for mothers of control children.

Suppose we were interested in testing H_0: $\lambda^{AB}=\lambda^{ABC}=0$ for the example. Then to get the "expected" frequencies under H_0 we would need to "fit" only to the AC and BC marginals. The model, Equation 9.6.35, under H_0 involves only superscripts AB, BC, A, B, and C, and fitting to the observed AB and BC marginal tables will insure that the A, B, and C margins "fit", as well.

Starting with the same $\hat{F}_{ijk}^{(0)}$s as before, that is, $\hat{F}_{ijk}^{(0)}=1, i,j=1,2; k=1,2,3$, we need $\hat{F}_{i+k}^{(0)}=\sum_{j=1}^{2}\hat{F}_{ijk}^{(0)}$ for the first cycle. The table of $\hat{F}_{i+k}^{(0)}$s is

	$k=1$	$k=2$	$k=3$
$i=1$	2	2	2
$i=2$	2	2	2

Fitting first to the AC marginal, we have $\hat{F}_{ijk}^{(1)}=\hat{F}_{ijk}^{(0)}f_{i+k}/\hat{F}_{i+k}^{(0)}$. The table of $\hat{F}_{ijk}^{(1)}$s is

$k=1$	$j=1$	$j=2$
$i=1$	$1\left(\frac{102}{2}\right)=51$	$1\left(\frac{102}{2}\right)=51$
$i=2$	$1\left(\frac{64}{2}\right)=32$	$1\left(\frac{64}{2}\right)=32$
$k=2$		
$i=1$	$1\left(\frac{67}{2}\right)=33.5$	$1\left(\frac{67}{2}\right)=33.5$
$i=2$	$1\left(\frac{46}{2}\right)=23$	$1\left(\frac{46}{2}\right)=23$
$k=3$		
$i=1$	$1\left(\frac{49}{2}\right)=24.5$	$1\left(\frac{49}{2}\right)=24.5$
$i=2$	$1\left(\frac{37}{2}\right)=18.5$	$1\left(\frac{37}{2}\right)=18.5$

For the next stage, fitting to the *BC* marginal table, we need $\hat{F}^{(1)}_{+jk} = \Sigma^2_{i=1}\hat{F}^{(1)}_{ijk}$:

	$k=1$	$k=2$	$k=3$
$j=1$	83	56.5	43
$j=2$	83	56.5	43

Then the values of $\hat{F}^{(2)}_{ijk} = \hat{F}^{(1)}_{ijk} f_{+jk} / \hat{F}^{(1)}_{+jk}$ are

$k=1$	$j=1$	$j=2$
$i=1$	$51\left(\frac{30}{83}\right)=18.433734$	$51\left(\frac{136}{83}\right)=83.566265$
$i=2$	$32\left(\frac{30}{83}\right)=11.566265$	$32\left(\frac{136}{83}\right)=52.433734$
$k=2$		
$i=1$	$33.5\left(\frac{42}{56.5}\right)=24.902654$	$33.5\left(\frac{71}{56.5}\right)=42.097345$
$i=2$	$23\left(\frac{42}{56.5}\right)=17.097345$	$23\left(\frac{71}{56.5}\right)=28.902654$
$k=3$		
$i=1$	$24.5\left(\frac{41}{43}\right)=23.360465$	$24.5\left(\frac{45}{43}\right)=25.639534$
$i=2$	$18.5\left(\frac{41}{43}\right)=17.639534$	$18.5\left(\frac{45}{43}\right)=19.360465$

This completes the first cycle. To start the next cycle would require "fitting" again to the *AC* marginal, for which we need $\hat{F}^{(2)}_{i+k} = \Sigma^2_{j=1}\hat{F}^{(2)}_{ijk}$:

	$k=1$	$k=2$	$k=3$
$i=1$	102.0000	67.0000	49.0000
$i=2$	64.0000	46.0000	37.0000

But the above table *is* the observed *AC* marginal table, and thus, there is no need for further iteration, since the multiplicative factors, $f_{i+k}/\hat{F}^{(2)}_{i+k}$, are all equal to one. Thus, the appropriate table of "expected" cell frequencies, using maximum likelihood estimation of the class probabilities under H_0, are those given in the table of $\hat{F}^{(2)}_{ijk}$s. The reason that the iteration automatically stops after one complete cycle here is that in this case it is possible to estimate the p_{ijk}s by the method of maximum likelihood *directly*.

For the general case of the $I \times J \times K$ contingency table, the likelihood function under H_0: $\lambda^{AB} = \lambda^{ABC} = 0$ is

$$L_0 = G \prod_{i,j,k} p_{ijk}^{f_{ijk}},$$

where G does not depend on the p_{ijk}s and where, under H_0,

$$p_{ijk}=\frac{p_{i+k}p_{+jk}}{p_{++k}},i=1,\ldots,I;j=1,\ldots,J;k=1,\ldots,K.$$

By solving linear equations by the standard method of using Lagrange multipliers, one can find the values of the parameters p_{i+k}, p_{+jk} and p_{++k} that maximize

$$\ln L_0=G'+\sum_i\sum_j\sum_k f_{ijk}\ln\frac{p_{i+k}p_{+jk}}{p_{++k}}$$

$$=G'+\sum_i\sum_k f_{i+k}\ln p_{i+k}+\sum_j\sum_k f_{+jk}\ln p_{+jk}-\sum_k f_{++k}\ln p_{++k},$$

subject to the following constraints on the parameters:

$$\sum_i\sum_k p_{i+k}-1=0,$$

$$\sum_j\sum_k p_{+jk}-1=0,$$

$$\sum_i p_{i+k}-p_{++k}=0,\ (k=1,\ldots,K),$$

and

$$\sum_j p_{+jk}-p_{++k}=0,\ (k=1,\ldots,K).$$

The solution is

$$\hat{p}_{i+k}=\frac{f_{i+k}}{n},\ (i=1,\ldots,I;\ k=1,\ldots,K),$$

$$\hat{p}_{+jk}=\frac{f_{+jk}}{n},\ (j=1,\ldots,J;\ k=1,\ldots,K),$$

and

$$\hat{p}_{++k}=\frac{f_{++k}}{n},\ (k=1,\ldots,K),$$

giving

$$\hat{p}_{ijk}=\frac{\hat{p}_{i+k}\hat{p}_{+jk}}{\hat{p}_{++k}}=\frac{f_{i+k}f_{+jk}}{nf_{++k}},\text{ all } i,j,k,$$

and

$$\hat{F}_{ijk}=n\hat{p}_{ijk}=\frac{f_{i+k}f_{+jk}}{f_{++k}}.$$

In the case of our example,

$$\hat{F}_{111}=n\hat{p}_{111}=\frac{f_{1+1}f_{+11}}{f_{++1}}=\frac{(102)(30)}{166}=18.4337,$$

where f_{1+1} can be found in the observed AC marginal table and f_{+11} in the observed BC marginal table; f_{++1} can be found in either the AC or the BC table. Similarly,

$$\hat{F}_{121}=n\hat{p}_{121}=\frac{f_{1+1}f_{+21}}{f_{++1}}=\frac{(102)(136)}{166}=83.5663,$$

etc. Note that these values are indentical to the values obtained after one cycle of iteration.

Because the iteration will automatically cease after each required marginal has been fitted once, it will, in general, not cause an enormous amount of unnecessary calculation if one uses the iterative method, even though the MLEs can be found directly.

The value of the χ^2 statistic for this example is

$$\chi^2 = \sum_{i=1}^{2} \sum_{j=1}^{2} \sum_{k=1}^{3} \frac{(f_{ijk} - \hat{F}_{ijk})^2}{\hat{F}_{ijk}} = 3.129.$$

Note that this is the same as the sum of the three χ^2s, one computed for each birth order, following Table 9.5.4. This is so because the above χ^2 statistic is algebraically identical to the sum of the three single-degree-of-freedom χ^2s in the previous case. Recall that the $\hat{F}_{ijk}$s are the corresponding $\hat{F}_{ijk}^{(2)}$s, because these completed the iteration. This χ^2 statistic has associated with it three degrees of freedom, corresponding to the three restrictions:

1. $\lambda_{11}^{AB} = 0$ (from which it follows that $\lambda_{12}^{AB} = \lambda_{21}^{AB} = \lambda_{22}^{AB} = 0$)
2. $\lambda_{111}^{ABC} = 0$ (from which it follows that $\lambda_{211}^{ABC} = \lambda_{121}^{ABC} = \lambda_{221}^{ABC} = 0$) and
3. $\lambda_{112}^{ABC} = 0$ (from which, along with $\lambda_{111}^{ABC} = 0$, it follows that the seven remaining three-factor interactions are also zero.)

Thus, because the significance probability is high ($.25 < p < .50$), we would not reject H_0 and we would conclude that, given the birth order, the two remaining variables, type of child and previous-loss history, are independent. Or, alternatively, within each birth order, the probability of previous-loss history of mothers of problem children has not been shown to be different from that of mothers of control children.

Note that if we wished to compare the log-linear results with the GSK result of Section 9.5.3, we would have to test for "no interaction" between type of child and birth order, that is, H_0: $\lambda^{AC} = \lambda^{ABC} = 0$. To do so, we proceed as outlined above for testing H_0: $\lambda^{AB} = \lambda^{ABC} = 0$, with AC replacing AB, and fitting to the BC and AB marginals, etc., we obtain a χ^2 value of 1.971, compared to the value 1.264 previously obtained, indicating that the two methods are not exactly equivalent.

We consider next various further hierarchical hypotheses. There is no set rule to follow here. Since $\lambda^{AB} = \lambda^{ABC} = 0$ has been shown to be a plausible hypothesis, we might consider next $\lambda^{A} = \lambda^{AB} = \lambda^{ABC} = 0$, or $\lambda^{AB} = \lambda^{AC} = 0$, or perhaps try to eliminate more than one set of parameters at one time, by testing next H_0: $\lambda^{AB} = \lambda^{BC} = \lambda^{AC} = \lambda^{ABC} = 0$. By testing any of these hierarchical hypotheses, we are investigating the possibility of eliminating some parameters from the model given by Equation 9.6.24. In order to interpret in terms of the class probabilities a more parsimonious model that might fit the data, we examine next the hierarchical hypothesis H_0: $\lambda^{AB} = \lambda^{AC} = \lambda^{ABC} = 0$, first in the general case of the three-way table, and then for our data.

Hierarchical Hypothesis H_0: $\lambda^{AB}=\lambda^{AC}=\lambda^{ABC}=0$. We have seen that $\lambda^{AB}=\lambda^{ABC}=0$ is equivalent to the conditional independence of variables A and B, given Variable C, that is, that

$$p_{ijk}=\frac{p_{i+k}p_{+jk}}{p_{++k}}.$$

Similarly $\lambda^{AC}=\lambda^{ABC}=0$ is equivalent to the conditional independence of variables A and C, given variable B, that is, that

$$p_{ijk}=\frac{p_{ij+}p_{+jk}}{p_{+j+}}.$$

It follows that

$$P(A=i|B=j)=\frac{p_{ij+}}{p_{+j+}}=\frac{p_{ijk}}{p_{+jk}}=\frac{p_{i+k}}{p_{++k}}.$$

Now,

$$\begin{aligned}P(A=i)&=\sum_{j=1}^{J}P(A=i,B=j)=\sum_{j=1}^{J}P(B=j)P(A=i|B=j)\\&=\sum_{j}P(B=j)\frac{p_{i+k}}{p_{++k}}=\frac{p_{i+k}}{p_{++k}}\sum_{j}P(B=j)=\frac{p_{i+k}}{p_{++k}}\\&=P(A=i|B=j),\end{aligned}$$

so that A and B are *unconditionally* independent. Similarly, it can be shown that A and C are also unconditionally independent. In addition, because

$$p_{ijk}=\frac{p_{i+k}}{p_{++k}}p_{+jk}=P(A=i)P(B=j,C=k),$$

we have the further result that the Variable A is independent of the *pair* of Variables, B and C. In terms of the example we have been considering, this hypothesis states that type of child and loss history are independent, given birth order and also that birth order and type of child are independent, given loss history. Together these imply that type of child is independent of loss history, type of child is independent of birth order, and type of child is independent of the pair of variables, birth order and loss history.

The maximum likelihood estimates of the p_{ijk}s could be found directly in this case, but it is simple to find the "expected" frequencies under the conditions imposed by H_0 using the iterative procedure. Under H_0,

$$v_{ijk}=\mu+\lambda_i^A+\lambda_j^B+\lambda_k^C+\lambda_{jk}^{BC},$$

and thus, we have only to "fit" to the BC and A marginals. The A marginals are:

Category of $A(i)$	f_{i++}
$i=1$	218
$i=2$	147
Total	365

For the BC marginals, see Table 9.6.4, and for the f_{ijk}s, see Table 9.5.3. Starting with $\hat{F}_{ijk}^{(0)}=1$ as before, we have, as the table of $\hat{F}_{ijk}^{(0)}$s:

	$k=1$		$k=2$		$k=3$	
	$j=1$	$j=2$	$j=1$	$j=2$	$j=1$	$j=2$
$i=1$	1	1	1	1	1	1
$i=2$	1	1	1	1	1	1

From this table we obtain the table of $\hat{F}_{+jk}^{(0)}$s:

	$k=1$	$k=2$	$k=3$
$j=1$	2	2	2
$j=2$	2	2	2

and, "fitting" to the BC marginals, we have, as the table of $\hat{F}_{ijk}^{(1)}$s:

$k=1$	$j=1$	$j=2$
$i=1$	$\frac{1}{2}(30)=15$	$\frac{1}{2}(136)=68$
$i=2$	$\frac{1}{2}(30)=15$	$\frac{1}{2}(136)=68$
$k=2$		
$i=1$	$\frac{1}{2}(42)=21$	$\frac{1}{2}(71)=35.5$
$i=2$	$\frac{1}{2}(42)=21$	$\frac{1}{2}(71)=35.5$
$k=3$		
$i=1$	$\frac{1}{2}(41)=20.5$	$\frac{1}{2}(45)=22.5$
$i=2$	$\frac{1}{2}(41)=20.5$	$\frac{1}{2}(45)=22.5$

To fit to the A marginals, we need the table of $\hat{F}_{i++}^{(1)}$:

i	$\hat{F}_{i++}^{(1)}$
1	182.5
2	182.5

Now, fitting to the A marginals, we have the table of $\hat{F}_{ijk}^{(2)}$s:

$k=1$	$j=1$	$j=2$
$i=1$	$\frac{15}{182.5}(218)=17.917808$	$\frac{68}{182.5}(218)=81.227397$
$i=2$	$\frac{15}{182.5}(147)=12.082191$	$\frac{68}{182.5}(147)=54.772602$

$k=2$	$j=1$	$j=2$
$i=1$	$\frac{21}{182.5}(218)=25.084931$	$\frac{35.5}{182.5}(218)=42.405479$
$i=2$	$\frac{21}{182.5}(147)=16.915068$	$\frac{35.5}{182.5}(147)=28.594520$
$k=3$		
$i=1$	$\frac{20.5}{182.5}(218)=24.487671$	$\frac{22.5}{182.5}(218)=26.876712$
$i=2$	$\frac{20.5}{182.5}(147)=16.512328$	$\frac{22.5}{182.5}(147)=18.123287$

To fit again to the BC marginals for the next cycle, we need the table of the $\hat{F}^{(2)}_{+jk}$s:

	$k=1$	$k=2$	$k=3$
$j=1$	30.0	42.0	41.0
$j=2$	136.0	71.0	45.0

Thus, we see that we have already recovered the BC marginals, and the "expected" frequencies based on maximum likelihood estimation of the p_{ijk}s are the $\hat{F}^{(1)}_{ijk}$s given in the table below:

	$k=1$		$k=2$		$k=3$	
	$j=1$	$j=2$	$j=1$	$j=2$	$j=1$	$j=2$
$i=1$	17.9	81.2	25.1	42.4	24.5	26.9
$i=2$	12.1	54.8	16.9	28.6	16.5	18.1

The χ^2 statistic for testing H_0 has the value

$$\chi^2=\frac{(20-17.9)^2}{17.9}+\frac{(82-81.2)^2}{81.2}+\ldots+\frac{(23-18.1)^2}{18.1}$$
$$=2.391,$$

to be compared with percentiles of the χ^2 distribution with five degrees of freedom. Because the significance probability is greater than .5, the test provides strong evidence that H_0 is true.

It might be instructive at this point to examine the marginal tables relevant to the three types of independence implied by H_0. For the independence of type of child and loss history, consider the AB marginal:

	Type of child (A)			
Loss category (B)	*Problem (i=1)*	*Control (i=2)*	*Total*	*% Problem*
Losses (j=1)	73	40	113	64.6%
No losses (j=2)	145	107	252	57.5%
Total	218	147	365	59.7%

The usual chi-square statistic for testing the above 2×2 table for independence has the value 1.618, which for one degree of freedom has an associated significance probability of about .20.

For the independence of type of child and birth order, consider the AC marginal:

	Type of Child (A)			
Birth Order (C)	Problem ($i=1$)	Control ($i=2$)	Total	% Problem
2 ($k=1$)	102	64	166	61.4%
3–4 ($k=2$)	67	46	113	59.3%
5+ ($k=3$)	49	37	86	57.0%
Total	218	147	365	59.7%

The value of the usual independence χ^2 statistic for the above table is less than 1, indicating support for the hypothesis that birth order and type of child are unconditionally independent.

For the independence of type of child on the one hand, and the *pair* of variables, birth order and loss history on the other, consider Table 9.6.6.

The value of the χ^2 statistic with five degrees of freedom for testing the above table for independence of the two classificatory variables is 3.667, which has a significance probability in excess of .5, supporting the hypothesis of independence of the variables type of child and the loss-history$\times$ birth-order combination.

We should point out that we are *not* suggesting an endless series of χ^2 tests. The purpose of presenting the three contingency tables above is solely to clarify the implications of support of the hierarchical hypothesis H_0: $\lambda^{AB}=\lambda^{AC}=\lambda^{ABC}=0$ by the data.

Having concluded that $\lambda^{AB}=\lambda^{AC}=\lambda^{ABC}=0$, we may proceed in various ways to examine additional sets of parameters that might be eliminated

TABLE 9.6.6 CROSS CLASSIFICATION OF 365 CHILDREN BY TYPE OF CHILD (*A*) AND LOSS HISTORY (*B*)$\times$BIRTH ORDER (*C*) CATEGORY

	Type of Child (A)			
Loss (B)$\times$ Birth Order (C)	Problem $i=1$	Control $i=2$	Total	%Problem
$L.^*(j=1)\times 2\ (k=1)$	20	10	30	66.7%
$NL.\ (j=2)\times 2\ (k=1)$	82	54	136	60.3%
$L.\ (j=1)\times 3\text{–}4\ (k=2)$	26	16	42	61.9%
$NL.\ (j=2)\times 3\text{–}4\ (k=2)$	41	30	71	57.7%
$L.\ (j=1)\times 5+\ (k=3)$	27	14	41	65.9%
$NL.\ (j=2)\times 5+\ (k=3)$	22	23	45	48.9%
Total	218	147	365	59.7%

* *L.* = Loss History, NL. = *No Loss History*

from the model (see Equation 9.6.24). Because we are limiting ourselves to the hierarchical scheme, there are still four sets of parameters that are eligible for elimination: the λ_i^As, the λ_{jk}^{BC}s the λ_k^Cs and the λ_j^Bs.

Because of the traditional interest in the hypothesis of unconditional independence of the three variables, we consider next the hierarchical hypothesis that is its equivalent.

Hierarchical Hypothesis H_0: $\lambda^{AB}=\lambda^{AC}=\lambda^{BC}=\lambda^{ABC}=0$, the Hypothesis of Unconditional Independence of the Three Variables A, B, and C. From our consideration of the consequences of $\lambda^{AB}=\lambda^{AC}=\lambda^{ABC}=0$, we know that the above hierarchical hypothesis is equivalent to the pairwise independence of the three variables A, B, and C. Further, we know that each variable is independent of the remaining pair of variables. Therefore,

$$\begin{aligned} p_{ijk} &= P(A=i,B=j,C=k)=P(A=i)P(B=j,C=k) \\ &= P(A=i)P(B=j)P(C=k), \\ &i=1,2,\ldots,I; \quad j=1,2,\ldots,J; \quad k=1,2,\ldots,K. \end{aligned}$$

Thus H_0 implies that the three variables are unconditionally independent.

On the other hand, assuming that A, B and C are unconditionally independent, that is, that

$$p_{ijk}=p_{i++}p_{+j+}p_{++k}, \text{ for all } i,j,k,$$

we have

$$v_{ijk}=\ln p_{i++}+\ln p_{+j+}+\ln p_{++k},$$

and therefore

$$v_{ij}=\frac{1}{K}\sum_K v_{ijk}=\ln p_{i++}+\ln p_{+j+}+\frac{1}{K}\sum_k \ln p_{++k},$$

and similarly,

$$v_{i.k}=\ln p_{i++}+\frac{1}{J}\sum_j \ln p_{+j+}+\ln p_{++k},$$

and

$$v_{.jk}=\frac{1}{I}\sum_i \ln p_{i++}+\ln p_{+j+}+\ln p_{++k}.$$

Further,

$$v_{i..}=\frac{1}{JK}\sum_k\sum_j v_{ijk}=\ln p_{i++}+\frac{1}{J}\sum_j \ln p_{+j+}+\frac{1}{K}\sum_k \ln p_{++k},$$

$$v_{.j.}=\frac{1}{I}\sum_i \ln p_{i++}+\ln p_{+j+}+\frac{1}{K}\sum_k \ln p_{++k},$$

and

$$v_{..k}=\frac{1}{I}\sum_i \ln p_{i++}+\frac{1}{J}\sum_j \ln p_{+j+}+\ln p_{++k},$$

and finally,

$$v_{\dots}=\frac{1}{IJK}\sum_k\sum_j\sum_i v_{ijk}=\frac{1}{I}\sum_i \ln p_{i++}+\frac{1}{J}\sum_j \ln p_{+j+}+\frac{1}{K}\sum_k \ln p_{++k}.$$

Substituting in Equations 9.6.26 and 9.6.27, we find

$$\lambda^{AB}=\lambda^{AC}=\lambda^{ABC}=0.$$

Thus, for the three-way table, unconditional independence of the three variables and zero two- and three-factor interactions are equivalent.

To test the hypothesis of independence of the three variables in the example, we need the expected frequencies for each of the six classes under maximum likelihood estimation of the class probabilities. These expected frequencies are easily seen to be

$$\hat{F}_{ijk}=\frac{f_{i++}f_{+j+}f_{++k}}{n^2}.$$

From the observed marginals we have,

$$f_{1++}=218$$
$$f_{2++}=147$$
$$f_{+1+}=113$$
$$f_{+2+}=252$$
$$f_{++1}=166$$
$$f_{++2}=113$$
$$f_{++3}=86$$

and

$$n=365.$$

We then have the following table of $\hat{F}_{ijk}$s under unconditional independence of the three variables, A, B and C:

$k=1$	$j=1$	$j=2$
$i=1$	30.7	68.5
$i=2$	20.7	46.2
$k=2$		
$i=1$	20.9	46.6
$i=2$	14.1	31.4
$k=3$		
$i=1$	15.9	35.5
$i=2$	10.7	23.9

The value of the χ^2 statistic formed by using the above values of the $\hat{F}_{ijk}$s in the formula

$$\sum_i\sum_j\sum_k\frac{(f_{ijk}-\hat{F}_{ijk})^2}{\hat{F}_{ijk}},$$

where the f_{ijk}s are the observed frequencies (See Table 9.3.1), is 29.408. There are seven degrees of freedom associated with this χ^2 statistic, and the significance probability is less than .001. We would therefore conclude that the three variables are not unconditionally independent.

Because we have already found evidence to support the hypothesis that $\lambda^{AB}=\lambda^{AC}=\lambda^{ABC}=0$, we might conclude that $\lambda^{BC}\neq 0$. In order to track down, in the data, an explanation for the conclusion that $\lambda^{BC}\neq 0$, let us see what we would have found if, after concluding that $\lambda^{AB}=\lambda^{ABC}=0$, instead of examining the hypothesis that $\lambda^{AB}=\lambda^{AC}=\lambda^{ABC}=0$, we had chosen to test H_0: $\lambda^{AB}=\lambda^{BC}=\lambda^{ABC}=0$.

Hierarachical Hypothesis H_0: $\lambda^{AB}=\lambda^{BC}=\lambda^{ABC}=0$. Proceeding in the same way as for testing H_0: $\lambda^{AB}=\lambda^{AC}=\lambda^{ABC}=0$, with BC replacing AC, we find, after having fit to the AC and to the B marginals, the following $\hat{F}_{ijk}$s:

$k=1$	$j=1$	$j=2$
$i=1$	31.57808	70.42192
$i=2$	19.81370	44.18630
$k=2$		
$i=1$	20.74247	46.25753
$i=2$	14.24110	31.75890
$k=3$		
$i=1$	15.16986	33.83014
$i=2$	11.45479	25.54521

The χ^2 statistic for testing this H_0 then has the value

$$\chi^2=\frac{(20-31.6)^2}{31.6}+\frac{(10-19.8)^2}{19.8}+\ldots+\frac{(23-25.5)^2}{25.5}$$
$$=29.556,$$

to be compared with the percentiles of the χ^2 distribution with five degrees of freedom. The significance probability is less than .001, indicating that some or all the conditions imposed by this hypothesis fail to hold. It might be instructive to examine the three kinds of independence implied by H_0 to see if we can isolate the reasons for the failure of H_0 to hold.

Implication 1: Type of child and loss category are independent. We have already examined the 2×2 table associated with these two variables and have found no evidence of a lack of independence.

Implication 2: Loss category and birth order are independent. To study this implication, we examine the following contingency table.

	Loss Category (B)			
Birth Order (C)	*Losses* ($j=1$)	*No Losses* ($j=2$)	*Total*	*% With Losses*
2 ($k=1$)	30	136	166	18.1%
3–4 ($k=2$)	42	71	113	37.2%
5+ ($k=3$)	41	45	86	47.7%
Total	113	252	365	31.0%

The value of the two degrees-of-freedom independence χ^2 statistic for the above table is 26.220, with a significance probability of less than .001. Clearly, loss category and birth order are not independent. This is not surprising, because we would expect that children with a high birth order come from larger families, in general, than those with a low birth order, and it is reasonable to assume that when there are more children the probability of losses is higher. Certainly, in our data, the proportion of mothers with losses increases with increasing birth order.

Implication 3: Loss category is independent of the pair of variables, type of child × birth order. To examine this aspect of H_0, we consider the following contingency table:

TABLE 9.6.7 CROSS-CLASSIFICATION OF 365 CHILDREN BY LOSS CATEGORY (*B*) AND TYPE OF CHILD (*A*)xBIRTH ORDER (*C*) CLASSIFICATION

Child Type $(A)\times$ *Birth Order* (C)	*Loss Category* (B) *Losses* $(j=1)$	*No Losses* $(j=2)$	*Total*	*% With Losses*
Problem $(i=1)\times 2$ $(k=1)$	20	82	102	19.6%
Control $(i=2)\times 2$ $(k=1)$	10	54	64	15.6%
Problem $(i=1)\times 3\text{–}4$ $(k=2)$	26	41	67	38.8%
Control $(i=2)\times 3\text{–}4$ $(k=2)$	16	30	46	34.8%
Problem $(i=1)\times 5+$ $(k=3)$	27	22	49	55.1%
Control $(i=2)\times 5+$ $(k=3)$	14	23	37	37.8%
Total	113	252	365	44.8%

The χ^2 value of 29.592 calculated for the above contingency table has, for five degrees of freedom, an associated significance probability less than .001, indicating a lack of independence between loss category on the one hand and type of child-birth order category on the other. An examination of the proportions with losses for the various type of child-birth order categories, confirms the results obtained for testing a previous hierarchical hypothesis, namely, H_0: $\lambda^{AB}=\lambda^{ABC}=0$. In that case we concluded that type of child and loss history are conditionally independent, given birth order. In the above contingency table we see that, within birth-order categories, the proportions with previous losses are remarkably similar for problem and for control children, except perhaps for birth order 5+.

From this discussion we conclude that λ^{BC} must remain in the model. Because we are restricting ourselves to hierarchical hypothesis, it follows that both λ^B and λ^C must also remain in the model. Therefore, there remains only one further hierarchical hypothesis that may be considered, H_0: $\lambda^A=\lambda^{AB}=\lambda^{AC}=\lambda^{ABC}=0$. We shall now briefly discuss the implications of this hypothesis for the general three-way table, and then consider its plausibility in the case of our data.

Hierarchical Hypothesis H_0: $\lambda^A=\lambda^{AB}=\lambda^{AC}=\lambda^{ABC}=0$. Under H_0 above, for all i, j, k,

$$v_{ijk} = \mu + \lambda_j^B + \lambda_k^C + \lambda_{jk}^{BC} = \mu - v_{...} + v_{.jk}.$$

Then $v_{ijk} = \ln p_{ijk}$ does not depend on i, and it follows that p_{ijk} is also independent of i. Therefore $P(A=i|B=j, C=k)$, the conditional probability that $A=i$, given that $B=j$ and $C=k$, is *constant*. In order to test this hypothesis, we have only to fit to the BC marginal, since that will automatically satisfy the B and C marginals. It is easily shown that the maximum likelihood estimator of p_{ijk} under H_0 is

$$\hat{p}_{ijk} = \frac{\hat{p}_{+jk}}{I} = \frac{f_{+jk}}{In} \quad (i=1,\ldots,I;\, j=1,\ldots,J;\, k=1,\ldots,K),$$

and therefore under H_0

$$\hat{F}_{ijk} = \frac{f_{+jk}}{I},$$

for all i, j, k. The table of $\hat{F}_{ijk}$s for testing H_0 is therefore:

$k=1$	$j=1$	$j=2$
$i=1$	$30/2=15$	$136/2=68$
$i=2$	$30/2=15$	$136/2=68$
$k=2$		
$i=1$	$42/2=21$	$71/2=35.5$
$i=2$	$42/2=21$	$71/2=35.5$
$k=3$		
$i=1$	$41/2=20.5$	$45/2=22.5$
$i=2$	$41/2=20.5$	$45/2=22.5$

The value of the χ^2 statistic for testing H_0 is

$$\chi^2 = \frac{(20-15)^2}{15} + \frac{(10-15)^2}{15} + \frac{(26-21)^2}{21} + \ldots + \frac{(23-25.5)^2}{25.5} = 17.326.$$

The associated number of degrees of freedom is six because the number of unknown parameters, excluding μ, under H_0 is five, namely, λ_1^A, λ_1^C, λ_2^C, λ_{11}^{BC}, and λ_{12}^{BC}, while under the full model, excluding μ there are eleven unknown parameters. Therefore H_0 is rejected with significance probability less than .005. Clearly the data show that the proportion of problem children is in general *not* close to that of control children within the various birth-order × loss-history categories.

We would conclude from the preceding analysis of our data that the model involving the fewest number of unknown parameters that "fits" the observed data is

$$v_{ijk} = \mu + \lambda_i^A + \lambda_j^B + \lambda_k^C + \lambda_{jk}^{BC}, \; (i=1,2;\, j=1,2;\, k=1,2,3).$$

The goodness of fit of this model was tested as H_0: $\lambda^{AB} = \lambda^{AC} = \lambda^{ABC} = 0$.

TABLE 9.6.8 HIERARCHICAL HYPOTHESES IN THE $I \times J \times K$ CONTINGENCY TABLE

Hypothesis	*Degrees of freedom*	*Number**	*Marginals to be fitted*	*Explanation of hypothesis*
1) $\lambda^{ABC}=0$	$(I-1)(J-1)(K-1)$	1	AB, AC, BC	The conditional two-factor interactions given the third variable are equal, that is, departure from conditional independence of each pair of variables, given the value of the third (as measured by the conditional second order interaction) is the same for all values of the third variable.
2) $\lambda^{AB}=\lambda^{ABC}=0$	$K(I-1)(J-1)$	3	AC, BC	Variables A and B conditionally independent, given the value of variable C.
3) $\lambda^{AB}=\lambda^{AC}=\lambda^{ABC}=0$	$(I-1)(JK-1)$	3	BC, A	Variables A and B *unconditionally* independent, as are variables A and C; A independent of the *pair*, BC.
4) $\lambda^{A}=\lambda^{AB}=\lambda^{AC}=\lambda^{ABC}=0$	$JK(I-1)$	3	BC	$P(A=i\|B=j,C=k)$ is constant, that is, the same for any i, in addition to #3 above.
5) $\lambda^{AB}=\lambda^{AC}=\lambda^{BC}=\lambda^{ABC}=0$	$IJK-I-J-K+2$	1	A, B, C	Variables A, B and C independent in the conventional sense, $P(A=i,B=j,C=k)=P(A=i)P(B=j)P(C=k)$ for all i,j,k.
6) $\lambda^{A}=\lambda^{AB}=\lambda^{AC}=\lambda^{BC}=\lambda^{ABC}=0$	$IJK-J-K+1$	3	B, C	Same as #5 above, and, in addition, $P(A=i\|B=j,C=k)$ is constant, all i.
7) $\lambda^{A}=\lambda^{B}=\lambda^{AB}=\lambda^{AC}=\lambda^{BC}=\lambda^{ABC}=0$	$K(IJ-1)$	3	C	Same as #6 and also $P(B=j\|A=i,C=k)$ constant (all j) and $P(A=i,B+j\|C=k)$ is also constant (all pairs i,j).
8) $\lambda^{A}=\lambda^{B}=\lambda^{C}=\lambda^{AB}=\lambda^{AC}=\lambda^{BC}=\lambda^{ABC}=0$	$IJK-1$	1	n	Equivalent to hypothesis of uniformity of distribution, that is, H_0: $p_{ijk}=1/IJK$, all i,j,k. Expected cell frequencies equal n/IJK, so fit to total n of observations.

*Number of hypotheses of the kind indicated.

Goodman (1970) has provided a list of hierarchical hypotheses for the three-way table, along with the interpretation of each one. Table 9.6.8 has been adapted from Goodman's Table 3. His 1970 paper also has a similar list for the case of the four-way table. We have previously discussed and illustrated the procedure for testing various hypotheses in the three-way case. A summary of the steps to be followed in testing a hierarchical hypothesis in the $I \times J \times K$ table follows.

9.6.3 Summary of Steps in Testing a Hierarchical Hypothesis

1. State the model. If there are no restrictions other than that $\Sigma_{ijk} p_{ijk} = 1$, the model for the $I \times J \times K$ contingency table is

$$v_{ijk} = \ln p_{ijk} = \mu + \lambda_i^A + \lambda_j^B + \lambda_k^C + \lambda_{ij}^{AB} + \lambda_{ik}^{AC} + \lambda_{jk}^{BC} + \lambda_{ijk}^{ABC},$$
$$i = 1,\ldots,I;\, j = 1,\ldots,\, J;\; k = 1,\ldots,K,$$

where the parameters of the model, the λs, are as they have been previously defined in equations 9.6.25 to 9.6.27, and μ, whose value is usually of no interest to the experimenter, is such that $\Sigma_{ijk} p_{ijk} = 1$ is satisfied. We have seen that, because of the definitions of the λs, $\Sigma_i \lambda_i^A = 0 = \Sigma_j \lambda_j^B = \Sigma_i \lambda_{ij}^{AB} = \Sigma_j \lambda_{ij}^{AB} = \Sigma_i \lambda_{ik}^{AC} = \Sigma_k \lambda_{ik}^{AC} = \Sigma_j \lambda_{jk}^{BC} = \Sigma_k \lambda_{jk}^{BC} = \Sigma_i \lambda_{ijk}^{ABC} = \Sigma_j \lambda_{ijk}^{ABC} = \Sigma_k \lambda_{ijk}^{ABC}$, and because μ is restricted as we have indicated, there are $I - 1 + J - 1 + K - 1 + (I-1)(J-1) + (I-1)(K-1) + (J-1)(K-1) + (I-1)(J-1)(K-1) = IJK - 1$ "free" parameters in the model. Thus, the reparameterization from the p_{ijk}s to the λs gives the same number of free parameters.

It may happen that one has reason to believe that, for example, one of the two-factor interactions, say λ^{BC}, is zero. The hierarchical scheme demands that then also $\lambda^{ABC} = 0$, and, in that case, the model would be

$$v_{ijk} = \mu + \lambda_i^A + \lambda_j^B + \lambda_k^C + \lambda_{ij}^{AB} + \lambda_{ik}^{AC},\ (i = 1,\ldots,I;\, j = 1,\ldots,J;\; k = 1,\ldots,K),$$

a model with $I - 1 + J - 1 + K - 1 + (I-1)(J-1) + (I-1)(K-1) = I(J + K - 1) - 1$ "free" parameters.

2. State the hypothesis. We restrict ourselves to hierarchical hypotheses so that, for example, $\lambda^{AB} = 0 \Rightarrow \lambda^{ABC} = 0$.

3. State the model if the conditions of the hypothesis hold. Thus, for example, if H_0: $\lambda^{AB} = 0 = \lambda^{ABC}$, then the model under H_0 would be

$$v_{ijk} = \mu + \lambda_i^A + \lambda_j^B + \lambda_k^C + \lambda_{ik}^{AC} + \lambda_{jk}^{BC},\ (i = 1,\ldots,I;\, j = 1,\ldots,J;\; k = 1,\ldots,K).$$

Count the number of free parameters in the restricted model (that is, restricted by H_0). In this case, there would be $I - 1 + J - 1 + K - 1 + (I-1)(K-1) + (J-1)(K-1) = K(I + J - 1) - 1$ free parameters. If the original model were the "full" model, the *reduction* in the number of free (unknown) parameters would be $IJK - 1 - [K(I + J - 1) - 1] = K(I-1)(J-1)$. Note that this reduction is equal to the number of new restrictions imposed by H_0, that is, $\lambda^{AB} = 0$ imposes $(I-1)(J-1)$ new restrictions, and $\lambda^{ABC} = 0$ imposes $(I-1)(J-1)(K-1)$ new restrictions, for a total of $(I-1)(J-1)(1 + K - 1) = K(I-1)(J-1)$.

4. Examine the restricted model (by H_0) for the highest order interactions still present (in the example, AC and BC), and for any lower-order interactions or main effects still in the model whose variable names do not appear in the highest order interactions noted. In the case of H_0 above, we have the three main effects associated with A, B, and C still in the model, but all three variable names are included in the pairs AC and BC. Use the iterative procedure to find the expected frequencies for all the $I \times J \times K$ cells, under maximum likelihood estimation, by fitting to the marginals of the highest order interaction terms bearing the names still included in the model, including lower order terms when necessary. In this case, we need only fit to the AC and BC marginals, since then the A, B, and C marginals will automatically fit. In those cases in which direct maximum likelihood estimates of the cell probabilities are easily computed, one may prefer to calculate these estimates, and denote them by $\hat{p}_{ijk}$. We can then calculate the expected cell frequencies under H_0, $\hat{F}_{ijk} = n\hat{p}_{ijk}$.

5. Calculate the χ^2 statistic

$$\sum_i \sum_j \sum_k \frac{(f_{ijk} - \hat{F}_{ijk})^2}{\hat{F}_{ijk}},$$

with associated degrees of freedom equal to the *reduction* in the number of free or unknown parameters under H_0 (or equivalently, the number of additional restrictions imposed by H_0 over those of the original model).

6. Accept or reject H_0 on the basis of the value of χ^2.

If the model we began with were not the full model, but, say

$$v_{ijk} = \mu + \lambda_i^A + \lambda_j^B + \lambda_k^C + \lambda_{ij}^{AB} + \lambda_{ik}^{AC},$$

a model with $I(J+K-1)-1$ free parameters, and if the hypothesis imposed the further restriction H_0: $\lambda^{AC} = 0$, the procedure would be as follows:

1. Calculate the χ^2 statistic associated with the above model, following the steps outlined above for *testing* this model as a hypothesis against the full model. This χ^2 statistic, denoted by $\chi^2_{R,F}$, would have associated with it

$$IJK - 1 - [I(J+K-1) - 1] = I(J-1)(K-1)$$

degrees of freedom.

2. Under H_0, the model becomes

$$v_{ijk} = \mu + \lambda_i^A + \lambda_j^B + \lambda_k^C + \lambda_{ij}^{AB}.$$

3. Fit the data to the marginals AB and C and compute the χ^2 statistic, say $\chi^2_{H_0,F}$, as though we were testing this hypothesis against the full model. Note that the restrictions imposed by this model number $(I-1)(K-1) + (J-1)(K-1) + (I-1)(J-1)(K-1) = (K-1)(IJ-1)$.

4. Compute the χ^2 statistic for testing H_0 against the restricted model, as

$$\chi^2_{H_0} = \chi^2_{H_0,F} - \chi^2_{R,F},$$

with associated degrees of freedom equal to the number of *additional* restrictions imposed by H_0 (over those imposed by the restricted model), that is, $(K-1)(IJ-1)-I(J-1)(K-1)=(I-1)(K-1)$, which is also equal to the reduction in the number of unknown parameters under H_0, compared with those under the restricted model.

9.6.4 Higher-Order Contingency Tables

In higher-order contingency tables, the variables and parameters of the log-linear model defined for the three-way table are analogously defined. For example, in a four-way table with Variables A, B, C, and D, with I, J, K, and L categories, respectively,

$$p_{ijkl} = P(A=i, B=j, C=k, D=1),$$

where again, $A=i$ means that variable A falls in the ith category of that variable, etc. Then

$$v_{ijkl} = \ln p_{ijkl}$$

$$v_{i...} = \frac{1}{JKL}\sum_j \sum_k \sum_l v_{ijkl}, \text{ etc.,}$$

$$v_{ij..} = \frac{1}{KL}\sum_k \sum_l v_{ijkl}, \text{ etc.,}$$

$$v_{ijk.} = \frac{1}{L}\sum_l v_{ijkl}, \text{ etc.,}$$

and

$$v_{....} = \frac{1}{IJKL}\sum_i \sum_j \sum_k \sum_l v_{ijkl}.$$

Here,

$$\lambda_i^A = v_{i...} - v_{....},$$

with the other main effects similarly defined.

$$\begin{aligned}\lambda_{ij}^{AB} &= v_{ij..} - v_{....} - (\lambda_i^A + \lambda_j^B)\\ &= v_{ij..} - v_{i...} - v_{.j..} + v_{....},\end{aligned}$$

is an example of a two-factor interaction,

$$\begin{aligned}\lambda_{ijk}^{ABC} &= v_{ijk.} - v_{....} - (\lambda_i^A + \lambda_j^B + \lambda_k^C + \lambda_{ij}^{AB} + \lambda_{ik}^{AC} + \lambda_{jk}^{BC})\\ &= v_{ijk.} - v_{ij..} - v_{i.k.} - v_{.jk.} + v_{i...} + v_{.j..} + v_{..k.} - v_{....},\end{aligned}$$

is an example of a three-factor interaction, and the four-factor interactions (of one kind only, that is, with only one set of superscripts), are

$$\begin{aligned}\lambda_{ijkl}^{ABCD} &= v_{ijkl} - v_{....} - (\lambda_i^A + \lambda_j^B + \lambda_k^C + \lambda_1^D + \lambda_{ij}^{AB} + \lambda_{ik}^{AC} + \lambda_{il}^{AB}\\ &\quad + \lambda_{jk}^{BC} + \lambda_{jl}^{BD} + \lambda_{kl}^{CD} + \lambda_{ijk}^{ABC} + \lambda_{ijl}^{ABD} + \lambda_{ikl}^{ACD} + \lambda_{jkl}^{BCD})\\ &= v_{ijkl} - v_{ijk.} - v_{ij.l} - v_{i.kl} - v_{.jkl} + v_{ij..}\\ &\quad + v_{i.k.} + v_{i..l} + v_{.jk.} + v_{.j.l} + v_{..kl} - v_{i...}\\ &\quad - v_{.j..} - v_{..k.} - v_{...l} + v_{....}.\end{aligned}$$

In dealing with the higher order tables, one must be careful of the interpretation of the various interactions. For example, $\lambda^{ABCD}=0$ means that the conditional three-factor interactions among three of the variables, given the fourth variable, are equal. As in the case of the analysis of variance, it becomes more difficult to interpret the interactions as the order, that is, the number of factors, increases.

The restriction $\lambda^{ABC}=0=\lambda^{ABCD}$ means that the conditional two-factor interactions between A and B, given C *and* D (that is, the pair) are equal, as are those between variables A and C, given B *and* D, and those between B and C, given A *and* D.

The interactions with which most readers are familiar are the two-factor interactions, particularly in the case of the two-way table, where we saw that $\lambda^{AB}=0$ means that the classificatory variables, A and B, are independent. In the four-way hierarchical case, $\lambda^{AB}=\lambda^{ABC}=\lambda^{ABCD}=0$ means that variables A and B are conditionally independent, given the *pair* C and D. In the general m-way table, by fixing $m-3$ of the variables, we can consider the resulting three-way table, given the "value" of those $m-3$ variables. That is, we may "lump" the $m-3$ variables together and consider them to be a single variable with the number of possible outcomes equal to the product of the number of possible outcomes for each of the $m-3$ variables. Then, what we have learned about the three-way case will hold conditionally, given the value of the new variable whose possible outcomes consist of all combinations of categories of the remaining $m-3$ variables. The important point to remember is that we must condition on *all* the remaining variables.

The consequences of this feature may be difficult to interpret. For example, if variables A and B are conditionally independent, given the pair C and D, it does not necessarily imply that A and B are conditionally independent given C (or D) alone. On the other hand, in the four-way case, if we sum over variable D in the original counts and find that A and B are conditionally independent, given C in the resulting three-way table, it does not imply that A and B are conditionally independent, given C and D. Thus, one must exercise some caution in the interpretation of various hierarchical hypotheses in terms of independence and conditional independence, which are the traditional concerns of interest in the analysis of contingency tables.

In general, the method outlined here offers an elegant unified approach that can be used to solve many problems of interest. It has the further advantage of focusing our attention on questions that may indeed be the appropriate ones to investigate, such as those concerning various types of conditional independence. The fact that one must exercise caution in interpreting results is not limited to the log-linear models and hypotheses we have discussed. However, there might be a temptation to be carried away by the neatness of the structure, without paying sufficient attention to the pertinence of the various models and hypotheses to a particular problem.

We have barely touched upon the subject of log-linear models here. Interested readers are referred to the literature cited for further study. In particular, there are some special cases of interest such as the $2 \times 2 \times 2$ table. Interesting and enlightening simplifications may be possible in such cases.

Finally, we have used the $I \times J \times K$ contingency table, a multinomial with IJK classes, as the focus of attention. Although we frequently deal with a sequence of multinomials rather than with a single one, the models we have discussed can still be used with appropriate interpretations of interactions in such situations.

9.7 COMPUTING

Some of the methods discussed in this chapter involve extensive calculations. For students who are familiar with APL, problems that involve successive operations on matrices can easily be solved using this mode. There are many computer programs that have been developed specifically for applying the types of analyses discussed in this chapter. A partial list of such programs is given below, along with the person or institution from which information concerning the documentation of the programs may be obtained. In some cases the programs themselves may be obtained from the same source.

GENCAT (A Computer Program For The Generalized Chi-Square Analysis Of Categorical Data Using Weighted Least Squares To Compute Wald Statistics)

J. Richard Landis, Ph.D.
Department of Biostatistics,
School of Public Health
University of Michigan
Ann Arbor, Michigan, 48109

ECTA (Everman's Contingency Table Analysis' Parameter Estimates And Tests)

Professor Leo A. Goodman
Department of Statistics
University of Chicago
1118 East 58th Street
Chicago, Illinois, 60637

CATLIN and *LINCAT*

Program Librarian
Department of Biostatistics,
School of Public Health
University of North Carolina
Chapel Hill, North Carolina, 27514

C-TAB (Analysis of Multidimensional Contingency Tables By Log-Linear Models)

International Educational Services
P. O. Box A 3650
Chicago, Illinois, 60690

MULTIQUAL (Log-Linear Analysis Of Nominal Or Ordinal Qualitative Data By The Method Of Maximum Likelihood)

National Educational Resources, Inc.
1525 East 53rd Street
Chicago, Illinois 60615

9.8 EXERCISES

9.1. Let X be the number of successes in n_1 independent binomial trials with probability of success p_1 on each trial, and let Y be the number of successes in n_2 independent trials (also independent of the n_1 other trials) with probability of success p_2 on each trial. We wish to test H_0: $p_1=p_2=p$, $(q=1-p)$.

(a)* Show that the maximum likelihood estimator of p, if H_0 is true, is

$$\hat{p}=\frac{X+Y}{n_1+n_2}\left(\hat{q}=1-\hat{p}=\frac{n_1+n_2-X-Y}{n_1+n_2}\right).$$

(b) Show that the Fisher χ^2 statistic

$$\chi^2_{(F)}=\frac{(X-n_1\hat{p})^2}{n_1\hat{p}}+\frac{(n_1-X-n_1\hat{q})^2}{n_1\hat{q}}+\frac{(Y-n_2\hat{p})^2}{n_2\hat{p}}+\frac{(n_2-Y-n_2\hat{q})^2}{n_2\hat{q}}$$

$$=\frac{(X-n_1\hat{p})^2}{n_1\hat{p}\hat{q}}+\frac{(Y-n_2\hat{p})^2}{n_2\hat{p}\hat{q}}$$

can be written

$$\frac{\left(\dfrac{X}{n_1}-\dfrac{Y}{n_2}\right)^2}{\hat{p}\hat{q}\left(\dfrac{1}{n_1}+\dfrac{1}{n_2}\right)}=\frac{[X(n_2-Y)-Y(n_1-X)]^2(n_1+n_2)}{n_1n_2(X+Y)(n_1-X+n_2-Y)}$$

(see Equation 9.5.2)

(c)* Show that the estimator of p that minimizes the Neyman χ^2 statistic, (if H_0 is true),

$$\chi^2_{(N)}=\frac{(X-n_1p)^2}{X}+\frac{(n_1-X-n_1q)^2}{n_1-X}+\frac{(Y-n_2p)^2}{Y}+\frac{(n_2-Y-n_2q)^2}{n_2-Y}$$

is

$$\tilde{p}=XY\frac{n_1^2(n_2-Y)+n_2^2(n_1-X)}{n_1^3Y(n_2-Y)+n_2^3X(n_1-X)}.$$

*The results should be noted by all students, but these proofs should only be attempted by those students with some knowledge of calculus.

(d) Show that the value of $\chi^2_{(N)}$ when $\tilde{p}$ is substituted for p in the expression for $\chi^2_{(N)}$ given in (c) above, can be written

$$\frac{\left(\frac{X}{n_1}-\frac{Y}{n_2}\right)^2}{\frac{X(n_1-X)}{n_1^3}+\frac{Y(n_2-Y)}{n_2^3}}.$$

(e) Comparing the expressions for the two χ^2 statistics in (b) and (d), you will see that they have the same numerators. Consider the square root of that numerator, namely, $X/n_1 - Y/n_2$, the difference between two independent observed proportions. If this difference, which is approximately normally distributed for large n_1 and n_2 is divided by an appropriate estimator of the standard error of the difference, the resulting statistic will have, for large n_1, n_2, approximately a standard normal distribution under H_0, and its square approximately a χ^2 distribution with one degree of freedom. The difference in the $\chi^2_{(F)}$ and $\chi^2_{(N)}$ statistics may be seen to be due to the use of different estimators of the standard error of $X/n_1 - Y/n_2$. How were these standard errors estimated in each case? Give an intuitive justification for each.

(f) The BNDD study on drug use and arrest charges cross-classified arrestees according to amphetamine use (current amphetamine user, nondrug user) and type of crime (person, property) with the following result, adapted from the data of Table VIII-21 of the report:

	Crime Against		
Drug Status	*Person*	*Property*	*Total*
Current Amphetamine User	61	77	138
Nondrug User	319	265	584
Total	380	342	722

i. Assuming that the model is two independent binomial distributions, compute the "usual" χ^2 statistic ($\chi^2_{(F)}$) for testing H_0: $p_1 = p_2$, where p_1 is the probability of a crime against the person among current amphetamine users who are arrested, and p_2 is the corresponding probability for nondrug users.

ii. Calculate the value of the Neyman χ^2 statistic.

iii. Test the same hypothesis with the help of a Wald χ^2 statistic.

iv. Compare the values of the three χ^2 statistics and comment.

(g) Interpret the results of the tests performed in (f), in terms of the data.

9.2. The Department of Mental Health Sciences of the Hahnemann Medical College and Hospital of Philadelphia is responsible for maintaining satellite mental health centers in greater Philadelphia. Three of these centers are: *Center City*, located in the political center of the city and serving a mixed racial and economic population; *Corinthian*, in the midst of a Black community consisting of mostly low-income families, with some pockets of blue-collar Blacks; and *Poplar*, a mixed Puerto Rican and Black community which is similar to Corinthian, but suffers somewhat worse economic conditions. The satellite clinics have patient rosters, and records are kept of patient visits to the centers. We are grateful to Dr. George Spivak, the Director of the Division of Research and Evaluation of the Hahnemann Community Mental Health Center for making available to us the following data on the number of patients, by age groups, seen during October 1977 in the three satellite centers.

Mental Health Center	*Age Group*	*Number of Patients*		
		Seen in Center	*Not Seen*	*Total Registered*
Center City	Adult	157	9	166
	Child	77	64	141
	Total	234	73	307
Corinthian	Adult	186	155	341
	Child	119	75	194
	Total	305	230	535
Poplar	Adult	130	18	148
	Child	65	80	145
	Total	195	98	293

(a) Using the data for the children only,
i. Perform the "usual" χ^2 test of homogeneity to determine if there is a significant difference in the proportion of children seen in the three centers among those registered.
ii. Test the same hypothesis with the help of a Wald statistic and compare the value of the χ^2 statistic obtained by this method with that in i. Comment.
(b) Is there evidence of age group-Health Center interaction? (Use the method suggested in Section 9.5.2). Interpret your result in terms of the data.
(c) For each center, calculate and interpret verbally the value of the odds ratio:

$$\frac{\text{proportion of adults seen}}{\text{proportion of adults not seen}} \Big/ \frac{\text{proportion of children seen}}{\text{proportion of children not seen}}$$

$$= \frac{\text{number of adults seen} \times \text{number of children not seen}}{\text{number of adults not seen} \times \text{number of children seen}}.$$

9.3. The Department of Health, Education and Welfare (1977) reported data on agreement between youths and their parents on health and behavioral questions cross-classified by sex, age and other demographic variables. Many interesting questions are examined based upon a sample of 6768 youths—3386 boys and 3382 girls.*

The following data represent a cross-classification of the youths by sex, expectation concerning educational goals of the parent for the youth, and the parent-youth agreement on goals. They are adapted from the data in Tables 23 and 24 of the report.

*These figures as well as the frequencies in the table that follows are based upon our interpretation of the details given in the report and may not be entirely accurate. For the purposes of this exercise, however, we will assume that they are correct.

Boys

Expectation of Parent for Youth	Parent-Youth Agreement			Row Total	Row Percent
	Youth Goal Higher	Goals the Same	Parents Goal Higher		
Quit school as soon as possible	48	81	–	129	3.8
Finish high school	287	508	48	843	24.9
Some college	306	598	176	1080	31.9
Finish college	208	443	206	857	25.3
Some postgraduate	64	251	162	477	14.1
Column Total	913	1881	592	3386	100.0
Column Percent	27.0	55.6	17.5	100.1	

Girls

Expectation of Parent for Youth	Parent-Youth Agreement			Row Total	Row Percent
	Youth Goal Higher	Goals the Same	Parents Goal Higher		
Quit school as Soon as possible	61	40	–	101	3.0
Finish high school	372	568	24	964	28.5
Some college	361	729	276	1366	40.4
Finish college	120	402	219	741	21.9
Some postgraduate	–	113	97	210	6.2
Column Total	914	1852	616	3382	100.0
Column Percent	27.0	54.8	18.2	100.0	

An examination of the marginal distributions shows that, although the degree of parent-youth agreement is remarkably similar for boys and girls, the parental educational expectations differ somewhat. While about the same proportion of parents expect at most a high school education (28.7% of parents of boys compared to 31.5% of parents of girls), there are differences in parental expectations at the higher educational levels according to the sex of the child. Relatively more parents of boys expect at least a college degree for their child than do parents of girls (39.4% compared to 28.1%, respectively).

There are many questions that may be of interest to educators. Consider the following presentation of the data with the expectations of parents collapsed into three categories: High school or less (High school −), Some college, Finish college or more (Finish college +).

Expectation of Parent for Youth	Sex of Youth	Parent-Youth Agreement			Total
		Youth Goal Higher	Goals the Same	Parent Goal Higher	
High School −	Boy	335	589	48	972
	Girl	433	608	24	1065
	Total	768	1197	72	2037
Some College	Boy	306	598	176	1080
	Girl	361	729	276	1366
	Total	667	1327	452	2446
Finish College +	Boy	272	694	368	1334
	Girl	120	515	316	951
	Total	392	1209	684	2285

(a) For the boys only, consider the 3×3 contingency table of expectation of parent for youth $\times$ parent-youth goal agreement. Are these two classifactory variables independent? Comment on your findings.

(b) Do the same for the girls only.

(c) Examine the two contingency tables carefully and comment on similarities and differences.

(d) Deleting those parent-youth pairs whose goals were the same, we have the following sex $\times$ goal-agreement frequencies:

Boys

	Parent-Youth Agreement					
Expectation of	*Youth Goal Higher*		*Parent Goal Higher*		*Total*	
Parent for Youth	*Number*	*Proportion*	*Number*	*Proportion*	*No.*	*Proportion*
High School −	335	.87467362	48	.12532637	383	1.0000000
Some College	306	.63485477	176	.36514522	482	1.0000000
Finish College +	272	.42500000	368	.57500000	640	1.0000000
Total	913	.60664451	592	.39335548	1505	1.0000000

Girls

	Parent-Youth Goal Agreement					
Expectation of	*Youth Goal Higher*		*Parent Goal Higher*		*Total*	
Parent for Youth	*Number*	*Proportion*	*Number*	*Proportion*	*No.*	*Proportion*
High School −	433	.94748358	24	.05251641	457	1.0000000
Some College	361	.56671899	276	.43328100	637	1.0000000
Finish College +	120	.27522935	316	.72477064	436	1.0000000
Total	914	.59738562	616	.40261437	1530	1.0000000

Note that among parent-youth pairs whose goals were in disagreement, the difference between the proportion of boys whose goals exceeded those of their parents and the corresponding proportion of girls is less than 1%.

Given that the parent-youth goals are not the same, let p_{ij} be the probability that a youth of sex i has educational goals exceeding those of a parent whose expectation is j; $(i=1,2;\ j=1,2,3)$, where

$i=1$ denotes a boy,

$i=2$ denotes a girl,

$j=1$ denotes a parental expectation of high school or less,

$j=2$ denotes a parental expectation of some college,

$j=3$ denotes a parental expectation of at least a college degree.

Assume that

$$p_{ij}=\mu+\alpha_i+\beta_j+(\alpha\beta)_{ij}, \quad (i=1,2;\ j=1,2,3),$$

where μ denotes a constant, α_i denotes the main effect of sex i, β_j denotes the main effect of parental expectation j, and $(\alpha\beta)_{ij}$ denotes the interaction of sex i and expectation j, and where

$$\sum_{i=1}^{2} \alpha_i = \sum_{j=1}^{3} \beta_j = \sum_{i=1}^{2} (\alpha\beta)_{ij} = \sum_{j=1}^{3} (\alpha\beta)_{ij} = 0.$$

Thus, the original six unknown parameters, p_{ij}, $(i=1,2;\ j=1,2,3)$ subject to the condition that $\Sigma_{j=1}^{3}\Sigma_{i=1}^{2} p_{ij} = 1$), have been replaced by six new parameters: μ, $\alpha_1(\alpha_2 = -\alpha_1)$, β_1, β_2, $(\beta_3 = -\beta_1 - \beta_2)$, $(\alpha\beta)_{11}$ and $(\alpha\beta)_{12}(\alpha\beta)_{13} = -(\alpha\beta)_{11} - (\alpha\beta)_{12}$; $(\alpha\beta)_{21} = -(\alpha\beta)_{11}$; $(\alpha\beta)_{22} = -(\alpha\beta)_{12}$; $(\alpha\beta)_{23} = -(\alpha\beta)_{13}$) and μ must be such that under the new parametrization, $\Sigma_{j=1}^{3}\Sigma_{i=1}^{2} p_{ij} = 1$.

i. By the method of weighted least squares, estimate the values of the new parameters; using the notation of Section 9.5.3, find the values of the matrices

$$\mathbf{Y}, \mathbf{X}, \hat{\mathbf{\Sigma}}_{\mathbf{Y}}^{-1}, \hat{\boldsymbol{\beta}}.$$

ii. Calculate the weighted least squares estimates of the p_{ij}s, by applying

$$\hat{\mathbf{p}} = \mathbf{X}\hat{\boldsymbol{\beta}},$$

and compare $\hat{\mathbf{p}}$ with $\mathbf{Y}$. Note that there is no hypothesis being tested here.

iii. Test the goodness of fit of the model

$$p_{ij} = \mu + \beta_j + (\alpha\beta)_{ij}, \quad (i=1,2;\ j=1,2,3),$$

which is equivalent to the hypothesis of no main effect for sex. In order to perform this test, the $\mathbf{Y}$ matrix will remain the same, as will $\hat{\mathbf{\Sigma}}_{\mathbf{Y}}^{-1}$. What are the values of the new matrices:

$\mathbf{X}_0 =$ The $\mathbf{X}$ matrix corresponding to H_0: $\alpha_i = 0$, $i = 1,2$,

$\hat{\boldsymbol{\beta}}_0 =$ the $\hat{\boldsymbol{\beta}}$ matrix corresponding to H_0,

and

$$\hat{\mathbf{p}}_0 = \mathbf{X}_0\hat{\boldsymbol{\beta}}_0,$$

the weighted least-squares estimate of $\mathbf{p}$ under the new model? Compare the new estimates with those calculated in part ii. Comment. Are you surprised at the outcome of the test? Can you think of an explanation?

9.4. Consider the sex × parental-expectation × goal-agreement data of Exercise 9.3(d). Let A_i represent sex i; $i=1$ for boys and $i=2$ for girls. Let B_j represent parental expectation j; $j=1$ for high school or less, $j=2$ for some college, and $j=3$ for college degree or more, and let C_k represent parent-youth goal agreement k; $k=1$ for youth goal higher than parental, $k=2$ for parental goal higher than youth.

Let $p_{ijk} = P(A=i, B=j, C=k)$, $(i=1,2;\ j=1,2,3;\ k=1,2)$ and let $\nu_{ijk} = \ln p_{ijk}$.

Consider the log-linear model

$$\nu_{ijk} = \mu + \lambda_i^A + \lambda_j^B + \lambda_k^C + \lambda_{ij}^{AB} + \lambda_{ik}^{AC} + \lambda_{jk}^{BC} + \lambda_{ijk}^{ABC},$$

$$(i = 1,2;\ j = 1,2,3;\ k = 1,2)$$

as defined in Section 9.6.2.

(a) Test the hypothesis H_0: $\lambda^{ABC} = 0$.

i. Display the table of expected frequencies under H_0.

ii. What is the value of the appropriate χ^2 statistic, its degrees of freedom and significance probability?

(b) Interpret verbally the outcome of the test in (a).

(c) Examine separately the 3×2 contingency tables of parental expectation $\times$ parent-youth goal agreement for boys and for girls, and relate your findings to your comments in (b).

9.5. In a study of the role of family structure in the mental health of children in a poor urban community in Chicago by Kellam, Ensminger and Turner (1977), the authors were interested in the relationships among many variables. In particular, they were interested in family structure beyond the categories "Mother-father present in the household" (the so-called nuclear family), and "Mother only present." They analyzed relationships between mental-status variables of the child and family-structure categories that included the traditional categories above but also included households in which the adult members were mother and grandmother, mother and stepfather, and others. Mental-health variables of the children included adaptational status ("adapting" or "maladapting," on the basis of the first-grade teacher's rating), and nervousness, ("not at all nervous," "a little nervous," and "pretty much or a lot nervous," on the basis of the child's self report in the third grade). Other variables such as self-reported sadness and income were also investigated. Some the the data reported in the study are given in the following table:

Family Type 1 (*Mother / Father*)

Third-grade self-assessed nervousness	*First-Grade Social Adaptational Status* 1 (*Adapting*)	2 (*Maladapting*)
1 (not at all nervous)	43	38
2 (a little nervous)	27	33
3 (pretty much or a lot nervous)	22	35

Family Type 2 (*Mother / Second Adult*)

Third-grade self-assessed nervousness	*First-Grade Social Adaptational Status* 1 (*Adapting*)	2 (*Maladapting*)
1 (not at all nervous)	5	8
2 (a little nervous)	14	9
3 (pretty much or a lot nervous)	4	13

Family Type 3 (*Mother Alone*)

Third-grade self-assessed nervousness	*First-Grade Social Adaptational Status* 1 (*Adapting*)	2 (*Maladapting*)
1 (not at all nervous)	11	35
2 (little nervous)	28	34
3 (pretty much or a lot nervous)	18	32

Let A_i represent family type i, $(i=1,2,3)$:
$i=1$ if mother and father are both present,
$i=2$ if mother and another adult are present,
$i=3$ if the mother is the only adult in the household.

Let B_j represent first grade adaptational status j, $(j=1,2)$:
$j=1$ if child is classified as adapting,
$j=2$ if child is classified as maladapting.

Let C_k represent third grade self-assessed nervousness, $(k=1,2,3)$:
$k=1$ if assessment is not at all nervous,
$k=2$ if assessment is a little nervous,
$k=3$ if assessment is pretty much or a lot nervous.

Let p_{ijk} be the probability of the joint occurrence of family type A_i, adaptational status B_j and self-assessed nervousness C_k, $(i=1,2,3; j=1,2; k=1,2,3)$; and let $\nu_{ijk}=\ln p_{ijk}$.

Consider the log-linear model

$$\nu_{ijk}=\mu+\lambda_i^A+\lambda_j^B+\lambda_k^C+\lambda_{ij}^{AB}+\lambda_{ik}^{AC}+\lambda_{jk}^{BC}+\lambda_{ijk}^{ABC},$$

$$(i=1,2,3; j=1,2; k=1,2,3).$$

(a) The authors tested the fit of this model with the last term left out of it, that is, they tested H_0: $\lambda^{ABC}=0$, and found the value of the χ^2 statistic to be 9.75, which for four degrees of freedom has a significance probability less than .044. The χ^2 statistic reported in the paper is not obtained in exactly the same manner as the one given in the text, but should be close in value to it. Compute the χ^2 statistic for testing H_0 and verify that it is in fact close to the one reported in the paper, and that it is significant at the .05 level. Present the table of expected values frequencies after three iterations based on maximum likelihood estimation of the p_{ijk}s.
(b) The results obtained above suggest that the conditional two-way tables for two of the variables, for fixed values of the third variable, do not show the same relationship. First fix the variable family type and examine all reasonable (hierarchical) log-linear hypotheses for the resulting two-way contingency tables. Then do the same fixing first-grade adaptational status. Discuss your results in terms of the data. Note that we could also consider the family type × first-grade adaptational status for each of the three levels of third-grade

nervousness, but we are not suggesting this because it seems less interesting.*

9.6. The National Academy of Sciences (1978), reported on a survey of a sample of Ph.D.s who had received their degrees between January 1934 and July 1976 and were residing in the United States in February 1977. The report includes demographic data such as sex, year (within time periods) of receipt of degree, and race, as well as employment status.

The data below are taken from estimates given in the report of the number of individuals in academic employment by sex, cohort (three time periods as indicated below were selected), and rank (the "no reported rank" category was deleted). The data given here are based upon Tables 1.7 and 2.7 of the report. Some fields listed in the report were excluded because of the small number of women Ph.D.s in these fields: earth science, engineering and agricultural science. Social sciences including the fields of archeology, linguistics, area studies, political science, public administration and international relations are included under "humanities," as in the report. Social scientists in other fields have been omitted because of possible overlap—some social science areas were included in the report in both the data on science/engineering (Table 1.7) and in the data on the humanities (Table 2.7).

Any inaccuracies in the data below are due to our own difficulties in interpreting the report rather than in the report itself. However, we hope that no serious distortions have been inadvertently introduced, and that the general trends in our adaptation of the data of the report are not seriously misleading.

With good intentions then, we present the number of individuals in academic employment by sex, cohort (year of degree 1950–59, 1960–69, 1970–74) rank (professor, associate professor, assistant professor, instructor-lecturer-other), and field (mathematics-computing, physics-chemistry, medical-biological sciences, psychology, humanities).

Males

Rank	*Field*				
Cohort 1950–59	*Math/ Computing*	*Physics/ Chemistry*	*Medical/ Bio. Sciences*	*Psych.*	*Humanities*
Professor	1417	4182	4854	2212	7713
Assoc. Prof.	155	554	883	286	638
Assist. Prof.	8	20	83	64	98
Instr.-Lect.	3	133	134	11	151
Cohort 1960–69					
Professor	2024	3778	4130	1811	8514
Assoc. Prof.	2165	4329	5122	2112	7474
Assist. Prof.	374	894	1294	269	1111
Instr.-Lect.	120	647	337	192	376

*These data were used in an assignment in a class in categorical data analysis in which the students were free to examine log-linear models of their own choosing. We are grateful to Steve Shuller, one of these students, for his analysis of the data, which suggested the structure of this exercise.)

Males					
Rank	*Field*				
Cohort 1970–74	*Math/ Computing*	*Physics/ Chemistry*	*Medical/ Bio. Sciences*	*Psych.*	*Humanities*
Professor	189	140	282	62	1101
Assoc. Prof.	1334	1608	1944	1145	6120
Assist. Prof.	1858	2806	4679	2157	7064
Instr.-Lect.	153	1264	1543	382	881

Females					
Rank	*Field*				
Cohort 1950–59	*Math/ Computing*	*Physics/ Chemistry*	*Medical/ Bio. Sciences*	*Psych.*	*Humanities*
Professor	44	163	333	190	603
Assoc. Prof.	8	60	168	83	263
Assist. Prof.	4	9	38	20	38
Instr.-Lect.	1	34	98	17	91
Cohort 1960–69					
Professor	99	80	318	230	1108
Assoc. Prof.	155	215	661	425	1584
Assist. Prof.	47	128	505	246	500
Instr.-Lect.	25	146	274	144	193
Cohort 1970–74					
Professor	6	7	45	16	230
Assoc. Prof.	55	44	257	267	1283
Assist. Prof.	215	204	1141	870	2556
Instr.-Lect.	52	181	654	196	689

Let A_i represent sex i, $(i=1,2)$:
$i=1$ for males,
$i=2$ for females.

Let B_j represent cohort j, $(j=1,2,3)$:
$j=1$ for 1950–59,
$j=2$ for 1960–69,
$j=3$ for 1970–74.

Let C_k represent rank k, $(k=1,2,3,4)$:
$k=1$ for professor,
$k=2$ for associate professor,
$k=3$ for assistant professor,
$k=4$ for instructor-lecturer-other.

Let D_l represent field l, $(l=1,2,3,4,5)$:
$l=1$ for mathematics/computing,
$l=2$ for physics/chemistry,
$l=3$ for medical/biological sciences,
$l=4$ for psychology,
$l=5$ for humanities.

Also let p_{ijkl} represent the joint probability of categories A_i, B_j, C_k, D_l and let

$$\nu_{ijkl} = \ln p_{ijkl} = \mu + \lambda_i^A + \lambda_j^B + \lambda_k^C + \lambda_l^D + \lambda_{ij}^{AB} + \lambda_{ik}^{AC} + \lambda_{il}^{AD} + \lambda_{jk}^{BC}$$
$$+ \lambda_{jl}^{BD} + \lambda_{kl}^{CD} + \lambda_{ijk}^{ABC} + \lambda_{ijl}^{ABD} + \lambda_{ikl}^{ACD} + \lambda_{jkl}^{BCD} + \lambda_{ijkl}^{ABCD}.$$

(a) Verify that there is strong evidence of four factor interaction, that is, that H_0: $\lambda^{ABCD} = 0$ is rejected with a very low significance probability. Write out the table of expected frequencies under H_0. The rejection of H_0 can be interpreted to mean that, keeping one of the four variables fixed, whatever is "going on" among the remaining three variables is not the same for each value of the fixed variable.
(b) We can keep any one of the four variables fixed and for each of its values examine the three-way contingency table of the other three variables. Consider separately the sex × rank × field tables only for the two cohorts 1960–69 and 1970–74, that is the ACD tables for $B = 2$ and for $B = 3$. Examine each of these for ACD interaction, and interpret your findings in terms of the data.

APPENDIX **A**

Matrix Algebra

Many of the mathematical expressions that occur in multivariate analysis would be extremely cumbersome if written in terms of ordinary algebra. Matrix algebra provides a means of writing such complex expressions in very simple terms. Therefore, we have used matrix algebra extensively, beginning in Chapter 4. This appendix is provided for the reader who is not familiar at all with matrix notation or who needs to review it in order to facilitate study of the material presented in Chapter 4 and the chapters that follow. Our presentation here is not exhaustive; rather we have tried to make it as concise as possible to facilitate its use as a handy reference. However, it includes all of the concepts and matrix operations used in this book. The reader who wishes to delve further into the subject might wish to consult the books by Horst (1963) or Searle (1966).

A.1 DEFINITIONS

Matrix. A matrix is a rectangular arrangement of numbers or of symbols that represent numbers. Typically, matrices are denoted by capital letters, printed in bold-face type, for example,

$$\mathbf{A}=\begin{bmatrix} 4 & 3 & 1 \\ 7 & 2 & 5 \end{bmatrix}.$$

Order. The order of a matrix is the number of rows and columns it has. A matrix having m rows and n columns is of order m by n (also written $m \times n$). Matrix **A** above is of order 2×3.

Element. Each of the numbers (or other appropriate symbols) that make up a matrix is called an element. The elements of matrix **A** above are 4, 3, 1, 7, 2, and 5. A particular element of a matrix is denoted by a symbol having two subscripts, the first to designate the row and the second the column. Thus, we might denote an element of **A** above by a_{ij}, that is, the element in the ith row and jth column of **A**. Then $a_{11}=4$, $a_{12}=3$, and so on.

Transpose. The transpose of a matrix is a new matrix formed by writing the rows of the original matrix as the columns of the new one. For example, the transpose of **A**, denoted by **A**′, would be

$$\mathbf{A}'=\begin{bmatrix} 4 & 7 \\ 3 & 2 \\ 1 & 5 \end{bmatrix}.$$

Square matrix. A matrix having an equal number of rows and columns is called a square matrix. Such a matrix is said to be of order n, because $n=m$. For example,

$$\mathbf{B}=\begin{bmatrix} 7 & 5 & 1 \\ 4 & 9 & 3 \\ 2 & 6 & 8 \end{bmatrix}$$

is a square matrix of order 3.

Principal diagonal. The elements in a square matrix that have equal row and column subscripts, that is, those for which $i=j$, constitute the principal diagonal of the matrix. For example, the principal diagonal of **B** above consists of the elements 7, 9, and 8, which we might express as b_{11}, b_{22}, and b_{33}, respectively. The sum of the diagonal elements of a square matrix is called the *trace* of the matrix. The trace of **B**, written tr(**B**), is equal to 24.

Diagonal matrix. A diagonal matrix is a square matrix in which all elements are zero except those in the principal diagonal. An example is

$$\mathbf{D}=\begin{bmatrix} 7 & 0 & 0 \\ 0 & 4 & 0 \\ 0 & 0 & 5 \end{bmatrix}.$$

Scalar. A scalar is a matrix of order 1×1, that is, a single number.

Scalar matrix. A scalar matrix is a diagonal matrix in which all of the elements of the principal diagonal are equal. A special case of a scalar matrix is the *identity matrix*, in which all diagonal elements are equal to one. An identity matrix is usually denoted by **I**, as in

$$\mathbf{I}=\begin{bmatrix} 1 & 0 & 0 \\ 0 & 1 & 0 \\ 0 & 0 & 1 \end{bmatrix}.$$

Sometimes an identity matrix is denoted by $\mathbf{I}_n$, with the subscript indicating the order. Thus, the above identity matrix would be denoted by $\mathbf{I}_3$.

Null matrix. A null matrix is a matrix in which all of the elements are zero. It is usually denoted by **0**.

Symmetric matrix. A symmetric matrix is a square matrix that is identical to its transpose, For example, if

$$\mathbf{C}=\begin{bmatrix} 7 & 0 & 4 \\ 0 & 5 & 1 \\ 4 & 1 & 2 \end{bmatrix},$$

then **C** is symmetric, that is, $\mathbf{C}=\mathbf{C}'$. Note that a diagonal matrix is a special case of a symmetric matrix. A square matrix that is symmetric is referred to merely as a symmetric matrix. A square matrix that is not symmetric is referred to as a square nonsymmetric matrix.

Column vector. A column vector is a matrix of order $n\times 1$, that is, a matrix having only one column. It is usually denoted by a small letter in boldface, for example,

$$\mathbf{a} = \begin{bmatrix} 3 \\ 4 \\ 8 \end{bmatrix}.$$

A column vector in which all elements are zero is called a *null vector*.

Row vector. A row vector is a matrix of order $1 \times m$, that is, a matrix having only one row. We shall denote it by the transpose of the corresponding column vector, for example,

$$\mathbf{a}' = [3 \quad 4 \quad 8].$$

Partitioned Matrix. A partitioned matrix is a matrix that has been divided (or partitioned) into smaller parts, each of which may be a matrix, a vector, or a scalar. For example, if a matrix, $\mathbf{R}$, of order 4×4, were defined as

$$\mathbf{R} = \left[\begin{array}{cc:cc} 1.0 & .2 & .3 & .5 \\ .2 & 1.0 & .4 & .1 \\ \hdashline .3 & .4 & 1.0 & .6 \\ .5 & .1 & .6 & 1.0 \end{array}\right],$$

it might be partitioned as indicated by the dashed lines to form four matrices, say, $\mathbf{R}_{11}$, $\mathbf{R}_{12}$, $\mathbf{R}_{21}$, and $\mathbf{R}_{22}$, each of order 2×2. Then the partitioned matrix may be written

$$\mathbf{R} = \left[\begin{array}{c:c} \mathbf{R}_{11} & \mathbf{R}_{12} \\ \hdashline \mathbf{R}_{21} & \mathbf{R}_{22} \end{array}\right].$$

Determinant of a matrix. If a matrix is square, such as $\mathbf{B}$ above, it has a determinant, which is denoted by

$$|\mathbf{B}| = \begin{vmatrix} 7 & 5 & 1 \\ 4 & 9 & 3 \\ 2 & 6 & 8 \end{vmatrix}.$$

A determinant is defined as a sum of terms; hence, unlike a matrix, it has a numerical value. In general, the number of terms in the sum is equal to the factorial of the order, n, of the determinant. Each term is the product of n elements, with only one element from each row and only one from each column. Thus, if we write $\mathbf{B}$ as

$$\mathbf{B} = \begin{bmatrix} b_{11} & b_{12} & b_{13} \\ b_{21} & b_{22} & b_{23} \\ b_{31} & b_{32} & b_{33} \end{bmatrix},$$

then $|\mathbf{B}|$ is the sum of $3! = 6$ terms, namely,

Number	*Term*
1	$b_{11}b_{22}b_{33}$
2	$b_{11}b_{23}b_{32}$
3	$b_{12}b_{21}b_{33}$
4	$b_{12}b_{23}b_{31}$
5	$b_{13}b_{21}b_{32}$
6	$b_{13}b_{22}b_{31}$

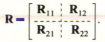

Note that the three row subscripts in each term are written in their natural order, that is, 1, 2, 3. However, in only one term, the first, are the column subscripts in natural order. The sign of each term can be determined as follows:

(1) Write down the column subscripts in the order in which they appear in the term as written, for example, for term 2, 1 3 2.

(2) Interchange pairs of column subscripts to achieve natural order, that is, 1 2 3. For term 2, natural order is achieved by interchanging the 2 and 3.

(3) If the number of interchanges performed in Step 2 is odd, the sign of the term is negative, if even, positive. For term 2, only one interchange is required, so the sign is negative.

Following these steps, the signs of all six terms are:

	Step 1	*Step* 2	*Step* 3	
Term	Original Order of Column Subscripts	Interchanges	Number of Interchanges Required	Sign
1	123	None	0	+
2	132	132---→123	1	−
3	213	213---→123	1	−
4	231	231--→132--→123	2	+
5	312	312--→132--→123	2	+
6	321	321---→123	1	−

Thus, the terms of $|\mathbf{B}|$ with their correct signs are:

Number	*Term*
1	$+b_{11}b_{22}b_{33}$
2	$-b_{11}b_{23}b_{32}$
3	$-b_{12}b_{21}b_{33}$
4	$+b_{12}b_{23}b_{31}$
5	$+b_{13}b_{21}b_{32}$
6	$-b_{13}b_{22}b_{31}$

Minors. If the ith row and the jth column of a determinant are eliminated, the remaining determinant is called the *minor* of the ijth element of the original determinant. For example, if the determinant of the matrix **B** is

$$|\mathbf{B}| = \begin{vmatrix} 7 & 5 & 1 \\ 4 & 9 & 3 \\ 2 & 6 & 8 \end{vmatrix},$$

the minor of the element $b_{12}=5$ is

$$m_{12} = \begin{vmatrix} 4 & 3 \\ 2 & 8 \end{vmatrix},$$

in which m_{12} denotes the minor of the element in the first row and second column of $|\mathbf{B}|$. If *any* number of *rows* of a determinant are eliminated, and if the *corresponding* columns (those for which $i=j$) are eliminated, the resulting determinant is called a *principal minor* of the original matrix. For the determinant $|\mathbf{B}|$ above, the principal minors and their numerical values (see Section A.2) are:

$$7, 9, 8, \begin{vmatrix} 9 & 3 \\ 6 & 8 \end{vmatrix} = 54, \begin{vmatrix} 7 & 1 \\ 2 & 8 \end{vmatrix} = 54, \begin{vmatrix} 7 & 5 \\ 4 & 9 \end{vmatrix} = 43,$$

and $|\mathbf{B}|=254$. Note that $|\mathbf{B}|$ is also a principal minor of the matrix $\mathbf{B}$. One of the properties of a Gramian matrix (defined below) is that its principal minors are all greater than or equal to zero.

Cofactors. The cofactor, e_{ij}, of element b_{ij} is defined as

$$e_{ij}=(-1)^{i+j}m_{ij}.$$

Thus, the cofactor of b_{ij} is the minor, e_{ij}, with a positive sign if $i+j$ is an even number and a negative sign if $i+j$ is an odd number. For example, the cofactor of element b_{33} in

$$|\mathbf{B}| = \begin{vmatrix} 7 & 5 & 1 \\ 4 & 9 & 3 \\ 2 & 6 & 8 \end{vmatrix}$$

is

$$e_{33} = \begin{vmatrix} 7 & 5 \\ 4 & 9 \end{vmatrix},$$

whereas the cofactor of element b_{21} is

$$e_{21} = -\begin{vmatrix} 5 & 1 \\ 6 & 8 \end{vmatrix}.$$

A.2 EVALUATING DETERMINANTS

A determinant may be evaluated, that is, its numerical value may be obtained, by finding the sum of its terms. For example, the value of

$$|\mathbf{B}| = \begin{vmatrix} 7 & 5 & 1 \\ 4 & 9 & 3 \\ 2 & 6 & 8 \end{vmatrix}$$

is $|\mathbf{B}| = (7)(9)(8) - (7)(3)(6) - (5)(4)(8) + (5)(3)(2) + (1)(4)(6) - (1)(9)(2) = 254$. Note that the terms of a determinant of order $n=2$, for example,

$$|\mathbf{A}| = \begin{vmatrix} a_{11} & a_{12} \\ a_{21} & a_{22} \end{vmatrix},$$

are $a_{11}a_{22}$ and $-a_{12}a_{21}$. Thus, $|\mathbf{A}| = a_{11}a_{22} - a_{12}a_{21}$, that is, the product of the elements of the principal diagonal minus the product of the off-diagonal elements.

A determinant may also be evaluated by the use of cofactors. The numerical value of $|\mathbf{B}|$ may be computed from

$$|\mathbf{B}| = \sum_{i=1}^{n} b_{ij}e_{ij} = \sum_{j=1}^{n} b_{ij}e_{ij}.$$

In other words, the numerical value of $|\mathbf{B}|$ is the weighted sum of the elements in row i (or column j), each element in row i (or column j) being weighted by its cofactor. Thus, using column 1 elements, we have

$$\begin{aligned}|\mathbf{B}| &= 7\begin{vmatrix} 9 & 3 \\ 6 & 8 \end{vmatrix} - 4\begin{vmatrix} 5 & 1 \\ 6 & 8 \end{vmatrix} + 2\begin{vmatrix} 5 & 1 \\ 9 & 3 \end{vmatrix} \\ &= (7)(9)(8)-(7)(3)(6)-(4)(5)(8)+(4)(1)(6)+(2)(5)(3)-(2)(1)(9) \\ &= 254.\end{aligned}$$

When a determinant is of order 3, a special procedure may be used for its evaluation. If we write

$$|\mathbf{B}| = \begin{vmatrix} b_{11} & b_{12} & b_{13} \\ b_{21} & b_{22} & b_{23} \\ b_{31} & b_{32} & b_{33} \end{vmatrix},$$

and repeat the first two columns following the third column, we obtain

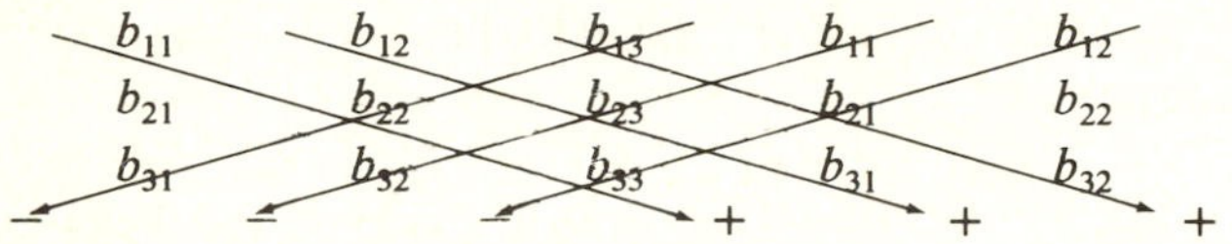

The diagonal lines indicate the multiplications to be performed, with the signs indicated below. Thus, $|\mathbf{B}| = b_{11}b_{22}b_{33} + b_{12}b_{23}b_{31} + b_{13}b_{21}b_{32} - b_{13}b_{22}b_{31} - b_{11}b_{23}b_{32} - b_{12}b_{21}b_{33}$. For the matrix

$$\mathbf{B} = \begin{bmatrix} 7 & 5 & 1 \\ 4 & 9 & 3 \\ 2 & 6 & 8 \end{bmatrix},$$

$$\begin{aligned}|\mathbf{B}| &= (7)(9)(8)+(5)(3)(2)+(1)(4)(6)-(1)(9)(2)-(7)(3)(6)-(5)(4)(8) \\ &= 254,\end{aligned}$$

which is the same result obtained earlier. We emphasize that this procedure may be applied only to determinants of order $n = 3$.

Singular matrix. A matrix whose determinant is equal to zero is said to be a singular matrix. A nonsingular matrix is one that has a nonzero determinant.

A.3 ALGEBRAIC OPERATIONS WITH MATRICES

Addition and Subtraction. Matrices *of the same order* may be added by summing corresponding elements to produce a new matrix. For example, suppose we have $\mathbf{A} = \begin{bmatrix} 2 & 3 & 4 \\ 1 & 6 & 8 \end{bmatrix}$ and $\mathbf{B} = \begin{bmatrix} 1 & 5 & 2 \\ 3 & 7 & 8 \end{bmatrix}$. The sum is $\mathbf{A}+\mathbf{B} = \begin{bmatrix} 2+1 & 3+5 & 4+2 \\ 1+3 & 6+7 & 8+8 \end{bmatrix} = \begin{bmatrix} 3 & 8 & 6 \\ 4 & 13 & 16 \end{bmatrix}$. The differences $\mathbf{A}-\mathbf{B}$ and

B − **A** may also be obtained. Thus,

$$\mathbf{A}-\mathbf{B}=\begin{bmatrix} 2-1 & 3-5 & 4-2 \\ 1-3 & 6-7 & 8-8 \end{bmatrix}=\begin{bmatrix} 1 & -2 & 2 \\ -2 & -1 & 0 \end{bmatrix}$$

and

$$\mathbf{B}-\mathbf{A}=\begin{bmatrix} -1 & 2 & -2 \\ 2 & 1 & 0 \end{bmatrix}.$$

Neither the sum $\mathbf{A}'+\mathbf{B}$ nor the differences $\mathbf{A}'-\mathbf{B}$ and $\mathbf{B}-\mathbf{A}'$, exist because $\mathbf{A}'$ and **B** are not of the same order. The determinant of the sum or difference of two matrices is not necessarily equal to the sum or difference of their determinants. For example, $\begin{vmatrix} 2 & 1 \\ 3 & 4 \end{vmatrix}=8-3=5$, and $\begin{vmatrix} 2 & 4 \\ 1 & 3 \end{vmatrix}=6-4=2$, so that the sum is $5+2=7$. However, the sum of the matrices is $\begin{bmatrix} 2 & 1 \\ 3 & 4 \end{bmatrix}+\begin{bmatrix} 2 & 4 \\ 1 & 3 \end{bmatrix}=\begin{bmatrix} 4 & 5 \\ 4 & 7 \end{bmatrix}$, and the determinant of the sum is $\begin{vmatrix} 4 & 5 \\ 4 & 7 \end{vmatrix}=28-20=8$, which is different from the sum of the determinants.

Multiplication. The product, say **AB**, of two matrices may be obtained only if the number of columns in **A** is equal to the number of rows in **B**. Thus, if **A** is of order $m \times n$ and **B** is of order $p \times q$, the product **AB** may be obtained only if $n=p$. If $n \neq p$, we say that the product **AB** does not exist. Similarly, the product **BA** exists only if $m=q$. Even if both **AB** and **BA** exist, however, **AB** is not necessarily equal to **BA**. In other words matrix multiplication is not *commutative*. For this reason, we must indicate the order of multiplication of **A** and **B**. For example, **AB** may be expressed either as **A** *postmultiplied* by **B** or as **B** *premultiplied* by **A**.

As an illustration, consider two matrices, **A**, of order 2×3, and **B**, of order 3×2. The product **AB** exists because **A** has 3 columns and **B** has 3 rows. The *ij*th element of **AB** is the sum of the products of corresponding elements in the *i*th row of **A** and the *j*th column of **B**. Thus, we have

$$\mathbf{AB}=\begin{bmatrix} a_{11} & a_{12} & a_{13} \\ a_{21} & a_{22} & a_{23} \end{bmatrix}\cdot\begin{bmatrix} b_{11} & b_{12} \\ b_{21} & b_{22} \\ b_{31} & b_{32} \end{bmatrix}$$

$$=\begin{bmatrix} a_{11}b_{11}+a_{12}b_{21}+a_{13}b_{31} & a_{11}b_{12}+a_{12}b_{22}+a_{13}b_{32} \\ a_{21}b_{11}+a_{22}b_{21}+a_{23}b_{31} & a_{21}b_{12}+a_{22}b_{22}+a_{23}b_{32} \end{bmatrix}.$$

Note that the order of the product is 2×2, that is the number of rows of **A** by the number of columns of **B**. In general, if $n=p$, the product $\mathbf{A}^{(m \times n)}\mathbf{B}^{(p \times q)}$, will be of order $m \times q$. Thus, the product $\mathbf{B}^{(3 \times 2)}\mathbf{A}^{(2 \times 3)}$, exists because **B** has 2 columns and **A** has 2 rows, and the product **BA** is of order 3×3. Note that **AB** and **BA** are obviously different, since **AB** is of order 2×2 and **BA** is of order 3×3. When both **AB** and **BA** exist, as in this case, the product having the lesser order (here **AB**) is sometimes called the *minor product*. The other product is called the *major product*.

Some other examples concerning the order of matrix products are given in the following table:

Order of **A**	*Order of* **B**	*Order of* **AB**	*Order of* **BA**
2×4	4×3	2×3	does not exist
5×5	5×5	5×5	5×5
3×1	1×3	3×3	1×1 (a scalar)
3×1	2×4	does not exist	does not exist
3×1	2×3	does not exist	2×1

As a numerical example, suppose we have $\mathbf{A}=\begin{bmatrix}2&0&3\\4&6&5\end{bmatrix}$ and $\mathbf{B}=\begin{bmatrix}1&2\\3&0\\5&4\end{bmatrix}$. Then

$$\mathbf{AB}=\begin{bmatrix}(2)(1)+(0)(3)+(3)(5) & (2)(2)+(0)(0)+(3)(4)\\(4)(1)+(6)(3)+(5)(5) & (4)(2)+(6)(0)+(5)(4)\end{bmatrix}$$

$$=\begin{bmatrix}17&16\\47&28\end{bmatrix}, \text{ and}$$

$$\mathbf{BA}=\begin{bmatrix}(1)(2)+(2)(4) & (1)(0)+(2)(6) & (1)(3)+(2)(5)\\(3)(2)+(0)(4) & (3)(0)+(0)(6) & (3)(3)+(0)(5)\\(5)(2)+(4)(4) & (5)(0)+(4)(6) & (5)(3)+(4)(5)\end{bmatrix}$$

$$=\begin{bmatrix}10&12&13\\6&0&9\\26&24&35\end{bmatrix}.$$

When **A** and **B** are both vectors, as in the third example in the above table, special terms are used to denote their products. Suppose we have $\mathbf{a}=\begin{bmatrix}4\\3\\6\end{bmatrix}$ and $\mathbf{b}=\begin{bmatrix}1\\3\\2\end{bmatrix}$. Then $\mathbf{a}'\mathbf{b}=(4)(1)+(3)(3)+(6)(2)=25$. This is called the *scalar product* of **a** and **b**. Note that $\mathbf{b}'\mathbf{a}$ also equals 25, that is, $\mathbf{a}'\mathbf{b}=\mathbf{b}'\mathbf{a}$. However, the product $\mathbf{ab}'=\begin{bmatrix}4&12&8\\3&9&6\\6&18&12\end{bmatrix}$, a matrix of order 3×3. This product is called the *matrix product* of **a** and **b**. Note that $\mathbf{ab}'=\mathbf{ba}'$.

When a product involves more than two matrices, for example, **ABC**, the multiplication is *associative*, that is, $\mathbf{A(BC)}=\mathbf{(AB)C}$. We assume, of course, that the indicated product exists. As an example, the product $\mathbf{A}^{(2\times3)}\mathbf{B}^{(3\times4)}\mathbf{C}^{(4\times1)}$ exists because **AB** exists and **BC** exists. However, **CAB** does not exist because **CA** does not exist. The product of three or more matrices is usually obtained by first finding the product of a pair (or pairs). Thus, to obtain **ABC**, we might first find **AB** and postmultiply by **C**. Alternatively, we might first find **BC** and premultiply by **A**.

Matrix multiplication is also *distributive*. Thus, assuming that the indicated products exist, $\mathbf{A(B+C)}=\mathbf{AB}+\mathbf{AC}$ and $\mathbf{(B+C)A}=\mathbf{BA}+\mathbf{CA}$.

We should mention a few special cases involving matrix multiplication:

1. When all of the elements of a matrix are multiplied by the same constant, say k, the process is sometimes called *scalar multiplication*. As an example, if $\mathbf{A}=\begin{bmatrix} 2 & 0 & 3 \\ 4 & 6 & 5 \end{bmatrix}$ and $k=2$, then

$$k\mathbf{A}=\begin{bmatrix} 4 & 0 & 6 \\ 8 & 12 & 10 \end{bmatrix}.$$

2. The product of a matrix and its transpose always exists and is always a symmetric matrix (which may be a scalar). Thus, $\mathbf{A}^{(m\times n)}\mathbf{A}'^{(n\times m)}$ is of order m and $\mathbf{A}'^{(n\times m)}\mathbf{A}^{(m\times n)}$ is of order n. If $m=n$, then $\mathbf{A}'\mathbf{A}=\mathbf{A}\mathbf{A}'$; otherwise, $\mathbf{A}'\mathbf{A}\neq\mathbf{A}\mathbf{A}'$.

3. The transpose of the product (if it exists) of two matrices is equal to the product of their tranposes in reverse order. Thus, $(\mathbf{AB})'=\mathbf{B}'\mathbf{A}'$. This can be generalized to several matrices, for example, $(\mathbf{ABC})'=\mathbf{C}'\mathbf{B}'\mathbf{A}'$.

4. A matrix product of the form $\mathbf{a}'\mathbf{B}\mathbf{a}$, in which $\mathbf{a}$ is an $n\times 1$ column vector and $\mathbf{B}$ is a square matrix of order n, is called a *quadratic form* of $\mathbf{a}$. Note that this product is a scalar because $\mathbf{a}'^{(1\times n)}\mathbf{B}^{(n\times n)}$ has order $1\times n$ and when postmultiplied by $\mathbf{a}^{(n\times 1)}$, the result is of order 1×1. This product is so named because each term of the sum contains either the square of an element of $\mathbf{a}$ or the product of two different elements of $\mathbf{a}$, each multiplied by a factor consisting of an element or elements of $\mathbf{B}$.

Division. In ordinary algebra an equation such as $ab=c$ $(a\neq 0)$ may be solved for b by dividing both sides of the equation by a to obtain $b=c/a$. Division by a is equivalent to multiplication by $a^{-1}=1/a$, the *reciprocal* of a, so that $(1/a)ab=(1/a)c$. Since $(1/a)a$ is equal to one, we have the desired result. In matrix algebra, the equation $\mathbf{AB}=\mathbf{C}$, in which $\mathbf{A}$ is a square nonsingular matrix, may be solved for $\mathbf{B}$ by computing the *inverse* of $\mathbf{A}$, denoted by $\mathbf{A}^{-1}$. The product $\mathbf{A}^{-1}\mathbf{A}=\mathbf{A}\mathbf{A}^{-1}$ is equal to the identity matrix, $\mathbf{I}$, defined earlier. $\mathbf{I}$ is equivalent to a 1 in ordinary algebra, so that if the product $\mathbf{IB}$ exists, then $\mathbf{IB}=\mathbf{B}$. The solution then is

$$\begin{aligned} \mathbf{AB}&=\mathbf{C} \\ \mathbf{A}^{-1}\mathbf{AB}&=\mathbf{A}^{-1}\mathbf{C} \\ \mathbf{IB}&=\mathbf{A}^{-1}\mathbf{C} \\ \mathbf{B}&=\mathbf{A}^{-1}\mathbf{C}. \end{aligned}$$

The product $\mathbf{A}^{-1}\mathbf{C}$ is thus analogous to $(1/a)c=c/a$, or, to "dividing" $\mathbf{C}$ by $\mathbf{A}$.

The inverse $\mathbf{A}^{-1}$ of a matrix $\mathbf{A}$ exists only if $\mathbf{A}$ is square and nonsingular, that is, if $|\mathbf{A}|\neq 0$. Under these assumptions, the inverse of $\mathbf{A}$ may be computed as follows:

1. Construct a new matrix, $\mathbf{E}$, with elements e_{ij}, each of which is the cofactor of a_{ij}.

2. Obtain the tranpose of $\mathbf{E}$, $\mathbf{E}'$. This matrix is called the *adjoint* of $\mathbf{A}$, denoted by adj($\mathbf{A}$).

3. Divide each element of adj($\mathbf{A}$) by $|\mathbf{A}|$. The result is $\mathbf{A}^{-1}$.

4. Verify the computations by the multiplication $\mathbf{A}^{-1}\mathbf{A}=\mathbf{A}\mathbf{A}^{-1}=\mathbf{I}$. If the computations of $\mathbf{A}^{-1}$ have been carried out to a sufficient number of significant digits (we suggest at least 8), then **I** should have elements that are accurate to within 5 places (or more) of 0 or 1. That is, zero elements should be $\pm.00000$ and elements of value one should be either 1.00001 or .99999, when rounded off to five places. If this degree of accuracy is not obtained, the computations are not correct.

As an example, suppose $\mathbf{A}=\begin{bmatrix} 1 & 2 & 0 \\ 0 & 3 & 1 \\ 2 & 2 & 4 \end{bmatrix}$. Then

1. $\mathbf{E}=\begin{bmatrix} 10 & 2 & -6 \\ -8 & 4 & 2 \\ 2 & -1 & 3 \end{bmatrix}$.
2. $\mathbf{E}'=\begin{bmatrix} 10 & -8 & 2 \\ 2 & 4 & -1 \\ -6 & 2 & 3 \end{bmatrix}$.
3. $|\mathbf{A}|=14$.

Then

$$\mathbf{A}^{-1}=\begin{bmatrix} 10/14 & -8/14 & 2/14 \\ 2/14 & 4/14 & -1/14 \\ -6/14 & 2/14 & 3/14 \end{bmatrix},$$

which in decimal form is

$$\mathbf{A}^{-1}=\begin{bmatrix} .71428571 & -.57142857 & .14285714 \\ .14285714 & .28571429 & -.07142857 \\ -.42857143 & .14285714 & .21428571 \end{bmatrix}.$$

4. $\mathbf{A}^{-1}\mathbf{A}=\begin{bmatrix} 1.00000 & .00000 & .00000 \\ .00000 & 1.00000 & .00000 \\ .00000 & .00000 & 1.00000 \end{bmatrix}$.

Again there are a few special cases to be mentioned concerning operations with the inverses of matrices. These are:

1. The inverse of a diagonal matrix, **D**, is a diagonal matrix, $\mathbf{D}^{-1}$, the elements of which are the reciprocals of the diagonal elements of **D**. For example, if $\mathbf{D}=\begin{bmatrix} 1 & 0 & 0 \\ 0 & 2 & 0 \\ 0 & 0 & 3 \end{bmatrix}$, then $\mathbf{D}^{-1}=\begin{bmatrix} 1 & 0 & 0 \\ 0 & 1/2 & 0 \\ 0 & 0 & 1/3 \end{bmatrix}$

2. The inverse of the product of a matrix by a scalar is equal to the reciprocal of the scalar multiplied by the inverse of the matrix. Thus, $(k\mathbf{A})^{-1}=(1/k)(\mathbf{A})^{-1}$.

3. If **A** and **B** are both nonsingular square matrices of order n, then $(\mathbf{AB})^{-1}=\mathbf{B}^{-1}\mathbf{A}^{-1}$.

A.4 SOLUTION OF A MATRIX EQUATION OF THE FORM $(A-\lambda_i I)v_i=0$

Many of the problems involving matrices with which we deal in this book involve the solutions for λ_i and $\mathbf{v}_i$ of a matrix equation of the form

$(\mathbf{A}-\lambda_i\mathbf{I})\mathbf{v}_i=\mathbf{0}$, in which $\mathbf{A}$ is a square symmetric matrix with $|\mathbf{A}|>0$, λ_i is the ith *characteristic root* or *eigenvalue* of $\mathbf{A}$, and $\mathbf{v}_i$ is the ith *characteristic vector* or *eigenvector*, corresponding to λ_i. The solution of such an equation for λ_i and $\mathbf{v}_i$ may be obtained through the following steps:

1. Write the determinantal equation, $|\mathbf{A}-\lambda\mathbf{I}|=0$. This is called the *characteristic equation* of $\mathbf{A}$. Solve this equation for the λ_is.

2. For each eigenvalue, λ_i, obtained in Step 1, write out the matrix $\mathbf{A}-\lambda_i\mathbf{I}$.

3. Compute adj$(\mathbf{A}-\lambda_i\mathbf{I})$ by the method described earlier.

4. Any column of adj$(\mathbf{A}-\lambda_i\mathbf{I})$ is an eigenvector associated with λ_i. The usual practice, however, is to "normalize" the vector by dividing its elements by the square root of their sum of squares. The resulting values are the normalized elements of $\mathbf{v}_i$. If desired to facilitate interpretation, the signs of all elements of $\mathbf{v}_i$ may be reversed by multiplying $\mathbf{v}_i$ by -1.

We illustrate the procedure with $\mathbf{A}=\begin{bmatrix} 6 & 2 \\ 2 & 3 \end{bmatrix}$:

1. $$|\mathbf{A}-\lambda\mathbf{I}|=\begin{vmatrix} 6-\lambda & 2 \\ 2 & 3-\lambda \end{vmatrix}=(6-\lambda)(3-\lambda)-4$$
$$=18-9\lambda+\lambda^2-4=\lambda^2-9\lambda+14=0.$$

Because $\lambda^2-9\lambda+14=(\lambda-7)(\lambda-2)$, the two characteristic roots of $\mathbf{A}$ are obtained by setting each of these factors equal to zero and solving for λ. Thus, $\lambda_1-7=0$ and $\lambda_2-2=0$, so that $\lambda_1=7$ and $\lambda_2=2$.

2. Using λ_1 we have $\mathbf{A}-\lambda_1\mathbf{I}=\begin{bmatrix} -1 & 2 \\ 2 & -4 \end{bmatrix}$.

3. The matrix of cofactors is then $\mathbf{E}=\begin{bmatrix} -4 & -2 \\ -2 & -1 \end{bmatrix}$, which is also the adjoint of $\mathbf{A}-\lambda_i\mathbf{I}$ because $\mathbf{E}$ is symmetric. Note that the columns of the adjoint are proportional to one another, that is, $-4/-2=-2/-1$.

4. Then $$\mathbf{v}_1=\begin{bmatrix} -4 \\ -2 \end{bmatrix}\frac{1}{\sqrt{(-4)^2+(-2)^2}}=\begin{bmatrix} -.89443 \\ -.44721 \end{bmatrix}.$$

Repeating Steps 2, 3, and 4, using λ_2, we have

$$\mathbf{v}_2=\begin{bmatrix} .22361 \\ -.44721 \end{bmatrix}.$$

It is useful to note that the products $\mathbf{v}_1'\mathbf{v}_2$ and $\mathbf{v}_2'\mathbf{v}_1$ are both equal to zero. In geometric terms, this means that the two eigenvectors are orthogonal, that is, uncorrelated.

The eigenvalues and eigenvectors of matrices have several important properties, which we shall state without proof:

1. If $\mathbf{v}_j$ and $\mathbf{v}_k$ are two eigenvectors associated with the eigenvalues λ_j and λ_k of a symmetric matrix, then $\mathbf{v}_j$ and $\mathbf{v}_k$ are orthogonal, that is, $\mathbf{v}_j'\mathbf{v}_k=0$.

2. The sum of the eigenvalues of a square matrix $\mathbf{A}$, that is, $\Sigma_{j=1}^{p}\lambda_j$, is equal to the trace of $\mathbf{A}$. As defined above, the trace of $\mathbf{A}$ is the sum of its diagonal elements.

3. The product of the eigenvalues of a square matrix **A** is equal to $|\mathbf{A}|$. Thus, if any of the eigenvalues of **A** are zero, then **A** is singular.

4. The *rank* of a square matrix **A** is equal to the number of its nonzero eigenvalues. The concept of rank has particular importance in factor analysis (see Chapter 8). In mathematical terms, the number of common factors is equal to the rank of the matrix of correlations among the variables. The problem of determining the number of common factors reduces, therefore, to that of deciding how many "nonzero" eigenvalues there are.

5. The rank of $\mathbf{C}=\mathbf{AB}$ is less than or equal to the smaller of the ranks of **A** or **B**.

6. If, for all vectors **a**, the quadratic form, $\mathbf{a}'\mathbf{Ba}$, is greater than or equal to zero, it is said to be *positive semidefinite*. If $\mathbf{a}'\mathbf{Ba}>0$ for all $\mathbf{a}\neq\mathbf{0}$, then $\mathbf{a}'\mathbf{Ba}$ is said to be *positive definite*. Although these terms properly apply to the quadratic form, they are also frequently used to describe **B** itself. In particular, if **B** is *symmetric* and positive semidefinite (or positive definite), it is called a *Gramian* matrix. As we stated earlier, all of the principal minor determinants of a Gramian matrix are greater than or equal to zero. In fact, any *symmetric* matrix is Gramian if all of its principal minor determinants are greater than or equal to zero.

It can be shown that if **B** is Gramian, then there exists a matrix **A**, such that $\mathbf{B}=\mathbf{AA}'$. In particular, the matrix of correlations,

$$\mathbf{R}=\begin{bmatrix} 1 & r_{12} & \cdots & r_{1p} \\ r_{21} & 1 & \cdots & r_{2p} \\ \vdots & \vdots & & \vdots \\ r_{p1} & r_{p2} & \cdots & 1 \end{bmatrix},$$

between p variables, is Gramian. This property of **R** is important in factor analysis because it enables us to write $\mathbf{R}=\mathbf{AA}'$, which Thurstone (1935, p. 70) called "the fundamental factor theorem." For a definition of **A** in this equation, see Chapter 8.

7. Given that λ and **v** are, respectively, an eigenvalue and associated eigenvector of matrix **A**,

a. $k\mathbf{A}$ (in which k is an arbitrary constant) has $k\lambda$ as an eigenvalue and **v** as the associated eigenvector.

b. $\mathbf{A}^x$ (in which x is any positive integer) has λ^x as an eigenvalue and **v** as the associated eigenvector. (Note: Because **A** is, by definition, a square matrix, it may be raised to any integral power; thus, $\mathbf{A}^x$ means "**A** multiplied by itself x times." For example, $\mathbf{A}^2=\mathbf{AA}$, $\mathbf{A}^3=\mathbf{AAA}$, and so on.)

c. $\mathbf{A}^{-1}$ (if it exists) has $1/\lambda$ as an eigenvalue and **v** as the associated eigenvector.

d. The matrix $\mathbf{A}+k\mathbf{I}$ (in which k is an arbitrary constant) has $(\lambda+k)$ as an eigenvalue and **v** as its associated eigenvector.

APPENDIX B
Some Elementary Differential Calculus

Our use of calculus in this book is limited to finding maxima and minima of ordinary algebraic expressions, of logarithmic functions, and of functions involving matrices. Although we could simply state here the formulas and rules for finding derivatives of such expressions, it seems more satisfying to try to provide some minimal insight into why these formulas and rules give us what we want. Because the *basic* idea of differentiation is quite simple, we have tried to provide a very elementary discussion of it. The formulas and rules presented later can then be used with some understanding of why they work.

B.1 THE LIMIT OF A FUNCTION OF *X*

As we shall see, the ***derivative*** of a function, $f(x)$, is really a particular type of limit, so we first have to discuss briefly the idea of a limit. Actually, the concept of a limit should be familiar to students who have considered, for example, the univariate normal distribution as the limiting form of the t distribution, as the sample size approaches infinity. This can be seen by referring to the table of the t distribution in Appendix C. The t distribution is never exactly equivalent to the normal, because the sample size cannot actually be infinite. However, we can still believe that the t distribution gets closer and closer to the normal form as the sample size becomes larger and larger. In fact, in considering the concept of a limit, we are interested in what happens as a variable *approaches* a particular value rather than in the situation in which the variable actually attains the specified value.

The limit, L, of a function, $f(x)$, as x approaches a constant, a, may be expressed as

$$\lim_{x\to a} f(x) = L.$$

Roughly speaking, this means that we can make $f(x)$ come as close to L as we want by making x come sufficiently close to, but not equal to, a. As we indicated, when we consider the limit of a function, we are not concerned with its value when $x = a$, but only with the limiting value, L, as x approaches a. As an example, we may write

$$\lim_{x\to 0} (x^2 + 1) = 1,$$

which means that (x^2+1) is arbitrarily near 1 if x is sufficiently close to 0, but not equal to 0. Regardless of how small, in absolute value, the difference, say ε, between (x^2+1) and 1 becomes, as x approaches 0, there is always a difference, δ, which in absolute value is less than ε. Thus, we are interested only in the value of the function when x is in the neighborhood of 0, not when $x=0$.

B.2 THE DERIVATIVE OF A FUNCTION OF *X*

Suppose we graph the relation $y=f(x)$, and obtain the curve given as Figure B.1. We now consider the point $P(x,y)$ and wish to determine the rate of increase of y with respect to x *at this point*. In other words, we wish to know the *slope* of a tangent to the curve at the point P. To obtain this value, we locate a point P', and denote the change in x between P and P' by Δx and the change in y between P and P' by Δy. The quotient $\Delta y/\Delta x$ is then the *average* rate of increase of y with respect to x between P and P'. We note that $\Delta y/\Delta x$ is also the slope of a line drawn to connect P and P'. It is not, however, the slope of the function *at P itself*.

FIGURE B.1 Graph of a Function, $y = f(x)$, Showing Average Rate of Increase of y With Respect to x Between Two Points, P and P'

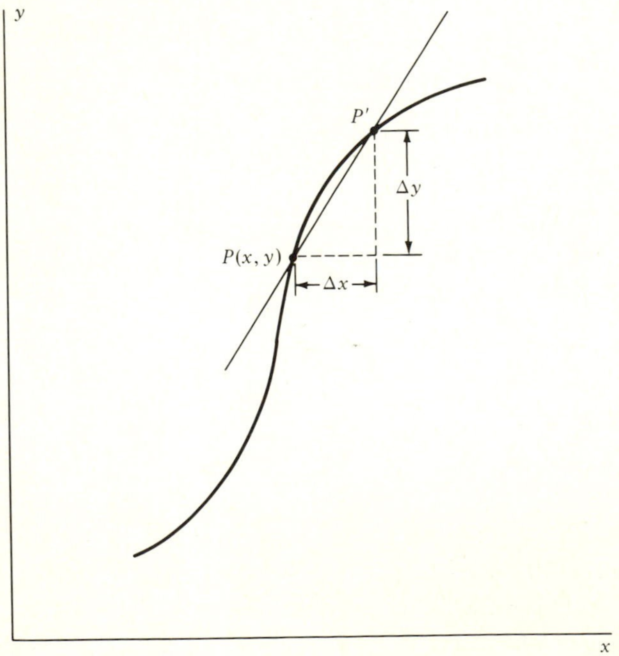

Now if we let Δx become smaller and smaller, we find that $\Delta y/\Delta x$ becomes closer and closer to representing the slope at P, that is, to being the slope of a tangent to the curve at P. Thus, we conclude that the slope at P is equal to the limit of $\Delta y/\Delta x$ as Δx approaches 0, that is,

$$\text{Slope at } P = \lim_{\Delta x \to 0} \Delta y/\Delta x\,.$$

This limit is called the *derivative* of $y=f(x)$ with respect to x*. It is commonly denoted by dy/dx, although other symbols, such as $f'(x)$ and y', are sometimes used.

An inspection of Figure B.1 shows that Δy is equal to the difference between the value of y at the point P' and the value of y at the point P. At P', $y=f(x+\Delta x)$ and at P, $y=f(x)$. Therefore, $\Delta y=f(x+\Delta x)-f(x)$ and we can express the derivative of y with respect to x as

$$\frac{dy}{dx} = \lim_{\Delta x \to 0} \frac{f(x+\Delta x)-f(x)}{\Delta x}\,.$$

The procedure for finding the derivative dy/dx, a process called *differentiation*, may be summarized as follows:

1. Obtain $\Delta y=f(x+\Delta x)-f(x)$.
2. Divide Δy by Δx to obtain $\Delta y/\Delta x=(f(x+\Delta x)-f(x))/\Delta x$.
3. Determine the limit of this quotient as $\Delta x \to 0$. This limit is the derivative, dy/dx.

As an example, we shall differentiate the function $y=x^2$ with respect to x. The graph of this function is shown in Figure B.2. The steps are:

1. $\Delta y=(x+\Delta x)^2-x^2=2x\Delta x+(\Delta x)^2$.
2. $\Delta y/\Delta x=2x+\Delta x$.
3. $dy/dx=\lim_{\Delta x \to 0}(2x+\Delta x)=2x$.

This means that we can find the slope of a tangent to the curve in Figure B.2 at *any* point $P(x,y)$ by substitution in $dy/dx=2x$. For example, if we wished to find the slope at the point ($x=1.5$, $y=(1.5)^2=2.25$), the result would be $dy/dx=2(1.5)=3$.

The procedure we have described may be applied to find the derivative of any function, if it exists. However, repeated applications of the procedure to functions of a particular form, for example, $y=x^n$, would soon reveal that a rule for differentiating such functions can be stated, so that carrying out the three steps in each case is not necessary. The rule is that the derivative of $y=x^n$ is

$$\frac{dy}{dx} = \frac{d}{dx}x^n = nx^{n-1}.$$

Thus, if $y=x^3$, $dy/dx=3x^2$; if $y=x^4$, $dy/dx=4x^3$; and so on.

Useful rules for other functions are:

1. If $y=kx^n$, $dy/dx=k(d/dx)x^n$. Thus, $(d/dx)2x^3=2(3x^2)=6x^2$.

*There are functions for which a limit, and hence a derivative, as defined here, does not exist. However, the definition of a derivative given is appropriate for all applications in this book.

FIGURE B.2 Graph of the Function $y = x^2$

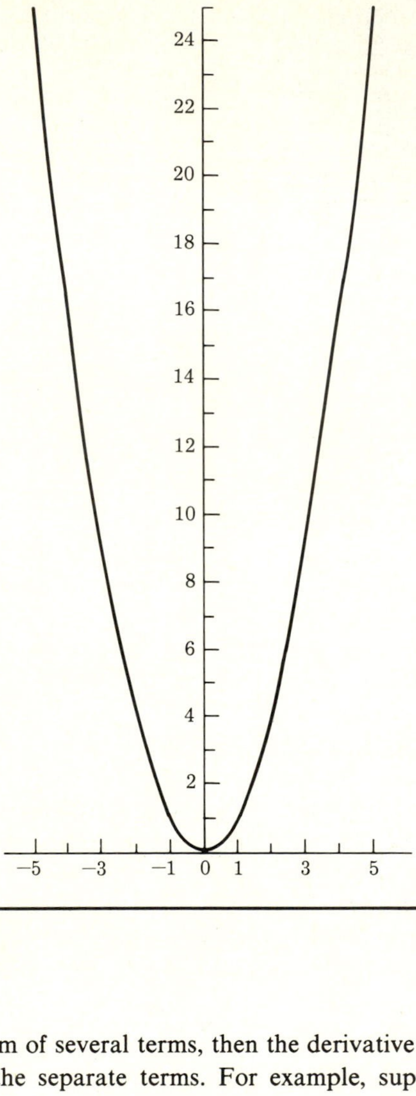

2. If y is the sum of several terms, then the derivative of y is the sum of the derivatives of the separate terms. For example, suppose $y=3x^2-2x$. Then $dy/dx=6x-2$.

3. The derivative of a constant, k, is equal to 0. That this is true can be seen by applying Rule 1 above. Thus, if we write $y=kx^0$, we see that $dy/dx=k(0)=0$.

4. The derivative of x with respect to itself is equal to 1, that is, $dx/dx=1$.

We state below without proof a number of formulas for differentiation, some of which have been used in this book. In each case, u and v represent functions of x, such as x^2, $3x^2-2x$, x^3-2x^2-1, and so on, whereas k, n, a, and $e = 2.71828\ldots$, are constants:

1. $\frac{d}{dx}(ku) = k\frac{du}{dx}$
2. $\frac{d}{dx}(u+v) = \frac{du}{dx} + \frac{dv}{dx}$
3. $\frac{d}{dx}(uv) = u\frac{dv}{dx} + v\frac{du}{dx}$
4. $\frac{d}{dx}\left(\frac{u}{v}\right) = \frac{v\frac{du}{dx} - u\frac{dv}{dx}}{v^2}$
5. $\frac{d}{dx}u^n = nu^{n-1}\frac{du}{dx}$
6. $\frac{d}{dx}\log_a u = \frac{\log_a e}{u}\frac{du}{dx}$
7. $\frac{d}{dx}\ln u = \frac{1}{u}\frac{du}{dx}$
8. $\frac{d}{dx}a^u = a^u \ln a\frac{du}{dx}$
9. $\frac{d}{dx}e^u = e^u\frac{du}{dx}$
10. $\frac{d}{dx}u^v = vu^{v-1}\frac{du}{dx} + u^v \ln u\frac{dv}{dx}$
11. $\frac{dy}{dx} = \frac{1}{\frac{dx}{dy}}$

B.3 PARTIAL DERIVATIVES

Sometimes we encounter functions that contain two different variables, say x and y, so that we have $z = f(x,y)$. In such cases, we may still obtain the derivative of z with respect to x *if we hold y constant*. Of course, we may also obtain the derivative of z with respect to y if we hold x constant. In either case the derivative is called a *partial derivative* and is usually denoted by

$$\frac{\partial z}{\partial x} \quad \text{or} \quad \frac{\partial z}{\partial y}.$$

The rules and formulas stated above can also be used for finding partial derivatives. For example, if $z = x^3 - 2x^2y + xy^2 + y^3$, then

$$\frac{\partial z}{\partial x} = 3x^2 - 2y(2x) + y^2 = 3x^2 - 4xy + y^2$$

and

$$\frac{\partial z}{\partial y} = -2x^2 + 2xy + 3y^2.$$

Note that in obtaining $\partial z/\partial x$, y was treated as a constant, and in obtaining $\partial z/\partial y$, x was treated as a constant.

B.4 FINDING RELATIVE MAXIMA AND MINIMA

The primary use made of differentiation in this book is in finding the value or values of a variable x for which $f(x)$ is a relative maximum or minimum*. Therefore, we shall discuss only this application of the process. There are many other applications that the interested reader may wish to pursue. Any good introductory college or advanced high school calculus textbook will describe such applications.

Suppose we wish to find the values of x for which the function $y = x^3 - x^2 - 8x + 6$ assumes its maximum or minimum values. The graph of a portion of this function is shown in Figure B.3. The point A is a maximum value and the point B is a minimum. In general, a function $y = f(x)$ has a maximum value at $x = a$ if y is larger when $x = a$ than when x is slightly

FIGURE B.3 Graph of the Function $y = x^3 - x^2 - 8x + 6$, Showing Maximum (A) and Minimum (B) Points and a Point of Inflection (P)

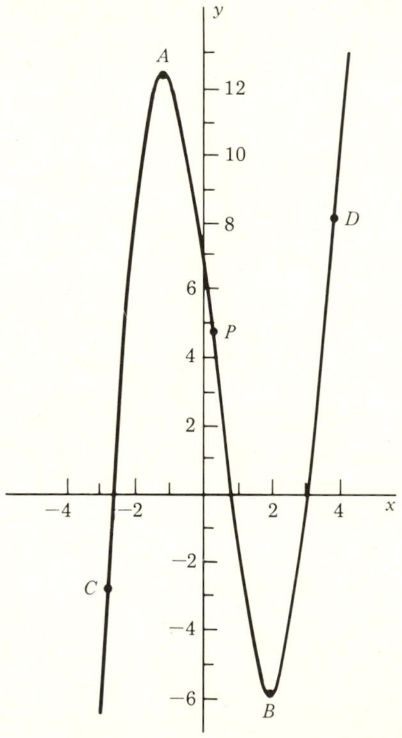

*For simplicity, we shall omit the adjective "relative" in the remainder of this section. It should be understood, however, that the discussion here applies to relative rather than absolute maxima and minima.

more or slightly less than a. Similarly, a function has a minimum at $x=b$ if y is smaller when $x=b$ than when x is slightly more or slightly less than b.

The values of x at A and B can be found through differentiation because the slopes of tangents to the curve at each of these points will be equal to zero. Thus, if we obtain

$$\frac{dy}{dx}=3x^2-2x-8,$$

and set it equal to zero, we have

$$3x^2-2x-8=0$$
$$(3x+4)(x-2)=0$$
$$x=2 \quad \text{and} \quad x=-4/3.$$

These results show that the tangent has slope equal to zero when $x=2$ and when $x=-4/3$. Although it is easy to see from Figure B.3 that the maximum point, A, corresponds to $x=-4/3$ and the minimum point, B, to $x=2$, we need to have a method for determining whether a particular result represents a maximum or minimum without graphing the function. Now if the point where $x=a$ is a maximum, then the slope (dy/dx) when x is slightly *less* than a will be *positive*, and the slope when x is slightly *more* than a will be *negative*. Similarly, if the point where $x=b$ is a minimum, then the slope when x is slightly *less* than b will be *negative*, and the slope when x is slightly *more* than b will be *positive*. In the preceding example, we may test each value, $x=2$ and $x=-4/3$, using these rules.

$x=2$. Because $dy/dx=3x^2-2x-8$, the value of the derivative when $x=1.9$ is -0.97 (negative) and when $x=2.1$ is 1.03 (positive). Therefore, $y=x^3-x^2-8x+6$ has a *minimum* at $x=2$.

$x=-4/3$. When $x=-1.4$ (slightly less than $-4/3$), $y=.68$ (positive), and when $x=-1.3$ (slightly more than $-4/3$), $y=-.33$ (negative). Therefore, y has a maximum at $x=-4/3$.

Alternatively, we can determine whether the point where $x=a$ (or b) is a maximum point or a minimum point by obtaining the *second derivative* of the original function. The second derivative is the derivative of the first derivative, and may be denoted by d^2y/dx^2. Because the second derivative is the derivative with respect to x of the function dy/dx, its value at $x=a$ is the *rate of change* of dy/dx. Therefore, where dy/dx is increasing, d^2y/dx^2 is positive, and where dy/dx is decreasing, d^2y/dx^2 is negative. A point at which the rate of change of dy/dx is equal to zero (that is, where $d^2y/dx^2=0$) is called a *point of inflection*.

For our example, in which $y=x^3-x^2-8x+6$, we found that $dy/dx=3x^2-2x-8$. Hence,

$$\frac{d^2y}{dx^2}=6x-2.$$

If we set d^2y/dx^2 equal to zero, we have

$$6x-2=0$$
$$x=1/3,$$

which indicates that a point of inflection occurs at $x=1/3$. This point is labeled P in Figure B.3. Now if we substitute $x=-4/3$ for x in d^2y/dx^2, we see that

$$6x-2=6(-4/3)-2=-10,$$

which indicates that the slope is decreasing at the point A in Figure B.3. In fact, in moving along the curve from point C to point P, the slope continually decreases, from a large positive value at C, to zero at A, and to a negative value between A and P. Thus, in this region, where d^2y/dx^2 is *negative*, the curve is concave *downward*, and point A is a *maximum*. A similar argument shows that the curve is concave *upward* between P and D, which means that B is a minimum. Thus, the rule is:

At a maximum point, $dy/dx=0$ and d^2y/dx^2 is negative.

At a minimum point, $dy/dx=0$ and d^2y/dx^2 is positive. Of course, if d^2y/dx^2 does not exist, then this rule cannot be used.

B.5 DIFFERENTIATION OF EXPRESSIONS INVOLVING MATRICES AND VECTORS

We frequently encounter expressions in multivariate analysis involving matrices and vectors, particularly the quadratic form, $\mathbf{a}'\mathbf{B}\mathbf{a}$, or functions of it. It is sometimes necessary, for example, in finding maxima or minima of such functions, to differentiate them with respect to $\mathbf{a}$ or $\mathbf{B}$. We show below how a formula for obtaining the derivative of $\mathbf{a}'\mathbf{B}\mathbf{a}$ with respect to $\mathbf{a}$ may be derived. We then state several rules that may be used for obtaining other derivatives of $\mathbf{a}'\mathbf{B}\mathbf{a}$.

Suppose that $\mathbf{a}'=[a_1 \quad a_2]$, $\mathbf{B}=\begin{bmatrix} b_{11} & b_{12} \\ b_{21} & b_{22} \end{bmatrix}$, and $\mathbf{a}=\begin{bmatrix} a_1 \\ a_2 \end{bmatrix}$. Then,

$$\begin{aligned} \mathbf{a}'\mathbf{B}\mathbf{a} &= \begin{bmatrix} a_1b_{11}+a_2b_{21} & a_1b_{12}+a_2b_{22} \end{bmatrix} \cdot \begin{bmatrix} a_1 \\ a_2 \end{bmatrix} \\ &= a_1^2b_{11}+a_1a_2b_{21}+a_1a_2b_{12}+a_2^2b_{22} \\ &= b_{11}a_1^2+(b_{21}+b_{12})a_1a_2+b_{22}a_2^2. \end{aligned}$$

We first take the partial derivatives of this function with respect to a_1 and a_2. The results are

$$\frac{\partial(\mathbf{a}'\mathbf{B}\mathbf{a})}{\partial a_1}=2b_{11}a_1+(b_{21}+b_{12})a_2$$

and

$$\frac{\partial(\mathbf{a}'\mathbf{B}\mathbf{a})}{\partial a_2}=(b_{21}+b_{12})a_1+2b_{22}a_2.$$

Then, arranging these results in the form of a column vector (because **a** is a column vector), we have

$$\begin{aligned}\frac{\partial(\mathbf{a}'\mathbf{B}\mathbf{a})}{\partial\mathbf{a}} &= \begin{bmatrix} 2b_{11}a_1+(b_{21}+b_{12})a_2 \\ (b_{21}+b_{12})a_1+2b_{22}a_2 \end{bmatrix} \\ &= \begin{bmatrix} 2b_{11} & (b_{21}+b_{12}) \\ (b_{21}+b_{12}) & 2b_{22} \end{bmatrix}\cdot\begin{bmatrix} a_1 \\ a_2 \end{bmatrix} \\ &= \left\{\begin{bmatrix} b_{11} & b_{12} \\ b_{21} & b_{22} \end{bmatrix}+\begin{bmatrix} b_{11} & b_{21} \\ b_{12} & b_{22} \end{bmatrix}\right\}\cdot\begin{bmatrix} a_1 \\ a_2 \end{bmatrix} \\ &= (\mathbf{B}+\mathbf{B}')\mathbf{a}.\end{aligned}$$

This result can be generalized to **a** of order $p\times 1$ and **B** of order $p\times p$. We give it as Rule 1 below along with several other rules, which may be used for obtaining partial derivatives of **a′Ba** with respect to **a′**, **B**, or **a**.

Rule 1. $\dfrac{\partial}{\partial\mathbf{x}}(\mathbf{x}'\mathbf{A}\mathbf{x})=(\mathbf{A}+\mathbf{A}')\mathbf{x}.$

Rule 2. $\dfrac{\partial}{\partial\mathbf{x}'}(\mathbf{x}'\mathbf{A}\mathbf{x})=\mathbf{x}'(\mathbf{A}+\mathbf{A}').$

When **A** is symmetric, Rules 1 and 2 reduce to:

Rule 1a. $\dfrac{\partial}{\partial\mathbf{x}}(\mathbf{x}'\mathbf{A}\mathbf{x})=2\mathbf{A}\mathbf{x}.$

Rule 2a. $\dfrac{\partial}{\partial\mathbf{x}'}(\mathbf{x}'\mathbf{A}\mathbf{x})=2\mathbf{x}'\mathbf{A}.$

Rule 3. $\dfrac{\partial}{\partial\mathbf{x}}(\mathbf{x}'\mathbf{A}\mathbf{y})=\dfrac{\partial}{\partial\mathbf{x}}(\mathbf{y}'\mathbf{A}\mathbf{x})=\mathbf{A}\mathbf{y}.$

Rule 4. $\dfrac{\partial}{\partial\mathbf{x}'}(\mathbf{x}'\mathbf{A}\mathbf{y})=\dfrac{\partial}{\partial\mathbf{x}'}(\mathbf{y}'\mathbf{A}\mathbf{x})=\mathbf{y}'\mathbf{A}.$

Rule 5. $\dfrac{\partial}{\partial\mathbf{A}}(\mathbf{x}'\mathbf{A}\mathbf{x})=\mathbf{x}\mathbf{x}'$ (**A** nonsymmetric).

When **A** is symmetric, Rule 5 becomes

Rule 5a. $\dfrac{\partial}{\partial\mathbf{A}}(\mathbf{x}'\mathbf{A}\mathbf{x})=2\mathbf{x}\mathbf{x}'-\mathbf{D}(x_i^2),$

in which the diagonal matrix,

$$\mathbf{D}(x_i^2)=\begin{bmatrix} x_1^2 & 0 & \cdots & 0 \\ 0 & x_2^2 & \cdots & 0 \\ \vdots & \vdots & & \vdots \\ 0 & 0 & \cdots & x_p^2 \end{bmatrix}.$$

B.6 THE USE OF LAGRANGE MULTIPLIERS

A type of problem one may encounter in multivariate analysis is that in which it is necessary to find a maximum or minimum value of a function when certain conditions or constraints have been imposed on the values of the variables involved. An example occurs in canonical correlation analysis where the problem is to find sets of weights on two linear combinations of variables that maximize their correlation. Because the number of such sets is infinite, the conditions are imposed in order to reach an unique solution, as well as to simplify the mathematics involved in the problem.

In general, the method of *Lagrange multipliers* is useful for finding values of variables, say x and y, that maximize or minimize a function, $f(x, y)$, when a condition $g(x,y)=0$, (or a set of conditions—see below) is imposed. To use the method, one first forms the function $F=f(x,y)+\lambda g(x,y)$, in which λ is the Lagrange multiplier. Partial differentiation of this function with respect to x, y, and λ yields expressions which, when set equal to zero, may be solved for x, y, and λ. The values of x and y obtained are those that maximize or minimize the function under the given constraint.

To clarify the procedure we use a simple example. Suppose we denote the length of a rectangle by x and its width by y, and wish to find the values of x and y that maximize the area, under the condition that $x+y=10$. Then $f(x,y)=xy$ and $g(x,y)=x+y-10=0$. Thus, $F=xy+\lambda(x+y-10)$. Taking partial derivatives with respect to x, y, and λ, and setting the resulting expressions equal to zero, we have

$$\frac{\partial F}{\partial x}=y+\lambda=0,$$

$$\frac{\partial F}{\partial y}=x+\lambda=0,$$

$$\frac{\partial F}{\partial \lambda}=x+y-10=0.$$

Solution yields the values $x=y=5$ and $\lambda=-5$. We conclude that the maximum area is achieved, under the condition $x+y=10$, when x and y are both equal to 5. The maximum area, of course, is $xy=25$.

Although the applications encountered in this book are somewhat more complicated than suggested by this example, the basic procedure is the same. In many applications, there are several variables involved and several conditions imposed so that, for p variables and m conditions ($m<p$), the function to be differentiated may be expressed as

$$F=f(x_1,x_2,\ldots,x_p)+\lambda_1 g_1(x_1,x_2,\ldots,x_p)+\cdots+\lambda_m g_m(x_1,x_2,\ldots,x_p).$$

Furthermore, in many applications, for example, canonical correlation, there is particular interest in the values of λ obtained, as well as in the values of the variables that maximize the function of interest.

APPENDIX C
Tables

TABLE C.1 AREAS AND ORDINATES OF THE STANDARD NORMAL DISTRIBUTION

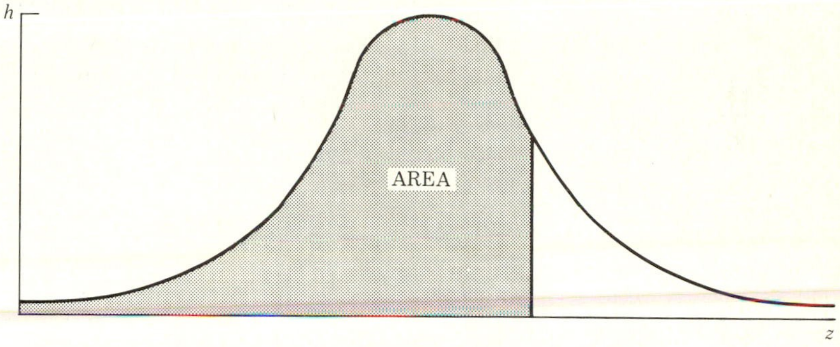

z	Area	h Ordinate	z	Area	h Ordinate
−3.00	.0013	.0044			
−2.99	.0014	.0046	−2.79	.0026	.0081
−2.98	.0014	.0047	−2.78	.0027	.0084
−2.97	.0015	.0048	−2.77	.0028	.0086
−2.96	.0015	.0050	−2.76	.0029	.0088
−2.95	.0016	.0051	−2.75	.0030	.0091
−2.94	.0016	.0053	−2.74	.0031	.0093
−2.93	.0017	.0055	−2.73	.0032	.0096
−2.92	.0018	.0056	−2.72	.0033	.0099
−2.91	.0018	.0058	−2.71	.0034	.0101
−2.90	.0019	.0060	−2.70	.0035	.0104
−2.89	.0019	.0061	−2.69	.0036	.0107
−2.88	.0020	.0063	−2.68	.0037	.0110
−2.87	.0021	.0065	−2.67	.0038	.0113
−2.86	.0021	.0067	−2.66	.0039	.0116
−2.85	.0022	.0069	−2.65	.0040	.0119
−2.84	.0023	.0071	−2.64	.0041	.0122
−2.83	.0023	.0073	−2.63	.0043	.0126
−2.82	.0024	.0075	−2.62	.0044	.0129
−2.81	.0025	.0077	−2.61	.0045	.0132
−2.80	.0026	.0079	−2.60	.0047	.0136

TABLE C.1 CONTINUED

z	Area	h Ordinate	z	Area	h Ordinate
−2.59	.0048	.0139	−2.09	.0183	.0449
−2.58	.0049	.0143	−2.08	.0188	.0459
−2.57	.0051	.0147	−2.07	.0192	.0468
−2.56	.0052	.0151	−2.06	.0197	.0478
−2.55	.0054	.0154	−2.05	.0202	.0488
−2.54	.0055	.0158	−2.04	.0207	.0498
−2.53	.0057	.0163	−2.03	.0212	.0508
−2.52	.0059	.0167	−2.02	.0217	.0519
−2.51	.0060	.0171	−2.01	.0222	.0529
−2.50	.0062	.0175	−2.00	.0228	.0540
−2.49	.0064	.0180	−1.99	.0233	.0551
−2.48	.0066	.0184	−1.98	.0239	.0562
−2.47	.0068	.0189	−1.97	.0244	.0573
−2.46	.0069	.0194	−1.96	.0250	.0584
−2.45	.0071	.0198	−1.95	.0256	.0596
−2.44	.0073	.0203	−1.94	.0262	.0608
−2.43	.0075	.0208	−1.93	.0268	.0620
−2.42	.0078	.0213	−1.92	.0274	.0632
−2.41	.0080	.0219	−1.91	.0281	.0644
−2.40	.0082	.0224	−1.90	.0287	.0656
−2.39	.0084	.0229	−1.89	.0294	.0669
−2.38	.0087	.0235	−1.88	.0301	.0681
−2.37	.0089	.0241	−1.87	.0307	.0694
−2.36	.0091	.0246	−1.86	.0314	.0707
−2.35	.0094	.0252	−1.85	.0322	.0721
−2.34	.0096	.0258	−1.84	.0329	.0734
−2.33	.0099	.0264	−1.83	.0336	.0748
−2.32	.0102	.0270	−1.82	.0344	.0761
−2.31	.0104	.0277	−1.81	.0351	.0775
−2.30	.0107	.0283	−1.80	.0359	.0790
−2.29	.0110	.0290	−1.79	.0367	.0804
−2.28	.0113	.0297	−1.78	.0375	.0818
−2.27	.0116	.0303	−1.77	.0384	.0833
−2.26	.0119	.0310	−1.76	.0392	.0848
−2.25	.0122	.0317	−1.75	.0401	.0863
−2.24	.0125	.0325	−1.74	.0409	.0878
−2.23	.0129	.0332	−1.73	.0418	.0893
−2.22	.0132	.0339	−1.72	.0427	.0909
−2.21	.0136	.0347	−1.71	.0436	.0925
−2.20	.0139	.0355	−1.70	.0446	.0940
−2.19	.0143	.0363	−1.69	.0455	.0957
−2.18	.0146	.0371	−1.68	.0465	.0973
−2.17	.0150	.0379	−1.67	.0475	.0989
−2.16	.0154	.0387	−1.66	.0485	.1006
−2.15	.0158	.0396	−1.65	.0495	.1023
−2.14	.0162	.0404	−1.64	.0505	.1040
−2.13	.0166	.0413	−1.63	.0516	.1057
−2.12	.0170	.0422	−1.62	.0526	.1074
−2.11	.0174	.0431	−1.61	.0537	.1092
−2.10	.0179	.0440	−1.60	.0548	.1109

z	Area	h Ordinate	z	Area	h Ordinate
−1.59	.0559	.1127	−1.09	.1379	.2203
−1.58	.0571	.1145	−1.08	.1401	.2227
−1.57	.0582	.1163	−1.07	.1423	.2251
−1.56	.0594	.1182	−1.06	.1446	.2275
−1.55	.0606	.1200	−1.05	.1469	.2299
−1.54	.0618	.1219	−1.04	.1492	.2323
−1.53	.0630	.1238	−1.03	.1515	.2347
−1.52	.0643	.1257	−1.02	.1539	.2371
−1.51	.0655	.1276	−1.01	.1562	.2396
−1.50	.0668	.1295	−1.00	.1587	.2420
−1.49	.0681	.1315	−0.99	.1611	.2444
−1.48	.0694	.1334	−0.98	.1635	.2468
−1.47	.0708	.1354	−0.97	.1660	.2492
−1.46	.0721	.1374	−0.96	.1685	.2516
−1.45	.0735	.1394	−0.95	.1711	.2541
−1.44	.0749	.1415	−0.94	.1736	.2565
−1.43	.0764	.1435	−0.93	.1762	.2589
−1.42	.0778	.1456	−0.92	.1788	.2613
−1.41	.0793	.1476	−0.91	.1814	.2637
−1.40	.0808	.1497	−0.90	.1841	.2661
−1.39	.0823	.1518	−0.89	.1867	.2685
−1.38	.0838	.1539	−0.88	.1894	.2709
−1.37	.0853	.1561	−0.87	.1922	.2732
−1.36	.0869	.1582	−0.86	.1949	.2756
−1.35	.0885	.1604	−0.85	.1977	.2780
−1.34	.0901	.1626	−0.84	.2005	.2803
−1.33	.0918	.1647	−0.83	.2033	.2827
−1.32	.0934	.1669	−0.82	.2061	.2850
−1.31	.0951	.1691	−0.81	.2090	.2874
−1.30	.0968	.1714	−0.80	.2119	.2897
−1.29	.0985	.1736	−0.79	.2148	.2920
−1.28	.1003	.1758	−0.78	.2177	.2943
−1.27	.1020	.1781	−0.77	.2206	.2966
−1.26	.1038	.1804	−0.76	.2236	.2989
−1.25	.1056	.1826	−0.75	.2266	.3011
−1.24	.1075	.1849	−0.74	.2296	.3034
−1.23	.1093	.1872	−0.73	.2327	.3056
−1.22	.1112	.1895	−0.72	.2358	.3079
−1.21	.1131	.1919	−0.71	.2389	.3101
−1.20	.1151	.1942	−0.70	.2420	.3123
−1.19	.1170	.1965	−0.69	.2451	.3144
−1.18	.1190	.1989	−0.68	.2483	.3166
−1.17	.1210	.2012	−0.67	.2514	.3187
−1.16	.1230	.2036	−0.66	.2546	.3209
−1.15	.1251	.2059	−0.65	.2578	.3230
−1.14	.1271	.2083	−0.64	.2611	.3251
−1.13	.1292	.2107	−0.63	.2643	.3271
−1.12	.1314	.2131	−0.62	.2676	.3292
−1.11	.1335	.2155	−0.61	.2709	.3312
−1.10	.1357	.2179	−0.60	.2743	.3332

TABLE C.1 CONTINUED

z	*Area*	*h* *Ordinate*	*z*	*Area*	*h* *Ordinate*
−0.59	.2776	.3352	−0.09	.4641	.3973
−0.58	.2810	.3372	−0.08	.4681	.3977
−0.57	.2843	.3391	−0.07	.4721	.3980
−0.56	.2877	.3410	−0.06	.4761	.3982
−0.55	.2912	.3429	−0.05	.4801	.3984
−0.54	.2946	.3448	−0.04	.4840	.3986
−0.53	.2981	.3467	−0.03	.4880	.3988
−0.52	.3015	.3485	−0.02	.4920	.3989
−0.51	.3050	.3503	−0.01	.4960	.3989
−0.50	.3085	.3521	0.00	.5000	.3989
−0.49	.3121	.3538	0.01	.5040	.3989
−0.48	.3156	.3555	0.02	.5080	.3989
−0.47	.3192	.3572	0.03	.5120	.3988
−0.46	.3228	.3589	0.04	.5160	.3986
−0.45	.3264	.3605	0.05	.5199	.3984
−0.44	.3300	.3621	0.06	.5239	.3982
−0.43	.3336	.3637	0.07	.5279	.3980
−0.42	.3372	.3653	0.08	.5319	.3977
−0.41	.3409	.3668	0.09	.5359	.3973
−0.40	.3446	.3683	0.10	.5398	.3970
−0.39	.3483	.3697	0.11	.5438	.3965
−0.38	.3520	.3712	0.12	.5478	.3961
−0.37	.3557	.3725	0.13	.5517	.3956
−0.36	.3594	.3739	0.14	.5557	.3951
−0.35	.3632	.3752	0.15	.5596	.3945
−0.34	.3669	.3765	0.16	.5636	.3939
−0.33	.3707	.3778	0.17	.5675	.3932
−0.32	.3745	.3790	0.18	.5714	.3925
−0.31	.3783	.3802	0.19	.5753	.3918
−0.30	.3821	.3814	0.20	.5793	.3910
−0.29	.3859	.3825	0.21	.5832	.3902
−0.28	.3897	.3836	0.22	.5871	.3894
−0.27	.3936	.3847	0.23	.5910	.3885
−0.26	.3974	.3857	0.24	.5948	.3876
−0.25	.4013	.3867	0.25	.5987	.3867
−0.24	.4052	.3876	0.26	.6026	.3857
−0.23	.4090	.3885	0.27	.6064	.3847
−0.22	.4129	.3894	0.28	.6103	.3836
−0.21	.4168	.3902	0.29	.6141	.3825
−0.20	.4207	.3910	0.30	.6179	.3814
−0.19	.4247	.3918	0.31	.6217	.3802
−0.18	.4286	.3925	0.32	.6255	.3790
−0.17	.4325	.3932	0.33	.6293	.3778
−0.16	.4364	.3939	0.34	.6331	.3765
−0.15	.4404	.3945	0.35	.6368	.3752
−0.14	.4443	.3951	0.36	.6406	.3739
−0.13	.4483	.3956	0.37	.6443	.3725
−0.12	.4522	.3961	0.38	.6480	.3712
−0.11	.4562	.3965	0.39	.6517	.3697
−0.10	.4602	.3970	0.40	.6554	.3683

z	Area	h Ordinate	z	Area	h Ordinate
0.41	.6591	.3668	0.91	.8186	.2637
0.42	.6628	.3653	0.92	.8212	.2613
0.43	.6664	.3637	0.93	.8238	.2589
0.44	.6700	.3621	0.94	.8264	.2565
0.45	.6736	.3605	0.95	.8289	.2541
0.46	.6772	.3589	0.96	.8315	.2516
0.47	.6808	.3572	0.97	.8340	.2492
0.48	.6844	.3555	0.98	.8365	.2468
0.49	.6879	.3538	0.99	.8389	.2444
0.50	.6915	.3521	1.00	.8413	.2420
0.51	.6950	.3503	1.01	.8438	.2396
0.52	.6985	.3485	1.02	.8461	.2371
0.53	.7019	.3467	1.03	.8485	.2347
0.54	.7054	.3448	1.04	.8508	.2323
0.55	.7088	.3429	1.05	.8531	.2299
0.56	.7123	.3410	1.06	.8554	.2275
0.57	.7157	.3391	1.07	.8577	.2251
0.58	.7190	.3372	1.08	.8599	.2227
0.59	.7224	.3352	1.09	.8621	.2203
0.60	.7257	.3332	1.10	.8643	.2179
0.61	.7291	.3312	1.11	.8665	.2155
0.62	.7324	.3292	1.12	.8686	.2131
0.63	.7357	.3271	1.13	.8708	.2107
0.64	.7389	.3251	1.14	.8729	.2083
0.65	.7422	.3230	1.15	.8749	.2059
0.66	.7454	.3209	1.16	.8770	.2036
0.67	.7486	.3187	1.17	.8790	.2012
0.68	.7517	.3166	1.18	.8810	.1989
0.69	.7549	.3144	1.19	.8830	.1965
0.70	.7580	.3123	1.20	.8849	.1942
0.71	.7611	.3101	1.21	.8869	.1919
0.72	.7642	.3079	1.22	.8888	.1895
0.73	.7673	.3056	1.23	.8907	.1872
0.74	.7704	.3034	1.24	.8925	.1849
0.75	.7734	.3011	1.25	.8944	.1826
0.76	.7764	.2989	1.26	.8962	.1804
0.77	.7794	.2966	1.27	.8980	.1781
0.78	.7823	.2943	1.28	.8997	.1758
0.79	.7852	.2920	1.29	.9015	.1736
0.80	.7881	.2897	1.30	.9032	.1714
0.81	.7910	.2874	1.31	.9049	.1691
0.82	.7939	.2850	1.32	.9066	.1669
0.83	.7967	.2827	1.33	.9082	.1647
0.84	.7995	.2803	1.34	.9099	.1626
0.85	.8023	.2780	1.35	.9115	.1604
0.86	.8051	.2756	1.36	.9131	.1582
0.87	.8078	.2732	1.37	.9147	.1561
0.88	.8106	.2709	1.38	.9162	.1539
0.89	.8133	.2685	1.39	.9177	.1518
0.90	.8159	.2661	1.40	.9192	.1497

TABLE C.1 CONTINUED

z	*Area*	*h* *Ordinate*	*z*	*Area*	*h* *Ordinate*
1.41	.9207	.1476	1.91	.9719	.0644
1.42	.9222	.1456	1.92	.9726	.0632
1.43	.9236	.1435	1.93	.9732	.0620
1.44	.9251	.1415	1.94	.9738	.0608
1.45	.9265	.1394	1.95	.9744	.0596
1.46	.9279	.1374	1.96	.9750	.0584
1.47	.9292	.1354	1.97	.9756	.0573
1.48	.9306	.1334	1.98	.9761	.0562
1.49	.9319	.1315	1.99	.9767	.0551
1.50	.9332	.1295	2.00	.9772	.0540
1.51	.9345	.1276	2.01	.9778	.0529
1.52	.9357	.1257	2.02	.9783	.0519
1.53	.9370	.1238	2.03	.9788	.0508
1.54	.9382	.1219	2.04	.9793	.0498
1.55	.9394	.1200	2.05	.9798	.0488
1.56	.9406	.1182	2.06	.9803	.0478
1.57	.9418	.1163	2.07	.9808	.0468
1.58	.9429	.1145	2.08	.9812	.0459
1.59	.9441	.1127	2.09	.9817	.0449
1.60	.9452	.1109	2.10	.9821	.0440
1.61	.9463	.1092	2.11	.9826	.0431
1.62	.9474	.1074	2.12	.9830	.0422
1.63	.9484	.1057	2.13	.9834	.0413
1.64	.9495	.1040	2.14	.9838	.0404
1.65	.9505	.1023	2.15	.9842	.0396
1.66	.9515	.1006	2.16	.9846	.0387
1.67	.9525	.0989	2.17	.9850	.0379
1.68	.9535	.0973	2.18	.9854	.0371
1.69	.9545	.0957	2.19	.9857	.0363
1.70	.9554	.0940	2.20	.9861	.0355
1.71	.9564	.0925	2.21	.9864	.0347
1.72	.9573	.0909	2.22	.9868	.0339
1.73	.9582	.0893	2.23	.9871	.0332
1.74	.9591	.0878	2.24	.9875	.0325
1.75	.9599	.0863	2.25	.9878	.0317
1.76	.9608	.0848	2.26	.9881	.0310
1.77	.9616	.0833	2.27	.9884	.0303
1.78	.9625	.0818	2.28	.9887	.0297
1.79	.9633	.0804	2.29	.9890	.0290
1.80	.9641	.0790	2.30	.9893	.0283
1.81	.9649	.0775	2.31	.9896	.0277
1.82	.9656	.0761	2.32	.9898	.0270
1.83	.9664	.0748	2.33	.9901	.0264
1.84	.9671	.0734	2.34	.9904	.0258
1.85	.9678	.0721	2.35	.9906	.0252
1.86	.9686	.0707	2.36	.9909	.0246
1.87	.9693	.0694	2.37	.9911	.0241
1.88	.9699	.0681	2.38	.9913	.0235
1.89	.9706	.0669	2.39	.9916	.0229
1.90	.9713	.0656	2.40	.9918	.0224

z	Area	h Ordinate	z	Area	h Ordinate
2.41	.9920	.0219	2.71	.9966	.0101
2.42	.9922	.0213	2.72	.9967	.0099
2.43	.9925	.0208	2.73	.9968	.0096
2.44	.9927	.0203	2.74	.9969	.0093
2.45	.9929	.0198	2.75	.9970	.0091
2.46	.9931	.0194	2.76	.9971	.0088
2.47	.9932	.0189	2.77	.9972	.0086
2.48	.9934	.0184	2.78	.9973	.0084
2.49	.9936	.0180	2.79	.9974	.0081
2.50	.9938	.0175	2.80	.9974	.0079
2.51	.9940	.0171	2.81	.9975	.0077
2.52	.9941	.0167	2.82	.9976	.0075
2.53	.9943	.0163	2.83	.9977	.0073
2.54	.9945	.0158	2.84	.9977	.0071
2.55	.9946	.0154	2.85	.9978	.0069
2.56	.9948	.0151	2.86	.9979	.0067
2.57	.9949	.0147	2.87	.9979	.0065
2.58	.9951	.0143	2.88	.9980	.0063
2.59	.9952	.0139	2.89	.9981	.0061
2.60	.9953	.0136	2.90	.9981	.0060
2.61	.9955	.0132	2.91	.9982	.0058
2.62	.9956	.0129	2.92	.9982	.0056
2.63	.9957	.0126	2.93	.9983	.0055
2.64	.9959	.0122	2.94	.9984	.0053
2.65	.9960	.0119	2.95	.9984	.0051
2.66	.9961	.0116	2.96	.9985	.0050
2.67	.9962	.0113	2.97	.9985	.0048
2.68	.9963	.0110	2.98	.9986	.0047
2.69	.9964	.0107	2.99	.9986	.0046
2.70	.9965	.0104	3.00	.9987	.0044

TABLE C.2 PERCENTILE POINTS OF CHI-SQUARE DISTRIBUTIONS*

	Percentile													
df	1	2	5	10	20	30	50	70	80	90	95	98	99	99.9
1	.0002	.0006	.0039	.0158	.0642	.148	.455	1.074	1.642	2.706	3.841	5.412	6.635	10.827
2	.0201	.0404	.103	.211	.446	.713	1.386	2.408	3.219	4.605	5.991	7.824	9.210	13.815
3	.115	.185	.352	.584	1.005	1.424	2.366	3.665	4.642	6.251	7.815	9.837	11.341	16.268
4	.297	.429	.711	1.064	1.649	2.195	3.357	4.878	5.989	7.779	9.488	11.668	13.277	18.465
5	.554	.752	1.145	1.610	2.343	3.000	4.351	6.064	7.289	9.236	11.070	13.388	15.086	20.517
6	.872	1.134	1.635	2.204	3.070	3.828	5.348	7.231	8.558	10.645	12.592	15.033	16.812	22.457
7	1.239	1.564	2.167	2.833	3.822	4.671	6.346	8.383	9.803	12.017	14.067	16.622	18.475	24.322
8	1.646	2.032	2.733	3.490	4.594	5.527	7.344	9.524	11.030	13.362	15.507	18.168	20.090	26.125
9	2.088	2.532	3.325	4.168	5.380	6.393	8.343	10.656	12.242	14.684	16.919	19.679	21.666	27.877
10	2.558	3.059	3.940	4.865	6.179	7.267	9.342	11.781	13.442	15.987	18.307	21.161	23.209	29.588
11	3.053	3.609	4.575	5.578	6.989	8.148	10.341	12.899	14.631	17.275	19.675	22.618	24.725	31.264
12	3.571	4.178	5.226	6.304	7.807	9.034	11.340	14.011	15.812	18.549	21.026	24.054	26.217	32.909
13	4.107	4.765	5.892	7.042	8.634	9.926	12.340	15.119	16.985	19.812	22.362	25.472	27.688	34.528
14	4.660	5.368	6.571	7.790	9.467	10.821	13.339	16.222	18.151	21.064	23.685	26.873	29.141	36.123
15	5.229	5.985	7.261	8.547	10.307	11.721	14.339	17.322	19.311	22.307	24.996	28.259	30.578	37.697
16	5.812	6.614	7.962	9.312	11.152	12.624	15.338	18.418	20.465	23.542	26.296	29.633	32.000	39.252
17	6.408	7.255	8.672	10.085	12.002	13.531	16.338	19.511	21.615	24.769	27.587	30.995	33.409	40.790
18	7.015	7.906	9.390	10.865	12.857	14.440	17.338	20.601	22.760	25.989	28.869	32.346	34.805	42.312
19	7.633	8.567	10.117	11.651	13.716	15.352	18.338	21.689	23.900	27.204	30.144	33.687	36.191	43.820
20	8.260	9.237	10.851	12.443	14.578	16.266	19.337	22.775	25.038	28.412	31.410	35.020	37.566	45.315
21	8.897	9.915	11.591	13.240	15.445	17.182	20.337	23.858	26.171	29.615	32.671	36.343	38.932	46.797
22	9.542	10.600	12.338	14.041	16.314	18.101	21.337	24.939	27.301	30.813	33.924	37.659	40.289	48.268
23	10.196	11.293	13.091	14.848	17.187	19.021	22.337	26.018	28.429	32.007	35.172	38.968	41.638	49.728
24	10.856	11.992	13.848	15.659	18.062	19.943	23.337	27.096	29.553	33.196	36.415	40.270	42.980	51.179
25	11.524	12.697	14.611	16.473	18.940	20.867	24.337	28.172	30.675	34.382	37.652	41.566	44.314	52.620
26	12.198	13.409	15.379	17.292	19.820	21.792	25.336	29.246	31.795	35.563	38.885	42.856	45.642	54.052
27	12.879	14.125	16.151	18.114	20.703	22.719	26.336	30.319	32.912	36.741	40.113	44.140	46.963	55.476
28	13.565	14.847	16.928	18.939	21.588	23.647	27.336	31.391	34.027	37.916	41.337	45.419	48.278	56.893
29	14.256	15.574	17.708	19.768	22.475	24.577	28.336	32.461	35.139	39.087	42.557	46.693	49.588	58.302
30	14.953	16.306	18.493	20.599	23.364	25.508	29.336	33.530	36.250	40.256	43.773	47.962	50.892	59.703

*If χ^2 is a chi-square variable with *df* greater than 30, then $z = \sqrt{2\chi^2} - \sqrt{2df-1}$ is very nearly normally distributed with mean 0 and standard deviation 1.

TABLE C.3 PERCENTILE POINTS OF *t* DISTRIBUTIONS

	*Percentile**												
df	55	60	65	70	75	80	85	90	95	97.5	99	99.5	99.95
1	.158	.325	.510	.727	1.000	1.376	1.963	3.078	6.314	12.706	31.821	63.657	636.619
2	.142	.289	.445	.617	.816	1.061	1.386	1.886	2.920	4.303	6.965	9.925	31.598
3	.137	.277	.424	.584	.765	.978	1.250	1.638	2.353	3.182	4.541	5.841	12.941
4	.134	.271	.414	.569	.741	.941	1.190	1.533	2.132	2.776	3.747	4.604	8.610
5	.132	.267	.408	.559	.727	.920	1.156	1.476	2.015	2.571	3.365	4.032	6.859
6	.131	.265	.404	.553	.718	.906	1.134	1.440	1.943	2.447	3.143	3.707	5.959
7	.130	.263	.402	.549	.711	.896	1.119	1.415	1.895	2.365	2.998	3.499	5.405
8	.130	.262	.399	.546	.706	.889	1.108	1.397	1.860	2.306	2.896	3.355	5.041
9	.129	.261	.398	.543	.703	.883	1.100	1.383	1.833	2.262	2.821	3.250	4.781
10	.129	.260	.397	.542	.700	.879	1.093	1.372	1.812	2.228	2.764	3.169	4.587
11	.129	.260	.396	.540	.697	.876	1.088	1.363	1.796	2.201	2.718	3.106	4.437
12	.128	.259	.395	.539	.695	.873	1.083	1.356	1.782	2.179	2.681	3.055	4.318
13	.128	.259	.394	.538	.694	.870	1.079	1.350	1.771	2.160	2.650	3.012	4.221
14	.128	.258	.393	.537	.692	.868	1.076	1.345	1.761	2.145	2.624	2.977	4.140
15	.128	.258	.393	.536	.691	.866	1.074	1.341	1.753	2.131	2.602	2.947	4.073
16	.128	.258	.392	.535	.690	.865	1.071	1.337	1.746	2.120	2.583	2.921	4.015
17	.128	.257	.392	.534	.689	.863	1.069	1.333	1.740	2.110	2.567	2.898	3.965
18	.127	.257	.392	.534	.688	.862	1.067	1.330	1.734	2.101	2.552	2.878	3.922
19	.127	.257	.391	.533	.688	.861	1.066	1.328	1.729	2.093	2.539	2.861	3.883
20	.127	.257	.391	.533	.687	.860	1.064	1.325	1.725	2.086	2.528	2.845	3.850
21	.127	.257	.391	.532	.686	.859	1.063	1.323	1.721	2.080	2.518	2.831	3.819
22	.127	.256	.390	.532	.686	.858	1.061	1.321	1.717	2.074	2.508	2.819	3.792
23	.127	.256	.390	.532	.685	.858	1.060	1.319	1.714	2.069	2.500	2.807	3.767
24	.127	.256	.390	.531	.685	.857	1.059	1.318	1.711	2.064	2.492	2.797	3.745
25	.127	.256	.390	.531	.684	.856	1.058	1.316	1.708	2.060	2.485	2.787	3.725
26	.127	.256	.390	.531	.684	.856	1.058	1.315	1.706	2.056	2.479	2.779	3.707
27	.127	.256	.389	.531	.684	.855	1.057	1.314	1.703	2.052	2.473	2.771	3.690
28	.127	.256	.389	.530	.683	.855	1.056	1.313	1.701	2.048	2.467	2.763	3.674
29	.127	.256	.389	.530	.683	.854	1.055	1.311	1.699	2.045	2.462	2.756	3.659
30	.127	.256	.389	.530	.683	.854	1.055	1.310	1.697	2.042	2.457	2.750	3.646
40	.126	.255	.388	.529	.681	.851	1.050	1.303	1.684	2.021	2.423	2.704	3.551
60	.126	.254	.387	.527	.679	.848	1.046	1.296	1.671	2.000	2.390	2.660	3.460
120	.126	.254	.386	.526	.677	.845	1.041	1.289	1.658	1.980	2.358	2.617	3.373
∞	.126	.253	.385	.524	.674	.842	1.036	1.282	1.645	1.960	2.326	2.576	3.291

*The lower percentiles are related to the upper percentiles tabulated above by the equation ${}_pt_n = -{}_{1-p}t_n$. Thus, the 10th percentile in the *t* distribution with 15*df* equals the negative of the 90th percentile in the same distribution, i.e., ${}_{10}t_{15} = -1.341$.

TABLE C.4 PERCENTILE POINTS OF *F* DISTRIBUTIONS

75th percentiles

n_2 \ n_1	1	2	3	4	5	6	7	8	9	10	12	15	20	24	30	40	60	120	∞
1	5.83	7.50	8.20	8.58	8.82	8.98	9.10	9.19	9.26	9.32	9.41	9.49	9.58	9.63	9.67	9.71	9.76	9.80	9.85
2	2.57	3.00	3.15	3.28	3.28	3.31	3.34	3.35	3.37	3.38	3.39	3.41	3.43	3.43	3.44	3.45	3.46	3.47	3.48
3	2.02	2.28	2.36	2.39	2.41	2.42	2.43	2.44	2.44	2.44	2.45	2.46	2.46	2.46	2.47	2.47	2.47	2.47	2.47
4	1.81	2.00	2.05	2.06	2.07	2.08	2.08	2.08	2.08	2.08	2.08	2.08	2.08	2.08	2.08	2.08	2.08	2.08	2.08
5	1.69	1.85	1.88	1.89	1.89	1.89	1.89	1.89	1.89	1.89	1.89	1.89	1.88	1.88	1.88	1.88	1.87	1.87	1.87
6	1.62	1.76	1.78	1.79	1.79	1.78	1.78	1.78	1.77	1.77	1.77	1.76	1.76	1.75	1.75	1.75	1.74	1.74	1.74
7	1.57	1.70	1.72	1.72	1.71	1.71	1.70	1.70	1.69	1.69	1.68	1.68	1.67	1.67	1.68	1.66	1.66	1.65	1.65
8	1.54	1.66	1.67	1.66	1.66	1.65	1.64	1.64	1.63	1.63	1.62	1.62	1.61	1.60	1.60	1.59	1.59	1.58	1.58
9	1.51	1.62	1.63	1.63	1.62	1.61	1.60	1.60	1.59	1.59	1.58	1.57	1.56	1.56	1.55	1.54	1.54	1.53	1.53
10	1.49	1.60	1.60	1.59	1.59	1.58	1.57	1.56	1.56	1.55	1.54	1.53	1.52	1.52	1.51	1.51	1.50	1.49	1.48
11	1.47	1.58	1.58	1.57	1.56	1.55	1.54	1.53	1.53	1.52	1.51	1.50	1.49	1.49	1.48	1.47	1.47	1.46	1.45
12	1.46	1.56	1.56	1.55	1.54	1.53	1.52	1.51	1.51	1.50	1.49	1.48	1.47	1.46	1.45	1.45	1.44	1.43	1.42
13	1.45	1.55	1.55	1.53	1.52	1.51	1.50	1.49	1.49	1.48	1.47	1.46	1.45	1.44	1.43	1.42	1.42	1.41	1.40
14	1.44	1.53	1.53	1.52	1.51	1.50	1.49	1.48	1.47	1.46	1.45	1.44	1.43	1.42	1.41	1.41	1.40	1.39	1.38
15	1.43	1.52	1.52	1.51	1.49	1.48	1.47	1.46	1.46	1.45	1.44	1.43	1.41	1.41	1.40	1.39	1.38	1.37	1.36
16	1.42	1.51	1.51	1.50	1.48	1.47	1.46	1.45	1.44	1.44	1.43	1.41	1.40	1.39	1.38	1.37	1.36	1.35	1.34
17	1.42	1.51	1.50	1.49	1.47	1.46	1.45	1.44	1.43	1.43	1.41	1.40	1.39	1.38	1.37	1.36	1.35	1.34	1.33
18	1.41	1.50	1.49	1.48	1.46	1.45	1.44	1.43	1.42	1.42	1.40	1.39	1.38	1.37	1.36	1.35	1.34	1.33	1.32
19	1.41	1.49	1.49	1.47	1.46	1.44	1.43	1.42	1.41	1.41	1.40	1.38	1.37	1.36	1.35	1.34	1.33	1.32	1.30
20	1.40	1.49	1.48	1.47	1.45	1.44	1.43	1.42	1.41	1.40	1.39	1.37	1.36	1.35	1.34	1.33	1.32	1.31	1.29
21	1.40	1.48	1.48	1.46	1.44	1.43	1.42	1.41	1.40	1.39	1.38	1.37	1.35	1.34	1.33	1.32	1.31	1.30	1.28
22	1.40	1.48	1.47	1.45	1.44	1.42	1.41	1.40	1.39	1.39	1.37	1.36	1.34	1.33	1.32	1.31	1.30	1.29	1.28
23	1.39	1.47	1.47	1.45	1.43	1.42	1.41	1.40	1.39	1.38	1.37	1.35	1.34	1.33	1.32	1.31	1.30	1.28	1.27
24	1.39	1.47	1.46	1.44	1.43	1.41	1.40	1.39	1.38	1.38	1.36	1.35	1.33	1.32	1.31	1.30	1.29	1.28	1.26
25	1.39	1.47	1.46	1.44	1.42	1.41	1.40	1.39	1.38	1.37	1.36	1.34	1.33	1.32	1.31	1.29	1.28	1.27	1.25
26	1.38	1.46	1.45	1.44	1.42	1.41	1.39	1.38	1.37	1.37	1.35	1.34	1.32	1.31	1.30	1.29	1.28	1.26	1.25
27	1.38	1.46	1.45	1.43	1.42	1.40	1.39	1.38	1.37	1.36	1.35	1.33	1.32	1.31	1.30	1.28	1.27	1.26	1.24
28	1.38	1.46	1.45	1.43	1.41	1.40	1.39	1.38	1.37	1.36	1.34	1.33	1.31	1.30	1.29	1.28	1.27	1.25	1.24
29	1.38	1.45	1.45	1.43	1.41	1.40	1.38	1.37	1.36	1.35	1.34	1.32	1.31	1.30	1.29	1.27	1.26	1.25	1.23
30	1.38	1.45	1.44	1.42	1.41	1.39	1.38	1.37	1.36	1.35	1.34	1.32	1.30	1.29	1.28	1.27	1.26	1.24	1.23
40	1.36	1.44	1.42	1.40	1.39	1.37	1.36	1.35	1.34	1.33	1.31	1.30	1.28	1.26	1.25	1.24	1.22	1.21	1.19
60	1.35	1.42	1.41	1.38	1.37	1.35	1.33	1.32	1.31	1.30	1.29	1.27	1.25	1.24	1.22	1.21	1.19	1.17	1.15
120	1.34	1.40	1.39	1.37	1.35	1.33	1.31	1.30	1.29	1.28	1.26	1.24	1.22	1.21	1.19	1.18	1.16	1.13	1.10
∞	1.32	1.39	1.37	1.35	1.33	1.31	1.29	1.28	1.27	1.25	1.24	1.22	1.19	1.18	1.16	1.14	1.12	1.08	1.00

TABLE C.4 CONTINUED

90th percentiles

n_2 \ n_1	1	2	3	4	5	6	7	8	9	10	12	15	20	24	30	40	60	120	∞
1	39.86	49.50	53.59	55.83	57.24	58.20	58.91	59.44	59.86	60.19	60.71	61.22	61.74	62.00	62.26	62.53	62.79	63.06	63.33
2	8.53	9.00	9.16	9.24	9.29	9.33	9.35	9.37	9.38	9.39	9.41	9.42	9.44	9.45	9.46	9.47	9.47	9.48	9.49
3	5.54	5.46	5.39	5.34	5.31	5.28	5.27	5.25	5.24	5.23	5.22	5.20	5.18	5.18	5.17	5.16	5.15	5.14	5.13
4	4.54	4.32	4.19	4.11	4.05	4.01	3.98	3.95	3.94	3.92	3.90	3.87	3.84	3.83	3.82	3.80	4.79	3.78	3.76
5	4.06	3.78	3.62	3.52	3.45	3.40	3.37	3.34	3.32	3.30	3.27	3.24	3.21	3.19	3.17	3.16	3.14	3.12	3.10
6	3.78	3.46	3.29	3.18	3.11	3.05	3.01	2.98	2.96	2.94	2.90	2.87	2.84	2.82	2.80	2.78	2.76	2.74	2.72
7	3.59	3.26	3.07	2.96	2.88	2.83	2.78	2.75	2.72	2.70	2.67	2.63	2.59	2.58	2.56	2.54	2.51	2.49	2.47
8	3.46	3.11	2.92	2.81	2.73	2.67	2.62	2.59	2.56	2.54	2.50	2.46	2.42	2.40	2.38	2.36	2.34	2.32	2.29
9	3.36	3.01	2.81	2.69	2.61	2.55	2.51	2.47	2.44	2.42	2.38	2.34	2.30	2.28	2.25	2.23	2.21	2.18	2.16
10	3.29	2.92	2.73	2.61	2.52	2.46	2.41	2.38	2.35	2.32	2.28	2.24	2.20	2.18	2.16	2.13	2.11	2.08	2.06
11	3.23	2.86	2.66	2.54	2.45	2.39	2.34	2.30	2.27	2.25	2.21	2.17	2.12	2.10	2.08	2.05	2.03	2.00	1.97
12	3.18	2.81	2.61	2.48	2.39	2.33	2.28	2.24	2.21	2.19	2.15	2.10	2.06	2.04	2.01	1.99	1.96	1.93	1.90
13	3.14	2.76	2.56	2.43	2.35	2.28	2.23	2.20	2.16	2.14	2.10	2.05	2.01	1.98	1.96	1.93	1.90	1.88	1.85
14	3.10	2.73	2.52	2.39	2.31	2.24	2.19	2.15	2.12	2.10	2.05	2.01	1.96	1.94	1.91	1.89	1.86	1.83	1.80
15	3.07	2.70	2.49	2.36	2.27	2.21	2.16	2.12	2.09	2.06	2.02	1.97	1.92	1.90	1.87	1.85	1.82	1.79	1.76
16	3.05	2.67	2.46	2.33	2.24	2.18	2.13	2.09	2.06	2.03	1.99	1.94	1.89	1.87	1.84	1.81	1.78	1.75	1.72
17	3.03	2.64	2.44	2.31	2.22	2.15	2.10	2.06	2.03	2.00	1.96	1.91	1.86	1.84	1.81	1.78	1.75	1.72	1.69
18	3.01	2.62	2.42	2.29	2.20	2.13	2.08	2.04	2.00	1.98	1.93	1.89	1.84	1.81	1.78	1.75	1.72	1.69	1.66
19	2.99	2.61	2.40	2.27	2.18	2.11	2.06	2.02	1.98	1.96	1.91	1.86	1.81	1.79	1.76	1.73	1.70	1.67	1.63
20	2.97	2.59	2.38	2.25	2.16	2.09	2.04	2.00	1.96	1.94	1.89	1.84	1.79	1.77	1.74	1.71	1.68	1.64	1.61
21	2.96	2.57	2.36	2.23	2.14	2.08	2.02	1.98	1.95	1.92	1.87	1.83	1.78	1.75	1.72	1.69	1.66	1.62	1.59
22	2.95	2.56	2.35	2.22	2.13	2.06	2.01	1.97	1.93	1.90	1.86	1.81	1.76	1.73	1.70	1.67	1.64	1.60	1.57
23	2.94	2.55	2.34	2.21	2.11	2.05	1.99	1.95	1.92	1.89	1.84	1.80	1.74	1.72	1.69	1.66	1.62	1.59	1.55
24	2.93	2.54	2.33	2.19	2.10	2.04	1.98	1.94	1.91	1.88	1.83	1.78	1.73	1.70	1.67	1.64	1.61	1.57	1.53
25	2.92	2.53	2.32	2.18	2.09	2.02	1.97	1.93	1.89	1.87	1.82	1.77	1.72	1.69	1.66	1.63	1.59	1.56	1.52
26	2.91	2.52	2.31	2.17	2.08	2.01	1.96	1.92	1.88	1.86	1.81	1.76	1.71	1.68	1.65	1.61	1.58	1.54	1.50
27	2.90	2.51	2.30	2.17	2.07	2.00	1.95	1.91	1.87	1.85	1.80	1.75	1.70	1.67	1.64	1.60	1.57	1.53	1.49
28	2.89	2.50	2.29	2.16	2.06	2.00	1.94	1.90	1.87	1.84	1.79	1.74	1.69	1.66	1.63	1.59	1.56	1.52	1.48
29	2.89	2.50	2.28	2.15	2.06	1.99	1.93	1.89	1.86	1.83	1.78	1.73	1.68	1.65	1.62	1.58	1.55	1.51	1.47
30	2.88	2.49	2.28	2.14	2.05	1.98	1.93	1.88	1.85	1.82	1.77	1.72	1.67	1.64	1.61	1.57	1.54	1.50	1.46
40	2.84	2.44	2.23	2.09	2.00	1.93	1.87	1.83	1.79	1.76	1.71	1.66	1.61	1.57	1.54	1.51	1.47	1.42	1.38
60	2.79	2.39	2.18	2.04	1.95	1.87	1.82	1.77	1.74	1.71	1.66	1.60	1.54	1.51	1.48	1.44	1.40	1.35	1.29
120	2.75	2.35	2.13	1.99	1.90	1.82	1.77	1.72	1.68	1.65	1.60	1.55	1.48	1.45	1.41	1.37	1.32	1.26	1.19
∞	2.71	2.30	2.08	1.94	1.85	1.77	1.72	1.67	1.63	1.60	1.55	1.49	1.42	1.38	1.34	1.30	1.24	1.17	1.00

TABLE C.4 CONTINUED

95th percentiles

n_2 \ n_1	1	2	3	4	5	6	7	8	9	10	12	15	20	24	30	40	60	120	∞
1	161.4	199.5	215.7	224.6	230.2	234.0	236.8	238.9	240.5	241.9	243.9	245.9	248.0	249.1	250.1	251.1	252.2	253.3	254.3
2	18.51	19.00	19.16	19.25	19.30	19.33	19.35	19.37	19.38	19.40	19.41	19.43	19.45	19.45	19.46	19.47	19.48	19.49	19.50
3	10.13	9.55	9.28	9.12	9.01	8.94	8.89	8.85	8.81	8.79	8.74	8.70	8.66	8.64	8.62	8.59	8.57	8.55	8.53
4	7.71	6.94	6.59	6.39	6.26	6.16	6.09	6.04	6.00	5.96	5.91	5.86	5.80	5.77	5.75	5.72	5.69	5.66	5.63
5	6.61	5.79	5.41	5.19	5.05	4.95	4.88	4.82	4.77	4.74	4.68	4.62	4.56	4.53	4.50	4.46	4.43	4.40	4.36
6	5.99	5.14	4.76	4.53	4.39	4.28	4.21	4.15	4.10	4.06	4.00	3.94	3.87	3.84	3.81	3.77	3.74	3.70	3.67
7	5.59	4.74	4.35	4.12	3.97	3.87	3.79	3.73	3.68	3.64	3.57	3.51	3.44	3.41	3.38	3.34	3.30	3.27	3.23
8	5.32	4.46	4.07	3.84	3.69	3.58	3.50	3.44	3.39	3.35	3.28	3.22	3.15	3.12	3.08	3.04	3.01	2.97	2.93
9	5.12	4.26	3.86	3.63	3.48	3.37	3.29	3.23	3.18	3.14	3.07	3.01	2.94	2.90	2.86	2.83	2.79	2.75	2.71
10	4.96	4.10	3.71	3.48	3.33	3.22	3.14	3.07	3.02	2.98	2.91	2.85	2.77	2.74	2.70	2.66	2.62	2.58	2.54
11	4.84	3.98	3.59	3.36	3.20	3.09	3.01	2.95	2.90	2.85	2.79	2.72	2.65	2.61	2.57	2.53	2.49	2.45	2.40
12	4.75	3.89	3.49	3.26	3.11	3.00	2.91	2.85	2.80	2.75	2.69	2.62	2.54	2.51	2.47	2.43	2.38	2.34	2.30
13	4.67	3.81	3.41	3.18	3.03	2.92	2.83	2.77	2.71	2.67	2.60	2.53	2.46	2.42	2.38	2.34	2.30	2.25	2.21
14	4.60	3.74	3.34	3.11	2.96	2.85	2.76	2.70	2.65	2.60	2.53	2.46	2.39	2.35	2.31	2.27	2.22	2.18	2.13
15	4.54	3.68	3.29	3.06	2.90	2.79	2.71	2.64	2.59	2.54	2.48	2.40	2.33	2.29	2.25	2.20	2.16	2.11	2.07
16	4.49	3.63	3.24	3.01	2.85	2.74	2.66	2.59	2.54	2.49	2.42	2.35	2.28	2.24	2.19	2.15	2.11	2.06	2.01
17	4.45	3.59	3.20	2.96	2.81	2.70	2.61	2.55	2.49	2.45	2.38	2.31	2.23	2.19	2.15	2.10	2.06	2.01	1.96
18	4.41	3.55	3.16	2.93	2.77	2.66	2.58	2.51	2.46	2.41	2.34	2.27	2.19	2.15	2.11	2.06	2.02	1.97	1.92
19	4.38	3.52	3.13	2.90	2.74	2.63	2.54	2.48	2.42	2.38	2.31	2.23	2.16	2.11	2.07	2.03	1.98	1.93	1.88
20	4.35	3.49	3.10	2.87	2.71	2.60	2.51	2.45	2.39	2.35	2.28	2.20	2.12	2.08	2.04	1.99	1.95	1.90	1.84
21	4.32	3.47	3.07	2.84	2.68	2.57	2.49	2.42	2.37	2.32	2.25	2.18	2.10	2.05	2.01	1.96	1.92	1.87	1.81
22	4.30	3.44	3.05	2.82	2.66	2.55	2.46	2.40	2.34	2.30	2.23	2.15	2.07	2.03	1.98	1.94	1.89	1.84	1.78
23	4.28	3.42	3.03	2.80	2.64	2.53	2.44	2.37	2.32	2.27	2.20	2.13	2.05	2.01	1.96	1.91	1.86	1.81	1.76
24	4.26	3.40	3.01	2.78	2.62	2.51	2.42	2.36	2.30	2.25	2.18	2.11	2.03	1.98	1.94	1.89	1.84	1.79	1.73
25	4.24	3.39	2.99	2.76	2.60	2.49	2.40	2.34	2.28	2.24	2.16	2.09	2.01	1.96	1.92	1.87	1.82	1.77	1.71
26	4.23	3.37	2.98	2.74	2.59	2.47	2.39	2.32	2.27	2.22	2.15	2.07	1.99	1.95	1.90	1.85	1.80	1.75	1.69
27	4.21	3.35	2.96	2.73	2.57	2.46	2.37	2.31	2.25	2.20	2.13	2.06	1.97	1.93	1.88	1.84	1.79	1.73	1.67
28	4.20	3.34	2.95	2.71	2.56	2.45	2.36	2.29	2.24	2.19	2.12	2.04	1.96	1.91	1.87	1.82	1.77	1.71	1.65
29	4.18	3.33	2.93	2.70	2.55	2.43	2.35	2.28	2.22	2.18	2.10	2.03	1.94	1.90	1.85	1.81	1.75	1.70	1.64
30	4.17	3.32	2.92	2.69	2.53	2.42	2.33	2.27	2.21	2.16	2.09	2.01	1.93	1.89	1.84	1.79	1.74	1.68	1.62
40	4.08	3.23	2.84	2.61	2.45	2.34	2.25	2.18	2.12	2.08	2.00	1.92	1.84	1.79	1.74	1.69	1.64	1.58	1.51
60	4.00	3.15	2.76	2.53	2.37	2.25	2.17	2.10	2.04	1.99	1.92	1.84	1.75	1.70	1.65	1.59	1.53	1.47	1.39
120	3.92	3.07	2.68	2.45	2.29	2.17	2.09	2.02	1.96	1.91	1.83	1.75	1.66	1.61	1.55	1.50	1.43	1.35	1.25
∞	3.84	3.00	2.60	2.37	2.21	2.10	2.01	1.94	1.88	1.83	1.75	1.67	1.57	1.52	1.46	1.39	1.32	1.22	1.00

TABLE C.4 CONTINUED

97.5th percentiles

n_2 \ n_1	1	2	3	4	5	6	7	8	9	10	12	15	20	24	30	40	60	120	∞
1	647.8	799.5	864.2	899.6	921.8	937.1	948.2	956.7	963.3	968.6	976.7	984.9	993.1	997.2	1001	1006	1010	1014	1018
2	38.51	39.00	39.17	39.25	39.30	39.33	39.36	39.37	39.39	39.40	39.41	39.43	39.45	39.46	39.46	39.47	39.48	39.49	39.50
3	17.44	16.04	15.44	15.10	14.88	14.73	14.62	14.54	14.47	14.42	14.34	14.25	14.17	14.12	14.08	14.04	13.99	13.95	13.90
4	12.22	10.65	9.98	9.60	9.36	9.20	9.07	8.98	8.90	8.84	8.75	8.66	8.56	8.51	8.46	8.41	8.36	8.31	8.26
5	10.01	8.43	7.76	7.39	7.15	6.98	6.85	6.76	6.68	6.62	6.52	6.43	6.33	6.28	6.23	6.18	6.12	607	6.02
6	8.81	7.26	6.60	6.23	5.99	5.82	5.70	5.60	5.52	5.46	5.37	5.27	5.17	5.12	5.07	5.01	4.96	4.90	4.85
7	8.07	6.54	5.89	5.52	5.29	5.12	4.99	4.90	4.82	4.76	4.67	4.57	4.47	4.42	4.36	4.31	4.25	4.20	4.14
8	7.57	6.06	5.42	5.05	4.82	4.65	4.53	4.43	4.36	4.30	4.20	4.10	4.00	3.95	3.89	3.84	3.78	3.73	3.67
9	7.21	5.71	5.08	4.72	4.48	4.32	4.20	4.10	4.03	3.96	3.87	3.77	3.67	3.61	3.56	3.51	3.45	3.39	3.33
10	6.94	5.46	4.83	4.47	4.24	4.07	3.95	3.85	3.78	3.72	3.62	3.52	3.42	3.37	3.31	3.26	3.20	3.14	3.08
11	6.72	5.26	4.63	4.28	4.04	3.88	3.76	3.66	3.59	3.53	3.43	3.33	3.23	3.17	3.12	3.06	3.00	2.94	2.88
12	6.55	5.10	4.47	4.12	3.89	3.73	3.61	3.51	3.44	3.37	3.28	3.18	3.07	3.02	2.96	2.91	2.85	2.79	2.72
13	6.41	4.97	4.35	4.00	3.77	3.60	3.48	3.39	3.31	3.25	3.15	3.05	2.95	2.89	2.84	2.78	2.72	2.66	2.60
14	6.30	4.86	4.24	3.89	3.66	3.50	3.38	3.29	3.21	3.15	3.05	2.95	2.84	2.79	2.73	2.67	2.61	2.55	2.49
15	6.20	4.77	4.15	3.80	3.58	3.41	3.29	3.20	3.12	3.06	2.96	2.86	2.76	2.70	2.64	2.59	2.52	2.46	2.40
16	6.12	4.69	4.08	3.73	3.50	3.34	3.22	3.12	3.05	2.99	2.89	2.79	2.68	2.63	2.57	2.51	2.45	2.38	2.32
17	6.04	4.62	4.01	3.66	3.44	3.28	3.16	3.06	2.98	2.92	2.82	2.72	2.62	2.56	2.50	2.44	2.38	2.32	2.25
18	5.98	4.56	3.95	3.61	3.38	3.22	3.10	3.01	2.93	2.87	2.77	2.67	2.56	2.50	2.44	2.38	2.32	2.26	2.19
19	5.92	4.51	3.90	3.56	3.33	3.17	3.05	2.96	2.88	2.82	2.72	2.62	2.51	2.45	2.39	2.33	2.27	2.20	2.13
20	5.87	4.46	3.86	3.51	3.29	3.13	3.01	2.91	2.84	2.77	2.68	2.57	2.46	2.41	2.35	2.29	2.22	2.16	2.09
21	5.83	4.42	3.82	3.48	3.25	3.09	2.97	2.87	2.80	2.73	2.64	2.53	2.42	2.37	2.31	2.25	2.18	2.11	2.04
22	5.79	4.38	3.78	3.44	3.22	3.05	2.93	2.84	2.76	2.70	2.60	2.50	2.39	2.33	2.27	2.21	2.14	2.08	2.00
23	5.75	4.35	3.75	3.41	3.18	3.02	2.90	2.81	2.73	2.67	2.57	2.47	2.36	2.30	2.24	2.18	2.11	2.04	1.97
24	5.72	4.32	3.72	3.38	3.15	2.99	2.87	2.78	2.70	2.64	2.54	2.44	2.33	2.27	2.21	2.15	2.08	2.01	1.94
25	5.69	4.29	3.69	3.35	3.13	2.97	2.85	2.75	2.68	2.61	2.51	2.41	2.30	2.24	2.18	2.12	2.05	1.98	1.91
26	5.66	4.27	3.67	3.33	3.10	2.94	2.82	2.73	2.65	2.59	2.49	2.39	2.28	2.22	2.16	2.09	2.03	1.95	1.88
27	5.63	4.24	3.65	3.31	3.08	2.92	2.80	2.71	2.63	2.57	2.47	2.36	2.25	2.19	2.13	2.07	2.00	1.93	1.85
28	5.61	4.22	3.63	3.29	3.06	2.90	2.78	2.69	2.61	2.55	2.45	2.34	2.23	2.17	2.11	2.05	1.98	1.91	1.83
29	5.59	4.20	3.61	3.27	3.04	2.88	2.76	2.67	2.59	2.53	2.43	2.32	2.21	2.15	2.09	2.03	1.96	1.89	1.81
30	5.57	4.18	3.59	3.25	3.03	2.87	2.75	2.65	2.57	2.51	2.41	2.31	2.20	2.14	2.07	2.01	1.94	1.87	1.79
40	5.42	4.05	3.46	3.13	2.90	2.74	2.62	2.53	2.45	2.39	2.29	2.18	2.07	2.01	1.94	1.88	1.80	1.72	1.64
60	5.29	3.93	3.34	3.01	2.79	2.63	2.51	2.41	2.33	2.27	2.17	2.06	1.94	1.88	1.82	1.74	1.67	1.58	1.48
120	5.15	3.80	3.23	2.89	2.67	2.52	2.39	2.30	2.22	2.16	2.05	1.94	1.82	1.76	1.69	1.61	1.53	1.43	1.31
∞	5.02	3.69	3.12	2.79	2.57	2.41	2.29	2.19	2.11	2.05	1.94	1.83	1.71	1.64	1.57	1.48	1.39	1.27	1.00

TABLE C.4 CONTINUED

99th percentiles

n_2 \ n_1	1	2	3	4	5	6	7	8	9	10	12	15	20	24	30	40	60	120	∞
1	4052	4999.5	5403	5625	5764	5859	5928	5982	6022	6056	6106	6157	6209	6235	6261	6287	6313	6339	6366
2	98.50	99.00	99.17	99.25	99.30	99.33	99.36	99.37	99.39	99.40	99.42	99.43	99.45	99.46	99.47	99.47	99.48	99.49	99.50
3	34.12	30.82	29.46	28.71	28.24	27.91	27.67	27.49	27.35	27.23	27.05	26.87	26.69	26.60	26.50	26.41	26.32	26.22	26.13
4	21.20	18.00	16.69	15.98	15.52	15.21	14.98	14.80	14.66	14.55	14.37	14.20	14.02	13.93	13.84	13.75	13.65	13.56	13.46
5	16.26	13.27	12.06	11.39	10.97	10.67	10.46	10.29	10.16	10.05	9.89	9.72	9.55	9.47	9.38	9.29	9.20	9.11	9.02
6	13.75	10.92	9.78	9.15	8.75	8.47	8.26	8.10	7.98	7.87	7.72	7.56	7.40	7.31	7.23	7.14	7.06	6.97	6.88
7	12.25	9.55	8.45	7.85	7.46	7.19	6.99	6.84	6.72	6.62	6.47	6.31	6.16	6.07	5.99	5.91	5.82	5.74	5.65
8	11.26	8.65	7.59	7.01	6.63	6.37	6.18	6.03	5.91	5.81	5.67	5.52	5.36	5.28	5.20	5.12	5.03	4.95	4.86
9	10.56	8.02	6.99	6.42	6.06	5.80	5.61	5.47	5.35	5.26	5.11	4.96	4.81	4.73	4.65	4.57	4.48	4.40	4.31
10	10.04	7.56	6.55	5.99	5.64	5.39	5.20	5.06	4.94	4.85	4.71	4.56	4.41	4.33	4.25	4.17	4.08	4.00	3.91
11	9.65	7.21	6.22	5.67	5.32	5.07	4.89	4.74	4.63	4.54	4.40	4.25	4.10	4.02	3.94	3.86	3.78	3.69	3.60
12	9.33	6.93	5.95	5.41	5.06	4.82	4.64	4.50	4.39	4.30	4.16	4.01	3.86	3.78	3.70	3.62	3.54	3.45	3.36
13	9.07	6.70	5.74	5.21	4.86	4.62	4.44	4.30	4.19	4.10	3.96	3.82	3.66	3.59	3.51	3.43	3.34	3.25	3.17
14	8.86	6.51	5.56	5.04	4.69	4.46	4.28	4.14	4.03	3.94	3.80	3.66	3.51	3.43	3.35	3.27	3.18	3.09	3.00
15	8.68	6.36	5.42	4.89	4.56	4.32	4.14	4.00	3.89	3.80	3.67	3.52	3.37	3.29	3.21	3.13	3.05	2.96	2.87
16	8.53	6.23	5.29	4.77	4.44	4.20	4.03	3.89	3.78	3.69	3.55	3.41	3.26	3.18	3.10	3.02	2.93	2.84	2.75
17	8.40	6.11	5.18	4.67	4.34	4.10	3.93	3.79	3.68	3.59	3.46	3.31	3.16	3.08	3.00	2.92	2.83	2.75	2.65
18	8.29	6.01	5.09	4.58	4.25	4.01	3.84	3.71	3.60	3.51	3.37	3.23	3.08	3.00	2.92	2.84	2.75	2.66	2.57
19	8.18	5.93	5.01	4.50	4.17	3.94	3.77	3.63	3.52	3.43	3.30	3.15	3.00	2.92	2.84	2.76	2.67	2.58	2.49
20	8.10	5.85	4.94	4.43	4.10	3.87	3.70	3.56	3.46	3.37	3.23	3.09	2.94	2.86	2.78	2.69	2.61	2.52	2.42
21	8.02	5.78	4.87	4.37	4.04	3.81	3.64	3.51	3.40	3.31	3.17	3.03	2.88	2.80	2.72	2.64	2.55	2.46	2.36
22	7.95	5.72	4.82	4.31	3.99	3.76	3.59	3.45	3.35	3.26	3.12	2.98	2.83	2.75	2.67	2.58	2.50	2.40	2.31
23	7.88	5.66	4.76	4.26	3.94	3.71	3.54	3.41	3.30	3.21	3.07	2.93	2.78	2.70	2.62	2.54	2.45	2.35	2.26
24	7.82	5.61	4.72	4.22	3.90	3.67	3.50	3.36	3.26	3.17	3.03	2.89	2.74	2.66	2.58	2.49	2.40	2.31	2.21
25	7.77	5.57	4.68	4.18	3.85	3.63	3.46	3.32	3.22	3.13	2.99	2.85	2.70	2.62	2.54	2.45	2.36	2.27	2.17
26	7.72	5.53	4.64	4.14	3.82	3.59	3.42	3.29	3.18	3.09	2.96	2.81	2.66	2.58	2.50	2.42	2.33	2.23	2.13
27	7.68	5.49	4.60	4.11	3.78	3.56	3.39	3.26	3.15	3.06	2.93	2.78	2.63	2.55	2.47	2.38	2.29	2.20	2.10
28	7.64	5.45	4.57	4.07	3.75	3.53	3.36	3.23	3.12	3.03	2.90	2.75	2.60	2.52	2.44	2.35	2.26	2.17	2.06
29	7.60	5.42	4.54	4.04	3.73	3.50	3.33	3.20	3.09	3.00	2.87	2.73	2.57	2.49	2.41	2.33	2.23	2.14	2.03
30	7.56	5.39	4.51	4.02	3.70	3.47	3.30	3.17	3.07	2.98	2.84	2.70	2.55	2.47	2.39	2.30	2.21	2.11	2.01
40	7.31	5.18	4.31	3.83	3.51	3.29	3.12	2.99	2.89	2.80	2.66	2.52	2.37	2.29	2.20	2.11	2.02	1.92	1.80
60	7.08	4.98	4.13	3.65	3.34	3.12	2.95	2.82	2.72	2.63	2.50	2.35	2.20	2.12	2.03	1.94	1.84	1.73	1.60
120	1.85	4.79	3.95	3.48	3.17	2.96	2.79	2.66	2.56	2.47	2.34	2.19	2.03	1.95	1.86	1.76	1.66	1.53	1.38
∞	6.63	4.61	3.78	3.32	3.02	2.80	2.64	2.51	2.41	2.32	2.18	2.04	1.88	1.79	1.70	1.59	1.47	1.32	1.00

TABLE C.4 CONTINUED

99.5th percentiles

n_2 \ n_1	1	2	3	4	5	6	7	8	9	10	12	15	20	24	30	40	60	120	∞
1	16211.	20000.	21615.	22500.	23056.	23437.	23715.	23925.	24091.	24224.	24426.	24630.	24836.	24940.	25044.	25148.	25253.	25359.	25465.
2	198.5	199.0	199.2	199.2	199.3	199.3	199.4	199.4	199.4	199.4	199.4	199.4	199.4	199.5	199.5	199.5	199.5	199.5	199.5
3	55.55	49.80	47.47	46.19	45.39	44.84	44.43	44.13	43.88	43.69	43.39	43.08	42.78	42.62	42.47	42.31	42.15	41.99	41.83
4	31.33	26.28	24.26	23.15	22.46	21.97	21.62	21.35	21.14	20.97	20.70	20.44	20.17	20.03	19.89	19.75	19.61	19.47	19.32
5	22.78	18.31	16.53	15.56	14.94	14.51	14.20	13.96	13.77	13.62	13.38	13.15	12.90	12.78	12.66	12.53	12.40	12.27	12.14
6	18.63	14.54	12.92	12.03	11.46	11.07	10.79	10.57	10.39	10.25	10.03	9.81	9.59	9.47	9.36	9.24	9.12	9.00	8.88
7	16.24	12.40	10.88	10.05	9.52	9.16	8.89	8.68	8.51	8.38	8.18	7.97	7.75	7.65	7.53	7.42	7.31	7.19	7.08
8	14.69	11.04	9.60	8.81	8.30	7.95	7.69	7.50	7.34	7.21	7.01	6.81	6.61	6.50	6.40	6.29	6.18	6.06	5.95
9	13.61	10.11	8.72	7.96	7.47	7.13	6.88	6.69	6.54	6.42	6.23	6.03	5.83	5.73	5.62	5.52	5.41	5.30	
10	12.83	9.43	8.08	7.34	6.87	6.54	6.30	6.12	5.97	5.85	5.66	5.47	5.27	5.17	5.07	4.97	4.86	4.75	4.64
11	12.23	8.91	7.60	6.88	6.42	6.10	5.86	5.68	5.54	5.42	5.24	5.05	4.86	4.76	4.65	4.55	4.44	4.34	4.23
12	11.75	8.51	7.23	6.52	6.07	5.76	5.52	5.35	5.20	5.09	4.91	4.72	4.53	4.43	4.33	4.23	4.12	4.01	3.90
13	11.37	8.19	6.93	6.23	5.79	5.48	5.25	5.08	4.94	4.82	4.64	4.46	4.27	4.17	4.07	3.97	3.87	3.76	3.65
14	11.06	7.92	6.68	6.00	5.56	5.26	5.03	4.86	4.72	4.60	4.43	4.25	4.06	3.96	3.86	3.76	3.66	3.55	3.44
15	10.80	7.70	6.48	5.80	5.37	5.07	4.85	4.67	4.54	4.42	4.25	4.07	3.88	3.79	3.69	3.58	3.48	3.37	3.26
16	10.58	7.51	6.30	5.64	5.21	4.91	4.69	4.52	4.38	4.27	4.10	3.92	3.73	3.64	3.54	3.44	3.33	3.22	3.11
17	10.38	7.35	6.16	5.50	5.07	4.78	4.56	4.39	4.25	4.14	3.97	3.79	3.61	3.51	3.41	3.31	3.21	3.10	2.98
18	10.22	7.21	6.03	5.37	4.96	4.66	4.44	4.28	4.14	4.03	3.86	3.68	3.50	3.40	3.30	3.20	3.10	2.99	2.87
19	10.07	7.09	5.92	5.27	4.85	4.56	4.34	4.18	4.04	3.93	3.76	3.59	3.40	3.31	3.21	3.11	3.00	2.89	2.78
20	9.94	6.99	5.82	5.17	4.76	4.47	4.26	4.09	3.96	3.85	3.68	3.50	3.32	3.22	3.12	3.02	2.92	2.81	2.69
21	9.83	6.89	5.73	5.09	4.68	4.39	4.18	4.01	3.88	3.77	3.60	3.43	3.24	3.15	3.05	2.95	2.84	2.73	2.61
22	9.73	6.81	5.65	5.02	4.61	4.32	4.11	3.94	3.81	3.70	3.54	3.36	3.18	3.08	2.98	2.88	2.77	2.66	2.55
23	9.63	6.73	5.58	4.95	4.54	4.26	4.05	3.88	3.75	3.64	3.47	3.30	3.12	3.02	2.92	2.82	2.71	2.60	2.48
24	9.55	6.66	5.52	4.89	4.49	4.20	3.99	3.83	3.69	3.59	3.42	3.25	3.06	2.97	2.87	2.77	2.66	2.55	2.43
25	9.48	6.60	5.46	4.84	4.43	4.15	3.94	3.78	3.64	3.54	3.37	3.20	3.01	2.92	2.82	2.72	2.61	2.50	2.38
26	9.41	6.54	5.41	4.79	4.38	4.10	3.89	3.73	3.60	3.49	3.33	3.15	2.97	2.87	2.77	2.67	2.56	2.45	2.33
27	9.34	6.49	5.36	4.74	4.34	4.06	3.85	3.69	3.56	3.45	3.28	3.11	2.93	2.83	2.73	2.63	2.52	2.41	2.29
28	9.28	6.44	5.32	4.70	4.30	4.02	3.81	3.65	3.52	3.41	3.25	3.07	2.89	2.79	2.69	2.59	2.48	2.37	2.25
29	9.23	6.40	5.28	4.66	4.26	3.98	3.77	3.61	3.48	3.38	3.21	3.04	2.86	2.76	2.66	2.56	2.45	2.33	2.21
30	9.18	6.35	5.24	4.62	4.23	3.95	3.74	3.58	3.45	3.34	3.18	3.01	2.82	2.73	2.63	2.52	2.42	2.30	2.18
40	8.83	6.07	4.98	4.37	3.99	3.71	3.51	3.35	3.22	3.12	2.95	2.78	2.60	2.50	2.40	2.30	2.18	2.06	1.93
60	8.49	5.79	4.73	4.14	3.76	3.49	3.29	3.13	3.01	2.90	2.74	2.57	2.39	2.29	2.19	2.08	1.96	1.83	1.69
120	8.18	5.54	4.50	3.92	3.55	3.28	3.09	2.93	2.81	2.71	2.54	2.37	2.19	2.09	1.98	1.87	1.75	1.61	1.43
∞	7.88	5.30	4.28	3.72	3.35	3.09	2.90	2.74	2.62	2.52	2.36	2.19	2.00	1.90	1.79	1.67	1.53	1.36	1.00

TABLE C.4 CONTINUED

99.9th percentiles

n_2 \ n_1	1	2	3	4	5	6	7	8	9	10	12	15	20	24	30	40	60	120	∞
1	4053*	5000*	5404*	5625*	5764*	5859*	5929*	5981*	6023*	6056*	6107*	6158*	6209*	6235*	6261*	6287*	6313*	6340*	6366*
2	998.5	999.0	999.2	999.2	999.3	999.3	999.4	999.4	999.4	999.4	999.4	999.4	999.4	999.5	999.5	999.5	999.5	999.5	999.5
3	167.0	148.5	141.1	137.1	134.6	132.8	131.6	130.6	129.9	129.2	128.3	127.4	126.4	125.9	125.4	125.0	124.5	124.0	123.5
4	74.14	61.25	56.18	53.44	51.71	50.53	49.66	49.00	48.47	48.05	47.41	46.76	46.10	45.77	45.43	45.09	44.75	44.40	44.05
5	47.18	37.12	33.20	31.09	29.75	28.84	28.16	27.64	27.24	26.92	26.42	25.91	25.39	25.14	24.87	24.60	24.33	24.06	23.79
6	35.51	27.00	23.70	21.92	20.81	20.03	19.46	19.03	18.69	18.41	17.99	17.56	17.12	16.89	16.67	16.44	16.21	15.99	15.75
7	29.25	21.69	18.77	17.19	16.21	15.52	15.02	14.63	14.33	14.08	13.71	13.32	12.93	12.73	12.53	12.33	12.12	11.91	11.70
8	25.42	18.49	15.83	14.39	13.49	12.86	12.40	12.04	11.77	11.54	11.19	10.84	10.48	10.30	10.11	9.92	9.73	9.53	9.33
9	22.86	16.39	13.90	12.56	11.71	11.13	10.70	10.37	10.11	9.89	9.57	9.24	8.90	8.72	8.55	8.37	8.19	8.00	7.81
10	21.04	14.91	12.55	11.28	10.48	9.92	9.52	9.20	8.96	8.75	8.45	8.13	7.80	7.64	7.47	7.30	7.12	6.94	6.76
11	19.69	13.81	11.56	10.35	9.58	9.05	8.66	8.35	8.12	7.92	7.63	7.32	7.01	6.85	6.68	6.52	6.35	6.17	6.00
12	18.64	12.97	10.80	9.63	8.89	8.38	8.00	7.71	7.48	7.29	7.00	6.71	6.40	6.25	6.09	5.93	5.76	5.59	5.42
13	17.81	12.31	10.21	9.07	8.35	7.86	7.49	7.21	6.98	6.80	6.52	6.23	5.93	5.78	5.63	5.47	5.30	5.14	4.97
14	17.14	11.78	9.73	8.62	7.92	7.43	7.08	6.80	6.58	6.40	6.13	5.85	5.56	5.41	5.25	5.10	4.94	4.77	4.60
15	16.59	11.34	9.34	8.25	7.57	7.09	6.74	6.47	6.26	6.08	5.81	5.54	5.25	5.10	4.95	4.80	4.64	4.47	4.31
16	16.12	10.97	9.00	7.94	7.27	6.81	6.46	6.19	5.98	5.81	5.55	5.27	4.99	4.85	4.70	4.54	4.39	4.23	4.06
17	15.72	10.66	8.73	7.68	7.02	9.56	6.22	5.96	5.75	5.58	5.32	5.05	4.78	4.63	4.48	4.33	4.18	4.02	3.85
18	15.38	10.39	8.49	7.46	6.81	6.35	6.02	5.76	5.56	5.39	5.13	4.87	4.59	4.45	4.30	4.15	4.00	3.84	3.67
19	15.08	10.16	8.28	7.26	6.62	6.18	5.85	5.59	5.39	5.22	4.97	4.70	4.43	4.29	1.14	3.99	3.84	3.68	3.51
20	14.82	9.95	8.10	7.10	6.46	6.02	5.69	5.44	5.24	5.08	4.82	4.56	4.29	4.15	4.00	3.86	3.70	3.54	3.38
21	14.59	9.77	7.94	6.95	6.32	5.88	5.56	5.31	5.11	4.95	4.70	4.44	4.17	4.03	3.88	3.74	3.58	3.42	3.26
22	14.38	9.61	7.80	6.81	6.19	5.76	5.44	5.19	4.99	4.83	4.58	4.33	4.06	3.92	3.78	3.63	3.48	3.32	3.15
23	14.19	9.47	7.67	6.69	6.08	5.65	5.33	5.09	4.89	4.73	4.48	4.23	3.96	3.82	3.68	3.53	3.38	3.22	3.05
24	14.03	9.34	7.55	6.59	5.98	5.55	5.23	4.99	4.80	4.64	4.39	4.14	3.87	3.74	3.59	3.45	3.29	3.14	2.97
25	13.88	9.22	7.45	6.49	5.88	5.46	5.15	4.91	4.71	4.56	4.31	4.06	3.79	3.66	3.52	3.37	3.22	3.06	2.89
26	13.74	9.12	7.36	6.41	5.80	5.38	5.07	4.83	4.64	4.48	4.24	3.99	3.72	3.59	3.44	3.30	3.15	2.99	2.82
27	13.61	9.02	7.27	6.33	5.73	5.31	5.00	4.76	4.57	4.41	4.17	3.92	3.66	3.52	3.38	3.23	3.08	2.92	2.75
28	13.50	8.93	7.19	6.25	5.66	5.24	4.93	4.69	4.50	4.35	4.11	3.86	3.60	3.46	3.32	3.18	3.02	2.86	2.69
29	13.39	8.85	7.12	6.19	5.59	5.18	4.87	4.64	4.45	4.29	4.05	3.80	3.54	3.41	3.27	3.12	2.97	2.81	2.64
30	13.29	8.77	7.05	6.12	5.53	5.12	4.82	4.58	4.39	4.24	4.00	3.75	3.49	3.36	3.22	3.07	2.92	2.76	2.59
40	12.61	8.25	6.60	5.70	5.13	4.73	4.44	4.21	4.02	3.87	3.64	3.40	3.15	3.01	2.87	2.73	2.57	2.41	2.23
60	11.97	7.76	6.17	5.31	4.76	4.37	4.09	3.87	3.69	3.54	3.31	3.08	2.83	2.69	2.55	2.41	2.25	2.08	1.89
120	11.38	7.32	5.79	4.95	4.42	4.04	3.77	3.55	3.38	3.24	3.02	2.78	2.53	2.40	2.26	2.11	1.95	1.76	1.54
∞	10.83	6.91	5.42	4.62	4.10	3.74	3.47	3.27	3.10	2.96	2.74	2.51	2.27	2.13	1.99	1.84	1.66	1.45	1.00

*Multiply these entries by 100.

TABLE C.5 CRITICAL VALUES OF THE SPEARMAN RANK CORRELATION COEFFICIENT, r', FOR TWO-TAILED AND ONE-TAILED PROBABILITIES, $\alpha(2)$ AND $\alpha(1)$, RESPECTIVELY

$\alpha(2)$	0.50	0.20	0.10	0.05	0.02	0.01	0.005	0.002	0.001
$\alpha(1)$	0.25	0.10	0.05	0.025	0.01	0.005	0.0025	0.001	0.005
n									
4	0.600	1.000	1.000						
5	0.500	0.800	0.900	1.000	1.000				
6	0.371	0.657	0.829	0.886	0.943	1.000	1.000		
7	0.321	0.571	0.714	0.786	0.893	0.929	0.964	1.000	1.000
8	0.310	0.524	0.643	0.738	0.833	0.881	0.905	0.952	0.976
9	0.267	0.483	0.600	0.700	0.783	0.833	0.867	0.917	0.933
10	0.248	0.455	0.564	0.648	0.745	0.794	0.830	0.879	0.903
11	0.236	0.427	0.536	0.618	0.709	0.755	0.800	0.845	0.873
12	0.224	0.406	0.503	0.587	0.671	0.727	0.776	0.825	0.860
13	0.209	0.385	0.484	0.560	0.648	0.703	0.747	0.802	0.835
14	0.200	0.367	0.464	0.538	0.622	0.675	0.723	0.776	0.811
15	0.189	0.354	0.443	0.521	0.604	0.654	0.700	0.754	0.786
16	0.182	0.341	0.429	0.503	0.582	0.635	0.679	0.732	0.765
17	0.176	0.328	0.414	0.485	0.566	0.615	0.662	0.713	0.748
18	0.170	0.317	0.401	0.472	0.550	0.600	0.643	0.695	0.728
19	0.165	0.309	0.391	0.460	0.535	0.584	0.628	0.677	0.712
20	0.161	0.299	0.380	0.447	0.520	0.570	0.612	0.662	0.696
21	0.156	0.292	0.370	0.435	0.508	0.556	0.599	0.648	0.681
22	0.152	0.284	0.361	0.425	0.496	0.544	0.586	0.634	0.667
23	0.148	0.278	0.353	0.415	0.486	0.532	0.573	0.622	0.654
24	0.144	0.271	0.344	0.106	0.476	0.521	0.562	0.610	0.642
25	0.142	0.265	0.337	0.398	0.466	0.511	0.551	0.598	0.630
26	0.138	0.259	0.331	0.390	0.457	0.501	0.541	0.587	0.619
27	0.136	0.255	0.324	0.382	0.448	0.491	0.531	0.577	0.608
28	0.133	0.250	0.317	0.375	0.440	0.483	0.522	0.567	0.598
29	0.130	0.245	0.312	0.368	0.433	0.475	0.513	0.458	0.589
30	0.128	0.240	0.306	0.362	0.425	0.467	0.504	0.549	0.580
31	0.126	0.236	0.301	0.356	0.418	0.459	0.496	0.541	0.571
32	0.124	0.232	0.296	0.350	0.412	0.452	0.489	0.533	0.563
33	0.121	0.229	0.291	0.345	0.405	0.446	0.482	0.525	0.554
34	0.120	0.225	0.287	0.340	0.399	0.439	0.475	0.517	0.547
35	0.118	0.222	0.283	0.335	0.394	0.433	0.468	0.510	0.539
36	0.116	0.219	0.279	0.330	0.388	0.427	0.462	0.504	0.533
37	0.114	0.216	0.275	0.325	0.383	0.421	0.456	0.497	0.526
38	0.113	0.212	0.271	0.321	0.378	0.415	0.450	0.491	0.519
39	0.111	0.210	0.267	0.317	0.373	0.410	0.444	0.485	0.513
40	0.110	0.207	0.264	0.313	0.368	0.405	0.439	0.479	0.507
41	0.108	0.204	0.261	0.309	0.364	0.400	0.433	0.473	0.501
42	0.107	0.202	0.257	0.305	0.359	0.395	0.428	0.468	0.495
43	0.105	0.199	0.254	0.301	0.355	0.391	0.423	0.463	0.490
44	0.104	0.197	0.251	0.298	0.351	0.386	0.419	0.458	0.484
45	0.103	0.194	0.248	0.294	0.347	0.382	0.414	0.453	0.479
46	0.102	0.192	0.246	0.291	0.343	0.378	0.410	0.448	0.474
47	0.101	0.190	0.243	0.288	0.340	0.374	0.405	0.443	0.469
48	0.100	0.188	0.240	0.285	0.336	0.370	0.401	0.439	0.465
49	0.098	0.186	0.238	0.282	0.333	0.366	0.397	0.434	0.460
50	0.097	0.184	0.235	0.279	0.329	0.363	0.393	0.430	0.456

α(2)	0.50	0.20	0.10	0.05	0.02	0.01	0.005	0.002	0.001
α(1)	0.25	0.10	0.05	0.025	0.01	0.005	0.0025	0.001	0.005
n									
52	0.095	0.180	0.231	0.274	0.323	0.356	0.386	0.422	0.447
54	0.094	0.177	0.226	0.268	0.317	0.349	0.379	0.414	0.439
56	0.092	0.174	0.222	0.264	0.311	0.343	0.372	0.407	0.432
58	0.090	0.171	0.218	0.259	0.306	0.337	0.366	0.400	0.424
60	0.089	0.168	0.214	0.255	0.300	0.331	0.360	0.394	0.418
62	0.087	0.165	0.211	0.250	0.296	0.326	0.354	0.388	0.411
64	0.086	0.162	0.207	0.246	0.291	0.321	0.348	0.382	0.405
66	0.084	0.160	0.204	0.243	0.287	0.316	0.343	0.376	0.399
68	0.083	0.157	0.201	0.239	0.282	0.311	0.338	0.370	0.393
70	0.082	0.155	0.190	0.235	0.278	0.307	0.333	0.365	0.388
72	0.081	0.153	0.195	0.232	0.274	0.303	0.329	0.360	0.382
74	0.080	0.151	0.193	0.229	0.271	0.299	0.324	0.355	0.377
76	0.078	0.149	0.190	0.226	0.267	0.295	0.320	0.351	0.372
78	0.077	0.147	0.188	0.223	0.264	0.291	0.316	0.346	0.368
80	0.076	0.145	0.185	0.220	0.260	0.287	0.312	0.342	0.363
82	0.075	0.143	0.183	0.217	0.257	0.284	0.308	0.338	0.359
84	0.074	0.141	0.181	0.215	0.254	0.280	0.305	0.334	0.355
86	0.074	0.139	0.179	0.212	0.251	0.277	0.301	0.330	0.351
88	0.073	0.138	0.176	0.210	0.248	0.274	0.298	0.327	0.347
90	0.072	0.136	0.174	0.207	0.245	0.271	0.294	0.323	0.343
92	0.071	0.135	0.173	0.205	0.243	0.268	0.291	0.319	0.339
94	0.070	0.133	0.171	0.203	0.240	0.265	0.288	0.316	0.336
96	0.070	0.132	0.169	0.201	0.238	0.262	0.285	0.313	0.332
98	0.069	0.130	0.167	0.199	0.235	0.260	0.282	0.310	0.329
100	0.068	0.129	0.165	0.197	0.233	0.257	0.279	0.307	0.326

Source: Zar, J. H., Significance testing of the Spearman rank correlation coefficient. *Journal of the American Statistical Association*, Sept. 1972, **67**, 578.

TABLE C.6 CRITICAL VALUES OF S FOR TESTING THE NULL HYPOTHESIS THAT KENDALL'S $\tau = 0$.*

One-Sided	*Two-Sided*	*Values of n* 4	5	6	7	8	9	10
$\alpha = .05$	$\alpha = .10$	6	8	11	13	16	18	21
.025	.05		10	13	15	18	20	23
.01	.02		10	13	17	20	24	27
.005	.01			15	19	22	26	29
.001	.002				21	24	30	35
.0005	.001				21	26	30	35
.0001	.0002					28	34	39
.00005	.0001					28	34	41

*Reject the null hypothesis if the absolute value of S is greater than or equal to the tabled value.

Source: adapted from Siegel, Sidney, *Nonparametric Statistics for the Behavioral Sciences*, New York, McGraw-Hill, 1956, Appendix Table Q; and Kendall, M. G., *Rank Correlation Methods*, Chas. Griffin and Co., 1948, Appendix Table 1.

TABLE C.7 CRITICAL VALUES OF s IN THE KENDALL COEFFICIENT OF CONCORDANCE

m	*n* 3†	4	5	6	7	*Additional values for n = 3* *m*	*s*
			Values at the .01 level of significance				
3			64.4	103.9	157.3	9	54.0
4		49.5	88.4	143.3	217.0	12	71.9
5		62.6	112.3	182.4	276.2	14	83.8
6		75.7	136.1	221.4	335.2	16	95.8
8	48.1	101.7	183.7	299.0	453.1	18	107.7
10	60.0	127.8	231.2	376.7	571.0		
15	89.8	192.9	349.8	570.5	864.9		
20	119.7	258.0	468.5	764.4	1,158.7		
			Values at the .01 level of significance				
3			75.6	122.8	185.6	9	75.9
4		61.4	109.3	176.2	265.0	12	103.5
5		80.5	142.8	229.4	343.8	14	121.9
6		99.5	176.1	282.4	422.6	16	140.2
8	66.8	137.4	242.7	388.3	579.9	18	158.6
10	85.1	175.3	309.1	494.0	737.0		
15	131.0	269.8	475.2	758.2	1,129.5		
20	177.0	364.2	641.2	1,022.2	1,521.9		

†Notice that additional critical values of s for $n = 3$ are given in the right-hand column of this table.

Source: adapted from Friedman, M., A comparison of alternative tests of significance for the problem of m rankings. *Annals of Mathematical Statistics*, 1940, **11**, 86–92.

TABLE C.8A TABLES FOR RAPID COMPUTATION OF THE TETRACHORIC CORRELATION (UNCORRECTED r_{tet} FROM *ab/cd*)

Uncorrected *r* from *ab/cd*																			
ad/bc	*r*	*ad/bc*	*r*	*ad/bc*	*r*	*ad/bc*	*r*	*ad/bc*	*r*	*ad/bc*	*r*	*ad/bc*	*r*	*ad/bc*	*r*	*ad/bc*	*r*	*ad/bc*	*r*
1.00	.000	1.50	.158	2.00	.267	2.50	.344	3.00	.408	4.25	.517	5.50	.590	8.00	.683	12.0	.762	24.5	.861
1.02	.007	1.52	.163	2.02	.270	2.52	.346	3.05	.414	4.30	.520	5.60	.595	8.10	.686	12.5	.769	25.0	.864
1.04	.014	1.54	.168	2.04	.274	2.54	.349	3.10	.419	4.35	.523	5.70	.600	8.20	.688	13.0	.776	26.0	.868
1.06	.021	1.56	.172	2.06	.278	2.56	.352	3.15	.425	4.40	.524	5.80	.604	8.30	.691	13.5	.782	27.0	.872
1.08	.028	1.58	.177	2.08	.281	2.58	.355	3.20	.430	4.45	.530	5.90	.609	8.40	.693	14.0	.788	28.0	.876
1.10	.035	1.60	.182	2.10	.285	2.60	.358	3.25	.435	4.50	.534	6.00	.613	8.50	.695	14.5	.794	29.4	.880
1.12	.042	1.62	.187	2.12	.288	2.62	.360	3.30	.440	4.55	.537	6.10	.617	8.60	.698	15.0	.799	30.0	.883
1.14	.049	1.64	.192	2.14	.291	2.64	.363	3.35	.444	4.60	.540	6.20	.621	8.70	.700	15.5	.804	31.0	.886
1.16	.056	1.66	.196	2.16	.294	2.66	.366	3.40	.449	4.65	.543	6.30	.625	8.80	.702	16.0	.808	32.0	.890
1.18	.063	1.68	.201	2.18	.298	2.68	.368	3.45	.454	4.70	.547	6.40	.629	8.90	.705	16.5	.812	33.0	.893
1.20	.070	1.70	.206	2.20	.301	2.70	.371	3.50	.458	4.75	.550	6.50	.633	9.00	.707	17.0	.817	34.0	.896
1.22	.077	1.72	.210	2.22	.304	2.72	.374	3.55	.463	4.80	.553	6.60	.637	9.20	.712	17.5	.821	35.0	.898
1.24	.084	1.74	.215	2.24	.307	2.74	.376	3.60	.467	4.85	.556	6.70	.641	9.40	.717	18.0	.825	36.0	.901
1.26	.090	1.76	.219	2.26	.310	2.76	.379	3.65	.471	4.90	.559	6.80	.645	9.60	.721	18.5	.829	37.0	.904
1.28	.096	1.78	.224	2.28	.312	2.78	.382	3.70	.475	4.95	.502	6.90	.648	9.80	.725	19.0	.832	38.0	.906
1.30	.102	1.80	.228	2.30	.315	2.80	.384	3.75	.479	5.00	.564	7.00	.651	10.00	.730	19.5	.835	39.0	.909
1.32	.108	1.82	.232	2.32	.318	2.82	.386	3.80	.483	5.05	.567	7.10	.655	10.20	.734	20.0	.838	40.0	.911
1.34	.114	1.84	.235	2.34	.321	2.84	.389	3.85	.487	5.10	.570	7.20	.658	10.40	.737	20.5	.840	42.0	.915
1.36	.121	1.86	.240	2.36	.324	2.86	.391	3.90	.492	5.15	.572	7.30	.661	10.60	.740	21.0	.843	44.0	.919
1.38	.127	1.88	.243	2.38	.327	2.88	.394	3.95	.496	5.20	.575	7.40	.664	10.80	.743	21.5	.846	46.0	.922
1.40	.133	1.90	.247	2.40	.330	2.90	.396	4.00	.500	5.25	.577	7.50	.668	11.00	.746	22.0	.849	48.0	.926
1.42	.139	1.92	.251	2.42	.333	2.92	.399	4.05	.503	5.30	.580	7.60	.671	11.20	.750	22.5	.852	50.0	.929
1.44	.144	1.94	.255	2.44	.335	2.94	.401	4.10	.507	5.35	.583	7.70	.674	11.40	.753	23.0	.854	55.0	.935
1.46	.149	1.96	.259	2.46	.338	2.98	.404	4.15	.510	5.40	.585	7.80	.677	11.60	.756	23.5	.856	60.0	.940
1.48	.153	1.98	.263	2.48	.341	2.98	.406	4.20	.514	5.45	.587	7.90	.680	11.80	.759	24.0	.859	70.0	.945

Source: Jenkins, W. L., An improved method for tetrachoric *r*. *Psychometrika*, September 1955, *20*, 254.

TABLE C.8B. TABLES FOR RAPID COMPUTATION OF THE TETRACHORIC CORRELATION (BASE CORRECTION)

Larger Split	*Base Correction*																							
	Smaller Split																							
n	10	11	32	13	14	15	16	17	18	19	20	22	24	26	28	30	32	34	36	38	40	42	44	46
80	225	217	210	204	197	190	184	178	173	168	163													
78	217	209	204	195	190	184	177	171	166	160	156	147												
76	208	201	195	189	182	176	169	163	158	152	148	139	132											
74	201	194	137	180	174	168	182	156	150	143	140	131	124	116										
72	195	187	180	173	166	160	154	148	142	137	132	123	116	109	101									
70	185	180	174	166	160	154	143	142	136	131	126	117	108	100	092	084								
68	182	174	165	160	154	148	142	136	130	124	120	109	100	091	084	076	070							
66	175	168	162	155	148	142	136	130	124	118	113	103	093	084	076	069	062	056						
64	169	162	155	148	141	134	128	122	116	111	105	096	086	077	070	063	056	050	045					
62	163	155	148	141	134	128	122	110	110	105	100	089	080	072	064	057	050	045	040	035				
60	157	150	142	135	129	122	115	110	104	099	094	083	074	066	058	052	045	040	035	030	025			
58	151	144	136	129	122	116	110	104	098	093	087	077	068	060	053	045	040	035	030	025	020	016		
56	145	137	130	122	116	110	104	098	092	086	081	072	063	055	048	041	036	030	025	020	016	013	010	
54	139	131	133	115	110	104	097	091	086	081	076	067	058	050	044	037	032	026	021	016	012	008	005	002
52	134	126	119	111	105	098	092	087	081	076	072	062	054	045	040	033	027	022	017	012	008	004	000	000
50	129	121	114	106	100	094	088	082	076	071	066	058	050	042	030	029	023	018	013	010	006	000	000	000
48	124	116	108	101	095	088	082	078	071	066	062	054	043	038	032	026	020	015	010	008	004	000	000	000
46	119	111	103	096	062	082	076	071	066	062	057	050	041	036	028	022	017	013	009	007	003	000	000	000
44	114	106	096	091	064	078	072	067	062	057	053	046	038	031	026	020	016	012	008	006	002	000	000	
42	110	102	094	086	080	073	068	062	058	053	049	042	035	029	023	018	014	012	008	006	001	000		
40	105	097	090	082	076	069	064	059	054	050	046	039	032	026	021	016	013	011	007	005	000			
38	101	093	066	078	072	066	060	055	051	047	043	036	030	024	020	016	012	011	007	004				
36	058	090	082	075	069	063	058	053	049	045	041	034	029	023	019	016	012	010	007					
34	095	087	078	071	066	060	056	051	047	043	039	033	026	023	018	015	011	010						
32	091	183	075	168	063	058	054	049	045	042	037	032	026	023	018	015	011							
30	088	080	072	065	061	056	052	048	043	040	036	031	026	022	018	015								
28	067	079	070	063	058	055	052	048	043	040	036	031	026	022	018									
26	086	078	069	063	058	055	051	048	043	040	036	031	026	022										
24	085	077	068	063	058	054	051	048	043	040	036	031	026											
22	085	077	068	063	058	054	051	048	043	040	036	031												
20	083	076	068	063	058	054	051	048	043	040	036													
18	086	076	068	063	058	054	051	048	043															
16	089	079	070	064	059	055	053																	
14	092	082	072	066	061																			
12	097	087	075																					
10	103																							

Source: Jenkins, W. L., An improved method for tetrachoric *r*. *Psychometrika*, September 1955, *20*, 254.

TABLE C.8C. TABLES FOR RAPID COMPUTATION OF THE TETRACHORIC CORRELATION (MULTIPLIERS FOR BASE CORRECTION)

	When larger split is .40 or less, use Difference between splits							*When larger split is more than .40, use Smaller split*					*Uncorrected*		*When larger split is .40 or less, use Difference between splits*							*When larger split is more than .40, use Smaller split*					
	00	05	10	15	20	25	30	10	20	30	40	50		*r*	00	05	10	15	20	25	30	10	20	30	40	50	
96	35	42	62	85	96	107	117	122	128	132	134	134	96	52	104	102	98	92	86	81	76	72	68	68	64	62	52
94	41	48	66	87	97	107	116	121	127	131	133	133	94	50	104	101	96	90	84	79	73	68	66	64	61	58	50
92	47	54	70	88	98	107	116	120	125	130	132	132	92	48	103	99	94	88	82	77	71	65	62	60	58	54	48
90	53	60	73	89	99	107	115	119	124	128	130	130	90	46	101	97	92	86	80	74	68	62	59	56	54	51	46
88	60	66	77	90	100	107	114	118	122	127	128	128	88	44	99	95	89	83	77	72	65	59	56	53	51	48	44
86	66	71	81	92	101	107	113	117	120	125	127	127	86	42	97	92	87	81	75	69	62	57	53	50	48	44	42
84	71	76	84	93	101	106	112	115	118	123	124	124	84	40	94	89	83	79	72	66	60	54	50	47	45	41	40
82	76	81	87	94	102	106	110	114	116	121	122	122	82	38	91	87	81	76	70	64	57	51	47	43	42	38	38
80	81	85	90	95	102	106	109	112	114	118	119	119	80	36	89	84	78	73	67	61	55	48	44	40	39	35	36
78	86	89	92	96	102	105	107	110	112	115	116	116	78	34	86	81	76	71	65	58	52	46	41	38	36	32	34
76	90	93	94	97	102	104	105	108	109	112	113	113	76	32	83	78	73	68	62	55	49	43	39	35	33	29	32
74	94	96	96	98	102	103	104	105	106	108	109	109	74	30	80	75	70	65	59	53	46	40	36	32	30	26	30
72	97	98	98	99	101	101	102	103	103	104	105	105	72	28	76	72	67	62	56	51	44	37	33	30	27	23	28
70	100	100	100	100	100	100	100	100	100	100	100	100	70	26	72	68	64	59	53	48	41	35	31	27	24	20	26
68	101	101	101	100	99	98	97	97	97	97	96	96	68	24	68	64	60	56	50	45	38	32	28	25	21	17	24
66	102	102	102	100	98	96	95	94	93	93	92	92	66	22	64	60	57	53	47	42	35	29	26	22	19	15	22
64	103	103	102	100	97	94	93	91	90	90	88	87	64	20	60	57	54	49	44	39	33	27	24	20	16	12	20
62	104	104	102	99	95	92	90	88	86	86	84	83	62	18	55	52	50	45	40	35	28	24	21	18	14	10	18
60	104	104	102	99	93	90	88	85	83	82	80	78	60	16	50	47	45	41	36	32	25	21	19	16	12	8	16
58	104	104	101	97	92	88	85	82	79	78	76	74	58	14	45	42	40	37	32	28	22	19	16	13	10	6	14
56	105	104	100	95	90	86	82	79	75	74	72	70	50	12	40	38	35	33	28	24	19	17	14	11	8	4	12
54	105	103	99	94	88	84	79	75	71	71	68	66	54	10	35	33	30	28	24	20	15	13	12	9	6	2	10

Source: Jenkins, W. L. An improved method for tetrachoric *r*. *Psychometrika*, September 1955, *20*, 254.

TABLE C.9a Graphs Showing Confidence Limits for the Population Correlation Coefficient, ρ, G Given the Sample Coefficient, r, Confidence Coefficient, $1 - 2\alpha = 0.95$

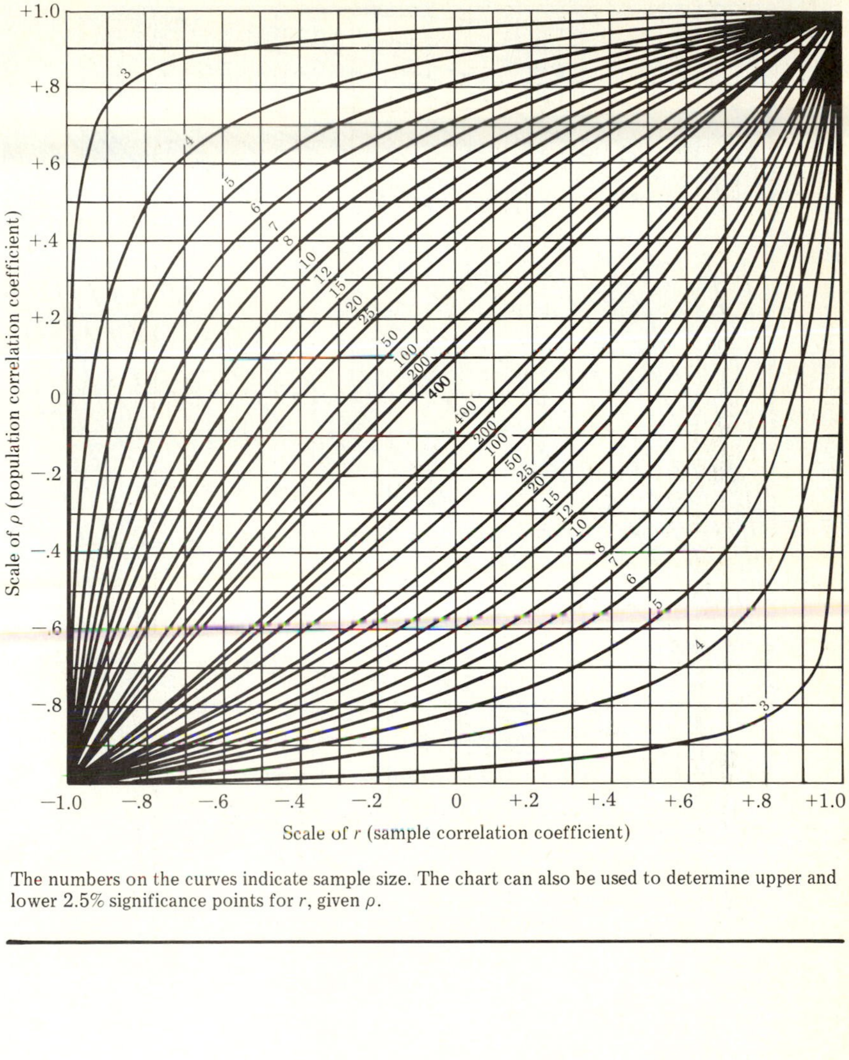

The numbers on the curves indicate sample size. The chart can also be used to determine upper and lower 2.5% significance points for r, given ρ.

TABLE C.9b Graphs Showing Confidence Limits for the Population Correlation Coefficient, ρ, Given the Sample Coefficient, r. Confidence Coefficient, $1 - 2\alpha = 0.99$

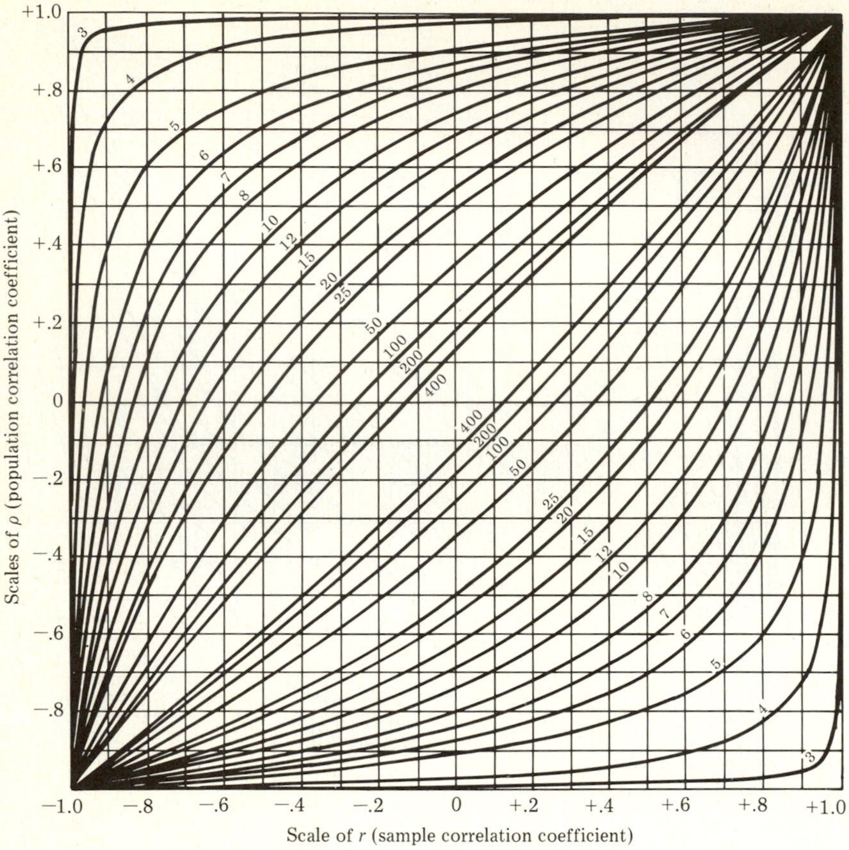

The numbers on the curves indicate sample size. The chart can also be used to determine upper and lower 0.5% significance points for r, given ρ.

TABLE C.10 THE TRANSFORMATION $Z = \tanh^{-1} r$ FOR THE CORRELATION COEFFICIENT

	r (3rd decimal)					*Proportional parts, for right side→*										*r (3rd decimal)*					
r	.000	.002	.004	.006	.008	1	2	3	4	5	6	7	8	9	10	.000	.002	.004	.006	.008	*r*
.00	.0000	.0020	.0040	.0060	.0080	1	3	4	5	7	8	9	11	12	13	.5493	.5520	.5547	.5573	.5600	.50
1	.0100	.0120	.0140	.0160	.0180	1	3	4	5	7	8	10	11	12	14	.5627	.5654	.5682	.5709	.5736	1
2	.0200	.0220	.0240	.0260	.0280	1	3	4	6	7	8	10	11	13	14	.5763	.5791	.5818	.5846	.5874	2
3	.0300	.0320	.0340	.0360	.0380	1	3	4	6	7	8	10	11	13	14	.5901	.5929	.5957	.5985	.6013	3
4	.0400	.0420	.0440	.0460	.0480	1	3	4	6	7	9	10	11	13	14	.6042	.6070	.6098	.6127	.6155	4
.05	.0500	.0520	.0541	.0561	.0581	1	3	4	6	7	9	10	12	13	14	.6184	.6213	.6241	.6270	.6299	.55
6	.0601	.0621	.0641	.0661	.0681	1	3	4	6	7	9	10	12	13	15	.6328	.6358	.6387	.6416	.6446	6
7	.0701	.0721	.0741	.0761	.0782	1	3	4	6	7	9	10	12	14	15	.6475	.6505	.6535	.6565	.6595	7
8	.0802	.0822	.0842	.0862	.0882	2	3	5	6	8	9	11	12	14	15	.6625	.6655	.6685	.6716	.6746	8
9	.0902	.0923	.0943	.0963	.0983	2	3	5	6	8	9	11	12	14	15	.6777	.6807	.6838	.6869	.6900	9
.10	.1003	.1024	.1044	.1064	.1084	2	3	5	6	8	9	11	13	14	16	.6931	.6963	.6994	.7026	.7057	.60
1	.1104	.1125	.1145	.1165	.1186	2	3	5	6	8	10	11	13	14	16	.7089	.7121	.7153	.7185	.7218	1
2	.1206	.1226	.1246	.1267	.1287	2	3	5	7	8	10	11	13	15	16	.7250	.7283	.7315	.7348	.7381	2
3	.1307	.1328	.1348	.1368	.1389	2	3	5	7	8	10	12	13	15	17	.7414	.7447	.7481	.7514	.7548	3
4	.1409	.1430	.1450	.1471	.1491	2	3	5	7	9	10	12	14	15	17	.7582	.7616	.7650	.7684	.7718	4
.15	.1511	.1532	.1552	.1573	.1593	2	4	5	7	9	11	12	14	16	18	.7753	.7788	.7823	.7858	.7893	.65
6	.1614	.1634	.1655	.1676	.1696	2	4	5	7	9	11	13	14	16	18	.7928	.7964	.7999	.8035	.8071	6
7	.1717	.1737	.1758	.1779	.1799	2	4	6	7	9	11	13	15	17	18	.8107	.8144	.8180	.8217	.8254	7
8	.1820	.1841	.1861	.1882	.1903	2	4	6	8	9	11	13	15	17	19	.8291	.8328	.8366	.8404	.8441	8
9	.1923	.1944	.1965	.1986	.2007	2	4	6	8	10	12	14	15	17	19	.8480	.8518	.8556	.8595	.8634	9
.20	.2027	.2048	.2069	.2090	.2111	2	4	6	8	10	12	14	16	18	20	.8673	.8712	.8752	.8792	.8832	.70
1	.2132	.2153	.2174	.2195	.2216	2	4	6	8	10	12	15	16	18	20	.8872	.8912	.8953	.8994	.9035	1
2	.2237	.2258	.2279	.2300	.2321	2	4	6	8	11	13	15	17	19	21	.9076	.9118	.9160	.9202	.9245	2
3	.2342	.2363	.2384	.2405	.2427	2	4	7	9	11	13	15	17	20	22	.9287	.9330	.9373	.9417	.9461	3
4	.2448	.2469	.2490	.2512	.2533	2	4	7	9	11	13	16	18	20	22	.9505	.9549	.9594	.9639	.9684	4
r	**.000**	**.002**	**.004**	**.006**	**.008**	**1**	**2**	**3**	**4**	**5**	**6**	**7**	**8**	**9**	**10**	**.000**	**.002**	**.004**	**.006**	**.008**	*r*
	r (3rd decimal)					*←Proportional parts, for left side*										*r (3rd decimal)*					

Interpolation

(1) $0 < r < 0.25$: find argument r_0 nearest to r and form $z = z(r_0) + \Delta r$ (where $\Delta r = r - r_0$), e.g. for $r = 0.2042$, $z = 0.2069 + 0.0002 = 0.2071$.

(2) $0.25 < r < 0.75$: find argument r_0 nearest to r and form $z = z(r_0) \pm P$, where P is the proportional part for $\Delta r = r - r_0$, e.g. for $r = 0.5146$, $z = 0.5682 + 0.0008 = 0.5690$; for $r = 0.5372$, $z = 0.6013 - 0.0011 = 0.6002$.

(3) $0.75 < r < 0.98$: use linear interpolation to get 3-decimal-place accuracy.

(4) $0.98 < r < 1$: form $z = -\frac{1}{2}\ln(1 - r) + 0.097 + \frac{1}{4}r$, with the help of table of natural logarithms.

TABLE C.10 CONTINUED

	− r (3rd decimal)					Proportional parts, for right side→										r (3rd decimal)					
r	.000	.002	.004	.006	.008	1	2	3	4	5	6	7	8	9	10	.000	.002	.004	.006	.008	r
.25	.2554	.2575	.2597	.2618	.2640	1	2	3	4	5	6	7	9	10	11	0.973	0.978	0.982	0.987	0.991	.75
6	.2661	.2683	.2704	.2726	.2747	1	2	3	4	5	6	8	9	10	11	0.996	1.001	1.006	1.001	1.015	6
7	.2769	.2790	.2812	.2833	.2855	1	2	3	4	5	6	8	9	10	11	1.020	1.025	1.030	1.035	1.040	7
8	.2877	.2899	.2920	.2942	.2964	1	2	3	4	5	7	8	9	10	11	1.045	1.050	1.056	1.061	1.066	8
9	.2986	.3008	.3029	.3051	.3073	1	2	3	4	5	7	8	9	10	11	1.071	1.077	1.082	1.088	1.093	9
.30	.3095	.3117	.3139	.3161	.3183	1	2	3	4	6	7	8	9	10	11	1.099	1.104	1.110	1.116	1.121	.80
1	.3205	.3228	.3250	.3272	.3294	1	2	3	4	6	7	8	9	10	11	1.127	1.133	1.139	1.145	1.151	1
2	.3316	.3339	.3361	.3383	.3406	1	2	3	4	6	7	8	9	10	11	1.157	1.163	1.169	1.175	1.182	2
3	.3428	.3451	.3473	.3496	.3518	1	2	3	5	6	7	8	9	10	11	1.188	1.195	1.201	1.208	1.214	3
4	.3541	.3564	.3586	.3609	.3632	1	2	3	5	6	7	8	9	10	11	1.221	1.228	1.235	1.242	1.249	4
.35	.3654	.3677	.3700	.3723	.3746	1	2	3	5	6	7	8	9	10	11	1.256	1.263	1.271	1.278	1.286	.85
6	.3769	.3792	.3815	.3838	.3861	1	2	3	5	6	7	8	9	10	12	1.293	1.301	1.309	1.317	1.325	6
7	.3884	.3907	.3931	.3954	.3977	1	2	3	5	6	7	8	9	10	12	1.333	1.341	1.350	1.358	1.367	7
8	.4001	.4024	.4047	.4071	.4094	1	2	4	5	6	7	8	9	11	12	1.376	1.385	1.394	1.403	1.412	8
9	.4118	.4142	.4165	.4189	.4213	1	2	4	5	6	7	8	9	11	12	1.422	1.432	1.442	1.452	1.462	9
.40	.4236	.4260	.4284	.4308	.4332	1	2	4	5	6	7	8	10	11	12	1.472	1.483	1.494	1.505	1.516	.90
1	.4356	.4380	.4404	.4428	.4453	1	2	4	5	6	7	8	10	11	12	1.528	1.539	1.551	1.564	1.576	1
2	.4477	.4501	.4526	.4550	.4574	1	2	4	5	6	7	9	10	11	12	1.589	1.602	1.616	1.630	1.644	2
3	.4599	.4624	.4648	.4673	.4698	1	2	4	5	6	7	9	10	11	12	1.658	1.673	1.689	1.705	1.721	3
4	.4722	.4747	.4772	.4797	.4822	1	2	4	5	6	7	9	10	11	12	1.738	1.756	1.774	1.792	1.812	4
.45	.4847	.4872	.4897	.4922	.4948	1	3	4	5	6	8	9	10	11	13	1.832	1.853	1.874	1.897	1.921	.95
6	.4973	.4999	.5024	.5049	.5075	1	3	4	5	6	8	9	10	11	13	1.946	1.972	2.000	2.029	2.060	6
7	.5101	.5126	.5152	.5178	.5204	1	3	4	5	6	8	9	10	12	13	2.092	2.127	2.165	2.205	2.249	7
8	.5230	.5256	.5282	.5308	.5334	1	3	4	5	7	8	9	10	12	13	2.298	2.351	2.410	2.477	2.555	8
9	.5361	.5387	.5413	.5440	.5466	1	3	4	5	7	8	9	11	12	13	2.647	2.759	2.903	3.106	3.453	9
r	.000	.002	.004	.006	.008	1	2	3	4	5	6	7	8	9	10	.000	.002	.004	.006	.008	r
	r (3rd decimal)					←Proportional parts, for left side										r (3rd decimal)					

Interpolation

(1) $0<r<0.25$: find argument r_0 nearest to r and form $z=z(r_0)+\Delta r$ (where $\Delta r=r-r_0$), e.g. for $r=0.2042$, $z=0.2069+0.0002=0.2071$.

(2) $0.25<r<0.75$: find argument r_0 nearest to r and form $z=z(r_0)\pm P$, where P is the proportional part for $\Delta r=r-r_0$, e.g. for $r=0.5146$, $z=0.5682+0.0008=0.5690$; for $r=0.5372$, $z=0.6013-0.0011=0.6002$.

(3) $0.75<r<0.98$: use linear interpolation to get 3-decimal-place accuracy.

(4) $0.98<r<1$: form $z=-\frac{1}{2}\ln(1-r)+0.097+\frac{1}{4}r$, with the help of table of natural logarithms.

APPENDIX **D**

Solution of a System of Linear Equations

As we indicated through examples given in Chapter 4, the solution of the normal equations in multiple regression analysis is relatively simple when the number of independent (predictor) variables is less than three. When there are three or more such variables, however, the computations become more difficult. One can either utilize a computer or use a computational method designed to organize the work so as to reduce computational labor and the possibility of error to a minimum. Several such methods have been proposed, one of which, due to Crout (1941), we present here.

We consider the case in which one wishes to obtain the vector of standardized regression coefficients **B**, given the matrix **R** of correlations among predictors (Xs), and the vector **c** of correlations between each predictor and the dependent variable, Y. Thus,

$$\mathbf{R}=\begin{bmatrix} 1 & r_{12} & \dots & r_{1p} \\ r_{21} & 1 & \dots & r_{2p} \\ \vdots & \vdots & & \vdots \\ r_{p1} & r_{p2} & \dots & 1 \end{bmatrix}, \quad \mathbf{B}=\begin{bmatrix} B_1 \\ B_2 \\ \vdots \\ B_p \end{bmatrix}, \quad \text{and } \mathbf{c}=\begin{bmatrix} r_{Y1} \\ r_{Y2} \\ \vdots \\ r_{Yp} \end{bmatrix},$$

and we wish to solve for **B** the matrix equation $\mathbf{RB}=\mathbf{c}$. Our illustration is based on a problem involving four predictor variables. Once the method is understood, it can easily be adapted for use with any number of predictors or with other equations of the same general form.

The steps are:

1. Construct the Original Matrix by inserting the matrix **R** and the vector **c** as indicated below and adding a column containing the sums of entries in each row.

Row	*Original Matrix* *Column* C_1	C_2	C_3	C_4	C_5	*Row Sums*
R_1	1 (*a*)	r_{12} (*e*)	r_{13} (*i*)	r_{14} (*m*)	r_{Y1} (*q*)	S_1
R_2	r_{21} (*b*)	1 (*f*)	r_{23} (*j*)	r_{24} (*n*)	r_{Y2} (*r*)	S_2
R_3	r_{31} (*c*)	r_{32} (*g*)	1 (*k*)	r_{34} (*o*)	r_{Y3} (*s*)	S_3
R_4	r_{41} (*d*)	r_{42} (*h*)	r_{43} (*l*)	1 (*p*)	r_{Y4} (*t*)	S_4
	←	—	**R**	→	← c →	

a. For the purpose of generality, the elements of the specific original matrix above will be denoted by the lower case alphabetic characters indicated in parentheses in the corresponding cells.

b. In applying the Crout Method to the solution of systems of simultaneous linear equations not involving either values of unity in the main diagonal of the basic square matrix and/or symmetric matrices, the procedure that follows may be used simply by appropriate substitution.

2. Prepare the Auxiliary Matrix:

	Column						
Row	C_1	C_2	C_3	C_4	C_5	Check	$\Sigma R_i + 1$
R_1	(1)	(5)	(9)	(13)	(17)	Check R_1	$\sum_2^5 R_1 + 1$
R_2	(2)	(6)	(10)	(14)	(18)	Check R_2	$\sum_3^5 R_2 + 1$
R_3	(3)	(7)	(11)	(15)	(19)	Check R_3	$\sum_4^5 R_3 + 1$
R_4	(4)	(8)	(12)	(16)	(20)	Check R_4	$\sum_5^5 R_4 + 1$

The number in parentheses in each cell merely identifies the position of the element to be computed. The entries in the Auxiliary Matrix are computed as follows:

a. Insert the elements in C_1 by copying corresponding elements from the Original Matrix. Thus:

$(1) = (a)$

$(2) = (b)$

$(3) = (c)$

$(4) = (d)$

b. Compute elements in R_1 as follows:

$(5) = (2)/(1) = (b)/(a)$

$(9) = (3)/(1) = (c)/(a)$

$(13) = (4)/(1) = (d)/(a)$

$(17) = (q)/(a)$

c. To obtain Check R_1, divide S_1 (from the Original Matrix) by element (1) in the Auxiliary Matrix. Thus, Check $R_1 = S_1/1$.

d. Obtain $\Sigma_2^5 R_1 = (5) + (9) + (13) + (17)$. Then $\Sigma_2^5 R_1 + 1$ should be equal to Check R_1.

e. Remaining elements in C_2 are obtained as follows:

$(6) = (f) - (5)(2)$

$(7) = (g) - (5)(3)$

$(8) = (h) - (5)(4)$

f. Remaining elements of R_2 are:

$(10)=(7)/(6)$

$(14)=(8)/(6)$

$(18)=[(r)-(17)(2)]/(6)$

g. Obtain Check $R_2=[S_2-(\text{Check } R_1)(2)]/(6)$.

h. Obtain $\Sigma_3^5 R_2=(10)+(14)+(18)$. Then $\Sigma_3^5 R_2+1$ should be equal to Check R_2.

i. Calculate the remaining elements in C_3.

$(11)=(k)-(9)(3)-(10)(7)$

$(12)=(l)-(9)(4)-(10)(8)$

j. To complete R_3, calculate

$(15)=(12)/(11)$

$(19)=[(s)-(17)(3)-(18)(7)]/(11)$

k. Obtain Check $R_3=[S_3-(\text{Check } R_1)(3)-(\text{Check } R_2)(7)]/(11)$

l. Obtain $\Sigma_4^5 R_3=(15)+(19)$. Then $\Sigma_4^5 R_3+1$ should be equal to Check R_3.

m. To complete R_4, calculate

$(16)=(p)-(13)(4)-(14)(8)-(15)(12)$

$(20)=[(t)-(17)(4)-(18)(8)-(19)(12)]/(16)$

n. Calculate Check R_4. Check $R_4=[S_4-(\text{Check } R_1)(4)-(\text{Check } R_2)(8)-(\text{Check } R_3)(12)]/(16)$

o. Obtain $\Sigma_5^5 R_4=(20)$. Then $\Sigma_5^5 R_4+1$ should be equal to Check R_4.

p. Compute the desired coefficients (in this case, elements of **B**) as follows:

$B_4=(20)$

$B_3=(19)-(15)B_4$

$B_2=(18)-(14)B_4-(10)B_3$

$B_1=(17)-(13)B_4-(9)B_3-(5)B_2$

q. Compute a final check by verifying that $\mathbf{RB}=\mathbf{c}$. For an example of the application of the Crout method to the data of Table 4.2.1, see Section 4.2.3.

APPENDIX E
Procedure for Multivariate Correction For Restriction in Range

We give here a step-by-step procedure for carrying out the computations required in the multivariate correction for restriction in range, discussed in Section 4.9. The procedure is presented in terms of the data of Table 4.8.1. We assume that the 1181 students in that example had been preselected on the basis of the GCT and ARI tests.

The steps are:

1. Construct the variance-covariance matrix as shown below:

Original Matrix					
Variables	C_1	C_2	C_3	C_4	C_Y
R_1	s_1^2	$r_{12}s_1s_2$	$r_{13}s_1s_3$	$r_{14}s_1s_4$	$r_Ys_1s_Y$
R_2	$r_{12}s_1s_2$	s_2^2	$r_{23}s_2s_3$	$r_{24}s_2s_4$	$r_Ys_2s_Y$
R_3	$r_{13}s_1s_3$	$r_{23}s_2s_3$	s_3^2	$r_{34}s_3s_4$	$r_Ys_3s_Y$
R_4	$r_{14}s_1s_4$	$r_{24}s_2s_4$	$r_{34}s_3s_4$	s_4^2	$r_Ys_4s_Y$
R_Y	$r_Ys_1s_Y$	$r_Ys_2s_Y$	$r_Ys_3s_Y$	$r_Ys_4s_Y$	s_Y^2

Based on the data of our example, we have:

Original Matrix							
		Explicit		*Incidental*			
	Variables	*GCT*	*ARI*	*MECH*	*CLER*	*GRADES*	
	GCT	33.735*	6.418	10.876	1.983	9.254	
Explicit	*ARI*	6.418	29.017	2.905	7.662	9.522	$\mathbf{V}_{sc}$
	MECH	10.876	2.905	56.822	2.174	22.286	
Incidental	*CLER*	1.983	7.662	2.174	53.022	5.706	$\mathbf{V}_{cc}$
	GRADES	9.254	9.521	22.286	5.706	54.383	

*These values do not agree exactly with those in Table 4.8.1 because of rounding. For example, the value $s_{GCT}^2 = 33.735$ was obtained from $(5.80818)^2$, which was rounded to 5.81 in Table 4.8.1.

2. Construct a partial variance-covariance Auxiliary Matrix.

a. Copy the first column from the Original Matrix.

b. Compute the remaining elements by the Crout method. (In this case, use the Crout method to obtain row and column entries for GCT and ARI, the variables subjected to explicit selection.)

Auxiliary Matrix

	Variables	*GCT*	*ARI*	*MECH*	*CLER*	*GRADES*
Explicit	X_1: *GCT*	33.735	0.19025*	0.32240	0.05878	0.27432
	X_2: *ARI*	6.418	27.796	0.03008	0.26209	0.27921
Incidental	X_3: *MECH*	10.876	0.836			
	X_4: *CLER*	1.983	7.285			
	Y: *GRADES*	9.254	7.761			

*Five decimal places are retained to avoid rounding error that may affect subsequent computations.

3. Construct the variance-covariance, $\mathbf{V}_{ss}$, of population values on variables subjected to explicit selection.

a. The population correlation matrix ($N = 30,300$) is:

Variables	*GCT*	*ARI*	*MECH*	*CLER*	μ	σ
X_1: *GCT*		.699	.511	.429	50.44	10.54
X_2: *ARI*			.416	.462	50.11	8.76
X_3: *MECH*				.143	49.84	8.95
X_4: *CLER*					49.87	8.77

It is necessary that $\mathbf{V}_{ss}$ be known only for the two predictors subjected to explicit selection, in this case, *GCT* and *ARI*.

b. $\mathbf{V}_{ss}$, calculated using five decimal places but rounding to three, is:

	GCT	*ARI*
X_1: *GCT*	$\sigma_1^2 = 111.195$	$\rho_{12}\sigma_1\sigma_2 = 64.536$
X_2: *ARI*	$\rho_{12}\sigma_1\sigma_2 = 64.536$	$\sigma_2^2 = 76.761$

4. Construct the Intermediate Matrix, **Q**:

$Q =$

	X_3	X_4	Y
X_1	a_{13}	a_{14}	a_{1Y}
X_2	a_{23}	a_{24}	a_{2Y}

a. Form the matrix **A**, by writing down the corresponding elements from the Auxiliary Matrix:

$A=$

	MECH	CLER	GRADES
GCT	0.32240	0.05878	0.27432
ARI	0.03008	0.26209	0.27921

b. To compute a_{13}, subtract from its corresponding element in the Auxiliary Matrix the cross-product of the element between it and the diagonal, and element a_{23}. Thus,

$$0.32240-[(0.19025)(0.03008)]=0.31668.$$

c. To compute a_{14}, subtract from its corresponding element in the Auxiliary Matrix the cross-product of the element to the right of the diagonal in the same row in the Auxiliary Matrix and element a_{24}. Thus,

$$0.05878-[(0.19025)(0.26209)]=0.00892.$$

d. To compute element a_{1Y}, subtract from its corresponding element in the Auxiliary Matrix the cross-product of the element to the right of the diagonal in the same row in the Auxiliary Matrix and the element a_{24}. Thus,

$$0.27432-[(0.19025)(0.27921)]=0.22120.$$

The Intermediate Matrix, **Q**, is then:

Q=

	MECH	CLER	GRADES
GCT	0.31668	0.00892	0.22120
ARI	0.03008	0.26209	0.27921

Note: The elements of the second row, that is $X_2=$ARI, have already been derived in developing the partial auxiliary matrix. Therefore, the second row in **Q** is the same as that in **A**.

5. Construct the matrix $\mathbf{V}_{EI}=\mathbf{V}_{ss}\mathbf{Q}$. This matrix involves variables subjected to explicit selection and variables subjected to incidental selection.

$\mathbf{V}_{EI}=$

	MECH	CLER	GRADES
GCT	37.154	17.906	42.615
ARI	22.746	20.694	35.708

6. Form the matrix $\mathbf{v}_{EI}$, the variance-covariance matrix of (X_1,X_2) and (X_3,X_4,Y). For our example,

$\mathbf{v}_{EI}=$

	MECH	CLER	GRADES
GCT	10.876	1.983	9.254
ARI	2.905	7.662	9.522

7. Compute the matrix product $(\mathbf{V}_{EI}-\mathbf{v}_{EI})'\mathbf{Q}$:

$(\mathbf{V}_{EI}-\mathbf{v}_{EI})'\mathbf{Q}=$

	MECH	*CLER*	*GRADES*
MECH	8.919	5.434	11.352
CLER	5.434	3.558	7.161
GRADES	11.352	7.161	14.691

8. Form the matrix $\mathbf{v}_{II}$, which is composed of the variance-covariance elements in the Original Matrix subjected to incidental selection. For this example:

$\mathbf{v}_{II}=$

	MECH	*CLER*	*GRADES*
MECH	56.822	2.174	22.286
CLER	2.174	53.022	5.706
GRADES	22.286	5.706	54.383

9. Compute $\mathbf{V}_{II}=\mathbf{v}_{II}+(\mathbf{V}_{EI}-\mathbf{v}_{EI})'\mathbf{Q}$:

$\mathbf{V}_{II}=$

	MECH	*CLER*	*GRADES*
MECH	65.741	7.608	33.638
CLER	7.608	56.580	12.867
GRADES	33.638	12.867	69.074

10. Extract the square roots of the elements of the principal diagonal of the $\mathbf{V}_{II}$ matrix. These are the estimates of the unrestricted standard deviations, $\hat{\sigma}_i$:

$$MECH\colon \hat{\sigma}_m=\sqrt{65.741}=8.10808$$
$$CLER\colon \hat{\sigma}_c=\sqrt{56.580}=7.52197$$
$$GRADES\colon \hat{\sigma}_g=\sqrt{69.074}=8.31108$$

11. Compute the reciprocals of the square roots of the elements in the principal diagonal of the $\mathbf{v}_{II}$ matrix:

$$MECH\colon 1/8.10000=0.12333$$
$$CLER\colon 1/7.52197=0.13294$$
$$GRADES\colon 1/8.31108=0.12032$$

12. Merge $\mathbf{V}'_{EI}$ and $\mathbf{V}_{II}$ to form a single table and multiply the entries in each row, respectively, by $1/\hat{\sigma}_m$, $1/\hat{\sigma}_c$ and $1/\hat{\sigma}_a$. The result is*:

	GCT	*ARI*	*MECH*	*CLER*	*GRADES*
MECH	4.582	2.805	8.108	0.938	4.149
CLER	2.380	2.751	1.011	7.522	1.711
GRADES	5.127	4.296	4.047	1.548	8.311
σ_i	10.54	8.76			
$\hat{\sigma}_i$			8.10808	7.52197	8.31108

*The entries denoted by σ_i and $\hat{\sigma}_i$ have been included for later use.

13. Correlations corrected for restriction in range caused by explicit selection on *GCT* and *ARI* are found by dividing each element of the *GCT* and *ARI* columns of the above matrix by their respective σ_is, and the elements of the remaining portion by their respective $\hat{\sigma}_i$s. Note that these values are given at the bottom of the table in Step 12. The corrected correlations are the entries in the following table:

	GCT	*ARI*	*MECH*	*CLER*	*GRADES*
MECH	0.435	0.320	1.000	0.125	0.490
CLER	0.226	0.314	0.125	1.000	0.206
GRADES	0.486	0.490	0.499	0.206	1.000

Hence the new inter-battary matrix corrected for restriction on *GCT* and *ARI* becomes:

	ARI	*MECH*	*CLER*	*GRADES*	*Mean*	*Standard Deviation*
GCT	.205	.435	.226	.486	56.72	10.54
ARI		.320	.314	.490	55.44	8.76
MECH			.125	.499	54.33	8.11
CLER				.206	50.56	7.52
GRADES					80.94	8.31

This new matrix of correlations yields an R^2 of .44, which is more than twice as large as that obtained using the uncorrected correlations. The multiple regression analysis for the *R* matrix in which the correlation coefficients subjected to incidental selection have been corrected for multivariate restriction in range is given below:

Original Matrix

	X_1	X_2	X_3	X_4	Y	
Variables	*GCT*	*ARI*	*MECH*	*CLER*	*GRADES*	ΣR_i
GCT	1.00000	.20500	.43500	.22600	.48600	2.35200
ARI	.20500	1.00000	.32000	.31400	.49000	2.32900
MECH	.43500	.32000	1.00000	.12500	.49900	2.37900
CLER	.22600	.31400	.12500	1.00000	.20600	1.87100
$\bar{X}$	56.72	55.44	54.33	50.56	80.94	
σ	10.54	8.76	8.11	7.52	8.31	

The Auxiliary Matrix is:

Auxiliary Matrix (Crout Method)

	X_1	X_2	X_3	X_4	Y		
Variables	*GCT*	*ARI*	*MECH*	*CLER*	*GRADES*	$\Sigma R_i + 1$	*Check*
GCT	1.00000	.20500	.43500	.22600	.48600	2.35200	2.35200
ARI	.20500	.95798	.23082	.26767	.40749	1.92785	1.92784
MECH	.43500	.23082	.75516	−.05006	.25628	1.20622	1.20622
CLER	.22600	.26767	−.03780	.87224	−.00369	0.99631	0.99631

and the standardized regression coefficients are:

$$B_1 = .30433$$
$$B_2 = .34682$$
$$B_3 = .25610$$
$$B_4 = -.00369.$$

The raw score form of the multiple regression equation becomes.

$$Y_i = 35.04 + .23994X_{1i} + .32900X_{2i} + .26242X_{3i} - .00400X_{4i}.$$

References

Anderson, T. W. *An introduction to multivariate statistical analysis*. New York: Wiley, 1958.

Anderson, T. W. and Rubin, H. Statistical inference in factor analysis. *Proceedings of the third Berkeley symposium on mathematical statistics and probability* (Volume 5). Berkeley: University of California Press, 1956. Pp. 111–150.

Armstrong, J. S. and Soelberg, P. On the interpretation of factor analysis. *Psychological Bulletin*, 1968, *70*, 361–364.

Bartlett, M. S. Contingency table interactions. *Journal of the Royal Statistical Society Supplement*, 1935, *2*, 248–252.

Bartlett, M. S. Multivariate analysis. *Journal of the Royal Statistical Society* (Series B), 1947, *9*, 176–197.

Bartlett, M. S. Tests of significance in factor analysis. *British Journal of Mathematical and Statistical Psychology*, 1950, *3*, 77–85.

Bartlett, M. S. A further note on tests of significance in factor analysis. *British Journal of Mathematical and Statistical Psychology*, 1951, *4*, 1–2.

Bayes, T. An essay towards solving a problem in the doctrine of chances. *Philosophical Transactions of the Royal Society*, *53*. (Reprinted in *Biometrika*, 1958, *45*, 293–315.)

Berger, A. and Gold, R. Z. Note on Cochran's *Q*-test for the comparison of correlated proportions. *Journal of the American Statistical Association*, 1973, *68*, 989–993.

Berkson, J. Application of minimum logit χ^2 estimate to a problem of Grizzle with a notation on the problem of no interaction. *Biometrics*, 1968, *24*, 75–95.

Bhapkar, V. P. Notes on analysis of categorical data. *Institute of Statistics* (Mimeo Series No. 477). Chapel Hill: University of North Carolina, 1966.

Birch, N. W. Maximum likelihood in three-way contingency tables. *Journal of the Royal Statistical Society* (Series B), 1963, *25*, 220–233.

Bishop, Y. M., Fienberg, S. E., and Holland, P. *Discrete multivariate analysis: Theory and practice*. Cambridge, MA: The Massachusetts Institute of Technology Press, 1975.

Bock, R. D. *Multivariate statistical methods in behavioral research*. New York: McGraw-Hill, 1975.

Bock, R. D. and Bargmann, R. E. Analysis of covariance structures. *Psychometrika*, 1966, *31*, 507–534.

Bryan, J. G. *A method for the exact determination of the characteristic equation of a matrix and applications to the discriminant function for more than two groups*. Unpublished doctoral dissertation, Harvard University, 1950.

Burt, C. *The factors of mind: An introduction to factor analysis in psychology*. New York: Macmillan, 1941.

Burt, C. The factorial study of temperamental traits. *British Journal of Mathematical and Statistical Psychology*, 1948, *1*, 178–203.

Burt, C. Tests of significance in factor analysis. *British Journal of Mathematical and Statistical Psychology*, 1952, *5*, 109–133.

Carroll, J. B. An analytical solution for approximating simple structure in factor analysis. *Psychometrika*, 1953, *18*, 23–28.

Carroll, J. B. Biquartimin criterion for rotation to oblique simple structure in factor analysis. *Science*, 1957, *126*, 1114–1115.

Carroll, J. B. *IBM 704 program for generalized analytic rotation solution in factor analysis.* Unpublished manuscript, Harvard University, 1960.

Cattell, R. B. The scree test for the number of factors. *Multivariate Behavioral Research*, 1966, *1*, 245–276.

Cattell, R. B. and Muerle, J. L. The "maxplane" program for factor rotation to oblique simple structure. *Educational and Psychological Measurement*, 1960, *20*, 569–590.

Chernoff, H. Large sample theory: Parametric case. *Annals of Mathematical Statistics*, 1956, *27*, 1–22.

Cochran, W. G. The comparison of percentages in matched samples. *Biometrika*, 1950, *37*, 256–266.

Cochran, W. G. Some methods for strengthening the common χ^2 tests. *Biometrics*, 1954, *10*, 417–451.

Cooley, W. W. and Lohnes, P. R. *Multivariate data analysis*. New York: Wiley, 1971.

Cox, D. R. *Analysis of binary data.* London: Metheren and Company, Ltd., 1970.

Cramér, H. *Mathematical methods of statistics*. Princeton, N.J.: Princeton University Press, 1946.

Crawford, C. *A general method of rotation for factor analysis*. Paper presented at the meeting of the Psychometric Society, April 1967.

Cronbach, L. J. and Gleser, G. C. Assessing profile similarity. *Psychological Bulletin*, 1953, *50*, 456–473.

Crout, P. D. *A short method for evaluating determinants and solving systems of linear equations with real or complex coefficients.* New York: American Institute of Electrical Engineers, 1941.

Danford, M. B. *Factor analysis and related techniques.* Unpublished doctoral dissertation, North Carolina State College, 1953.

Darlington, R. B. Multiple regression in psychological research and practice. *Psychological Bulletin*, 1968, *79*, 161–182.

Darlington, R. B., Weinberg, S. L., and Wahlberg, H. J. Canonical variate analysis and related techniques. *Review of Educational Research*, 1973, *43*, 433–454.

Davidoff, M. D. and Goheen, H. W. Table for rapid determination of the tetrachoric correlation coefficient. *Psychometrika*, 1953, *18*, 115–121.

Draper, N. and Smith, H. *Applied regression analysis*. New York: Wiley, 1966.

Dunlap, J. W. and Cureton, E. E. On the analysis of causation. *Journal of Educational Psychology*, 1930, *21*, 657–679.

Eber, H. W. Toward oblique simple structure: Maxplane. *Multivariate Behavioral Research*, 1966, *1*, 112–125.

Eckerman, W., Bates, J. D., Rachel, J. V., and Poole, W. K. *Drug usage and arrest charges: A study of drug usage and arrest charges among arrestees in six metropolitan areas of the United States*. Washington, D.C.: U.S. Department of Justice, Bureau of Narcotics and Dangerous Drugs, December 1971.

Ferguson, G. The concept of parsimony in factor analysis. *Psychometrika*, 1954, *19*, 281–290.

Fienberg, S. E. *The analysis of cross-classified categorical data*. Cambridge, MA: The Massachusetts Institute of Technology Press, 1977.

Fisher, R. A. On an absolute criterion for fitting frequency curves. *Messenger of Mathematics*, 1912, *41*, 155–160.

Fisher, R. A. On the "probable error" of a coefficient of correlation deduced from a small sample. *Metron*, 1921, *1*(Part 4), 1–32.

Fisher, R. A. On the interpretation of χ^2 from contingency tables and the calculation of *P*. *Journal of the Royal Statistical Society*, 1922, *85*, 87–94.

Fisher, R. A. The use of multiple measurements in taxonomic problems. *Annals of Eugenics*, 1936, *7*, 179–188.

Fisher, R. A. The statistical utilization of multiple measurements. *Annals of Eugenics*, 1938, *8*, 376–386.

Fisher, R. A. *Statistical methods for research workers* (12th ed.). Edinburgh: Oliver & Boyd, 1954.

Francis, J. G. F. The QR transformation (Parts I and II). *The Computer Journal*, 1961–1962, *4*, 265–271; 332–345.

Freeman, H. *Introduction to statistical inference*. Reading, Mass.: Addison-Wesley, 1963.

Galton, F. *Natural inheritance*. New York: Macmillan, 1889.

Geisser, S. Posterior odds for multivariate normal classifications. *Journal of the Royal Statistical Society*, 1964, *26*, 69–76.

Gleason, T. C. On redundancy in canonical analysis. *Psychological Bulletin*, 1976, *83*, 1004–1006.

Gnedenko, B. V. *Theory of probability* (4th ed.). New York: Chelsea Publishing Co., 1968.

Goodman, L. The multivariate analysis of qualitative data: Interactions among multiple classifications. *Journal of the American Statistical Association*, 1970, *65*, 226–256.

Grizzle, J. E. A new method of testing hypotheses and estimating parameters for the logistic model. *Biometrics*, 1961, *17*, 372–385.

Grizzle, J. E., Starmer, C. F., and Koch, G. C. Analysis of categorical data by linear models. *Biometrics*, 1969, *25*, 489–503.

Gulliksen, H. *Theory of mental tests.* New York: Wiley, 1950.

Guttman, L. Image theory for the structure of quantitative variates. *Psychometrika*, 1953, *18*, 277–296.

Guttman, L. Some necessary conditions for common-factor analysis. *Psychometrika*, 1954, *19*, 149–161.

Guttman, L. "Best possible" estimates of communalities. *Psychometrika*, 1956, *21*, 273–85.

Haberman, S. J. *The analysis of frequency data*. Chicago: The University of Chicago Press, 1974.

Hakstian, A. R. and Abell, R. A. A further comparison of oblique factor transformation methods. *Psychometrika*, 1974, *39*, 429–444.

Harman, H. H. *Modern factor analysis* (3rd ed.). Chicago: University of Chicago Press, 1976.

Harman, H. H. and Jones, W. H. Factor analysis by minimizing residuals (minres). *Psychometrika*, 1966, *31*, 351–368.

Harris, C. W. Some Rao-Guttman relationships. *Psychometrika*, 1962, *27*, 247–263.

Harris, C. W. and Kaiser, H. F. Oblique factor analytic solutions by orthogonal transformations. *Psychometrika*, 1964, *29*, 347–362.

Harris, R. J. *A primer of multivariate statistics*. New York: Academic Press, 1975.

Hays, W. L. *Statistics for the social sciences* (2nd ed.). New York: Holt, Rinehart & Winston, 1973.

Hendrikson, A. E. and White, P. D. PROMAX: A quick method for rotation to oblique simple structure. *British Journal of Mathematical and Statistical Psychology*, 1964, *17*, 65–70.

Holzinger, K. and Swineford, F. The bi-factor method. *Psychometrika*, 1937, *2*, 41–54.

Horst, P. *Matrix algebra for social scientists*. New York: Holt, Rinehart & Winston, 1963.

Hotelling, H. The generalization of Student's ratio. *Annals of Mathematical Statistics*, 1931, *2*, 360–378.

Hotelling, H. Analysis of a complex of statistical variables into principal components. *Journal of Educational Psychology*, 1933, *24*, 417–441; 498–520.

Hotelling, H. The most predictable criterion. *Journal of Educational Psychology*, 1935, *26*, 139–142.

Hotelling, H. Relations between two sets of variates. *Biometrika*, 1936, *28*, 321–377.

Hotelling, H. The selection of variates for use in prediction with some comments on the general problem of nuisance parameters. *Annals of Mathematical Statistics*, 1940, *11*, 271–283.

Jacobi, C. G. J. Über ein leichtes Verfahren die in der Theorie der Säcularstörungen vorkommenden Gleichungen numerisch aufzulösen. *J. Reine Angewandte Mathematik*, 1846, *30*, 51–94.

Jenkins, W. L. An improved method for tetrachoric *r*. *Psychometrika*, 1955, *20*, 253–258.

Jennrich, R. I. and Sampson, P. F. Rotation for simple loadings. *Psychometrika*, 1966, *31*, 313–323.

Jöreskog, K. G. On the statistical treatment of residuals in factor analysis. *Psychometrika*, 1962, *27*, 335–354.

Jöreskog, K. G. Testing a simple structure hypothesis in factor analysis. *Psychometrika*, 1966, *31*, 165–178.

Jöreskog, K. G. Some contributions to maximum likelihood factor analysis. *Psychometrika*, 1967, *32*, 443–482.

Jöreskog, K. G. A general approach to confirmatory maximum likelihood factor analysis. *Psychometrika*, 1969, *34*, 183–202.

Jöreskog, K. G. A general method for analysis of covariance structures. *Biometrika*, 1970, *57*, 239–251.

Kaiser, H. *The varimax method of factor analysis*. Unpublished Ph.D. dissertation, University of California at Berkeley, 1956.

Kaiser, H. The varimax criterion for analytic rotation in factor analysis. *Psychometrika*, 1958, *23*, 187–200.

Kaiser, H. The application of electronic computers to factor analysis. *Educational and Psychological Measurement*, 1960, *20*, 141–151.

Kaiser, H. A note on Guttman's lower bound for the number of common factors. *British Journal of Statistical Psychology*, 1961, *14*, 1.

Kaiser, H. Psychometric approaches to factor analysis. *Proceedings of the 1964 Invitational Conference on Testing Problems*. Princeton, N.J.: Educational Testing Service, 1965, pp. 37–45.

Kaiser, H. A second-generation Little Jiffy. *Psychometrika*, 1970, *35*, 401–415.

Kaiser, H. and Caffrey, J. Alpha factor analysis. *Psychometrika*, 1965, *30*, 1–14.

Kellam, S. J., Ensminger, M., and Turner, R. J. Family structure and the mental health of children. *Archives of General Psychiatry*, 1977, *34*, 1012–1022.

Kendall, M. G. *The advanced theory of statistics* (Volume 1). London: Griffin, 1952.

Kendall, M. G. Review of Uppsala symposium on psychological factor analysis. *Journal of the Royal Statistical Society* (Series A), 1954, *107*, 462–483.

Kendall, M. G. *Rank correlation methods* (3rd ed.). London: Griffin, 1962.

Kendall, M. G. and Lawley, D. N. The principles of factor analysis. *Journal of the Royal Statistical Society* (Series B), 1956, *18*, 83–84.

Kendall, M. G. and Stuart, A. *The advanced theory of statistics* (Volume 2). New York: Hafner, 1974.

Kendall, M. G. and Stuart, A. *The advanced theory of statistics* (Volume 3). New York: Hafner, 1976.

Kerlinger, F. N. *Foundations of behavioral research* (2nd ed.), New York: Holt, Rinehart & Winston, 1973.

Kerlinger, F. N. and Pedhazur, E. J. *Multiple regression in behavioral research*. New York: Holt, Rinehart & Winston, 1973.

Killion, R. A. and Zahn, D. A. *A bibliography of contingency table literature: 1900 to 1973* (FSU Statistics Report M-300). The Florida State University Department of Statistics, Tallahassee, Florida, August 1974.

Lawley, D. N. The estimation of factor loadings by the method of maximum likelihood. *Proceedings of the Royal Society of Edinburgh*, 1940, *60*, 64–82.

Lord, F. M. and Novick, M. R. *Statistical theories of mental test scores*. Reading, MA: Addison-Wesley, 1968.

Lubin, A. Linear and non-linear discriminating functions. *British Journal of Mathematical and Statistical Psychology*, 1950, *3*, 90–103.

Mahalonobis, P. C. On the generalized distance in statistics. *Proceedings of the National Institute of Science, India*, 1936, *12*, 49–55.

Maxwell, A. E. Statistical methods in factor analysis. *Psychological Bulletin*, 1959, *56*, 228–235.

McNemar, Q. On the number of factors. *Psychometrika*, 1942, *7*, 9–18.

McNemar, Q. Note on the sampling error of the difference between correlated proportions or percentages. *Psychometrika*, 1947, *12*, 153–157.

McNemar, Q. *Psychological statistics* (4th ed.). New York: Wiley, 1969.

Mehrens, W. A. and Lehman, I. J. *Measurement and evaluation in education and psychology*. New York: Holt, Rinehart & Winston, 1973.

Mendenhall, W. *Introduction to linear models and the design and analysis of experiments*. Belmont, CA: Duxbury Press, 1968.

Merenda, P. F., Novack, H. S. and Bonaventura, E. Relationships between the California Test of Mental Maturity and the Stanford Achievement Test Battery at the primary levels. *Educational and Psychological Measurement*, 1972, *32*, 1079–1087.

Merenda, P. F., Novack, H. S., and Bonaventura, E. Multivariate analysis of the California Test of Mental Maturity, Primary Forms. *Psychological Reports*, 1976, *38*, 487–493.

Miller, J. K. *The development and application of bimultivariate correlation: A measure of statistical association between multivariate measurement sets.* Unpublished doctoral dissertation, State University of New York at Buffalo, 1969.

Miller, J. K. In defense of the general canonical correlation index: Reply to Nicewander and Wood. *Psychological Bulletin*, 1975, *82*, 207–209.

Mosteller, F. Association and estimation in contingency tables. *Journal of the American Statistical Association*, 1968, *63*, 1–28.

Mulaik, S. A. Are personality factors raters' conceptual factors? *Journal of Consulting Psychology*, 1964, *28*, 506–511.

Mulaik, S. A. *The foundations of factor analysis*. New York: McGraw-Hill, 1972.

National Research Council. *Science, engineering, and humanities doctorates in the United States—1977 profile*. Washington, DC: National Academy of Sciences, 1978.

Neuhaus, J. O. and Wrigley, C. The quartimax method: An analytical approach to orthogonal simple structure. *British Journal of Mathematical and Statistical Psychology*, 1954, *7*, 81–91.

Neyman, J. Contribution to the theory of the χ^2 test. *Proceedings of the Berkeley Symposium on Mathematical Statistics and Probability*. Berkeley, CA: University of California Press, 1949.

Nicewander, W. A. and Wood, D. A. Comments on "A general canonical correlation index." *Psychological Bulletin*, 1974, *81*, 92–94.

Nicewander, W. A. and Wood, D. A. On the mathematical bases of the general canonical correlation index: Rejoinder to Miller. *Psychological Bulletin*, 1975, *82*, 210–212.

Osgood, C. E. *Method and theory in experimental psychology*. Oxford: Oxford University Press, 1953.

Osgood, C. E., Suci, G. J., and Tannenbaum, P. H. *The measurement of meaning*. Urbana, IL: University of Illinois Press, 1957.

Pearson, K. On the criterion that a given system of deviations from the probable in the case of a correlated system of variables is such that it can be reasonably supposed to have arisen from random sampling. *Philosophical Magazine*, 1900, *5*, 157.

Pearson, K. On the correlation of characters not quantitatively measured. *Philosophical Transactions of the Royal Society* (Series A), 1901, *195*, 1–47. [a]

Pearson, K. On lines and planes of closest fit to systems of points in space. *Philosophical Magazine*, 1901, *6*, 559–572. [b]

Rao, C. R. The utilization of multiple measurements in problems of biological classification. *Journal of the Royal Statistical Society* (B), 1948, *10*, 159–193.

Rao, C. R. *Advanced statistical methods in biometric research*. New York: Wiley, 1952.

Rao, C. R. Estimation and tests of significance in factor analysis. *Psychometrika*, 1955, *20*, 93–111.

Rao, C. R. *Linear statistical inference and its applications* (2nd ed.). New York: Wiley, 1973.

Rao, C. R. and Slater, P. Multivariate analysis applied to differences between neurotic groups. *British Journal of Mathematical and Statistical Psychology*, 1949, *2*, 17–29.

Reynolds, H. T. *The analysis of cross-classifications*. Glencoe, IL: Free Press, 1977.

Rulon, P. J., Tiedeman, D. V., Tatsuoka, M. S., and Langmuir, C. R. *Multivariate statistics for personnel classification*. New York: Wiley, 1967.

Saunders, D. R. An analytical method for rotation to orthogonal simple structure. *Research Bulletin 53–10*. Princeton, NJ: Educational Testing Service, 1953.

Saunders, D. R. The rationale for an "oblimax" method of transformation in factor analysis. *Psychometrika*, 1961, *26*, 317–324.

Saunders, D. R. Trans-varimax. *American Psychologist*, 1962, *17*, 395.

Schatzoff, M. *Exact distributions of Wilks' likelihood ratio criterion and comparisons with competitive tests*. Unpublished doctoral dissertation. Harvard University, 1964.

Schatzoff, M. Exact distributions of Wilks' likelihood ratio criterion. *Biometrika*, 1966, *53*, 347–358.

Searle, S. R. *Matrix algebra for the biological sciences*. New York: Wiley, 1966.

Searle, S. R. *Linear models*. New York: Wiley, 1971.

Sherin, R. J. A matrix formulation of Kaiser's varimax criterion. *Psychometrika*, 1966, *31*, 535–538.

Snider, J. G. and Osgood, C. E. (Eds.). *Semantic differential technique: A sourcebook*. Chicago: Aldine, 1969.

Sorenson, W. W. Test of mechanical principles as a suppressor variable for the prediction of effectiveness on a mechanical repair job. *Journal of Applied Psychology*, 1966, *50*, 348–352.

Spearman, C. The proof and measurement of association between two things. *American Journal of Psychology*, 1904, *15*, 72–101. [a]

Spearman, C. General intelligence, objectively determined and measured. *American Journal of Psychology*, 1904, *15*, 201–293. [b]

Spearman, C. *The abilities of man*. New York: Macmillan, 1927.

Spearman, C. and Holzinger, K. Note on the sampling error of tetrad differences. *British Journal of Psychology*, 1925, *16*, 86–88.

Stewart, D. and Love, W. A general canonical correlation index. *Psychological Bulletin*, 1968, *70*, 160–163.

Super, D. E. Career development. In J. Davitz and S. Ball (Eds.), *Psychology of the educational process*. New York: McGraw-Hill, 1970.

Super, D. E. (Ed.). *Measuring vocational maturity for counseling and evaluation*. Washington: American Personnel and Guidance Association, 1974.

Tatsuoka, M. *The relationship between canonical correlation and discriminant analysis*. Cambridge, MA: Educational Research Corporation, 1953.

Tatsuoka, M. *Multivariate analysis*. New York: Wiley, 1971.

Thorndike, R. M. and Weiss, D. J. Stability of canonical components. *Proceedings of the American Psychological Association*, 1970, *78*, 107–108.

Thurstone, L. L. Multiple factor analysis. *Psychological Review*, 1931, *38*, 406–427.

Thurstone, L. L. *The vectors of mind*. Chicago: The University of Chicago Press, 1935.

Thurstone, L. L. Primary mental abilities. *Psychometric Monographs* (No. 1). Chicago: University of Chicago Press, 1938.

Thurstone, L. L. *Multiple factor analysis*. Chicago: University of Chicago Press, 1947.

Tiedeman, D. V., Bryan, J. G., and Rulon, P. J. *The utility of the airman classification battery for assignment of airmen to eight air force specialties*. Cambridge, MA: Educational Research Corporation, 1953.

Timm, N. H. *Multivariate analysis*. Belmont, CA: Brooks-Cole, 1975.

United States Department of Health, Education and Welfare. The association of health attitudes and perceptions of youths 12–17 years of age with those of their parents: United States, 1966–1970. *Vital and Health Statistics, Series 11* (Data from the National Health Survey). Publication No. (HRA) 77–1643, March 1977.

Velicer, W. F. Determining the number of components from the matrix of partial correlations. *Psychometrika*, 1976, *41*, 321–327.

Wald, A. Tests of statistical hypotheses concerning several parameters when the number of observations is large. *Transactions of the American Mathematical Society*, 1943, *54*, 426–482.

Ward, J. and Jennings, E. *Introduction to linear models*. Englewood Cliffs, NJ: Prentice-Hall, 1973.

Wechsler, H., Dorsey, J. L., and Bovey, J. D. A follow-up study of residents in internal medicine, pediatrics, and obstetrics-gynecology training programs: Implications for the supply of primary care physicians. *New England Journal of Medicine*, 1978, *298*, #1.

Weiss, D. J. Canonical correlation analysis in counseling psychology. *Journal of Counseling Psychology*, 1972, *19*, 241–252.

Wilks, S. Certain generalizations in the analysis of variance. *Biometrika*, 1932, *24*, 471–494.

Winer, B. J. *Statistical principles in experimental design* (2nd ed.). New York: McGraw-Hill, 1971.

Wold, H. Some artificial experiments in factor analysis. *Uppsala Symposium on Psychological Factor Analysis*. Uppsala, Sweden: Almqvist and Wiksell, 1953.

Young, G. Maximum likelihood estimation and factor analysis. *Psychometrika*, 1941, *6*, 49–54.

Zar, J. H. Significance testing of the Spearman rank correlation coefficient. *Journal of the American Statistical Association*, 1972, *67*, 578.

Index